Goal-Directed Decision Making

Goal-Directed Decision Making
Computations and Neural Circuits

Edited by

RICHARD MORRIS
AARON BORNSTEIN
AMITAI SHENHAV

ELSEVIER

ACADEMIC PRESS

An imprint of Elsevier

Academic Press is an imprint of Elsevier
125 London Wall, London EC2Y 5AS, United Kingdom
525 B Street, Suite 1650, San Diego, CA 92101, United States
50 Hampshire Street, 5th Floor, Cambridge, MA 02139, United States
The Boulevard, Langford Lane, Kidlington, Oxford OX5 1GB, United Kingdom

Library of Congress Cataloging-in-Publication Data
A catalog record for this book is available from the Library of Congress

British Library Cataloguing-in-Publication Data
A catalogue record for this book is available from the British Library

ISBN: 978-0-12-812098-9

For information on all Academic Press publications visit our website at
https://www.elsevier.com/books-and-journals

Working together
to grow libraries in
developing countries

www.elsevier.com • www.bookaid.org

Publisher: Nikki Levy
Acquisition Editor: Joslyn Chaiprasert-Paguio
Editorial Project Manager: Gabriela Capille
Production Project Manager: Anusha Sambamoorthy
Cover Designer: Greg Harris

Typeset by TNQ Technologies

CONTENTS

CONTRIBUTORS

Bernard W. Balleine
Decision Neuroscience Lab, School of Psychology, UNSW Sydney, Kensington, NSW, Australia

Jesus Bertran-Gonzalez
Decision Neuroscience Laboratory, School of Psychology, University of New South Wales, Sydney, NSW, Australia

Rahul Bhui
Departments of Psychology and Economics & Center for Brain Science, Harvard University, Cambridge, MA, United States

Matthew M. Botvinick
DeepMind, London, United Kingdom
Gatsby Computational Neuroscience Unit, UCL, London, United Kingdom

Regina M. Carelli
Department of Psychology and Neuroscience, University of North Carolina, Chapel Hill, NC, United States

Luke J. Chang
Department of Psychological and Brain Sciences, Dartmouth College, Hanover, NH, United States

Anne G.E. Collins
Department of Psychology, University of California, Berkeley, Berkeley, CA, United States

Laura H. Corbit
Department of Psychology, The University of Toronto, Toronto, ON, Canada

Etienne Coutureau
CNRS, INCIA, UMR 5287, Bordeaux, France
Université de Bordeaux, INCIA, UMR 5287, Bordeaux, France

Fiery A. Cushman
Department of Psychology, Harvard University, Cambridge, MA, United States

Sanne de Wit
Department of Clinical Psychology, University of Amsterdam, Amsterdam, The Netherlands
Amsterdam Brain and Cognition, University of Amsterdam, Amsterdam, The Netherlands

Anthony Dickinson
Department of Psychology, University of Cambridge, Cambridge, United Kingdom

Oriel FeldmanHall
Department of Cognitive, Linguistic & Psychological Sciences, Brown University, Providence, RI, United States

Teri M. Furlong
Neuroscience Research Australia, Randwick, NSW, Australia

Samuel J. Gershman
Department of Psychology, Harvard University, Cambridge, MA, United States
Center for Brain Science, Harvard University, Cambridge, MA, United States

Catherine A. Hartley
Department of Psychology, New York University, New York, NY, United States
Center for Neural Science, New York University, New York, NY, United States

Wouter Kool
Department of Psychology, Harvard University, Cambridge, MA, United States

Vincent Laurent
Decision Neuroscience Laboratory, School of Psychology, University of New South Wales, Sydney, NSW, Australia

Mimi Liljeholm
Cognitive Sciences, University of California, Irvine, Irvine, CA, United States

Elliot A. Ludvig
Department of Psychology, University of Warwick, Coventry, United Kingdom

Kevin J. Miller
Princeton Neuroscience Institute, Princeton University, Princeton, NJ, United States

Richard W. Morris
School of Psychology, University of New South Wales (NSW), Sydney, NSW, Australia
School of Medicine, University of Sydney, Sydney, NSW, Australia
Centre for Translational Data Science, University of Sydney, Sydney, NSW, Australia

Travis M. Moschak
Department of Psychology and Neuroscience, University of North Carolina, Chapel Hill, NC, United States

Yael Niv
Princeton Neuroscience Institute & Department of Psychology, Princeton University, Princeton, NJ, United States

Shauna L. Parkes
CNRS, INCIA, UMR 5287, Bordeaux, France
Université de Bordeaux, INCIA, UMR 5287, Bordeaux, France

Omar D. Pérez
Division of the Humanities and Social Sciences, California Institute of Technology, California, United States

Giovanni Pezzulo
Institute of Cognitive Sciences and Technologies, National Research Council, Rome, Italy

Hillary A. Raab
Department of Psychology, New York University, New York, NY, United States

A. David Redish
Department of Neuroscience, University of Minnesota, Minneapolis, MN, United States

Brandy Schmidt
Department of Neuroscience, University of Minnesota, Minneapolis, MN, United States

Geoffrey Schoenbaum
National Institute on Drug Abuse, National Institute of Health, Baltimore, MD, United States
Department of Anatomy and Neurobiology, University of Maryland School of Medicine, Baltimore, MD, United States
Solomon H. Snyder Department of Neuroscience, The John Hopkins University, Baltimore, MD, United States

Nicolas W. Schuck
Max Planck Research Group NeuroCode, Max Planck Institute for Human Development, Berlin, Germany
Princeton Neuroscience Institute & Department of Psychology, Princeton University, Princeton, NJ, United States

Melissa J. Sharpe
National Institute on Drug Abuse, National Institute of Health, Baltimore, MD, United States
Princeton Neuroscience Institute, Princeton University, Princeton, NJ, United States
School of Psychology, UNSW Australia, Sydney, NSW, Australia

Amitai Shenhav
Department of Cognitive, Linguistic, and Psychological Sciences, Brown Institute for Brain Science, Brown University, Providence, RI, United States

Alec Solway
Virginia Tech Carilion Research Institute, Roanoke, VA, United States

Elizabeth A. West
Department of Psychology and Neuroscience, University of North Carolina, Chapel Hill, NC, United States

Andrew M. Wikenheiser
National Institute on Drug Abuse Intramural Research Program, Baltimore, MD, United States

Robert Wilson
Department of Psychology, University of Arizona, Tucson, AZ, United States

PREFACE

How do we decide? Through the early part of the 20th century, the dominant view of animal behavior was based on the idea that it is driven by reinforcement of learned associations between a stimulus and a response: The more often an animal is rewarded after responding a particular way in the presence of a stimulus, the more likely it is to make that same response the next time it encounters that stimulus. This idea that all learned behavior is underpinned by stimulus–response (S-R) associations (habits), acquired by a subtle process of reinforcement, was turned on its head by subsequent research that showed that animals can also use internal models of their environment to flexibly guide behavior across different environments.

These observations, and the associated theories of what became known as *goal-directed* behavior, not only transformed our understanding of behavior but also formed the basis for entirely new fields of study within psychology, neuroscience, and computer science. Even so, exactly how we represent what Edward Tolman called the "causal texture of our environment"—and, consequently, how we use it to decide between actions— remains fundamentally unresolved.

Recent years have seen a renewed convergence of cross-disciplinary efforts to understand goal-directed decisions, unearthing surprising results and providing new frameworks for understanding what has been found. This volume represents the breadth and depth of this pursuit, collecting observations and ideas from the anatomical to the algorithmic.

The first section begins with a review by **Dickinson and Peréz** of how the axioms of goal-directed learning, which Dickinson laid out 30 years ago, have guided research into decision-making over that time, as well how both habitual and goal-directed systems exist alongside each other and arbitrate decision-making. **Liljeholm** further develops these ideas by showing how constructs from computer science, such as instrumental divergence, can enrich our understanding of goal-directed learning and even answer why arbitration occurs in the way it does. **Solway and Botvinick** address an oft-neglected issue in goal-directed learning, namely how evidence can be integrated over time to support decision-making. **Bhui** proposes an innovative approach to thinking about goal-directed decision-making using the economic theory of case-based reasoning, in which past experiences can serve as analogies for novel choice problems. **Collins** describes how advances in our understanding of hierarchical learning can enrich S-R models to flexibly represent environmental structure, and so challenge our notions about the boundaries between S-R and goal-directed learning. **Schmidt, Wikenheiser, and Redish** then delve into how rodents navigate structured environments by simulating among multiple possible trajectories. **Kool, Cushman, and Gershman** end this section by returning to the issue of arbitration between decision-making systems, and review computational principles that may help us understand how both competition and cooperation between systems can occur.

The second section surveys what is known about the neural substrates of decision-making. **Coutureau and Parkes** review the crucial involvement of different subregions of the prefrontal cortex in goal-directed learning, proposing that decision-making is mediated by cortico–cortical as well as cortico–subcortical pathways. **West, Moschak, and Carelli** review the role of subcortical regions in flexible decision-making, highlighting the importance of interactions between the nucleus accumbens core and shell. **Bertram-Gonzalez and Laurent** dive deeper into the nucleus accumbens shell and propose that the shell silently learns the necessary contingencies during Pavlovian conditioning that enable future action selection. **Sharpe and Schoenbaum** provide an important reappraisal of the reward prediction error account of phasic dopamine signals in the ventral tegmental area and cast it as a more general teaching signal for nonrewarding events as well. Finally, **Schuck, Wilson, and Niv** present a novel theory of the role of the orbitofrontal cortex in mapping the partially observable state representation necessary for goal-directed learning.

The third section presents the application of decision-making science to understanding child development (**Raab and Hartley**), social behavior (**FeldmanHall and Chang**), disorders of compulsivity (**de Wit**), as well as drug addiction (**Furlong and Corbit**) and psychosis (**Morris**). For instance, **FeldmanHall and Chang** introduce a new model of goal-directed behavior within social environments, describing how emotional signals are used to monitor the degree to which one's actions facilitate or impede their social goals.

The final chapters take a step back to offer a broad perspective on how competing theories and approaches to studying goal-directed decision-making can be reconciled and further honed moving forward. **Miller, Ludvig, Pezzulo, and Shenhav** discuss existing tensions between psychological and computational classifications that distinguish habitual and goal-directed behavior and elaborate on recent models that relieve this tension. Finally, **Balleine** reexamines what we mean by "reward" and "reinforcement" and outlines a provocative account of how they might be distinguished according to their learning rules, neural circuits and bases in motivation.

The merger of ideas and techniques from psychology, computer science, and neuroscience is bearing fruit at a remarkable pace. Still, much is yet to be discovered. The entries in this volume reflect this fast-moving state of affairs by serving both as syntheses of what is known, as well as guides to what is yet unknown. We look forward to seeing what the future will bring.

Richard W. Morris
Aaron M. Bornstein
Amitai Shenhav
March 2018

CHAPTER 1

Actions and Habits: Psychological Issues in Dual-System Theory

Anthony Dickinson[1], Omar D. Pérez[2]

[1]Department of Psychology, University of Cambridge, Cambridge, United Kingdom; [2]Division of the Humanities and Social Sciences, California Institute of Technology, California, United States

Over a quarter of a century ago, Heyes and Dickinson (1990) offered two behavioral criteria for the attribution of intentionality to animal action that have subsequently served to characterize goal-directed behavior (de Wit & Dickinson, 2009). These criteria are rooted in the folk psychology of action. If asked to explain why a hungry rat is pressing a lever for food pellets, the folk psychologist tells us that the rat *desires* access to the food pellets and *believes* that pressing the lever will yield access. This account is deceptively simple in that it has two explicit mental entities, the belief and desire, and an implicit process for deploying these entities in the control of behavior. In spite of this apparent simplicity, the belief—desire account has psychologically important features that motivate our conception of goal-directed action.

A philosopher of mind may well point out that beliefs and desires are particular types of mental entities, propositional attitudes, which have two important features. First, the content of the belief or desire, such as believing that "lever pressing causes access to food pellets" or desiring that "there is access to food pellets," is a representation of an event, state, or relationship with a propositional-like structure. Second, this propositional content stands in a relationship, or attitude, to the event or state of affairs that is represented. A belief represents a supposed state of affairs in the world and therefore has a world-to-representation fit in that it can be either true or false of the world (Searle, 1983). By contrast, a desire has a representation-to-world fit in that its content represents a state of affairs that is currently not true but that the agent wishes it to be so. Therefore, a desire's fit to the world is one of fulfillments in that the content of the desire can either be fulfilled or unfulfilled.

The third, and often implicit, component of the belief—desire account is a process of practical inference that takes the belief and desire as it arguments to yield an intention to act. As our aim was to marshal beliefs and desires in a psychological account, rather than a philosophical analysis of action, we choose to present the content and the practical inference in a programming language, PROLOG, which was designed, at least in part, to simulate cognitive processes and is reasonably transparent with respect to

Goal-Directed Decision Making
ISBN 978-0-12-812098-9, https://doi.org/10.1016/B978-0-12-812098-9.00001-2

content. So a minimal program to generate lever pressing for food pellets might take the following form:

cause(lever-press,access(food-pellet),g).	—belief
access(food-pellet,δ).	—desire
perform(A,g∗δ):-cause(A,access(O),g),access(O,δ).	—practical inference

This little program is similar to that offered in Heyes and Dickinson (1990) except for the addition of the parameters g and $δ$, which serve to quantify the believed strength of the causal relationship between lever pressing and access to the food pellets and the strength of the desire for this goal, respectively. Therefore, g represents the believed reliability or rate with which lever pressing will cause access to the food pellets. If we ran this little program to determine which action this impoverished agent intends to perform, it would return an intention to lever-press in the form *perform(lever-press)* with the will to execute the intention determined by the product of g and $δ$. If either of these parameters is zero, there will be a lack of will to execute the intention. Henceforth, we shall ignore the quantification of beliefs and desire except where necessary.

This psychological account of action is both causal and rational. It is causal in the sense that it is the interaction of the content of the belief and desire in the process of practical inference that determines the content of the intention, and hence the particular goal-directed action performed. Second, and importantly, the practical inference process yields a rational action. If the belief *cause(lever-press,access(food-pellet))* is true and the intention *perform(A)* is executed then, other things being equal, the desire *access(food-pellet)* must, *of necessity*, be fulfilled. The psychological rationality of lever pressing within the present context may appear obvious, and possibly trivial, but, as we shall discuss below, the issue of whether representational content can cause responses that are nonrational with respect to that content is a matter of dispute.

This belief—desire account led Heyes and Dickinson to offer two behavioral criteria for determining whether a particular behavior is an intentional or goal-directed action: the desire (goal) and belief (instrumental) criteria. We shall discuss each in turn.

DESIRE CRITERION

A straightforward prediction of the belief—desire account is that, if following lever-press training under the desire for access to food pellets, this desire is reduced, the animal's propensity to press the lever should immediately decrease without any further training. Reducing the desire for access to the food pellets by setting the $δ$ parameter in the little PROLOG program to zero yields an intention to lever-press but without any will to execute it. When Adams (1980) first attempted to assess the status of lever pressing for food pellets in Dickinson's lab by removing his rats' desire for these pellets, he could find no evidence that lever pressing was goal-directed. Having trained the rats to press a lever for access to the food pellets, he then reduced the desire for these pellets by conditioning a food aversion to the pellets in the absence of an opportunity to lever-press. When once again given the opportunity to press the lever (in the absence of the food pellets), the rate of pressing was totally unaffected by whether or not the food pellets were desired at the time.

We were surprised that our rats were insensitive to outcome devaluation, given that many years before Tolman and Gleitman (1949) had convincingly demonstrated that devaluing one of the goal boxes in an E-maze by associating it with electric shocks induced an immediate reluctance by their rats to take the turn leading to the devalued goal. Therefore, we suspected that it was the type of action and training that might be critical in determining whether performance was sensitive to removal of a desire through goal or outcome devaluation.

Adams (1980) trained his rats to press the lever on a variable interval schedule, which models a resource that depletes and then regenerates with time, such as nectar, in that the schedule specifies the average time interval that has to elapse before the next reward becomes available for collection. By contrast, foraging in a nondepleting source is modeled by ratio schedules, which specifies the probability with which each action yields an outcome. Therefore, there is a more direct causal connection between the action and the outcome under a ratio contingency in that it does not involve an additional causal process, outcome regeneration with time. When we switched to training on a variable ratio schedule, we found that lever pressing for food by hungry rats could be goal-directed, at least by the desire criterion. Our rats pressed the lever at a reduced rate following the removal of the desire for the food pellets by aversion conditioning when tested extinction so that the devalued pellets were not presented (Adams & Dickinson, 1981). Subsequently, we established that the type of training schedule is critical in determining whether or not an action is goal-directed. Ratio-trained lever pressing for food is more sensitive to outcome devaluation than interval-trained responding even when either the probability of an outcome following a lever press or the overall outcome rates are matched (Dickinson, Nicholas, & Adams, 1983).

On the basis of these early studies, the desire criterion as implemented by the outcome revaluation test has been widely accepted as a canonical assay for the goal-directed status of an action not only in rodents but also in monkeys (Rhodes & Murray, 2013), children (Klossek, Russell, & Dickinson, 2008), and adult humans (Valentin, Dickinson, & O'Doherty, 2007). Furthermore, these studies established that whether or not an action meets the desire criterion depends upon the conditions of training. We have already noted that training on ratio contingency is more likely to establish a goal-directed action than is the equivalent interval training. Moreover, in his doctoral research, Adams (1982) discovered that overtraining rats on a ratio schedule can render performance autonomous of the current value of the outcome in a devaluation test, a finding subsequently replicated with humans (Tricomi, Balleine, & O'Doherty, 2009). The status of an instrumental action also depends upon whether or not the agent has a choice between this target action and another action yielding a different outcome during training. The target action retains its goal-directed status following choice training even though equivalent single-action training by both rats (Kosaki & Dickinson, 2010) and children (Klossek, Yu, & Dickinson, 2011) is sufficient to establish autonomy of the current goal value. Finally, it is not just the training conditions that determine whether instrumental behavior is goal-directed—subjecting both rats (Dias-Ferreira et al., 2009) and humans (Schwabe & Wolf, 2009, 2010) to a stressful experience prior to either training or testing renders performance insensitive to outcome devaluation.

BELIEF CRITERION

Although most of the early outcome revaluation studies failed to find any effect of revaluation following interval training, an exception is an irrelevant incentive study. Krieckhaus and Wolf (1968) trained thirsty rats to lever-press for either a sodium or potassium solution or for water before revaluing the sodium relative to the other solutions by inducing a sodium appetite. When tested in extinction, the rats trained with sodium solution (the outcome relevant to the current sodium appetite) pressed more in the extinction test than those trained with either the potassium solution or water, thereby demonstrating that lever pressing met the desire criterion for goal-directedness. However, Dickinson was concerned about the apparent discrepancy between this finding and Adams's (1980; see above) failure to observe an outcome devaluation effect using aversion conditioning following interval training, as Krieckhaus and Wolf trained their rats on an interval schedule.

At issue is whether the induction of a desire for sodium impacted upon lever pressing through a belief that this action yields access to a sodium solution. Unless there are plausible grounds for ascribing such a belief to their rats, this revaluation effect does not warrant a belief–desire account. This issue can be addressed by varying whether or not lever pressing causes access to the sodium solution, while equating exposure to the sodium. A belief–desire account requires that rats trained with sodium contingent upon lever pressing should respond more under the sodium appetite than those trained with noncontingent sodium. To examine this issue, thirsty rats lever-pressed for either a sodium solution or water using the ratio schedule previously employed by Adams and Dickinson (1981) to demonstrate the goal-directed status of rodent lever pressing (Dickinson & Nicholas, 1983). Under this schedule, one group of thirsty rats, Group Na(w), pressed for the sodium solution (Na) on a ratio schedule while water (W) was presented noncontingently, whereas for a second group, Group W(na), these contingencies were reversed so that these rats pressed for water while receiving the sodium solution noncontingently. This training enabled us to assess whether the outcome revaluation effect observed by Krieckhaus and Wolf (1968) under a sodium appetite met the belief criterion, which requires that the outcome revaluation effect is mediated by the instrumental contingency. Only Group Na(w), for which the lever pressing produced the sodium solution, could have acquired the belief *cause(lever-press, access(sodium-solution))*, and therefore this group should have responded more than Group W(na) under the sodium appetite if lever pressing was goal-directed.

Importantly, performance during the outcome revaluation test under the sodium appetite was unaffected by whether lever pressing had been trained with water or sodium solution as the outcome. This null result contrasts with the fact we replicated the basic outcome revaluation effect under the sodium appetite. In a second pair of groups, Groups K(w) and W(k), we replaced the sodium solution with a potassium solution and found that both these groups pressed less under the sodium appetite than Groups Na(w) and W(na). What this result shows is that the revaluation effect depended upon experiencing sodium in the training context but, importantly, not upon learning that lever pressing yields sodium or, in other words, learning about the causal relationship between action and outcome. Consequently, this irrelevant incentive effect does not meet the belief criterion, which requires that the outcome revaluation effect is mediated by the instrumental contingency. To the extent that a goal-directed action is conceived of as an action that supports a belief—desire explanation, the revaluation of the sodium outcome did not identify the instrumental action as goal-directed.

The importance of deploying the desire criterion in concert with the belief criterion is also illustrated by a more recent study by Jonkman, Kosaki, Everitt, and Dickinson (2010). We extensively trained rats to lever-press for food pellets on an interval schedule, a training regime expected to render performance autonomous of the current value of the pellets. However, devaluing the pellets by aversion conditioning yielded a substantial outcome devaluation effect in a subsequent extinction test but, once again, this effect was mediated, at least in part, by an association between the context, rather than lever pressing, and the pellets. Extinguishing the context—pellets association between the devaluation treatment and the extinction test markedly reduced the outcome devaluation effect. So again we must have concerns about whether this devaluation effect meets the belief criterion because the outcome devaluation effect depended upon the association between the context and outcome rather than upon knowledge about the instrumental contingency between the lever pressing and the food pellets.

Some have argued that the belief criterion is too stringent in that it excludes behavior that we might want to characterize as goal-directed in terms of its manifest functional properties. For example, Carruthers has argued that the belief criterion is not necessary for the attribution of belief—desire psychology to an animal. He asks of a bird flying toward a food source "why should we not say that the animal behaves as it does because it *wants* something and *believes* that the desired thing can be found at a certain represented location on the (mental) map" (Carruthers, 2004) p. 211; see also (Allen & Bekov, 1995;

Heyes & Dickinson, 1995) and leave the causation of the behavior implicit, for example, in the activation of a "flying-in-that-direction schemata." In essence, what Carruthers is arguing is that we should allow a representational psychology in which a belief with a content that makes no reference to an action should be capable of causing that action as a response to entertaining the belief.

The problem with filling the explanatory gap between a belief and action with some implicit causation is the absence of any account of how the content of the belief determines the action selected. When applied to the Dickinson and Nicholas (1983) study in which rats that received noncontingent sodium, Group W(na), pressed the lever more under a sodium appetite than the animals that received noncontingent potassium, Group W(k), by appealing to some implicit causal process that deployed Group W(na)'s belief that sodium was available in the training context to generate lever pressing. This is indeed a deeply mysterious implicit process as throughout training lever pressing had produced water but never sodium solution—in any reasonable sense of the term, the sodium solution could not have functioned as a psychological goal of this action.

This example illustrates a cardinal feature of an intentional account of behavior, such as that offered by belief–desire psychology. An intentional account must specify not only the content of the beliefs and desires but also the process that deploys this content to generate the behavior. Moreover, if the content of these propositional attitudes is to have a causal role, the process must respect the fact that those beliefs and desires have representational content and therefore truth and fulfillment values, respectively. As a consequence, the process deploying these intentional states must not only cause the action but also be rational with respect to the representational content of these states. As we pointed out in our introduction to belief–desire psychology, this is the case for practical inference in that if the belief is true and action performed, *of necessity* the desire must be fulfilled. This rationality criterion is clearly not met by the process that caused lever pressing in Group W(na) in the Dickinson and Nicholas (1983) study.

There is little point arguing in general about the appropriate characterization of behavior as "goal-directed." From a functional or biological point view, an invigoration of the predominant behavior in an environment associated with sodium while under a sodium appetite may well enhance the likelihood of encountering salt and in this sense be goal-directed. But from a psychological perspective, a failure to mark out behavior mediated by a belief about its causal effects just serves to conflate goal-directed control with other importantly distinct processes. One such process is the source of instrumental behavior that is unaffected by outcome revaluation following interval or extended training.

HABITS

Folk psychology recognizes another form of instrumental behavior, habits, which are elicited directly by the stimulus context without thought for their consequences. In case of habits, we explain our behavior, not by an appeal to beliefs and desires, but rather to the fact that the response is one that we have regularly performed in this situation in the past. Within academic psychology, habits have been traditionally explained by the stimulus-response/reinforcement mechanisms envisaged by Thorndike's law of effect (Thorndike, 1911) whereby, for example, the outcome of a lever press, a food pellet, just serves to strengthen or reinforce an association between the stimulus context and the lever press response without encoding any information about the outcome itself. Others have argued that simple stimulus-response contiguity is sufficient for habit learning (Guthrie, 1959). In both cases, however, the loss of desire for the food pellets in a revaluation test cannot impact directly on lever pressing, which is simply elicited when the agent is replaced in the training context after the devaluation treatment.

This conception of a habit emphasizes the fact that its performance is autonomous of the current value of the outcome that reinforced the habit in the first place. However, it is often assumed that the canonical feature of a habit is that its performance makes minimal demands on general cognitive resources, a view that prioritizes *automaticity* over *autonomy*. This priority has recently been challenged by Economides, Kurth-Nelson, Lübbert, Guitart-Masip, and Dolan (2015) who argue that automaticity may not differentiate goal-directed behavior from habitual behavior, at least after more extensive training. Rather than using the standard outcome revaluation test as the assay of goal-directed status, they employed a decision task developed by Daw, Gershman, Seymour, Dayan, and Dolan (2011). In this task each of the two choice options leads to one outcome with a high probability and the other outcome with the complementary, low probability. The critical feature of the task is that high probability outcome for one choice is also the low probability outcome for the other option with the relative value of the two outcomes changing slowly throughout each session of training. Consequently, having received a low probability outcome with the relatively higher value, the optimal strategy is to switch choice options because the other option is more likely to yield this currently high-valued outcome. The implementation of this strategy appears to require knowledge of the choice—outcome contingencies of the two options, and the goal-directed status of this strategy has been validated against the standard outcome revaluation test (Friedel et al., 2014; Gillan, Otto, Phelps, & Daw, 2015).

What Economides et al. (2015) reported is that goal-directed performance on this decision task becomes unaffected by the addition of an independent cognitive load leading them to conclude that goal-directed performance can be automatic, thereby endorsing autonomy of the current outcome value, rather than automaticity, as the cardinal feature of habitual behavior. It is possible, however, for optimal choices in this decision task to become habitual if the low probability but high-valued outcome in the context of one choice option can act as a stimulus associated habitually with the choice of the other option on the next trial. This switch is more likely to have been reinforced by the receipt of the higher value outcome than is repeating the same choice. Whatever the merits of this habitual account, the result highlights the fact that automaticity cannot be taken as a primary feature distinguishing habitual from goal-directed control so that behavioral autonomy remains the canonical marker of the habitual.

The motivation of habits

The fact that the current incentive value of a reinforcer does not exert direct control over habitual responding through its role as the instrumental outcome does not mean that this value has no impact on this form of responding. Recall that both the Krieckhaus and Wolf (1968) and Jonkman et al. (2010) demonstrated an outcome revaluation effect that, on further analysis, failed to meet the belief criterion and turned out to be mediated by the context—outcome association. Such Pavlovian associations endow a stimulus, be it a complex stimulus such as context or a simple discrete stimulus, with the capacity to exert a general motivational influence over habitual responding.

By using the Pavlovian—instrumental transfer (PIT) paradigm, Dawson and Dickinson demonstrated the generality of this form of motivation (Dickinson & Dawson, 1987). They trained hungry rats on three types of trial. In one type, a lever was inserted, and pressing it was reinforced on an interval schedule with food pellets. In the other trial types, the lever remained retracted, and one of two stimuli was presented. Food pellets were freely presented during one stimulus and sugar water during the other. The experiment then concluded with a transfer test in which, for the first time, the lever was inserted during trials in which one of the stimuli was present, although pressing was never reinforced during the test. This test allowed us to assess the influence of the Pavlovian stimuli on instrumental responding without the response having been trained in the presence of the stimuli. Moreover, to vary the motivational relevance of these stimuli, half of the rats were tested hungry as in training and half thirsty. Both the stimuli associated with the pellets and sugar water were relevant to the hunger state, whereas only the sugar—water stimulus was relevant under thirst.

The amount of transfer respected the motivational relevance of the stimuli. For animals tested under the training state of hunger, if anything, the pellet stimulus elicited most lever presses. By contrast, thirsty rats pressed more in the presence of the sugar—water stimulus than during the pellet stimulus. The theoretical significance of this transfer effect lies with its generality in that it was the stimulus associated with the sugar water that had the greatest impact on performance under thirst even though lever pressing had been trained with the pellets. Therefore, although this motivational influence is sensitive to the relevance of the Pavlovian reinforcers to the animal's current motivational state, the transfer is not mediated by a goal-directed component of the instrumental behavior and so, by default, must have operated through habitual responding. This conclusion is bolstered by the fact that training regimes that render performance impervious to outcome devaluation, such as extensive training (Holland, 2004) and interval as opposed to ratio training (Wiltgen et al., 2012), also render performance more sensitive to the motivational influence of Pavlovian stimuli.

The motivational influence manifest in the general PIT effect also resolves an outstanding issue arising from our discussion of the belief criterion. In that discussion, we noted a number if instances in which the training context potentiates instrumental performance through its association with a motivationally relevant reinforcer. The general PIT effect provides a ready explanation of this potentiation by assuming that the Pavlovian context → reinforcer association is acquired concurrently with the instrumental learning and thereby exerts a motivational influence on the habitual component throughout training.

In summary, the residual responding observed following outcome devaluation is best characterized as habitual in that it is not a product of a desire for the outcome. Even so, habitual behavior can be sensitive to the motivational relevance of the outcome and responds appropriately to shifts in motivational state, such as that between hunger and thirst. However, this sensitivity reflects the association of the outcome with the training context rather than with the instrumental action.

Outcome expectations and habits

In a classic PIT study, Lovibond (1983) reported different transfer profiles when thirsty rabbits were responding for a sucrose solution on interval and ratio schedules. Following interval training, the enhanced instrumental responding was confined to the Pavlovian stimulus and was positively related to the response rate immediately prior to its onset. This profile points to the motivational influence just discussed, which in classical Hullian fashion interacts in a multiplicative fashion with habit strength as registered by the prestimulus responding (Hull, 1952). In contrast, the Pavlovian stimulus was effective in restarting ratio responding at times when the rabbit had stopped performing during the prestimulus period, a reinstatement that then outlasted the stimulus itself. However, the Pavlovian stimulus had less impact on ratio performance if the rabbit was already reliably responding prior to its onset, an effect recently replicated with humans (Colagiuri & Lovibond, 2015). On the assumption that the ratio responding was goal-directed, this profile suggests that the Pavlovian stimulus served to prime rather than motivate instrumental action.

In contrast to the motivation of habits manifested in general PIT, another form of transfer, (outcome-)specific PIT, illustrates how this priming operates. During Pavlovian training, two stimuli (Ss) are associated with different outcomes (Os), such as food pellet and a sugar solution, which are also used to reinforce two different responses (Rs) during separate instrumental training. Although the stimuli and responses have never been trained together, each stimulus preferentially enhances the response trained with the same outcome during a transfer test. Classically, specific PIT is interpreted in terms of an associative, rather than motivational process. The Pavlovian training sets up S → O associations, whereas the instrumental training results in O → R associations. Amalgamating these two associations at test enables the stimulus to select the response with the common outcome through an S → O → R chain. Put more colloquially, the stimulus elicits an expectation of the outcome, which in turn activates its response.

Although this associative analysis is generally accepted, how the outcome expectation comes to control its response remains an issue. In a standard specific PIT study, the two instrumental responses are trained separately and sometimes in different stimulus contexts. As a consequence, the contextual stimuli should come to activate a "representation" or expectation of the instrumental outcome through Pavlovian learning, so that the instrumental response is reinforced in the presence of this activated outcome representation. The activated outcome representation can therefore function as stimulus capable of eliciting the instrumental response through a habitual S-R/reinforcement mechanism, a function that should then transfer to an outcome representation activated by the Pavlovian stimulus (Trapold & Overmier, 1972). Therefore, just as in the case of general PIT, habitual responding can appear to be goal-directed in the sense that it is mediated by a representation of the outcome.

This reinforcement account of specific PIT was challenged by a complex and sophisticated analysis by Rescorla and Colwill. By assessing the relative control exerted by an activated representation of a reward when it functioned as a stimulus for a response or the outcome of that response, they found that the outcome function always exerted greater control (Rescorla, 1992; Rescorla & Colwill, 1989). However, perhaps the simplest way of addressing this issue is by training the two instrumental responses concurrently in the same stimulus context. We shall illustrate this point with a study by Watson, Wiers, Hommel, and De Wit (2014).

Human participants were trained to press one of two keys (Rc) for a chocolate outcome (Oc) and the other (Rp) for a popcorn outcome (Op) before receiving Pavlovian pairings of one abstract stimulus (Sc) with chocolate and another (Sp) with popcorn. During the transfer test, the participants preferentially performed the response trained with the same outcome as the stimulus indicating that they had acquired Sc → Oc → Rc and Sp → Op → Rp associative chains. The crucial feature of the instrumental training in this study was that both the response options were available concurrently in a common stimulus context without any information about which would be the next available outcome. Moreover, the outcomes became available in a random order. Given this training, each outcome representation should have been equally associated with both responses, thereby vitiating the S-R/reinforcement account of the selective transfer observed by Watson et al.

Ideomotor theory offers an alternative explanation of the genesis of O → R associations. This theory has its origins in the 19th century accounts for voluntary action (Stock & Stock, 2004) and, although largely neglected during the 20th century, the last decade has seen a renaissance of interest (Shin, Proctor, & Capaldi, 2010). When applied to instrumental behavior, ideomotor theory also argues that responding is a product of an S → O → R chain but differs from S-R/reinforcement account in the source of the O → R link. The central idea is that the O → R association is generated by the pairings of the response and the outcome brought about by the instrumental contingency so that the direction of the association is backward with respect to the causal (and temporal) sequence of the response and outcome. As a result, according to this account, the stimulus context retrieves a memory of the outcome available in this context in the past through Pavlovian learning, which in turn activates the response that produced the outcome through ideomotor learning, thereby generating the S → O → R chain. Because the origin of the O → R associations lies in experience with the instrumental R−O contingencies, the ideomotor account provides an explanation of the selective transfer observed by Watson et al. (2014).

Although there are good reasons for believing in the reality of the ideomotor learning (de Wit & Dickinson, 2016; Shin et al., 2010), it fulfills neither the desire nor the belief criteria for goal-directed behavior in spite of the fact that responses are selected by a representation of their outcomes. With respect to the belief criterion, the issue is theoretical. Whereas a lever press → food pellet association could represent the belief *cause(leverpress,access(food-pellet))*, given that the mechanism for deploying the association respects the fact that it is a representation (see below), it surely is perverse to argue that a backward food pellet → lever press association represents the fact the lever press causes access to food pellets in that the direction of the association is the opposite of the direction of causation.

The problem for the desire criterion is empirical. As specific PIT is an assay of the S → O → R chain, this form of transfer should be sensitive to motivational variables if it is to fulfill the desire criterion. However, in contrast to general PIT, this form of transfer is often impervious to the motivation status of the outcome. For example, the magnitude of the specific PIT was unaffected by devaluing one of the food outcomes by prefeeding in both human participants (Watson et al., 2014) and rodents. Moreover, in the latter case, the insensitivity of specific PIT to outcome revaluation contrasted with a reduction in the general form in the same experiment (Corbit, Janak, & Balleine, 2007), thereby confirming an earlier report by Rescorla (1994) that specific PIT is unaffected by devaluation of the outcome by aversion conditioning in rats. The resistance of specific PIT to outcome devaluation by aversion conditioning is also observed in humans, at least as assessed by the difference in responding to the Pavlovian stimuli associated with the same and different outcomes of biological relevance (Eder & Dignath, 2016). In summary, specific PIT fails to meet the desire criterion in that the effect is insensitive to outcome value, at least when the outcomes are of biological relevance.[1]

Although interactions between Pavlovian and instrumental learning may well endow behavior with a veneer of goal-directedness, they do so by engaging forms of habit learning rather than a belief–desire psychology that we argue characterizes goal-directed behavior, at least at the psychological level. General PIT reflects the fact that the concurrent activation of a reward representation exerts a motivational influence over ongoing habitual behavior. By contrast, an activated representation of the sensory properties of an outcome generates specific PIT by priming or eliciting the habitual component of instrumental behavior.

DUAL-SYSTEM THEORIES OF INSTRUMENTAL LEARNING

The claim that Pavlovian conditioning can motivate or control instrumental performance is referred to as two-process theory (Rescorla & Solomon, 1967; Trapold & Overmier, 1972). As we have discussed, however, instrumental behavior itself involves two systems, the goal-directed and the habitual, which we shall refer to as the *dual-system theory* of instrumental behavior to distinguish it from the classic two-process account of Pavlovian–instrumental interactions. Dual-system theories, while endorsing a version of the stimulus-response account of habit learning, differ in their conception of goal-directed learning. Ever since Tolman (1959) recast the belief–desire account of folk psychology into the jargon of academic psychology, cognitive and social psychologists have offered a plethora of expectancy-value theories of goal-directed action. However, these theories do not offer an integrated account of both goal-directed and habitual responding and therefore lie outside the scope of this chapter.

Over the last decade or so, research on human goal-directed and habitual behavior has become increasingly couched in terms of computational reinforcement theory, which has its origins in machine learning and offers a normative account of both goal-directed and habitual behavior (Dolan & Dayan, 2013; Sutton & Barto, 1998). According to dual-system reinforcement theory (Daw, Niv, & Dayan, 2005), goal-directed behavior is controlled by *model-based* computations in which the agent uses *state* prediction errors to learn a model of the state transitions produced by the instrumental contingencies. This model therefore functions like the instrumental belief of folk psychology. Model-based control contrasts with a less computationally

---

[1] An exception to this generalization may well be transfer procedures that involve purely symbolic outcomes that we suspect engages inferential processes that lie outside the scope of the basic belief–desire psychology, which is the focus of this review. For example, Allman et al. reported that transfer is sensitive to changes in the values of fictitious currencies in a stock market task (Allman, DeLeon, Cataldo, Holland, & Johnson, 2010).

demanding *model-free control*, which uses *reward* prediction errors to learn habit-like responses.

Although the reinforcement theory assumes that both systems learn concurrently, an additional computational process is required to determine whether the model-based or the model-free system controls the behavioral output. Daw et al. (2005) originally suggested that the arbitration between the two systems or controllers is based on the uncertainty of the utilities predicted by each system. The selected controller is the one that yields the most certain prediction as assessed by the state and reward prediction errors generated by the two systems. Moreover, they demonstrated by simulation that this arbitration, when biased against the model-based system as a result of its inherent computational cost, can predict, at least at a qualitative level, both the development of behavioral autonomy with extensive training (Adams, 1982) and the maintained sensitivity to outcome devaluation following extensive choice training between two actions yielding different outcomes (e.g., Klossek et al., 2011; Kosaki & Dickinson, 2010). Since this initial research, elaborations of this arbitration have been proposed and investigated within the context of human decision tasks (see Chapter 7).

The most problematic finding for computational reinforcement theory arises from our original contrast between the effects of ratio and interval training. Recall that training on an interval contingency is more likely to engender habitual control than ratio training even when the probability of the outcome on the interval schedule is the same as or higher than that on the ratio schedule (Dickinson et al., 1983). The prediction error learning algorithms of reinforcement theory are sensitive to the probability of outcomes but not to temporal variations in the likelihood that a response will produce an outcome. Therefore, as the arbitration is based on the state and reinforcement prediction errors generated by the model-based and model-free processes, respectively, there is no reason why interval training should yield more uncertain predictions than matched ratio training.

Rate correlational theory

Prior to our discovery that ratio-trained behavior is more likely to be goal-directed than interval-based responding, Baum suggested that the elevated response rate maintained by ratio contingencies could be explained by a correlational-based law of effect (Baum, 1973). Whereas response strength is increased by temporally contiguous pairings of response and reward according to Thorndike's (1911) law of effect, Baum argued that responding is determined by the correlation between the rate of responding and the rate of the reward or outcome as assessed across a series of time samples. By remembering the number of responses and outcomes that fell within each sample, the correlation between the response and outcome rates can be computed across the time samples. This correlational law of effect readily explains the impact of the primary variables of instrumental conditioning, response—outcome contiguity, and contingency. As the temporal interval between a response and the contingent reward lengthens, the likelihood that the response and its reward fall in different time samples increases, thereby reducing the experienced correlation. Similarly, degrading the contingency by delivering rewards independently of responding will also reduce the experienced correlation between response and reward rates.

Ratio and interval training

With respect to the ratio—interval contrast, a critical feature of the interval contingency is that the maximum rate of reward is set by the scheduled average interreward interval, so that variations in responding have little impact on the reward rate once the response rate is sufficient to collect the rewards almost as soon as they become available. As a consequence, the correlation for an interval contingency is likely to be low. By contrast, given that the response rate varies sufficiently on a ratio schedule, there will always be a positive correlation between response and reward rates with the strength of the correlation being determined by the probability of reward. Therefore, the response rate—outcome rate correlation should be higher for the ratio than for the interval contingency even when the reward probability or the reward rate is matched across the two contingencies.

Given this property of the rate correlation, Dickinson (1985) suggested that the goal-directed component of a dual-system theory is determined by the currently experienced rate correlation. Specifically, within the framework of belief—desire psychology, the claim is that the g parameter of the instrumental belief, which represents the believed strength of the causal relationship between action and outcome, is determined by the experienced rate correlation. Direct evidence for this claim comes from the fact that the rate correlation generated across 10-s samples during training on ratio and interval contingencies has been positively related to human performance (Pérez et al., 2016) and judgments of the causal effectiveness of an action (Tanaka, Balleine, & O'Doherty, 2008). Dickinson (1985) also suggested that the habit learning follows Thorndike's law of effect in which increments in S-R strength are produced by contiguous reinforcement of the response. Finally, in contrast to computational reinforcement theory, Dickinson, Balleine, Watt, Gonzalez, and Boakes (1995) assumed that the outputs of the goal-directed and habit systems simply summate to determine the current response rate.

The rationale for this simple summation assumption is illustrated by the lever-press rates displayed in Fig. 1.1. The ratio and interval data come from the outcome devaluation test of Dickinson et al. (1983). In this experiment, rats received limited lever-press training for food pellets before aversion conditioning to devalue the pellets. As we have already noted, a devaluation effect was observed following ratio but not interval training. For present purposes, however, the important point is that devaluation of the ratio reward reduced responding to the level produced by the interval groups. Because the rate correlation should be close to zero for the interval groups, this level of responding

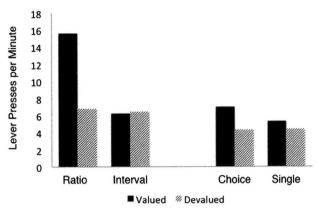

Figure 1.1 Mean lever presses per minute during an extinction test following either the devaluation of the outcome by aversion conditioning (Devalued) or following no devaluation (Valued). The rates following ratio and interval training are taken from Experiment 1 of Dickinson et al. (1983), whereas the rates following choice and single-response training are taken from the choice and NCT1 groups, respectively, of Experiment 2 of Kosaki and Dickinson (2010).

represents the habit strength produced by the number and probability of the reinforcing outcome experienced during training. As the ratio and interval groups received the same number and probability of reinforcement, we should expect the habit strength to have been similar in the two training conditions. In accord with this analysis, having removed the goal-directed component by devaluation, the residual habitual ratio responding was similar to the level of interval responding. Therefore, the valued ratio performance appears to be the sum of the goal-directed and habitual components.

Fig. 1.2 illustrates simulations of the rate correlational dual-system theory (see Appendix for details) for training on a random interval schedule (top panel) and on ratio schedule under which the probability of the outcome was yoked to the probability generated by performance on the interval schedule (bottom panel). From the outset of training, the probability of a response generated by the goal-directed learning (pg) is greater for ratio than for interval training with the interval pg rapidly dropping to near zero with further training. As a result, the overall probability of responding (p) converges more rapidly with the probability generated by the habit learning (ph) under interval training than under ratio training, thereby explaining why Dickinson et al. (1983) could observe an outcome devaluation effect following limited ratio training but not following equivalent interval training when the probabilities of the outcome per response were matched.

Extended training

The bottom panel of Fig. 1.2 shows that the goal-directed pg also declines with extended ratio training with the result that the overall summed p converges on the habit ph, thereby explaining why overtraining often produces behavioral autonomy following further ratio

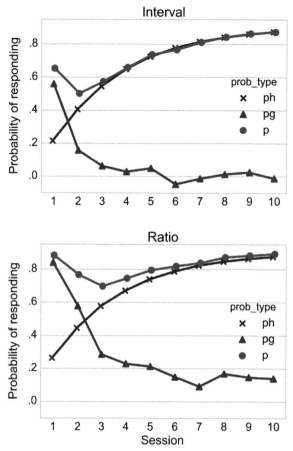

Figure 1.2 Mean probability of responding per time unit (*p*) on a random interval (top panel) and yoked ratio schedule (bottom panel) derived from 30 simulations of the rate correlation dual-system theory. Whereas outcomes became available on average once every 10 time units on the interval schedule, the average probability of an outcome for a response for each simulation on the ratio schedule was matched to that generated by a master simulation of the interval schedule for each session. Also displayed are the probabilities of a response per unit time generated by the goal-directed (*pg*) and the habit learning (*ph*). During pretraining (not shown), outcomes were delivered on a ratio 1 schedule for the first session and a random ratio 5 schedule for the next two sessions. All sessions terminated after the delivery of the 30 outcomes. See Appendix for details of the simulated theory.

training (e.g., Adams, 1982). Of course, there always remains an actual positive ratio contingency but, as there is little variation in responding across the time samples when the response probability *p* is high due to the acquisition of habit strength *ph*, the agent no longer experiences a rate correlation. Thus, stereotyped responding generates behavioral autonomy according to this theory.

Choice training

Finally, the remaining data in Fig. 1.1 show performance during the devaluation test of an experiment by Kosaki and Dickinson (2010), which illustrates that rats remain sensitive to outcome devaluation following choice training even though the equivalent single-response training yields behavioral autonomy. The important feature of these data is the fact that the summation effect is again observed in that outcome devaluation following choice training reduces responding to the level observed following single-response training. The reason for persistent goal-directed responding is because the choice prevents the development of stereotyped rates of responding. When the animal responds at one of the two choice options, there will be a reduction in number of the other response and its outcome in the current time samples relative to those samples in which this latter response is chosen. Therefore, even when responding is at asymptote with strong and equal habit strengths for each response option, there will still be a variation in the number of each of the two responses and their outcomes across the time samples. These variations ensure that the agent continues to experience a positive rate correlation for each response option so that goal-directed control should persist. When we always have to make a decision about which action to take, then behavior never becomes purely habitual.

Avoidance

The empirical and theoretical development of dual-system theory has focused almost exclusively on rewarded behavior to the neglect of instrumental avoidance. Not only can instrumental action be deployed to gain resources but also to avoid an aversive event or state, which is referred to as a negative reinforcer in the jargon of behavioral psychology because the reinforcement operates through a negative contingency with the action. Baum (1973) discussed avoidance within the context of his correlational law of effect in that an agent experiences a negative correlation between the action and the reinforcer under an avoidance contingency. If experience of a negative rate correlation generates a belief that the action prevents the reinforcer and if the agent desires that it should not occur, then folk psychology provides a straightforward account of avoidance. More formally, recall that within our PROLOG description of practical inference, the product of g and δ determines the will to execute the intention inferred from the respective belief and desire. If we assume that $access(O,\delta)$ represents the content of a desire to avoid O when δ has a negative value, the will to execute the intention will be positive when pg also has a negative value under an avoidance contingency.

Whether or not avoidance behavior meets the desire criterion for goal-directedness, at least as assessed by the reinforcer revaluation paradigm, has received scant attention. Recently, however, A. Fernando, G. Urcelay, A. Mar, A. Dickinson, and T. Robbins (2014) trained rats to press a lever to avoid a foot-shock before revaluing the shock by presenting it noncontingently under the influence of an analgesic dose of morphine. A subsequent test of lever pressing in the absence of the shock revealed that the revaluation treatment had reduced avoidance responding, thereby demonstrating that avoidance met the desire criterion with respect to foot-shock.

Furthermore, there is evidence that the habit system also contributes to avoidance. In the procedure employed by Fernando et al., each lever press not only canceled the next scheduled shocks but also turned on an auditory feedback stimulus to facilitate the acquisition of avoidance. A second set of studies confirmed that this stimulus augmented avoidance (A. B. P. Fernando, G. P. Urcelay, A. C. Mar, A. Dickinson, & T. W. Robbins, 2014), and it is usually assumed that such explicit feedback stimuli play the same role as intrinsic action-generated feedback stimuli but in a more salient form. Although the process by which a feedback stimulus becomes a conditioned reinforcer is beyond the scope of this chapter, it is usually assumed that its reinforcing function derives from its role as a safety signal predicting the absence of the aversive reinforcer. More pertinent, however, is the fact that such a stimulus provides a potential source of positive reinforcement for an avoidance habit being not only contingent but also contiguous with the avoidance response.

Direct evidence that the feedback stimulus served to reinforce an avoidance habit comes from the insensitivity of performance to revaluation of this stimulus. Fernando et al. used the same revaluation procedure, as they had employed successfully for the foot-shock by giving noncontingent presentations of the feedback stimulus under morphine. Although this revaluation treatment enhanced the ability of the feedback stimulus to reinforce the avoidance, it had no detectable effect in an extinction test immediately following the revaluation. Therefore, the feedback stimulus appears to act as a positive reinforcer of the avoidance habit, so that in the case of intrinsic feedback stimuli habitual responding becomes self-reinforcing.

Taken together, these studies suggest that instrumental avoidance is also under dual-system control in which responding is goal-directed with respect to avoidance of the primary negative reinforcer, the shock, at the same time as being habitual with respect to response feedback stimuli. It should be noted, however, that the goal-directed status of avoidance has only been evaluated against the desire criterion, and whether or not the effects of revaluing the primary reinforcer (foot-shock) is mediated by knowledge of the instrumental avoidance contingency, as required by the belief criterion, remains to be determined.

System integration

In summary, rate correlational theory, when assimilated into a dual-system framework, provides a principled account of goal-directed and habitual control. However, there are outstanding issues. Our version of dual-system theory assumes that the outputs of the goal-directed and habitual systems summate in generating behavior but fail to offer commensurate psychologies for the two systems that would allow for such summation. We appeal to an intentional psychology involving the process of practical inference to explain goal-directed action, whereas habitual responding is attributed to a mechanistic psychology in which the process of excitation (and inhibition) operates through associative connections. Dickinson (2012) has suggested that this disjunction might be resolved by an associative account of practical inference within the processing architecture of an associative—cybernetic model. This model has its origins in Thorndike's ideational theory of instrumental action (Thorndike, 1931), which Sutton and Barto (1981) subsequently simulated 50 years later. Thorndike considered a hungry animal at the choice point in T-maze in which it has learned that a right turn leads to food and a left turn to shock. The animal resolves this choice by imagining turning right, which in turn leads to the associative activation of a food representation. The positive evaluation of the food then feeds back to augment the activation of the right turn sufficiently to produce this response. In contrast, contemplation of a left turn activates a shock representation with its evaluative feedback inhibiting this response.

Within this framework, Dickinson (2012) suggested that the initial activation of the response representation constitutes a latent intention that, in turn, retrieves the belief about the causal consequences of action in the form of the association between the response and outcome representations. The desire for the outcome is then implemented by an activated association between the outcome representation and a motivational mechanism with the feedback process completing the practical inference to generate an executable intention. Dickinson and Balleine (1993) integrated this associative—cybernetic system with an S–R/reinforcement process for habits to generate a fully mechanistic dual-system psychology. However, our purpose here is not to describe and evaluate the associative—cybernetic model, which has been extensively presented in the literature (Balleine & Ostlund, 2007; de Wit & Dickinson, 2009; Dickinson, 1994) but rather to use this model to illustrate the necessity for an integration of the intentional and mechanistic psychologies in the control of behavior.

LOOKING TO THE FUTURE

Goal-directedness is a profligate concept in psychology, having been ascribed to behavior ranging from target-directed responses, such as preparatory grasping of objects, through instrumental action that yields access to currently valued resources, to future planning, such as food caching and subscribing to a pension plan. This chapter focuses on the second, instrumental sense of the term that, when integrated into a dual-system theory, has become a focus of research during the last decade or so. Although Dickinson (1985) argued for the rate correlational account of goal-directed learning within the framework of dual-system control over 30 years ago, more than half of the citations to this paper have occurred within the last 5 years.

Although most recent research has focused on the neurobiology rather than psychology of dual-system control (Balleine & O'Doherty, 2010; Dolan & Dayan, 2013), the belief and desire criteria, derived from the practical inference account of folk psychology, still provide empirical benchmarks for identifying an important class of adaptive behavior that is goal-directed in terms of the underlying psychological processes rather than just at the functional or descriptive level. Whether the beliefs and desires that underpin these criteria should be regarded metaphorically, at least in the case of animals, remains a contentious issue. Dickinson addressed this issue by arguing that the acquisition of human beliefs or judgments about the causal status of an action are governed by the same learning processes as animal instrumental conditioning (Dickinson & Shanks, 1995), a finding compatible with the claim that goal-directed instrumental learning yields propositional-like representations in both humans and other animals. Correspondingly, in their hedonic interface theory Dickinson and Balleine (2009) argued that goal values are learned and represented abstractly and therefore could also operate through a desire with propositional-like content.

There has always been an inferential asymmetry in the assignment of control based on the outcome revaluation paradigm. Whereas sensitivity to revaluation indicates goal-directed control, it is the failure to detect a revaluation effect that marks habitual responding. It is therefore reassuring for dual-system theories that the two forms of control have been doubly dissociated by manipulations of corticostriatal systems (Balleine & O'Doherty, 2010). Goal-directed and habitual control have also been dissociated motivationally with the latter being more sensitive to the general motivating effects of the stimulus context brought about by Pavlovian conditioning. Indeed, this form of motivation may endow habits with a veneer of goal-directedness in that they can be sensitive to outcome revaluation, thereby meeting the desire criterion. This veneer is deepened by the fact that a sensory representation of the outcome can also act as a stimulus capable of priming or eliciting the response through an $S \rightarrow O \rightarrow R$ chain generated by S-R/reinforcement and/or ideomotor learning. However, both the motivational and stimulus effects fail the belief criterion in that they do not operate through a causal representation of the instrumental action—outcome contingency.

An influential theoretical advance in the last decade or so is the distinction between model-based and model-free control within the framework of computational reinforcement theory. However, the state prediction error learning rules of the model-based account fail to capture important aspects of goal-directed control. The propensity of interval contingencies to establish habitual control is problematic for reinforcement theory and suggests a form of goal-directed learning based upon the rate correlation. Moreover, the interaction between goal-directed and habitual control often appears to be cooperative rather than competitive.

Finally, although the desire criterion has generally been accepted as an assay of goal-directed status, there are problematic cases. Consider food caching. If caching is regarded as the instrumental action and recovery of the food as the outcome, the value of the food at the time of action can be dissociated from its value at the time of the outcome. The desire criterion specifies that it is the value of the outcome at the time of caching that should control this action if it is goal-directed, whereas from a functional perspective, it is the value of the food at the time of recovery, when it is needed, that should be critical. Dickinson with Clayton et al. examined this issue using jays, which are omnivorous cachers (Cheke & Clayton, 2012; Correia, Dickinson, & Clayton, 2007). The idea behind these experiments was to manipulate the relative values of two foods at caching and at recovery to determine which value controls the caching. For example, in one experiment the jays cached two different types of food, pine seeds and dog kibble, in the morning followed in the afternoon by the opportunity to recover their caches. To manipulate the relative values of these two foods, some of the jays were prefed the pine seeds prior to caching in the morning to reduce their value relative to the dog kibbles through specific satiety. In accord with the desire criterion, these jays cached less seeds than kibble. However, in the afternoon, these jays were prefed the kibbles prior to recovery, so that the seeds were now the more valuable food. At issue was the caching preference next morning when the birds were again prefed the seeds. If motivational control reflects the current desires, as the desire criterion requires, the jays should again have cached more of the kibble having just been prefed seeds. By contrast, the biological function of caching predicts that caching should reflect the relative values of the foods at recovery, and therefore a switch in caching preference from the kibble to the seeds is predicted on this second caching episode. In accord with this prediction, the jays cached relatively more seeds than kibbles in spite of having just been prefed the seeds. Thus, the jays preferentially cached the food items that were valuable at the time of recovery rather than those that were valuable when they were actually engaged in caching. It is not that the relative desires of jays had switched on the second caching episode because they continued to eat more of the nonprefed food, kibble in this case, at the same time as preferentially caching the prefed food, the seeds.

It is, of course, the case that food caching is a specialized biological adaptation, although this experiment shows that the choice of which food to cache is instrumental in the sense that it is sensitive to the outcomes at the time of recovery. Moreover, caching appears similar to many forms of our own future planning, such as stocking up the larder for next week's meals. It may well be that such future planning engages psychological processes that differ from those mediating goal-directed action as defined by the desire criterion. For example, it has been suggested that future planning involves episodic memory and it is well-established that the recovery searches by the jays are controlled by an episodic-like memory for the caching episode (Clayton & Dickinson, 1998). The integration or differentiation of goal-directed behavior and future planning remains an important research issues in the study of purposive behavior.

APPENDIX: SIMULATION OF RATE CORRELATION DUAL-SYSTEM THEORY

The simulation of the goal-directed learning assumes that the agent segments time into a series of time samples, each 20 time units in length where a unit is the time required to execute a single response. The number of responses and outcomes in each new sample is registered in a working memory with a capacity of 20 successive samples so that as the memory cycles each new sample replaces the oldest one. The probability of a response per time unit generated by goal-directed learning after the kth memory cycle [$pg(k)$] is equal to the Pearson's correlation coefficient [$r(k)$] between the number of responses and outcomes in each sample calculated across the samples in the current memory cycle weighted with the correlation obtained in the previous cycle [$r(k-1)$], so that $pg(k) = \theta r(k) + [1 - \theta]r(k-1)$, where $0 < \theta < 1$.

The probability of a response in each time unit (t) generated by the habit learning [$ph(t)$] is determined by a reinforcement learning algorithm that is sensitive to the contiguity between responses and outcomes. If a response was reinforced, the change in the probability of responding (Δph) generated by the habit learning was given by $\Delta ph = \alpha[1 - ph(t)]$, where α is the learning rate for reinforced responses ($0 < \alpha < 1$), the asymptotic value of ph is 1, and $ph(t)$ is the current value of ph. Conversely, if a response was performed but not reinforced, the probability changed according to $\Delta ph = -\beta[ph(t)]$, where β is a learning rate for nonreinforced responses ($0 < \beta < 1$). Therefore, the value of the probability generated by habit learning in each cycle is given by the final value accrued by ph during the cycle. The values chosen for the parameters were $\theta = 0.5$, $\alpha = 0.01$ and $\beta = \alpha/100$.

Finally, the model assumes that the outputs of the two systems summate so that the probability of a response in each cycle k is given by $p(k) = pg(k) + ph(k) - pg(k) \cdot ph(k)$, where $ph(k)$ is the value of ph accumulated throughout the simulation.

REFERENCES

Adams, C. D. (1980). Post-conditioning devaluation of an instrumental reinforcer has no effect on extinction performance. *Quarterly Journal of Experimental Psychology, 32*(3), 447–458. https://doi.org/10.1080/14640748008401838.

Adams, C. D. (1982). Variations in the sensitivity of instrumental responding to reinforcer devaluation. *Quarterly Journal of Experimental Psychology: Comparative and Physiological Psychology, 34B*(2), 77–98.

Adams, C. D., & Dickinson, A. (1981). Instrumental responding following reinforcer devaluation. *Quarterly Journal of Experimental Psychology: Comparative and Physiological Psychology, 33B*, 109–122.

Allen, C., & Bekov, M. (1995). Cognitive ethology and the intentionality of animal behaviour. *Mind and Lanuage, 10*(4), 313–328.

Allman, M. J., DeLeon, I. G., Cataldo, M. F., Holland, P. C., & Johnson, A. W. (2010). Learning processes affecting human decision making: An assessment of reinforcer-selective Pavlovian-to-instrumental transfer following reinforcer devaluation. *Journal of Experimental Psychology: Animal Behavior Processes, 36*(3), 402–408. https://doi.org/10.1037/a001787.

Balleine, B. W., & O'Doherty, J. P. (2010). Human and rodent homologies in action control: Corticostriatal determinants of goal-directed and habitual action. *Neuropsychopharmacology, 35*, 48—69. https://doi.org/10.1038/npp.2009.131.

Balleine, B. W., & Ostlund, S. B. (2007). Still at the choice-point. Action selection and initiation in instrumental conditioning. *Annals of the New York Academy of Sciences, 1104*, 147—171.

Baum, W. M. (1973). The correlation-based law of effect. *Journal of the Experimental Analysis of Behavior, 20*(1), 137—153.

Carruthers, P. (2004). On being simple minded. *American Philosophical Quarterly, 41*(3), 205—220.

Cheke, L. G., & Clayton, N. S. (2012). Eurasian jays (*Garrulus glandarius*) overcome their current desires to anticipate two distinct future needs and plan for them appropriately. *Biology Letters, 8*(2), 171—175. https://doi.org/10.1098/rsbl.2011.0909.

Clayton, N. S., & Dickinson, A. (1998). Episodic-like memory during cache recovery by scrub-jays. *Nature, 395*, 272—274.

Colagiuri, B., & Lovibond, P. F. (2015). How food cues can enhance and inhibit motivation to obtain and consume food. *Appetite, 84*, 79—87. https://doi.org/10.1016/j.appet.2014.09.023.

Corbit, L. H., Janak, P. H., & Balleine, B. W. (2007). General and outcome-specific forms of Pavlovian-instrumental transfer: The effect of shifts in motivational state and inactivation of the ventral tegmental area. *European Journal of Neuroscience, 26*, 3141—3149.

Correia, S. P. C., Dickinson, A., & Clayton, N. S. (2007). Western scrub-jays anticipate future needs independently of their current motivational state. *Current Biology, 17*, 856—861. https://doi.org/10.1016/j.cub.2007.03.063.

Daw, N. D., Gershman, S. J., Seymour, B., Dayan, P., & Dolan, R. J. (2011). Model-based influences on humans' choices and striatal prediction errors. *Neuron, 69*(6), 1204—1215. https://doi.org/10.1016/j.neuron.2011.02.027.

Daw, N. D., Niv, Y., & Dayan, P. (2005). Uncertainty-based competition between prefrontal and dorsolateral striatal systems for behavioral control. *Nature Neuroscience, 8*, 1704—1711.

Dias-Ferreira, E., Sousa, J. C., Melo, I., Morgado, P., Mesquita, A. R., Cerqueira, J. J., et al. (July 2009). Chronic stress causes frontostriatal reorganization and affects decision-making. *Science, 31*, 621—625.

Dickinson, A. (1985). Actions and habits: The development of behavioural autonomy. *Philosophical Transactions of the Royal Society (London), B, 308*, 67—78.

Dickinson, A. (1994). Instrumental conditioning. In N. J. Mackintosh (Ed.), *Animal learning and cognition* (pp. 45—79). San Diego, CA: Academic Press.

Dickinson, A. (2012). Associative learning and animal cognition. *Philosophical Transactions of the Royal Society B, 367*, 2733—2742. https://doi.org/10.1098/rstb.2012.0220.

Dickinson, A., & Balleine, B. (1993). Actions and responses: The dual psychology of behaviour. In N. Eilan, R. A. McCarthy, & M. W. Brewer (Eds.), *Problems in the philosophy and psychology of spatial representation* (pp. 277—293). Oxford: Blackwell.

Dickinson, A., & Balleine, B. (2009). Hedonics: The cognitive-motivational interface. In M. L. Kringelbach, & K. C. Berridge (Eds.), *Pleasures of the brain* (pp. 74—84). New York: New York: Oxford University Press.

Dickinson, A., Balleine, B., Watt, A., Gonzalez, F., & Boakes, R. A. (1995). Motivational control after extended instrumental training. *Animal Learning and Behavior, 23*, 197—206.

Dickinson, A., & Dawson, G. R. (1987). Pavlovian processes in the motivational control of instrumental performance. *Quarterly Journal of Experimental Psychology, 39B*, 201—213.

Dickinson, A., & Nicholas, D. J. (1983). Irrelevant incentive learning during instrumental conditioning: The role of the drive-reinforcer and response-reinforcer relationships. *The Quarterly Journal of Experimental Psychology Section B: Comparative and Physiological Psychology, 35*(3), 249—263. https://doi.org/10.1080/14640748308400909.

Dickinson, A., Nicholas, D. J., & Adams, C. D. (1983). The effect of the instrumental training contingency on susceptiblity to reinforcer devaluation. *Quarterly Journal of Experimental Psychology, 35B*, 35—51.

Dickinson, A., & Shanks, D. R. (1995). Instrumental action and causal representation. In D. Sperber, D. Premack, & A. J. Premack (Eds.), *Causal cognition* (pp. 5—25). Oxford: Clarendon Press.

Dolan, R. J., & Dayan, P. (2013). Goals and habits in the brain. *Neuron, 80*, 312–325. https://doi.org/10.1016/j.neuron.2013.09.007.

Economides, M., Kurth-Nelson, Z., Lübbert, A., Guitart-Masip, M., & Dolan, R. J. (2015). Model-based reasoning in humans becomes automatic with training. *PLoS Computational Biology, 11*(9), e1004463.

Eder, A. B., & Dignath, D. (2016). Cue-elicited food seeking is eliminated with aversive outcomes following outcome devaluation. *Quarterly Journal of Experimental Psychology, 69*(3), 574–588. https://doi.org/10.1080/17470218.2015.1062527.

Fernando, A., Urcelay, G., Mar, A., Dickinson, A., & Robbins, T. (2014). Free-operant avoidance behavior by rats after reinforcer revaluation using opioid agonists and d-amphetamine. *Journal of Neuroscience, 34*(18), 6286–6293. https://doi.org/10.1523/JNEUROSCI.4146-13.2014.

Fernando, A. B. P., Urcelay, G. P., Mar, A. C., Dickinson, A., & Robbins, T. W. (2014). Safety signals as instrumental reinforcers during free-operant avoidance. *Learning and Memory, 21*, 488–497. https://doi.org/10.1101/lm.034603.114.

Friedel, E., Koch, S. P., Wendt, J., Heinz, A., Deserno, L., & Schlagenhauf, F. (2014). Devaluation and sequential decisions: Linking goal-directed and model-based behavior. *Frontiers in Human Neuroscience, 8*, 587. https://doi.org/10.3389/fnhum.2014.00587.

Gillan, C. M., Otto, A. R., Phelps, E. A., & Daw, N. D. (2015). Model-based learning protects against forming habits. *Cognitive, Affective, & Behavioral Neuroscience, 15*, 523–536. https://doi.org/10.3758/s13415-015-0347-6.

Guthrie, E. R. (1959). Association by contiguity. In S. Koch (Ed.), *Psychology: A study of a science* (Vol. 2, pp. 158–195). New York: McGraw-Hill.

Heyes, C., & Dickinson, A. (1990). The intentionality of animal action. *Mind and Language, 5*, 87–104.

Heyes, C., & Dickinson, A. (1995). Folk psychology won't go away: Response to Allen and Bekoff. *Mind and Language, 10*(4), 329–332.

Holland, P. C. (2004). Relations between Pavlovian-instrumental transfer and reinforcer devaluation. *Journal of Experimental Psychology: Animal Behavior Processes, 30*, 104–117.

Hull, C. L. (1952). *A behavior system*. New Haven: Yale University Press.

Jonkman, S., Kosaki, Y., Everitt, B. J., & Dickinson, A. (2010). The role of contextual conditioning in the effect of reinforcer devaluation on instrumental performance by rats. *Behaviour Processes, 83*, 276–281. https://doi.org/10.1016/j.beproc.2009.12.017.

Klossek, U. M. H., Russell, J., & Dickinson, A. (2008). The control of instrumental action following outcome devaluation in young children aged between 1 and 4 years. *Journal of Experimental Psychology: General, 137*, 39–51.

Klossek, U. M. H., Yu, S., & Dickinson, A. (2011). Choice and goal-directed behavior in preschool children. *Learning and Behavior, 39*, 350–357. https://doi.org/10.3758/s13420-011-0030-x.

Kosaki, Y., & Dickinson, A. (2010). Choice and contingency in the development of behavioral autonomy during instrumental conditioning. *Journal of Experimental Psychology: Animal Behavior Processes, 36*(3), 334–342. https://doi.org/10.1037/a0016887.

Krieckhaus, E. E., & Wolf, G. (1968). Acquisition of sodium by rats: Interaction of innate mechanisms and latent learning. *Journal of Comparative and Physiological Psychology, 65*, 197–201.

Lovibond, P. F. (1983). Facilitation of instrumental behavior by a Pavlovian appetitive conditioned stimulus. *Journal of Experimental Psychology: Animal Behavior Processes, 9*, 225–247.

Pérez, O., Aitken, M. R. F., Zhukovsky, P., Soto, F. A., Urcelay, G. P., & Dickinson, A. (2016). Human instrumental performance in ratio and interval contingencies: A challenge for associative theory. *The Quarterly Journal of Experimental Psychology*, 1–33. https://doi.org/10.1080/17470218.2016.1265996.

Rescorla, R. A. (1992). Response-outcome versus outcome-response associations in instrumental learning. *Animal Learning and Behavior, 20*(3), 223–232.

Rescorla, R. A. (1994). Transfer of instrumental control mediated by a devalued outcome. *Animal Learning and Behavior, 22*, 27–33.

Rescorla, R. A., & Colwill, R. M. (1989). Associations with anticipated and obtained outcomes in instrumental learning. *Animal Learning and Behavior, 17*(3), 291–303.

Rescorla, R. A., & Solomon, R. L. (1967). Two-process learning theory: Relationship between Pavlovian conditioning and instrumental learning. *Psychological Review, 74*, 151–182.

Rhodes, S. E. V., & Murray, E. A. (2013). Differential effects of amygdala, orbital prefrontal cortex, and pre-limbic cortex lesions on goal-directed behavior in rhesus macaques. *Journal of Neuroscience, 33*(8), 3380–3389. https://doi.org/10.1523/JNEUROSCI.4374-12.2013.

Schwabe, L., & Wolf, O. T. (2009). Stress prompts habit behavior in humans. *Journal of Neuroscience, 29*(22), 7191–7198.

Schwabe, L., & Wolf, O. T. (2010). Socially evaluated cold pressor stress after instrumental learning favors habits over goal-directed action. *Psychoneuroendocrinology, 35*(7), 977–986. https://doi.org/10.1016/j.psyneuen.2009.12.010.

Searle, J. R. (1983). *Intentionality: An essay in the philosophy of mind*. Cambridge: Cambridge University Press.

Shin, Y. K., Proctor, R. W., & Capaldi, E. J. (2010). A review of contemporary ideomotor theory. *Psychological Bulletin, 136*(6), 943–947. https://doi.org/10.1037/a0020541.

Stock, A., & Stock, C. (2004). A short history of ideo-motor action. *Psychological Research, 68*, 176–188.

Sutton, R. S., & Barto, A. G. (1981). An adaptive network that constructs and uses an internal model of its world. *Cognition and Brain Theory, 4*, 217–246.

Sutton, R. S., & Barto, A. G. (1998). *Reinforcement learning*. Cambridge, MA: The MIT Press.

Tanaka, S. C., Balleine, B. W., & O'Doherty, J. P. (2008). Calculating consequences: Brain systems that encode the causal effects of actions. *Journal of Neuroscience, 28*(26), 6750–6755.

Thorndike, E. L. (1911). *Animal intelligence: Experimental studies*. New York: Macmillan.

Thorndike, E. L. (1931). *Human learning*. New York: Century.

Tolman, E. C. (1959). Principles of purposive behavior. In S. Koch (Ed.), *Psychology: A study of a science* (Vol. 2, pp. 92–157). New York: McGraw-Hill.

Tolman, E. C., & Gleitman, H. (1949). Studies in learning and motivation: I. Equal reinforcements in both end-boxes, followed by shock in one end-box. *Journal of Experimental Psychology, 39*, 810–819.

Trapold, M. A., & Overmier, J. B. (1972). The second learning process in instrumental learning. In A. H. Black, & W. F. Prokasy (Eds.), *Classical conditioning II: Current research and theory* (pp. 427–452). New York: Appleton-Century-Crofts.

Tricomi, E., Balleine, B. W., & O'Doherty, J. P. (2009). A specific role for posterior dorsolateral striatum in human habit learning. *European Journal of Neuroscience, 29*, 2225–2232.

Valentin, V. V., Dickinson, A., & O'Doherty, J. P. (2007). Determining the neural substrates of goal-directed learning in the human brain. *Journal of Neuroscience, 27*, 4019–4026.

Watson, P., Wiers, R. W., Hommel, B., & De Wit, S. (2014). Working for food you don't desire. Cues interfere with goal-directed food-seeking. *Appetite, 79*, 139–148. https://doi.org/10.1016/j.appet.2014.04.005.

Wiltgen, B. J., Sinclair, C., Lane, C., Barrows, F., Molina, M., & Chabanon-Hicks, C. (2012). The effect of ratio and interval training on Pavlovian-instrumental transfer in mice. *PLoS One, 7*(10), e48227. https://doi.org/10.1371/journal.pone.0048227.

de Wit, S., & Dickinson, A. (2009). Associative theories of goal-directed behaviour: A case for animal–human translational models. *Psychological Research, 73*(4), 463–476.

de Wit, S., & Dickinson, A. (2016). Ideomotor mechanism of goal-directed behavior. In T. S. Braver (Ed.), *Motivation and cognitive control* (pp. 123–142). New York: Psychology Press.

CHAPTER 2

Instrumental Divergence and Goal-Directed Choice

Mimi Liljeholm
Cognitive Sciences, University of California, Irvine, Irvine, CA, United States

INTRODUCTION

An essential aspect of flexible choice is that alternative actions yield distinct consequences: If all available actions have identical, or highly similar, outcome distributions, such that selecting one action over another does not significantly alter the probability of any given outcome state, an agent's ability to exert control over its environment is considerably impaired. Conversely, when alternative actions produce distinct outcome states, discrimination and selection between actions allow an agent to flexibly obtain the currently most desired outcome. Since subjective outcome utilities are constantly changing, such flexible control is essential for reward maximization and, thus, may have intrinsic value, serving to motivate and reinforce specific decisions, as well as to generally justify the processing cost of goal-directed computations. In this chapter, I discuss work investigating the role of *instrumental divergence*—the degree to which actions differ with respect to their outcome probability distributions—in goal-directed choice.

Formal theories of goal-directed decisions postulate that the agent generates a "cognitive map" of stochastic relationships between actions and states such that, for each action in a given state, a probability distribution is specified over possible outcome states. These transition probabilities are then combined with current estimates of outcome utilities in order to generate action values—the basis of goal-directed choice (Daw, Niv, & Dayan, 2005; Doya, Samejima, Katagiri, & Kawato, 2002). Although computationally expensive (Keramati, Dezfouli, & Piray, 2011; Otto, Raio, Chiang, Phelps, & Daw, 2013; Otto, Skatova, Madlon-Kay, & Daw, 2014), the "on-the-fly" binding of outcome probabilities with utilities offers adaptive advantage over more automatic action selection, which uses cached values based on reinforcement history (Sutton & Barto, 1998). There are, however, situations in which the processing cost of goal-directed computations does not yield the return of flexible control.

As an illustration, consider the scenario in Fig. 2.1A, which shows two available actions, A1 and A2, with bars representing the transition probabilities of each action into three potential outcome states, O1, O2, and O3. Here, the goal-directed approach prescribes that the agent retrieves each transition probability, estimates the current utility of each outcome, computes the product of each utility and associated probability, sums

Goal-Directed Decision Making
ISBN 978-0-12-812098-9, https://doi.org/10.1016/B978-0-12-812098-9.00002-4

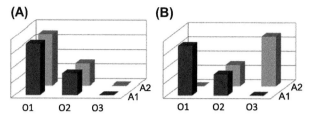

Figure 2.1 Probability distributions over three potential outcomes (O1, O2, and O3) for two available actions (A1 and A2) across which instrumental divergence is zero (1A) or high (1B).

across the resulting value distribution for each action, and, finally, compares the two action values. Of course, given equivalent costs, actions that have identical outcome distributions, as in Fig. 2.1A, will inevitably have the same value, eliminating the need for costly goal-directed computations. However, critically, this lack of instrumental divergence also eliminates the power of choice: Selecting A1 over A2, or vice versa, does not alter the probability of any given outcome state.

Now consider the scenario in Fig. 2.1B, in which the probability distribution of A2 has been reversed across the three outcomes, yielding high instrumental divergence. Note that if the utilities of O1 and O3 are the same, then according to conventional accounts of economic choice, from reinforcement learning (RL) theory to rational choice theory and prospect theory, all actions depicted in Fig. 2.1 have the same *expected utility*. Consequently, there should be no preference for the scenario depicted in Fig. 2.1B over that in Fig. 2.1A. And yet, if one considers the dynamic nature of subjective outcome utilities, the two scenarios clearly differ. To appreciate the significance of this difference, imagine that O1 and O3 represent food and water, respectively, and that at the point of choosing between the two scenarios, you are as hungry as you are thirsty. However, having committed, for example, to Fig. 2.1B, you might find that after a large meal without a drop to drink, your desire for O3 is suddenly greater than that for O1. A few hours later, having thoroughly quenched your thirst, you may again prefer O1. Unlike the scenario illustrated in Fig. 2.1A, the instrumental contingencies in Fig. 2.1B allow you to produce the currently desired outcome as preferences change, by switching between actions. Thus, even when expected utilities are presently the same, the possibility that outcome utilities may subsequently change renders the flexible control afforded by high instrumental divergence essential for long-term reward maximization. As such, high instrumental divergence may have intrinsic utility, eliciting a significant preference for the environment in Fig. 2.1B over that in Fig. 2.1A.

Theories of instrumental behavior distinguish between the goal-directed decisions described above, which are motivated by the probability and current utility of their consequences, and habitual actions, which are rigidly and automatically elicited by the stimulus environment based on their reinforcement history (Balleine & Dickinson, 1998). Although considerable evidence has substantiated this theoretical distinction, and in spite of its far-reaching implications, ranging from the structuring of economic policies to the treatment of compulsive pathology, very little is still known about what factors induce the use of one instrumental strategy over the other. As noted, when instrumental divergence is zero, the greater processing cost of goal-directed computations does not yield the return of flexible control, suggesting that a less resource-intensive, habitual, action selection strategy might be optimal. One possibility, therefore, is that a lack of instrumental divergence (i.e., a failure to map alternative actions to distinct outcome states) results in a degradation of goal-directed performance, eliciting a greater reliance on habitual control.

In this chapter, I will review behavioral and neural support for the role of instrumental divergence in goal-directed decision-making. I will begin by formalizing instrumental divergence as the information-theoretic distance between outcome distributions associated with available action alternatives, relating this novel decision variable to, and dissociating it from, a range of motivational and cognitive factors. I will then review recent work addressing the intrinsic utility of instrumental divergence, including its relevance to psychopathology, and, finally, discuss the potential role of instrumental divergence in the arbitration between goal-directed and habitual decision strategies.

AN INFORMATION-THEORETIC FORMALIZATION OF INSTRUMENTAL DIVERGENCE

Conceptually, instrumental divergence is simply the difference between outcome distributions associated with alternative actions. This concept can be formalized as the Jensen–Shannon (JS) divergence of instrumental outcome probability distributions. Let P_1 and P_2 be the respective outcome probability distributions for two available actions, O be the set of possible outcomes, and $P(o)$ be the probability of a particular outcome, o. The instrumental (JS) divergence is

$$ID = \frac{1}{2}\sum_{o \in O}\log\left(\frac{P_1(o)}{P_*(o)}\right)P_1(o) + \frac{1}{2}\sum_{o \in O}\log\left(\frac{P_2(o)}{P_*(o)}\right)P_2(o), \qquad (2.1)$$

where

$$P_* = \frac{1}{2}(P_1 + P_2).$$

Note that instrumental divergence is defined here with respect to the sensory rather than motivational features of outcome states. Since subjective outcome utilities may change from one moment to the next (e.g., due to sensory satiety), a measure of divergence based on outcome utilities would be inherently unstable. Thus, a definition in terms of nonvalenced sensory features is critical for the broad, organizing, role of instrumental divergence posited here, which includes guiding the organism toward high-agency environments and signaling the need to switch to a habitual decision strategy. Note also that instrumental divergence is defined on distributions associated with *available* action alternatives: If P_1 and P_2 were outcome distributions associated with different cues, or with any other events not subject to the agents volition, their divergence, although relevant to the predictability of the outcome, would not be instrumental and, consequently, would have no implications for flexible instrumental control.

While JS divergence is only one of many distance measures, it has several advantages, including its symmetry and generality: It applies to nominal and numerical, discrete and continuous random variables, and it intuitively generalizes to any arbitrary finite number of probability distributions (Lin, 1991), allowing for comparisons of multiple action alternatives. JS divergence is also intimately related to Shannon entropy, a decision variable frequently shown to influence economic choice (Abler, Herrnberger, Grön, & Spitzer, 2009; Erev & Barron, 2005; Holt & Laury, 2005), that is greatest when the distribution over outcomes is uniform. Given a set of available actions, A, where $p(o|a)$ and $p(o,a)$ are, respectively, the conditional and joint probabilities of outcome o, the Shannon entropy is

$$H = -\sum_{a \in A} \sum_{o \in O} p(o,a) \log p(o|a). \tag{2.2}$$

In spite of the close relationship between the two measures (JS divergence is simply the symmetrized relative entropy), they have dramatically different implications: While Shannon entropy reflects uncertainty about the state of the outcome variable given performance of a particular action, or given a set of available actions as in Eq. (2.2), JS divergence, as applied here, reflects the degree to which discrimination and selection between available actions increases the controllability of the outcome. As discussed in the next section, these closely related information-theoretic variables elicit neural activity in distinct brain regions.

NEURAL CORRELATES OF INSTRUMENTAL DIVERGENCE

A large literature has identified neural signals scaling with trial-by-trial estimates of goal-directed action values (i.e., the expected utility of available response options) (Gläscher, Hampton, & O'doherty, 2008; Rangel & Hare, 2010; Wunderlich, Dayan, & Dolan, 2012; Wunderlich, Rangel, & O'Doherty, 2009). While instrumental divergence is not a measure of the value of performing a particular action (since it is defined with respect to sensory rather than motivational outcome features), it may improve the efficacy of such estimates by identifying instances in which a goal-directed decision strategy yields flexible control over outcomes. This marker of flexible control can then be used to guide the organism toward high-agency environments, to signal that a transition to habitual performance might be advantageous, or to restrict searches of the state—action space. Given these important characteristics, one might expect a neural signature of instrumental divergence to be present during human goal-directed performance, dissociable from the well-established neural correlates of action values. In this section, I will review prominent formal accounts of goal-directed action values, highlighting their relevance to, and recently demonstrated neural dissociation from, instrumental divergence.

Formal accounts of goal-directed action values: Early accounts of goal-directed performance formalized the strength of the action—reward relationship as the difference between two conditional probabilities; the probability of gaining a target reward (r), given that a specific action (a) is performed and the probability of gaining the reward in the absence of that action ($\sim a$) (Hammond, 1980):

$$\Delta P = p(r|a) - p(r|\sim a). \tag{2.3}$$

Sensitivity to this "instrumental contingency" is a defining property of goal-directed actions that has been reliably demonstrated in humans (Chatlosh, Neunaber, & Wasserman, 1985; Liljeholm, Tricomi, O'Doherty, & Balleine, 2011; Shanks & Dickinson, 1991) as well as rodents (Balleine & Dickinson, 1998; Hammond, 1980). Instrumental divergence can be characterized as a generalization of the instrumental contingency rule, extending the contrast over multiple actions and sensory-specific outcomes. The representational change achieved by this simple extension is profound; while the instrumental contingency is a signed measure of the relative advantage of performing a particular action, instrumental divergence is a symmetric measure of the degree to which discrimination and selection between actions alters the probabilities of potential outcome states (i.e., the degree of flexible instrumental control).

A more recent formal framework that represents the full sensory-specific outcome distributions of alternative actions is model-based RL (e.g., Daw et al., 2005). Specifically, for each action available in the current state, and for all possible outcome states, model-based RL maintains separate representations of the probability of transitioning into a possible subsequent state, given that a particular action is performed in the current state, $T(s,a,s')$, and the reward associated with that subsequent state, $R(s')$. Transition probabilities and rewards are dynamically combined, at each choice point, to yield action values:

$$Q(s, a) = \sum_{s'} T(s, a, s') * \left[r(s') + \gamma \max_{d'} Q(s', d') \right], \tag{2.4}$$

where $Q(s',d')$ is the, recursively defined, value of an action performed in the subsequent state and γ is a discount parameter. The transition probabilities may be presumed to be known, or may be incrementally acquired based on trial-by-trial feedback, using a state prediction error:

$$T(s, a, s') = T(s, a, s') + \eta(1 - T(s, a, s')), \tag{2.5}$$

where η is the learning rate. Note that, although sensory-specific transition probabilities are explicitly estimated and represented, they are used solely in the service of generating action values, through their combination with outcome utilities. In contrast, the argument set forth in this chapter is that sensory-specific transition probabilities are also used to estimate instrumental divergence, which is in turn used to guide the deployment of goal-directed processes.

Neural correlates of motivational and information-theoretic variables: As a first step in evaluating the representation of instrumental divergence in goal-directed processes, Liljeholm, Wang, Zhang, and O'Doherty (2013) used fMRI to investigate a neural signal scaling with instrumental divergence, and the dissociability of such a signal from the effects of other motivational and information-theoretic variables. On each trial in their choice task, participants selected between two available actions, given a set of food treats potentially produced by those actions (see Fig. 2.2A). The probability distributions over food treats, including a "no treat" outcome, for four distinct action alternatives were trained to criterion prior to the choice task, and this procedure was repeated in each of three consecutive blocks, using different probabilities and food treats in each block, to ensure sufficient variance. The subject-specific utilities of food treats were assessed using evaluative pleasantness ratings and a standard, incentive compatible, Becker–DeGroot–Marschak auction (Becker, DeGroot, & Marschak, 1964).

Liljeholm et al. (2013) modeled the BOLD response during the choice period of each trial as a function of the instrumental divergence between the actions available on the trial and the values of those actions derived using model-based RL. Consistent with previous work (Gläscher et al., 2008; Wunderlich et al., 2012, 2009), they found that the value of the chosen action scaled with activity in the ventromedial prefrontal cortex: In contrast, instrumental divergence correlated with activity in the right supramarginal gyrus of the inferior parietal lobule (IPL)—a region previously implicated in the planning, execution, and observation of goal-directed actions (Fincham, Carter, Van Veen, Stenger, & Anderson, 2002; Liljeholm, Molloy, & O'Doherty, 2012; Liljeholm et al., 2011). Importantly, the effect of instrumental divergence in the IPL was also dissociable from other information-theoretic and motivational variables, such as the entropy of outcome distributions for chosen actions, which scaled with activity in the dorsolateral prefrontal cortex (DLPFC), and the summed utility of potential food treats, which elicited activity in the insula and ventral striatum (see Fig. 2.2B). A Bayesian model selection analysis ruled out additional competing variables, such as the difference between reward probabilities associated with available action alternatives (i.e., the absolute value of ΔP) and the overall probability of reward on each trial, as sources of the IPL activity. It should be noted that a BOLD signal scaling with instrumental divergence says very little about how a distributed neural code of instrumental divergence may be implemented—an important question for future work. Nonetheless, the identification of a neural signal scaling with instrumental divergence during instrumental choice performance supports

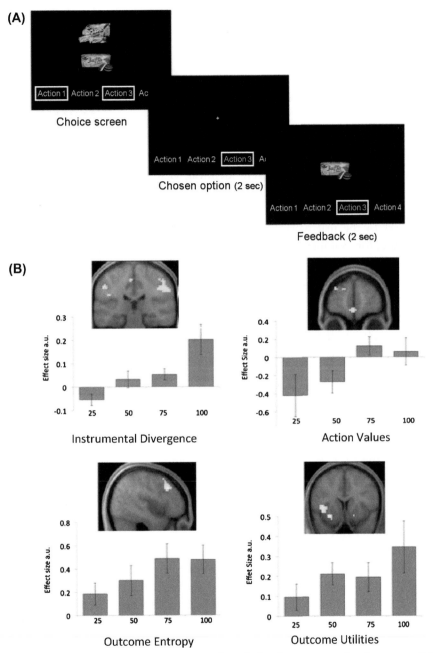

Figure 2.2 (A) Illustration of a trial in the choice task, with the choice screen showing two available actions and the food treats potentially produced by those actions. (B) Parametric modulation, during the choice screen period of each trial, of activity in the IPL by instrumental divergence (top left), of ventromedial prefrontal cortex activity by the value of the chosen action (top right), of DLPFC activity by the outcome entropy for the chosen action (bottom left), and of anterior insula and ventral striatum by the summed utility of potential outcomes (bottom right). *(Task and results from Liljeholm et al. (2013).)*

the notion that this variable may play an important role in decision-making. In subsequent sections, I discuss more direct, behavioral evidence for an influence of instrumental divergence on goal-directed choice.

INSTRUMENTAL DIVERGENCE AND THE INTRINSIC UTILITY OF CONTROL

Imagine that you had to commit, for some duration, to one of the two environments illustrated in Fig. 2.1A and B, and that, at the time of making your decision, the subjective utilities of O1 and O3 were identical, yielding identical expected values for all actions across environments. If outcome utilities were static, the high instrumental divergence afforded by the probability distributions in Fig. 2.1B would be of little consequence. In the real world, however, subjective outcome utilities are constantly changing, due, for example, to sensory-specific satiety or changes in motivational states. Given this dynamic nature of subjective utilities, the high instrumental divergence in Fig. 2.1B is essential for long-term reward maximization and, as such, may have intrinsic utility, serving to motivate and reinforce decisions that guide the agent toward environments that enable flexible instrumental control. In this section, I review recent research investigating the intrinsic utility of instrumental divergence, its dissociability from related constructs, such as outcome diversity and free choice, and its role in psychopathology.

An experimental test of the utility of flexible instrumental control: A recent study by Mistry and Liljeholm (2016) investigated the intrinsic utility of flexible instrumental control using a novel paradigm, illustrated in Fig. 2.3, in which participants choose between environments with either high or low instrumental divergence. Specifically, participants assumed the role of a gambler in a casino, playing a set of four slot machines (i.e., alternative actions, respectively labeled A1—A4) that yielded three differently colored tokens, each worth a particular amount of money, with different probabilities. In each of several gambling rounds, participants were required to first select a "room" in which only two slot machines were available, and they were restricted to playing on those two machines on subsequent trials within that round. Critically, the two slot machines available in a room had either identical probability distributions over token outcomes, yielding zero divergence (as in Fig. 2.1A), or symmetrically opposite distributions, yielding relatively high divergence (as in Fig. 2.1B). The measure of interest, thus, was the decision at the beginning of each block (top of Fig. 2.3), between a high- versus zero-divergence room.

While the probabilities with which each slot machine yielded each colored token were fixed throughout the task, and pretrained to criterion prior to gambling, the monetary values assigned to different token colors changed intermittently and unpredictably (about every fourth gambling round on average). In addition to mimicking

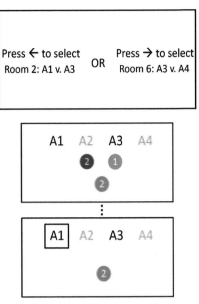

Figure 2.3 Illustration of task used by Mistry and Liljeholm (2016), showing the choice screen at the onset of a round (top) and the choice (middle) and feedback (bottom) screens on a trial within the round.

changes in the utilities of natural rewards, these changes in monetary values served to vary expected monetary utilities across gambling rooms, confirming that participants were sensitive to monetary payoffs, and to pit conventional currency against the utility of flexible control. Thus, while in some rounds, room options differed only in terms of instrumental divergence, in other rounds, expected monetary payoffs also differed across rooms, in either the same or opposite direction of instrumental divergence.

When participants choose between two gambling rooms with identical expected monetary payoffs but different levels of instrumental divergence, they choose the high-divergence room about 70% of the time, confirming a strong preference for flexible instrumental control all else being equal. Of primary interest, however, is how participants responded when high instrumental divergence was pitted against monetary gain. Here, alternative formal predictions may be generated using model-based RL agents that either do or do not consider the utility of flexible control. Specifically, the term $r(s')$ in Eq. (2.4) can be defined solely in terms of monetary reward, $r(s') = m(s')$, or in terms of both monetary reward and instrumental divergence, $r(s') = m(s') + w*ID(s')$, where w is a free parameter accounting for individual differences in the perceived utility of flexible control. Using a softmax distribution to translate action values into action probabilities, and fitting free parameters to choice data by minimizing negative log likelihood, we can derive a prediction, by each model, regarding the proportion of choosing the room with a greater expected monetary payoff when

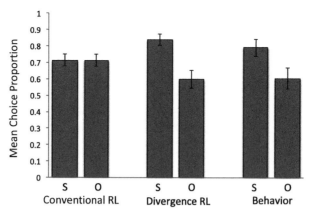

Figure 2.4 RL predictions and mean choice proportions from Experiment 1 by Mistry and Liljeholm (2016). Mean proportion of selecting the room with a greater expected monetary payoff when instrumental divergence differed across rooms in either the same (S) or opposite (O) directions, for a conventional RL model, an RL model that considers the utility of instrumental divergence, and behavioral choices. Error bars = SEM. *RL*, reinforcement learning.

instrumental divergence differs across rooms in either the same or opposite direction. Fig. 2.4 shows these predictions, for each model, together with the actual proportion of choices made by participants.

The conventional RL agent is, of course, likely to select the room with a greater monetary payoff whether the instrumental divergence of that room is zero or relatively high. In contrast, the divergence RL agent is significantly more likely to select the room with greater expected monetary payoff when that room also has relatively high instrumental divergence than when it has zero divergence. This was also the case with participants' choice behavior: The preference for a room with greater expected monetary payoff was significantly reduced when that room had zero instrumental divergence, and the alternative room, associated with a lower expected monetary payoff, had relatively high instrumental divergence. As a result, the divergence RL agent provides a significantly better fit to behavior than does the conventional model. Note that, when instrumental divergence differed across rooms in the opposite direction of monetary reward, the utility of flexible control was directly pitted against that of monetary gain. The reduction in preference, thus, shows a willingness to incur a monetary loss for access to high instrumental divergence. A parametric search for the exact trade-off between instrumental divergence and monetary reward, and investigation of the common neural value-scale mediating such trade-offs, is an important avenue for future work.

Instrumental divergence, perceptual outcome diversity, and free choice: The idea of "portfolio diversification"—mixing a wide variety of investments in order to reduce the impact of a single poorly performing source—is fundamental to theories of risk management. Ayal and Zakay (2009) conducted a series of psychological experiments in which participants choose among various "betting pools," where the perceptual diversity of betting options varied while the expected monetary gain was held constant. They found a significant preference for the most perceptually diverse pool and further, that the effort to maximize perceptual diversity sometimes led participants to prefer alternatives with lower expected monetary gain (see Schwartenbeck et al., 2015 for similar results). In the study by Mistry and Liljeholm described earlier, the perceptual diversity of obtainable outcomes was greater in high-divergence rooms than in zero-divergence rooms. Specifically, in zero-divergence rooms, there was a high probability of obtaining a blue token, a relatively low probability of obtaining a red token, and a zero probability of obtaining a green token (with specific token colors counterbalanced across participants). In contrast, in high-divergence rooms, participants were able to obtain blue, red, *and* green tokens by switching between actions across trials. Consequently, even when the expected monetary gain of high- and zero-divergence rooms was identical, the perceptual diversity of obtainable outcomes was greater in high-divergence rooms than in zero-divergence rooms.

Now, consider a scenario in which a computer algorithm chooses between the actions in a particular gambling room, selecting each action equally often by alternating across trials. Given such absence of voluntary choice, the high-divergence room no longer yields flexible instrumental control. Indeed, in the absence of free choice, neither the high- nor zero-divergence condition can be considered instrumental. However, such a computer algorithm would still yield greater perceptual diversity in high-divergence rooms than in zero-divergence rooms. Consequently, if choices were driven by a desire to maximize perceptual diversity, rather than instrumental divergence, they should not differ depending on whether the participant or an alternating computer algorithm chooses between the actions in a room. In a second study, Mistry and Liljeholm used an "autoplay" option, in which the computer selected between the two actions available in a room, to rule out perceptual outcome diversity as the source of a preference for flexible instrumental control. Specifically, in each block, one room option was always self-play—participants choose freely between actions available in the selected room—and the other option was always autoplay—a computer algorithm alternated between actions across trials—as indicated by labels printed below options on the room-choice screen. Instrumental divergence was either the same (high or zero) or different across room options.

The results of the study are shown in Fig. 2.5. When choosing between a high-divergence and a zero-divergence room (left two bars), participants preferred the high-divergence room when it was self-play (while the zero-divergence room was autoplay) but had no preference when the high-divergence room was autoplay (and the zero-divergence room was self-play). Since the high-divergence room was always associated with greater perceptual diversity, these results suggest that preferences were instead driven, as hypothesized, by instrumental divergence. The self-play versus autoplay

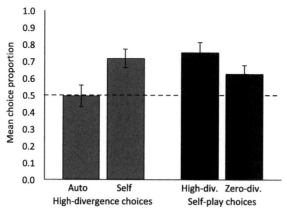

Figure 2.5 Mean choice proportions from Experiment 2 reported by Mistry and Liljeholm (2016). Mean proportions of high-divergence choices over zero-divergence choices (left) for blocks in which the high-divergence option was autoplay (Auto) versus blocks in which the high-divergence option was self-play (Self), and mean proportions of self-play choices over autoplay choices (right) for blocks in which both options had high divergence (High-div.) versus blocks in which both options had zero divergence (Zero-div.). *Dashed lines* indicate chance performance. Error bars = SEM.

manipulation is also related to a well-established preference for free choice over forced choice demonstrated across species, from pigeons to primates, including humans (Catania & Sagvolden, 1980; Leotti & Delgado, 2011, 2014; Suzuki, 1999). In Mistry and Liljeholm's second study, when instrumental divergence was held constant across self-play and autoplay options, participants choose the self-play over autoplay room significantly more often when both rooms had high instrumental divergence than when both rooms had zero instrumental divergence (right two bars in Fig. 2.5), suggesting that the value of choice depends less on whether a decision is voluntarily made and more on the extent to which decisions have a meaningful impact on future states.

Instrumental divergence and psychopathology: An aberrant experience of instrumental control, or "sense of agency" (SOA), is a common characteristic of various psychiatric disorders (Haggard, Martin, Taylor-Clarke, Jeannerod, & Franck, 2003; Keeton, Perry-Jenkins, & Sayer, 2008; Maeda et al., 2012; Martin & Penn, 2002; Peterson & Seligman, 1984; Seligman, Abramson, Semmel, & Von Baeyer, 1979; Voss et al., 2010; Werner, Trapp, Wüstenberg, & Voss, 2014): Schizophrenic individuals, in particular, differ from healthy controls in their self versus external attributions of events, as well as in the degree of intentional binding—a perceived compression of the time interval between an action and its consequence (Haggard et al., 2003; Maeda et al., 2012; Martin & Penn, 2002; Voss et al., 2010; Werner et al., 2014). While operational definitions of agency and volition differ across such findings, they share some fundamental limitations: First, they often conflate the estimation or representation of an action—outcome contingency with the subjective experience of volitional control (e.g., by manipulating outcome entropy or contiguity). Second, they tend to focus exclusively on cognitive or perceptual judgments, thus failing to address motivational aspects of SOA. In contrast, instrumental divergence provides a novel measure of agency that varies independently of outcome contiguity and predictability, and without eliminating volition, thus disambiguating the contribution of basic instrumental processes, such as simple contingency learning, to the apparent dysregulation of agency in schizophrenia. Moreover, unlike previous assessments of SOA, our task assessing a preference for high instrumental divergence can dissociate motivational aspects of flexible instrumental control from purely cognitive representations, at both behavioral and neural levels.

In a recent, unpublished, study, to begin to address the nature of aberrant SOA in schizophrenia, particularly with respect to its role in motivated behavior, we used the Oxford—Liverpool Inventory of Feelings and Experiences (O-LIFE) (Mason, Claridge, & Jackson, 1995) to relate individual differences in schizotypy to performance on the task used in the second study by Mistry and Liljeholm, discussed in the previous section. The O-LIFE questionnaire measures four dimensions of schizotypy—unusual experiences, cognitive disorganization, introvertive anhedonia, and impulsive nonconformity. We found that scores on the dimensions of unusual experiences and introvertive anhedonia, phenomenologically related, respectively, to positive and negative symptoms of schizophrenia, predicted a preference for high instrumental divergence. Specifically, as illustrated in Fig. 2.6, scores on both of these dimensions were significantly, negatively, correlated with the proportion of high-divergence choices over zero-divergence choices when the high-divergence option was self-play and with the proportion of self-play over autoplay choices when both options had high divergence. In contrast, there was no significant correlation between any schizotypy dimension and the proportion of choices for options that did not involve high instrumental divergence - i.e., rooms with high divergence but auto-play, or with zero divergence and self-play; indeed, neither of these latter choice proportions deviated significantly from chance, suggesting a complete lack of preference for either perceptual diversity or self-play in the absence of instrumental divergence. Moreover, no schizotypy dimension predicted preferences for greater monetary pay-offs, specifically implicating the utility of agency as a target for modulation in schiotypy.

The finding that schizotypal traits in healthy individuals modulate a preference for high instrumental divergence suggests that effects of instrumental divergence might also be significantly altered in clinical populations, potentially accounting for aspects of behavioral pathology in schizophrenia. Notably, the supramarginal gyrus of the IPL, implicated in neural computations of instrumental divergence by Liljeholm et al. (2013), has been frequently shown to differ volumetrically across schizophrenic and neurotypical individuals (Buchanan et al., 2004; Goldstein et al., 1999; Peng et al., 1994; Pol et al., 2001; Zhou et al., 2007), highlighting a possible anatomical basis

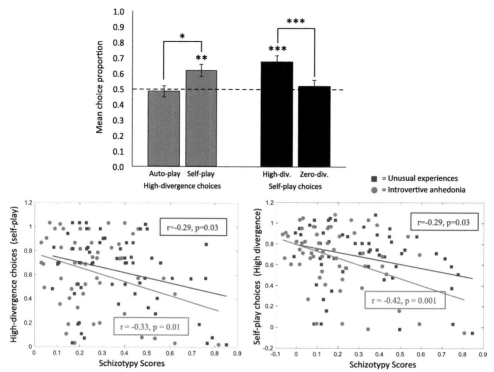

Figure 2.6 Results from a preliminary study (n = 60) assessing the relationship between schizotypal traits and a preference for high instrumental divergence. Top: Mean choice proportions for the same conditions as those listed in Fig. 2.5. Error bars = SEM. *Dashed line* indicates chance performance. * = $P < .05$, ** = $P < .005$, *** = $P < .0001$. Bottom: Residual plots of choice proportions and schizotypy scores (points scored out of total possible), adjusted for the number of training blocks to criterion on action–outcome probabilities, and for the order of Oxford–Liverpool Inventory of Feelings and Experience administration (i.e., before or after the gambling task).

for any differences in cognitive or motivational representations of instrumental divergence. Future research will be aimed at assessing whether individuals diagnosed with schizophrenia differ from healthy controls in their behavioral preference for high instrumental divergence and in underlying neural value computations.

INSTRUMENTAL DIVERGENCE AS A BOUNDARY CONDITION ON GOAL-DIRECTEDNESS

Unlike goal-directed decisions, habitual performance is insensitive to the current utility of action outcomes: an inflexibility that has been argued to result from a model-free RL process, in which instrumental responses come to be rigidly elicited by the stimulus environment based on their reinforcement history (Daw et al., 2005). Specifically, for a given action performed in a particular state, the model-free action value is updated as

$$Q(s, a) \leftarrow Q(s, a) + \alpha[(r(s') + \gamma Q(s', a')) - Q(s, a)], \tag{2.6}$$

where α is the learning rate and remaining terms are defined as for Eq. (2.4). Critically, the value of an action is updated *following* its execution in a particular state, only to be stored and not retrieved until the agent reenters that state. Consequently, model-free action selection reflects only past reinforcement, without regard for the current utility of future states. The flexibility of goal-directed decisions, while computationally expensive (Keramati et al., 2011; Otto et al., 2013, 2014), offers a clear adaptive advantage over such cached retrieval. However, as noted, when instrumental divergence is very low, the greater processing cost of goal-directed computations does not yield the return of flexible instrumental control, suggesting that a less resource-intensive, habitual, strategy might be optimal. One possibility, therefore, is that instrumental divergence serves as a boundary condition on the deployment of goal-directed behavior, increasing reliance on "fast and frugal" habits in environments that inherently impede flexibility.

In this section, I review evidence from the rodent literature suggesting that reliance on a goal-directed versus habitual strategy might depend on the degree of instrumental divergence. I then discuss a recent human neuroimaging study assessing the role of instrumental divergence in biasing behavior toward goal-directed versus habitual control, and sketch a formal description of such arbitration.

Instrumental divergence and behavior—reward correlations: In a series of seminal paper, Dickinson et al. demonstrated that goal-directed sensitivity to current outcome values depends on both the extent and nature of instrumental training: First, extensive but not moderate training produced insensitivity to outcome devaluation (Adams, 1982). Second, and more intriguingly, even moderate training resulted in devaluation insensitive performance if animals were trained on an interval schedule, in which reward delivery depends on the time elapsed since the last reward, but not if they were trained on a ratio schedule, in which the delivery of reward depends on the number of responses since the last reward (Dickinson, Nicholas, & Adams, 1983). Another clue to what factors may influence goal-directed and habitual arbitration came with a couple of demonstrations by Colwill and Rescorla (1990), showing that, contrary to the reports by Dickinson et al., animals remained sensitive to outcome devaluation in spite of being extensively trained on an interval schedule. A critical difference in methods was that, while Dickinson et al. used a single lever yielding a single outcome, Colwill and Rescorla trained animals on two different instrumental responses, each yielding a distinct sensory-specific reward. Holland (2004) directly contrasted these two procedures, demonstrating that, indeed, performance remains sensitive to outcome devaluation in spite of extensive training on an interval schedule when different instrumental responses yield distinct sensory-specific outcomes but not when alternative responses yield the same outcome.

Dickinson (1985) suggested that the critical factor arbitrating between goal-directed and habitual performance might be the correlation between variations in response performance and variations in obtained outcomes. Dickinson noted that animals tended to show relatively large variations in performance across early training sessions but exhibited a consistently high rate of responding with extended training. He concluded that, rather than the extent of training per se, it was the reduced variation in performance during late training stages, and the resulting reduction in the behavior—reward correlation, that was responsible for devaluation insensitive performance. This framework also predicts the differences between ratio and interval schedules: On an interval schedule, no amount of responding will yield reward until a particular interval has passed; once the interval has passed, a single reward is delivered given a response, whether that response was preceded by 1, 10, or 100 responses. In other words, variations in response rates have virtually no impact on the reward rate. In contrast, on a ratio schedule, the number of obtained rewards increases linearly with the number of responses. As with response—reward correlations, instrumental divergence, defined over the quantitative variables of response and reward rate, increases across interval and ratio schedules. Moreover, instrumental divergence is greater whenever qualitatively different instrumental responses yield distinct sensory-specific outcomes, as in the studies by Colwill and Rescorla (1990) and Holland (2004). Thus, the notion that instrumental divergence arbitrates between goal-directed and habitual performance is an extension of Dickinson's "behavior—reward correlation" theory to the case of multiple actions and sensory-specific outcomes.

A study assessing the role of instrumental divergence in strategy arbitration: In a recent neuroimaging study, Liljeholm, Dunne, & O'doherty (2015) employed a task aimed at encouraging goal-directed versus habitual responding using environments with high versus zero instrumental divergence. Specifically, in this task (illustrated in Fig. 2.7A), participants had to maintain the balance of a virtual system of fluid-filled beakers, using four distinct actions across four abstract cues, in order to avoid incurring a rapidly cumulating monetary loss. As long as all beakers had sufficient fluid, system balance was maintained and yielded continuous monetary reward. However, on each trial, one of the beakers would be emptied causing "system imbalance" and monetary loss until the participant refilled the beaker by performing a particular instrumental action. The emptying of a beaker was accompanied by the onset of one of four abstract cues. In a high-divergence condition, each action deterministically and uniquely regulated a particular beaker, so that there was no overlap between the sensory-specific outcome probability distributions of the four actions. Conversely, in the zero-divergence condition, each abstract cue signaled that a particular action would be effective in regulating any beaker that needed to have its fluid refilled at the moment, regardless of the identity of that beaker: Consequently, across trials, while each action was paired with a specific antecedent cue, it was decorrelated from the refilling of any particular beaker, generating a complete overlap of sensory-specific outcome probability distributions associated with alternative actions.

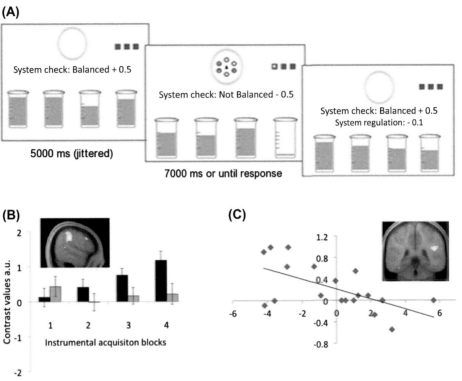

Figure 2.7 (A) Illustration of a trial in the beaker regulation task. (B) Increases in IPL activity across acquisition blocks in the high-divergence condition but not the zero-divergence condition. (C) Correlation between differences in IPL activity (x-axis) and differences in devaluation insensitivity (y-axis) across high- and zero-divergence conditions. *IPL*, inferior parietal lobule. *(Task and results from Liljeholm et al. (2015).)*

Following acquisition of the system-balancing task, one of the beakers was "devalued": Specifically, participants were instructed, as well as given the opportunity to passively observe across several trials, that one of the beakers was no longer relevant for system balance, which would be maintained, and continue to yield points, even when the liquid in this beaker dropped below threshold. Having correctly identified the devalued beaker, participants were allowed to again regulate the beaker system in a final test phase. Defining any response aimed at refilling the now devalued beaker as habitual (i.e., as devaluation insensitive), Liljeholm et al. found significantly greater habitual test performance in the zero-divergence than in the high-divergence condition. At the neural level, during the initial acquisition phase, activity in the supramarginal gyrus of the IPL, the region found by Liljeholm et al. (2013) to encode instrumental divergence, increased across blocks of acquisition in the high-divergence, but not the zero-divergence, condition (see Fig. 2.7B). Moreover, in the test phase, differences in IPL activity across high- and zero-divergence conditions predicted behavioral differences in habitual performance (Fig. 2.7C). It should be noted that the task employed by Liljeholm et al. (2015) was quite complex, with several factors (e.g., the strong stimulus control in the zero-divergence condition or the threat of cumulative loss) potentially contributing to neural and behavioral effects. Nonetheless, the results provide compelling initial evidence for a role of instrumental divergence in the arbitration between decision strategies, while also corroborating previous work implicating the IPL in a neural representation of instrumental divergence.

Formalizing arbitration by instrumental divergence: While a fully specified computational account of arbitration between goal-directed and habitual control by instrumental divergence is beyond the scope of this chapter, a preliminary sketch might characterize the probability of deploying a goal-directed versus habitual strategy as a logistic function of instrumental divergence, *ID*, such that

$$P(Q_{MB}(s, a)) = \frac{1}{1 + \exp^{-A(ID-B)}}$$

and

$$P(Q_{MF}(s, a)) = 1 - P(Q_{MB}(s, a))$$

(2.7)

where $P(Q_{MB}(s,a))$ and $P(Q_{MF}(s,a))$ are probabilities of using model-based and model-free action values respectively, B is a free parameter indicating the value of instrumental divergence at which the two control strategies are equally likely, and A specifies the strength of the bias toward a particular control strategy as instrumental divergence deviates from the indifference point set by B. Reasonable constraints on B would be the lower, 0, and upper, log(n), bounds of instrumental divergence, where n is the number of actions (i.e., outcome distributions) being considered. This simple rule predicts an increased reliance on goal-directed, over habitual, behavioral control with increasing levels of instrumental divergence.

OPEN QUESTIONS AND CONCLUDING REMARKS

In this chapter, I have reviewed some empirical evidence for the role of instrumental divergence—a formal index of flexible instrumental control—in goal-directed choice. In particular, I have addressed the utility of instrumental divergence, operationally defined as a preference for high-divergence environments, and the use of instrumental divergence as a boundary condition on the deployment of goal-directedness. At the neural level, I have discussed two studies implicating the supramarginal gyrus of the IPL—a region previously linked to a range of goal-directed processes—in the representation of instrumental divergence. While this recent work offers compelling preliminary evidence for the importance of instrumental divergence as a psychological construct, several critical questions remain open, many of which have been noted throughout this chapter. In this section, I will focus on two issues fundamental to the representation and implementation of instrumental divergence.

First, at the core of the proposal set forth in this chapter is the notion that instrumental divergence has intrinsic utility, serving both to justify the processing cost of goal-directed computations and to motivate decisions that guide the organism toward high-agency environments. But where exactly does this utility come from? In Instrumental divergence and the intrinsic utility of control section, I modeled the utility of instrumental divergence by including it as a reward surrogate in a model-based RL algorithm. This approach makes two assumptions: First, instrumental divergence is an explicitly represented variable and second, the apparent utility is directly attached to this variable, either a priori or through experience. An alternative possibility is that the agent assumes that subjective utilities may change over time, computing the values of future states and actions over a set of possible configurations of subjective utilities. Returning to the example provided in the introduction, given a choice between the scenarios depicted in Fig. 2.1A and B, respectively, and given that the subjective utilities of O1 and O3 are the same at the time of choosing, if the agent considers an array of possible changes in those subjective utilities, computing model-based action values over all possibilities, then the high-divergence scenario depicted in Fig. 2.1B would likely yield the greatest expected utility, since it allows the agent to select, for each hypothesized future utility configuration, the action that yields the outcome with greatest hypothetical utility. Thus, it is possible that an influence of instrumental divergence on choice preferences, such as that demonstrated by Mistry and Liljeholm (2016), could emerge in the absence of any explicit representation of instrumental divergence.

The nature of its apparent utility notwithstanding, if instrumental divergence is an explicitly represented variable, as suggested by the neural correlates identified by Liljeholm et al. (2013), another fundamental question is how exactly this construct is implemented neurally. In other words, is there a distributed neural code that carries information about the extent to which alternative actions differ with respect to their outcome distributions, analogous to the computation specified in Eq. (2.1)? A possible solution to this problem might be a neural network that discriminates between actions based on their outcome distributions. Specifically, initial layers in the network might retrieve the sensory-specific outcome features associated with distinct action alternatives, and those outcome features would then serve as inputs to subsequent layers that identify individual actions: The greater the decoding of action identities by the output layer of this network, the greater the instrumental divergence of considered action alternatives.

In conclusion, in addition to a range of open questions regarding its specific effects on decision-making, more fundamental aspects of instrumental divergence, such as the computational basis of its apparent utility, and the architecture of its neural implementation, must also be addressed by a comprehensive account. Clearly, assessment of the role of instrumental divergence in goal-directed choice is still in its infancy. Nonetheless, the initial findings reviewed in this chapter—ranging from a behavioral influence on choice preferences and devaluation sensitivity to neural signaling in a region frequently implicated in goal-directed control—promise exciting possibilities.

ACKNOWLEDGMENTS

The writing of this chapter was funded by a CAREER grant from the National Science Foundation (1654187) awarded to Mimi Liljeholm. The author thanks Richard Morris for helpful discussion.

REFERENCES

Abler, B., Herrnberger, B., Grön, G., & Spitzer, M. (2009). From uncertainty to reward: BOLD characteristics differentiate signaling pathways. *BMC Neuroscience, 10*(1), 154.

Adams, C. D. (1982). Variations in the sensitivity of instrumental responding to reinforcer devaluation. *The Quarterly Journal of Experimental Psychology, 34*(2), 77—98.

Ayal, S., & Zakay, D. (2009). The perceived diversity heuristic: The case of pseudodiversity. *Journal of Personality and Social Psychology, 96*(3), 559.

Balleine, B. W., & Dickinson, A. (1998). Goal-directed instrumental action: Contingency and incentive learning and their cortical substrates. *Neuropharmacology, 37*(4), 407—419.

Becker, G. M., DeGroot, M. H., & Marschak, J. (1964). Measuring utility by a single-response sequential method. *Systems Research and Behavioral Science, 9*(3), 226—232.

Buchanan, R. W., Francis, A., Arango, C., Miller, K., Lefkowitz, D. M., McMahon, R. P., ... Pearlson, G. D. (2004). Morphometric assessment of the heteromodal association cortex in schizophrenia. *American Journal of Psychiatry, 161*(2), 322—331.

Catania, A. C., & Sagvolden, T. (1980). Preference for free choice over forced choice in pigeons. *Journal of the Experimental Analysis of Behavior, 34*(1), 77—86.

Chatlosh, D. L., Neunaber, D. J., & Wasserman, E. A. (1985). Response-outcome contingency: Behavioral and judgmental effects of appetitive and aversive outcomes with college students. *Learning and Motivation, 16*(1), 1—34.

Colwill, R. M., & Rescorla, R. A. (1990). Effect of reinforcer devaluation on discriminative control of instrumental behavior. *Journal of Experimental Psychology: Animal Behavior Processes, 16*(1), 40.

Daw, N. D., Niv, Y., & Dayan, P. (2005). Uncertainty-based competition between prefrontal and dorsolateral striatal systems for behavioral control. *Nature Neuroscience, 8*(12), 1704—1711.

Dickinson, A. (1985). Actions and habits: The development of behavioural autonomy. *Philosophical Transactions of the Royal Society of London B: Biological Sciences, 308*(1135), 67—78.

Dickinson, A., Nicholas, D. J., & Adams, C. D. (1983). The effect of the instrumental training contingency on susceptibility to reinforcer devaluation. *The Quarterly Journal of Experimental Psychology, 35*(1), 35—51.

Doya, K., Samejima, K., Katagiri, K. I., & Kawato, M. (2002). Multiple model-based reinforcement learning. *Neural Computation, 14*(6), 1347—1369.

Erev, I., & Barron, G. (2005). On adaptation, maximization, and reinforcement learning among cognitive strategies. *Psychological Review, 112*(4), 912.

Fincham, J. M., Carter, C. S., Van Veen, V., Stenger, V. A., & Anderson, J. R. (2002). Neural mechanisms of planning: A computational analysis using event-related fMRI. *Proceedings of the National Academy of Sciences, 99*(5), 3346—3351.

Gläscher, J., Hampton, A. N., & O'doherty, J. P. (2008). Determining a role for ventromedial prefrontal cortex in encoding action-based value signals during reward-related decision making. *Cerebral Cortex2, 19*(2), 483—495.

Goldstein, J. M., Goodman, J. M., Seidman, L. J., Kennedy, D. N., Makris, N., Lee, H., ... Tsuang, M. T. (1999). Cortical abnormalities in schizophrenia identified by structural magnetic resonance imaging. *Archives of General Psychiatry, 56*(6), 537–547.

Haggard, P., Martin, F., Taylor-Clarke, M., Jeannerod, M., & Franck, N. (2003). Awareness of action in schizophrenia. *Neuroreport, 14*(7), 1081–1085.

Hammond, L. J. (1980). The effect of contingency upon the appetitive conditioning of free-operant behavior. *Journal of the Experimental Analysis of Behavior, 34*(3), 297–304.

Holland, P. C. (2004). Relations between Pavlovian-instrumental transfer and reinforcer devaluation. *Journal of Experimental Psychology: Animal Behavior Processes, 30*(2), 104.

Holt, C. A., & Laury, S. K. (2005). Risk aversion and incentive effects: New data without order effects. *The American Economic Review, 95*(3), 902–904.

Keeton, C. P., Perry-Jenkins, M., & Sayer, A. G. (2008). Sense of control predicts depressive and anxious symptoms across the transition to parenthood. *Journal of Family Psychology, 22*(2), 212.

Keramati, M., Dezfouli, A., & Piray, P. (2011). Speed/accuracy trade-off between the habitual and the goal-directed processes. *PLoS Computational Biology, 7*(5), e1002055.

Leotti, L. A., & Delgado, M. R. (2011). The inherent reward of choice. *Psychological Science, 22*(10), 1310–1318.

Leotti, L. A., & Delgado, M. R. (2014). The value of exercising control over monetary gains and losses. *Psychological Science, 25*(2), 596–604.

Liljeholm, M., Dunne, S., & O'doherty, J. P. (2015). Differentiating neural systems mediating the acquisition vs. expression of goal-directed and habitual behavioral control. *European Journal of Neuroscience, 41*(10), 1358–1371.

Liljeholm, M., Molloy, C. J., & O'Doherty, J. P. (2012). Dissociable brain systems mediate vicarious learning of stimulus–response and action–outcome contingencies. *Journal of Neuroscience, 32*(29), 9878–9886.

Liljeholm, M., Tricomi, E., O'Doherty, J. P., & Balleine, B. W. (2011). Neural correlates of instrumental contingency learning: Differential effects of action–reward conjunction and disjunction. *Journal of Neuroscience, 31*(7), 2474–2480.

Liljeholm, M., Wang, S., Zhang, J., & O'Doherty, J. P. (2013). Neural correlates of the divergence of instrumental probability distributions. *Journal of Neuroscience, 33*(30), 12519–12527.

Lin, J. (1991). Divergence measures based on the Shannon entropy. *IEEE Transactions on Information Theory, 37*(1), 145–151.

Maeda, T., Kato, M., Muramatsu, T., Iwashita, S., Mimura, M., & Kashima, H. (2012). Aberrant sense of agency in patients with schizophrenia: Forward and backward over-attribution of temporal causality during intentional action. *Psychiatry Research, 198*(1), 1–6.

Martin, J. A., & Penn, D. L. (2002). Attributional style in schizophrenia: An investigation in outpatients with and without persecutory delusions. *Schizophrenia Bulletin, 28*(1), 131–141.

Mason, O., Claridge, G., & Jackson, M. (1995). New scales for the assessment of schizotypy. *Personality and Individual Differences, 18*(1), 7–13.

Mistry, P., & Liljeholm, M. (2016). Instrumental divergence and the value of control. *Scientific Reports, 6*.

Otto, A. R., Raio, C. M., Chiang, A., Phelps, E. A., & Daw, N. D. (2013). Working-memory capacity protects model-based learning from stress. *Proceedings of the National Academy of Sciences, 110*(52), 20941–20946.

Otto, A. R., Skatova, A., Madlon-Kay, S., & Daw, N. D. (2014). Cognitive control predicts use of model-based reinforcement learning. *Journal of Cognitive Neuroscience, 27*(2), 319–333.

Peng, L. W., Lee, S., Federman, E. B., Chase, G. A., Barta, P. E., & Pearlson, G. D. (1994). Decreased regional cortical gray matter volume in schizophrenia. *American Journal of Psychiatry, 151*(6), 843.

Peterson, C., & Seligman, M. E. (1984). Causal explanations as a risk factor for depression: Theory and evidence. *Psychological Review, 91*(3), 347.

Pol, H. E. H., Schnack, H. G., Mandl, R. C., van Haren, N. E., Koning, H., Collins, D. L., & Kahn, R. S. (2001). Focal gray matter density changes in schizophrenia. *Archives of General Psychiatry, 58*(12), 1118–1125.

Rangel, A., & Hare, T. (2010). Neural computations associated with goal-directed choice. *Current Opinion in Neurobiology, 20*(2), 262–270.

Schwartenbeck, P., FitzGerald, T. H., Mathys, C., Dolan, R., Kronbichler, M., & Friston, K. (2015). Evidence for surprise minimization over value maximization in choice behavior. *Scientific Reports, 5*.

Seligman, M. E., Abramson, L. Y., Semmel, A., & Von Baeyer, C. (1979). Depressive attributional style. *Journal of Abnormal Psychology, 88*(3), 242.

Shanks, D. R., & Dickinson, A. (1991). Instrumental judgment and performance under variations in action-outcome contingency and contiguity. *Memory & Cognition, 19*(4), 353–360.

Sutton, R. S., & Barto, A. G. (1998). *Reinforcement learning: An introduction* (Vol. 1, No. 1). Cambridge: MIT Press.

Suzuki, S. (1999). Selection of forced-and free-choice by monkeys (*Macaca fascicularis*). *Perceptual and Motor Skills, 88*(1), 242–250.

Voss, M., Moore, J., Hauser, M., Gallinat, J., Heinz, A., & Haggard, P. (2010). Altered awareness of action in schizophrenia: A specific deficit in predicting action consequences. *Brain, 133*(10), 3104–3112.

Werner, J. D., Trapp, K., Wüstenberg, T., & Voss, M. (2014). Self-attribution bias during continuous action-effect monitoring in patients with schizophrenia. *Schizophrenia Research, 152*(1), 33–40.

Wunderlich, K., Dayan, P., & Dolan, R. J. (2012). Mapping value based planning and extensively trained choice in the human brain. *Nature Neuroscience, 15*(5), 786–791.

Wunderlich, K., Rangel, A., & O'Doherty, J. P. (2009). Neural computations underlying action-based decision making in the human brain. *Proceedings of the National Academy of Sciences, 106*(40), 17199–17204.

Zhou, S. Y., Suzuki, M., Takahashi, T., Hagino, H., Kawasaki, Y., Matsui, M., … Kurachi, M. (2007). Parietal lobe volume deficits in schizophrenia spectrum disorders. *Schizophrenia Research, 89*(1), 35–48.

CHAPTER 3

The Temporal Dynamics of Reward-Based Goal-Directed Decision-Making

Alec Solway[1], Matthew M. Botvinick[2,3]
[1]Virginia Tech Carilion Research Institute, Roanoke, VA, United States; [2]DeepMind, London, United Kingdom; [3]Gatsby Computational Neuroscience Unit, UCL, London, United Kingdom

The focus of this chapter is on the temporal dynamics of reward-based goal-directed decision-making, specifically, on models that describe how decision-making *evolves* during an individual episode. Unlike static choice models, which focus just on the choice itself, dynamic models describe the time course of the decision process. Behaviorally, this means that such models also make predictions about reaction times. Neurally, rather than describing static latent quantities that have to be computed and represented on the way to a decision, such models provide more comprehensive constraints on the components of the decision process and the time course of their interaction. Although reward-based goal-directed choice has a long history of study, surprisingly little work has been conducted on dynamic models of the decision process. Luckily, decision dynamics have a rich history of study in other cognitive domains, such as perceptual decision-making and memory retrieval (Luce, 1986), and a large body of work exists on which researchers of reward-based choice can rely. This literature is extensive, and includes a wide range of models, but most share a common focal point: the idea of *evidence integration*. We begin by presenting the general framework within the context of one such particularly successful model, the Ratcliff drift-diffusion model (Ratcliff & McKoon, 2008).

THE DRIFT-DIFFUSION MODEL

Here we provide a brief high-level description of the model, shown schematically in Fig. 3.1. Mathematical detail can be found in Ratcliff (1978). The model describes binary choice—a decision between two options. The state of the decision process is a single quantity representing the total amount of relative evidence collected so far in favor of one option versus the other. It is initialized to a reference value, labeled "z" in the figure, and samples of competing evidence then gradually arrive over time (the general version of the drift-diffusion model describes a continuous time process, but the model can be conceptualized in discrete time), which the model integrates, resulting in momentary fluctuations in the state variable. The state variable is biased to travel in a particular

Goal-Directed Decision Making
ISBN 978-0-12-812098-9, https://doi.org/10.1016/B978-0-12-812098-9.00003-6

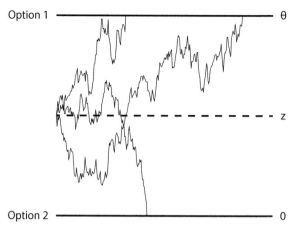

Figure 3.1 The drift-diffusion model, a model of one-step binary choice. The state of the decision is represented by a point particle that begins its journey at "z" and is biased to drift toward one of two boundaries, representing the two options under consideration. Each step is also subject to noise. A decision is made when the particle hits one of the boundaries, and the amount of time taken is the reaction time for the decision. The figure displays several sample decision paths.

direction and by a particular magnitude on average (the drift piece), but the distance traveled is subject to noise (the diffusion piece). A decision is made when the amount of overall evidence accrued hits one of two decision boundaries, which correspond to the available options. The path of this evidence integration process represents the path of the decision, and the time taken to reach one of the boundaries is the reaction time.

The state variable, or the evidence integrator, in such models may be prescribed a probabilistic interpretation (Bogacz, Brown, Moehlis, Holmes, & Cohen, 2006; Gold & Shadlen, 2001, 2002) and thought of as a quantity proportional to the log likelihood or posterior ratio of the two options under consideration. The log likelihood ratio for momentary evidence e is equal to $\log \frac{Pr(e|h_1)}{Pr(e|h_2)}$, where h_1 and h_2 are hypotheses corresponding to each of the two available options being the correct one (e.g., in perceptual decision-making; in reward-based choice, presented later, the hypotheses correspond to one option being valued greater than the other). If each sample of evidence is independent, their log likelihood ratios are summed to compute the log likelihood ratio for the entire sequence of evidence, analogous to the evidence integrator. If the prior over the two hypotheses is not uniform, this information can be incorporated by adding an appropriate bias to the overall log likelihood ratio, analogous to adjusting the reference point "z." Finally, an optimal decision rule known as the sequential probability ratio test stipulates that a decision should be made when the likelihood ratio exceeds one of two decision boundaries, exactly as described above. This perspective serves as partial motivation for a more direct application of iterative probabilistic computation in a model of multistep choice presented in a later section (Botvinick & An, 2009; Solway & Botvinick, 2012).

The drift-diffusion model has four classes of parameters: those related to the drift rate (the average bias in evidence toward one or the other boundary), the location of one of the boundaries (the second boundary can be fixed to an arbitrary value, e.g., 0), the initial starting point for the evidence relative to the boundaries, and the nondecision time, which includes the amount of time taken to process the stimulus through the visual system and to generate a motor response. Like the location of the second boundary, within-trial noise affecting momentary fluctuations in evidence is not identifiable, given the other parameters, and is fixed to a constant value. There is also across-trial variability: the drift rate has a Gaussian distribution across trials, and the starting point and nondecision components are uniformly distributed within a given range.

The model stipulates that during decision-making, new evidence in favor of each option is gradually made available, but no single sample of evidence is conclusive. The drift rate represents the average strength of evidence for choosing one option over the other. In perceptual decision-making tasks, part of the uncertainty is inherent to the stimulus itself. For example, a canonical task in this domain asks participants to judge the direction in which a subset of noisy dots are moving (Newsome, Britten, & Movshon, 1989) over time. On each trial, some percentage of dots coherently move up or down (or in other versions, left or right), while the remaining dots move at random. Because of the randomness inherent in the motion of the dot display, it is difficult to use a single pair of frames from this stimulus to make a decision. For example, on a trial with 5% coherence, 5% of the dots move in concert, while the remaining 95% appear at random. Multiple frames are required to judge the direction in which a subset of dots regularly move together. As the coherence level is increased, more evidence is presented in each pair of frames, resulting in a larger drift rate when the drift-diffusion model is fit to data from this task (Ratcliff & McKoon, 2008).

The boundary separation represents the amount of caution employed by the decision-maker. Boundaries that are further away from the reference point take longer to reach, and such decision episodes average over more evidence in the process. However, the increased accuracy that results comes at the cost of additional deliberation time. The integration process may be metabolically costly, and the time spent can potentially be used to gather other rewards.

Depending on situational context, it may be optimal to accept a lower level of accuracy for an individual trial in order to quickly proceed to the next trial. Setting the location of the boundary thus amounts to a speed/accuracy trade-off decision, and it should be possible to find the *optimal* speed/accuracy trade-off for a given decision context (Bogacz et al., 2006; Gold & Shadlen, 2002). Related to the boundary separation, the initial value for the relative evidence reflects the initial inclination toward one or the other option and is also an important variable in setting the optimal decision strategy. Whether and how the brain adjusts these two parameters in an optimal fashion is the subject of ongoing work (Bogacz et al., 2006; Bogacz, Hu, Holmes, & Cohen, 2010; Simen et al., 2009; Simen, Cohen, & Holmes, 2006).

The interpretation of the nondecision time is straightforward: This component of the model represents the sum of the amount of time taken for the stimulus information to travel through the sensory system and the amount of time taken to physically generate the appropriate motor response. Depending on the decision context, it may also include the time necessary to convert the decision into the appropriate action (Wunderlich, Rangel, & O'Doherty, 2010). Most models of temporal dynamics, including the drift-diffusion model, focus on the decision process itself, and the inner workings of this component of the model are left unspecified. It captures, in an abstract fashion, the residual time to reach a decision, without describing the process that unfolds during that time.

Although this may seem like an unsatisfactory aspect of this class of decision models, it is also an advantage, in that the decision process can be studied independent of stimulus-response modalities. If these two aspects of processing can really be separated, such a model can make predictions across domains. This appears to be the case. For example, although developed in the context of recognition memory (Ratcliff, 1978), the drift-diffusion model has been widely applied to data from perceptual decision-making (e.g., Gold & Shadlen, 2007; Ratcliff, 2002; Ratcliff & McKoon, 2008), as well as to lexical decision-making (Wagenmakers, Ratcliff, Gomez, & McKoon, 2008). Given the model's success in explaining data across decision domains, it is natural to begin studying the temporal dynamics of reward-based goal-directed choice using the same set of principles.

SIMPLE BINARY CHOICE

One line of work has imported the drift-diffusion model verbatim to study reward-based binary choice problems where participants choose between food items or other types of products. This work has demonstrated that the model can successfully account for mean accuracy and correct-and-error reaction times for such decisions, as well as the entire shape of the reaction time distributions (Milosavljevic, Malmaud, Huth, Koch, & Rangel, 2010). A key feature of these results is that the best-fitting drift rate is proportional to the difference in value (inferred from self-reported ratings) between the two items in question. This finding is intuitive: the larger the difference in value between two options, the more (noisy) the evidence is conveyed per unit time in favor of one option over the other. In this way, value difference in reward-based decision-making is analogous to the amount of stimulus (un)certainty in perceptual decision-making.

These initial findings have been expanded in several ways. In conjunction with eye tracking, it has been shown that drift rate is also influenced by overt differences in attention, with enhanced processing for items that are attended to (Krajbich & Rangel, 2011; Krajbich, Armel, & Rangel, 2010). A simple variant of the drift-diffusion model can also account for decisions with more than two choices (Krajbich & Rangel, 2011). In this version, evidence for each item is accrued by independent integrators, and a decision is made when the difference between the largest integrator and the other integrators exceeds a threshold. In the case of binary choice, this model is analogous to the drift-diffusion model, where the relative evidence represents the difference between two independent integrators. Some work has begun mapping out the neural circuits implementing these computations in the context of reward-based choice. A neural implementation of a simple variant of the drift-diffusion model, in tandem with dynamic causal modeling, was used to show that noisy reward information represented by the ventromedial prefrontal cortex is passed on to evidence accumulators in the dorsomedial prefrontal cortex and the intraparietal sulcus, which in turn inform motor-related areas to execute the appropriate action (Hare, Schultz, Camerer, O'Doherty, & Rangel, 2011).

In reward-based decision-making tasks, like in memory retrieval, the source of the decision noise and what is being "integrated over" is less clear. In such tasks, the stimuli themselves are usually not noisy: For example, a task may ask participants to choose between two clearly displayed chocolate bars. However, the rewards associated with the stimuli may be noisy. One possibility is that values are constructed by randomly sampling previous experiences with the object in question from memory (e.g., instances of the person eating the type of chocolate bar displayed). A second possibility, not mutually exclusive from the first, is that values are affected by a random sampling of the features associated with each object and its corresponding value (Milosavljevic et al., 2010). These explanations are speculative at present, and much work remains to be done to clarify the contributions to decision noise in reward-based choice (Shadlen & Shohamy, 2016).

Although the drift-diffusion model can explain data on reward-based choice, it may be a suboptimal decision strategy in this domain in the form described above. In perceptual decision-making, the relative reward associated with picking one option over the other is independent of decision difficulty (participants are encouraged to be equally accurate in all experimental conditions). However, in reward-based decision-making tasks, decision difficulty is inversely proportional to relative reward. Here, by definition, difficult decisions involve choosing between items that are close in value, and there is little to gain by being accurate compared to easier decisions, where values are farther apart (Oud et al., 2016). Under some circumstances, the standard drift-diffusion model can be made optimal by adjusting (collapsing) the decision boundary over time (Tajima, Drugowitsch, & Pouget, 2016)[1]. Whether and how the brain is able to approximate this optimum is an open question.

---

[1] Collapsing decision bounds also come into play in explaining optimal behavior in perceptual decision-making when trial difficulty is mixed (Drugowitsch, Moreno-Bote, Churchland, Shadlen, & Pouget, 2012).

The drift-diffusion model is only one of many evidence integration models, and it is not the only one to have been applied to reward-based choice. A related perspective comes from work on decision field theory (Busemeyer & Townsend, 1993; Diederich, 1997). This model was developed specifically in the context of reward-based goal-directed choice and has been applied to a richer array of decision problems, including problems with probabilistic outcomes and outcomes with multiple attributes. At the core of the model is a similar evolving noisy preference state. It has been successful not only at explaining the relationship between choice and reaction time data, but its dynamic nature has also shed light on choice data that has challenged some static models (Busemeyer & Townsend, 1993). The breadth of this work is beyond the scope of this introductory review.

While an understanding of one-step choice is slowly beginning to take shape, less work has been conducted on multistep decision problems. It is of course an unfortunate state of affairs, because real-life decisions seldom involve a single step of action. This is true for both of major decisions, for example, deciding what kind of career to pursue, and less important, more proximal decisions, such as how to plan a weekend trip. In examining the temporal dynamics of multistep decisions, there is no work in other decision domains to fall back on, because multiple steps of actions are less inherent to decisions in other domains. We next discuss some initial approaches to this problem, each from a somewhat different perspective, but tied together by the principle of gradual evidence accumulation.

MULTISTEP DRIFT-DIFFUSION MODELS

Simple binary choice can be represented by a shallow decision tree like the one shown in Fig. 3.2A. The participant starts in the same initial root state at the start of the trial, and deterministically transitions to one of two other states depending on the action they choose, collecting reward along the way. A simple way to extend this paradigm to

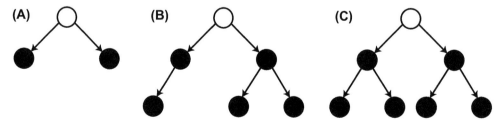

Figure 3.2 Decision trees whose temporal dynamics have been studied. The *open circle* represents the starting state, each *closed circle* represents a potential future state associated with reward, and each *arrow* represents a deterministic action. (A) Simple reward-based binary choice (e.g., Krajbich et al., 2010; Milosavljevic et al., 2010). (B) Experiment 1 of Solway and Botvinick (2015). (C) Experiment 2 of Solway and Botvinick (2015).

multiple steps of action is to add one additional choice point on one side of this decision tree, as shown in Fig. 3.2B. Choosing "left" results in a deterministic transition and reward, and at the second stage, the participant is forced to choose a particular second item, with only one option available. Choosing "right" results in a different deterministic transition and reward, but the participant is also then asked to make a second decision between two other items. A simple further extension of this paradigm asks participants to make a second decision on both sides of the decision tree, as shown in Fig. 3.2C.

Solway and Botvinick (2015) ran two experiments, each asking participants to solve one of these two types of decision trees. The stimuli consisted of various products, such as DVDs and board games, which participants rated on a five-point scale at the start of the experiment. At the first stage, participants were able to see the entire structure of the decision tree. Making a choice at the first stage removed the opposite side of the tree, and participants were asked to then make a second selection, even if only one option was available.

In one-step decision problems, competition is between individual items. The drift-diffusion model and its variants implement this competition, keeping track of the evidence accrued in favor of each item. The difficulty of choice (i.e., of this competition) can be indexed by a function of the disparity in value among items. In binary choice, this function is usually just the absolute difference between the values (ratings) of the two items. In multialternative choice, a variety of different functions can be used. One useful option is the difference between the maximum value and the average of the remaining values (Krajbich & Rangel, 2011). Decisions that are more difficult based on these definitions typically have slower reaction times and lower (subjective) accuracy.

This perspective can be translated to the multistep case by reasoning in terms of each complete branch of the decision tree, i.e., each path from the root node at the top to a leaf node at the bottom, instead of individual items. For example, the decision tree in Fig. 3.2B has three such paths, and the decision tree in Fig. 3.2C has four. Rather than competition among individual items, in multistep choice, there is competition between the paths through the decision tree. It is straightforward to extend the above definition of decision difficulty to the multistep case, with one potential measure being the difference in value between the maximum-valued branch and the average of the remaining branches. Fig. 3.3A and D shows (in blue) first-stage accuracy and reaction time as a function of this measure of difficulty for the decision tree with three branches from Solway and Botvinick (2015). Fig. 3.4A and D plots the same information for the decision tree with four branches. Accuracy increases and reaction time decreases for decisions that are considered less difficult under this definition, similar to the one-step case.

Fig. 3.3B,E and 3.4B,E display accuracy and reaction times for the second-stage choice as a function of the absolute difference in value between the remaining items (forced choice trials for the decision tree with three branches are not included). The

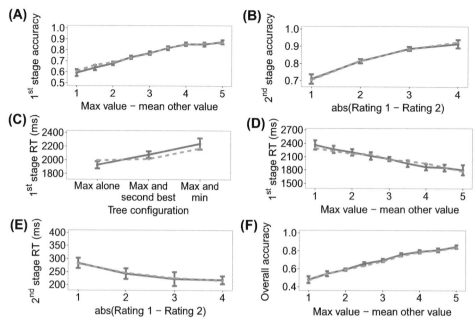

Figure 3.3 Results of Experiment 1 from Solway and Botvinick (2015). Empirical data appear in blue with *solid lines*, and the winning model in orange with *dashed lines*. Bars represent within-subject confidence intervals (Morey, 2008). In the figure and in the following description, *value* refers to the sum of the ratings along one path of the decision tree. (A) First-stage choice accuracy as a function of the difference between the maximum value and average of the other two values. A trial is considered correct if the first-stage choice does not rule out the optimal path. (B) Second-stage choice accuracy as a function of the absolute difference between the ratings of the items remaining at the second-stage. A trial is considered correct if the higher rated item is selected. Only trials where a second-stage choice had to be made are included. (C) First-stage reaction time for correct trials. A trial is considered correct if the best overall path was selected. (D) Second-stage reaction time for correct trials. A trial is considered correct if the best overall path was selected. (E) First-stage reaction time for correct trials, as defined in (C) and (D), as a function of the paths that appear together in the tree. For example, "Max and second best" means that the two paths with the two largest values were grouped on one side (pressing left or pressing right at the first stage, depending on the paths' location, would leave both of them in play), and the smaller valued path was on the other side by itself. (F) Overall choice accuracy, taking both stages into account. *(From Solway and Botvinick (2015).)*

data reveal accuracy and reaction time effects at the second stage as well, although the absolute value of reaction time is much smaller than what is seen in one-step choice. Together, these aspects of the data suggest that deliberation continues after the first-stage decision is made.

A model describing the temporal dynamics of multistep choice must capture the two properties discussed so far: (1) competition between the branches of the decision tree and (2) deliberation that begins at earlier decision points continues at later decision points. A simple extension of the drift-diffusion model that captures these properties, and also turns

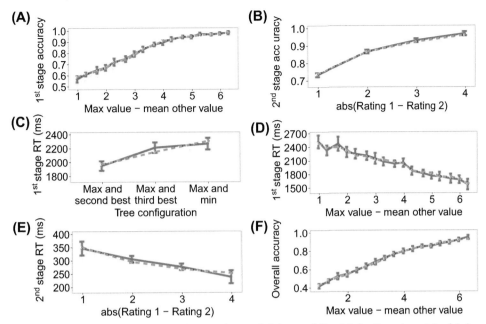

Figure 3.4 Results of Experiment 2. The panels parallel those of Fig. 3.3 for Experiment 1 of Solway and Botvinick (2015). *(From Solway and Botvinick (2015).)*

out to be the most parsimonious explanation of the data among several alternatives (Solway & Botvinick, 2015), is displayed schematically in Fig. 3.5. The number of evidence integrators equals the number of paths through the decision tree. During deliberation, noisy reward information associated with each item is sampled and contributes to the evidence for the corresponding branch. A first-stage decision is made when a sufficient amount of evidence has accrued on behalf of one path relative to the others, that is, when the magnitude of the largest integrator exceeds that of the remaining integrators by a threshold amount. The deliberation process then continues at the next stage, with the surviving integrators continuing from where they left off until a second threshold is reached. The model's predictions are overlaid in orange in Fig. 3.3 and 3.4, where it can be seen that the model captures the accuracy and reaction time effects at both decision stages.

Solway and Botvinick (2015) also tested a number of alternative multistep decision models. We briefly describe the three that are, perhaps, of most interest. The first alternative assumes that decisions are conducted using backward induction: The best course of action at each state at the bottom of the decision tree is determined first, and the best set of options are then moved up and integrated with reward information above. The backward induction model provided a worse account of the data than the model described above—human decision-making is more parallel in nature and simultaneously takes

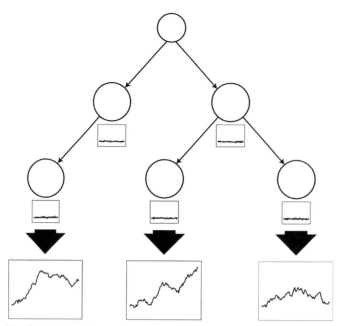

Figure 3.5 Model of Solway and Botvinick (2015). Evidence in favor of each item is generated independently for each branch and contributes to a corresponding evidence integrator. A first-stage decision is made when the magnitude of the largest integrator exceeds that of the remaining integrators by a threshold amount. The deliberation process then continues to the next stage, with the surviving integrators continuing from where they left off until a second threshold is reached.

into account items at multiple level of a decision tree. Besides providing a worse quantitative fit, the model is also mechanistically awkward. As described above, the data suggest that decision-making continues at lower levels of the tree after decisions are made at higher levels. The model thus has to reason bottom–up on the first pass, and then again top–down during a second pass, all while maintaining evidence accrued in favor of items ruled out at the bottom of the first pass in addition to the chosen items.

Another model alternative assumed that if an item is shared between multiple complete branches of the tree (e.g., the top left and right items in Fig. 3.2B and C), then during each iteration of the deliberation process a single noisy sample of reward is generated for each such item and contributes to all (in this case, both) of the branches in which it participates. This seems like the most parsimonious assumption, however, in the winning model described above, this actually is not the case. Instead, in the model above, a separate sample of reward is generated for each item in each branch. Besides providing a worse quantitative fit, there are qualitative differences in fit that suggest the simpler model's mechanisms provide a poor account of the data. In particular, second-stage accuracy is overestimated without additional deliberation at the second stage, in turn requiring the model to perform no additional deliberation, and predicting a flat

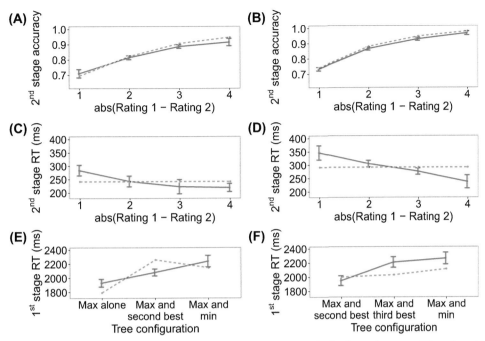

Figure 3.6 Simulation of both experiments (Experiment 1 appears in the first column, and Experiment 2 in the second) from Solway and Botvinick (2015) using a version of the model with correlated noise. Because noise is correlated at the top level, the amount of noise that distinguishes individual branches on each side of the tree is effectively halved, driving up second-stage accuracy. This in turn predicts a flat second-stage reaction time curve. The reduction in noise at the first-stage differentially affects different tree configurations, also resulting in a poor fit to this aspect of the data. *(From Solway and Botvinick (2015).)*

second-stage reaction time curve, contrary to the data (Fig. 3.6). Mechanistically, this difference arises from noise being shared at the top level. In the winning model with uncorrelated noise, noise for the top level item can help push the evidence for either branch in which it participates above threshold, creating more opportunity for the inferior branch to win the competition and bringing accuracy in line with what is seen in the data. In contrast, in the correlated noise model, the amount of noise that distinguishes individual branches on each side of the tree is effectively halved, driving up second-stage accuracy.

Although possibly surprising at first, the finding that the winning model samples items separately for each path may have an intuitive explanation: Rather than sampling branches in parallel, individuals may apply a serial sampling procedure, navigating down the tree one branch at a time (see David Redish's Chapter 6). Such an explanation seems especially likely with larger decision trees. Eye tracking might be able to answer whether this is the case in the present context.

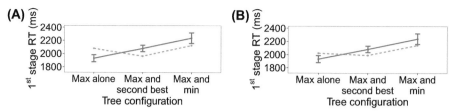

Figure 3.7 Simulation of Experiment 1 from Solway and Botvinick (2015) using a version of the model with (A) pruning and (B) pruning and correlated noise. The pruning mechanism predicts that first-stage decisions should be fastest when the best and second best paths appear on the same side of the decision tree, contrary to the data. *(From Solway and Botvinick (2015).)*

Finally, the third model variant of interest assumed that participants "prune" the tree and make a first-step decision when the evidence for the worst item on one side is larger than the evidence for the best item on the other side by a threshold amount. In addition to providing a worse quantitative fit, this model predicts that for the tree with three branches (Fig. 3.2B), decisions should be fastest when the best and second best branches are on the same side, also contrary to the data (Fig. 3.7). The explanation for this is intuitive: When the maximum branch is by itself, the pruning mechanism cannot fire at all, but when the maximum and second best branches are together, the evidence for the second best item will sometimes overweigh the worst branch on the other side, triggering a first-stage decision.

In short, the model that best accounts for data on multistep decisions is based on two principles: (1) the branches of the decision tree representing the decision compete independently, with evidence accruing on behalf of each and (2) the competition continues further down the tree from where it left off at the level above. The model, like most models, aims to capture key properties of the data rather than to provide a complete account of decision-making in every possible context. The current version of the model certainly has limitations, perhaps especially with regard to how much larger decision trees (in terms of breadth and/or depth) are handled (e.g., Huys et al., 2012, 2015), and future work will need to incorporate ideas from static choice models of large decision contexts into dynamic models of deliberation.

Some work is also being done to understand multistep decision-making using the language of decision field theory (Hotaling, 2013; Hotaling & Busemeyer, 2012). The general idea is similar to the multistep drift-diffusion model described earlier, in which decisions are made by simulating paths down the decision tree. Although it also offers a process level account of temporal dynamics, this work has so far focused more on choice than reaction time data. The model has been shown to account for a multistep version of the payoff variability effect, where outcomes with larger variance displayed multiple levels down a decision tree, result in more random decisions at higher levels of the tree. Other work has also begun exploring attentional biases by modeling how particular trajectories are overweighed or underweighed, and clustering participants based on model parameters (e.g., into planners and nonplanners), although this work appears to be in early stages.

A BAYESIAN PERSPECTIVE ON MULTISTEP CHOICE

Extending previous work by Botvinick and An (2009), Solway and Botvinick (2012) describe a somewhat different perspective on multistep choice. The motivation for this line of work is twofold. First, as described above, the drift-diffusion model has a probabilistic interpretation, with the evolving evidence proportional to the log likelihood or posterior ratio of the two options under consideration. We will relate this idea to the current model more specifically below. Second, there is a growing literature modeling numerous aspects of cognition, such as (again) perception, in Bayesian terms. Bayesian models specify a joint probability distribution among a set of component variables, and given values for a subset of these variables, the Bayesian framework describes how to compute the posterior probability distribution over the remaining variables conditional on the given set. For example, one class of perception models posit that object recognition is Bayesian in nature: the brain learns and maintains a generative model describing how latent object features give rise to percepts and uses Bayesian principles to compute the posterior probabilities of different features generating a given frame (Dayan, Hinton, Neal, & Zemel, 1995; Kersten, Mamassian, & Yuille, 2004).

Many such models can be usefully represented as a Bayesian network, like the one shown in Fig. 3.8. In this example, the components **Π** and **S** may be thought of as "generating" the component **A** using a probabilistic function. It should be noted,

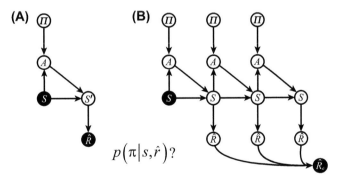

Figure 3.8 The planning-as-inference model. The components of the environment and their relationships are represented as a Bayesian network (probabilistic model). The nodes labeled **S** represent states, the nodes **A** represent actions, the nodes R̂ represent immediate one-step reward, and the single node labeled R̂c represents the cumulative reward for an entire multistep episode. Policy nodes, labeled **Π**, dictate what action to perform in each state at each time step. Planning proceeds by iteratively computing $p(\Pi|s_1, \hat{R}_c = 1)$, using the posterior from each round as the prior on the next round. *(Adapted from Solway and Botvinick (2012).)*

however, that although the model is framed in generative terms, the underlying relationship is not necessarily causal. That is, it is not necessarily the case that $\mathbf{\Pi}$ and $\mathbf{S}$ give rise to $\mathbf{A}$ (although that could be the case). Instead, the model describes conditional probabilities; how certain quantities rely on other quantities.

Solway and Botvinick (2012) detail a Bayesian perspective on goal-directed choice based on a generative model of action—outcome and outcome—reward contingencies. The structure of the model is motivated by another modeling framework called *model-based reinforcement learning*. Reinforcement learning (RL) is a framework for learning and decision-making whose algorithms can be divided into two broad groups: model-free and model-based (Dolan & Dayan, 2013; Sutton & Barto, 1998). Model-free RL learns directly, through trial and error, the actions that are reward-maximizing for a given stimulus. Model-based RL, on the other hand, makes decisions by first learning a model of the environment and then converting the model into a decision. The structure of the model is defined broadly and can be used to represent a very wide range of decision problems. It consists of two parts: a state transition function that dictates for each state/action pair the probability of transitioning to each other state in the environment (the action—outcome contingencies) and a reward function that describes for each such transition how much reward or punishment to expect (the outcome—reward contingencies). Taken together, this information can be used to generate the sort of decision trees we discussed previously, rooted at the current state of the environment.

The RL framework has a long history and successful track record for describing a range of behavioral and neural data (Dolan & Dayan, 2013). Model-based RL, in particular, has been used as a computational substrate for goal-directed control, but most of this work has focused on static choice models. Solway and Botvinick (2012) describe a dynamic perspective on model-based choice based on an iterative Bayesian inference procedure. The decision model is displayed schematically in Fig. 3.8 and incorporates the two pieces of the world model stipulated by model-based RL[2]. The nodes labeled $\mathbf{S}$ represent states, the nodes $\mathbf{A}$ represent actions, the nodes $\hat{\mathbf{R}}$ represent immediate one-step reward, and the single node labeled $\hat{\mathbf{R}}_c$ represents the cumulative reward for an entire multistep episode. Policy nodes, labeled $\mathbf{\Pi}$, dictate what action to perform in each state at each time step. Goal-directed choice requires computing the optimal policies, i.e., the actions at each state and time step that maximize the cumulative long-term reward. Continuous reward in this framework is transformed to a probabilistic scale, such that reward nodes are Bernoulli and take on values of 0 or 1, with magnitude encoded in the probability of "success" (i.e., observing a 1). Although this sounds like a technical nuisance, it is crucial for querying the model, as will be described shortly.

---

[2] Although the reward function is sometimes defined in terms of state/action/state triplets, here for simplicity it is defined just in terms of states.

Fig. 3.8A displays a slice of the model for a single step of action. This structure can be concatenated ad infinitum to model multiple steps, as shown in Fig. 3.8B. The state for the first time step is shaded to signify that it is "given," that is, the individual knows where they currently are. The goal then is to determine the policies (i.e., what action to take at each step and in each state, signified by the Π nodes) that maximize reward. This can be done in two ways. First, the decision-maker can set Π to each possible set of values in turn, compute $p(\hat{R}_c|s_1, \Pi)$ for each setting, and choose the policies that maximize this value. However, there is also another way to query the model. Rather than reasoning forward from possible policies to implied rewards, Bayesian inference can be used to reason backward, conditioning on reward, and computing the posterior over the policies that give rise to it. That is, rather than computing $p(\hat{R}_c|s_1, \Pi)$ for each setting of Π, we compute $p(\Pi|s_1, \hat{R}_c = 1)$. This is exactly analogous to the sort of reasoning posited by Bayesian models of perception, which work backward to map latent features to percepts. In goal-directed decision-making, we instead have a model of *reward* and reason backward to the policies that maximize it.

In order to compute optimal policies under this scheme, the inference process has to be performed iteratively over several rounds, each time using the value of $p(\Pi|s_1, \hat{R}_c = 1)$ computed on the previous round as the prior for the next round. Such an approach is necessary for two reasons. First, the action that is optimal at each step depends on the actions taken at other steps. These restrictions are mutually constraining, and several iterations are required for the policies to settle in to the optimum. Second, as we have seen before, single instances of reward information may be noisy, and this noise can be muted by averaging over multiple samples. Herein also lies the parallel between the present framework and models of evidence integration. The evolving posterior over the policy nodes may be viewed as evidence for performing a particular action (at a particular state and time step). As described previously, the drift-diffusion model has a similar probabilistic interpretation, with the evidence proportional to the log posterior ratio of the available action policies conditional on the "observations," which in reward-based decision-making, are noisy samples of reward. Reaction time can be modeled by setting a threshold on these values, and rendering a decision when the probability of a policy exceeds the threshold amount.

The model is demonstrated to make appropriate decisions in a number of different classical and modern experimental contexts, such as experiments on outcome revaluation and contingency degradation, which are naturally framed within the RL framework by setting and adjusting the appropriate parts of the reward and transition function. If inference is performed using an algorithm called loopy belief propagation (for details, see Pearl, 1988), the model offers an explanation for the existence and interaction of neurons coding a number of different decision variables, each previously studied in isolation, including offered and chosen values, action values, policies, and other variables.

CONCLUDING REMARKS

In this chapter, we have outlined in broad strokes the computational work to date on models of deliberation in reward-based goal-directed decision-making. Unlike static choice models, such models make predictions not only about the sort of choices that people make but also about the trajectory of the decision and the time it takes to make the decision. These models borrow from a rich literature on perceptual decision-making and memory retrieval (Luce, 1986) and share the general premise that deliberation involves accruing (averaging over) noisy evidence on behalf of each option. For one-step choice problems, two popular accounts of the data have come from the drift-diffusion model (Ratcliff & McKoon, 2008) and decision field theory (Busemeyer & Townsend, 1993). More recently, there has been a growing interest in expanding our understanding of decision dynamics to include multistep choice problems, which are obviously prevalent in everyday life. There are three perspectives on this question: one extending the drift-diffusion model to the multistep case (Solway & Botvinick, 2015), another providing a Bayesian inference perspective (Solway & Botvinick, 2012), and the third using the framework of decision field theory (Hotaling, 2013; Hotaling & Busemeyer, 2012). None of these models describe the entirety of the decision process; instead, each aims to capture key properties of the data. Future work will need to offer a more integrated account of decision-making by using data that go beyond the boundaries where each model breaks down.

REFERENCES

Bogacz, R., Brown, E., Moehlis, J., Holmes, P., & Cohen, J. D. (2006). The physics of optimal decision making: A formal analysis of models of performance in two-alternative forced-choice tasks. *Psychological Review, 113*(4), 700–765.

Bogacz, R., Hu, P. T., Holmes, P. J., & Cohen, J. D. (2010). Do humans produce the speed–accuracy trade-off that maximizes reward rate? *The Quarterly Journal of Experimental Psychology, 63*(5), 863–891.

Botvinick, M. M., & An, J. (2009). Goal-directed decision making in prefrontal cortex: A computational framework. In D. Koller, D. Schuurmans, Y. Bengio, & L. Bottou (Eds.), *Vol. 21. Advances in neural information processing systems* (pp. 169–176).

Busemeyer, J. R., & Townsend, J. T. (1993). Decision field theory: A dynamic-cognitive approach to decision making in an uncertain environment. *Psychological Review, 100*(3), 432–459.

Dayan, P., Hinton, G. E., Neal, M. R., & Zemel, R. S. (1995). The Helmholtz machine. *Neural Computation, 7*(5), 889–904.

Diederich, A. (1997). Dynamic stochastic models for decision making under time constraints. *Journal of Mathematical Psychology, 41*(3), 260–274.

Dolan, R. J., & Dayan, P. (2013). Goals and habits in the brain. *Neuron, 80*(2), 312–325.

Drugowitsch, J., Moreno-Bote, R., Churchland, A. K., Shadlen, M. N., & Pouget, A. (2012). The cost of accumulating evidence in perceptual decision making. *Journal of Neuroscience, 32*(11), 3612–3628.

Gold, J. I., & Shadlen, M. N. (2001). Neural computations that underlie decisions about sensory stimuli. *Trends in Cognitive Sciences, 5*(1), 10–16.

Gold, J. I., & Shadlen, M. N. (2002). Banburismus and the brain: Decoding the relationship between sensory stimuli, decisions, and reward. *Neuron, 36*(2), 299–308.

Gold, J. I., & Shadlen, M. N. (2007). The neural basis of decision making. *Annual Review of Neuroscience, 30*, 535–574.

Hare, T. A., Schultz, W., Camerer, C. F., O'Doherty, J. P., & Rangel, A. (2011). Transformation of stimulus value signals into motor commands during simple choice. *Proceedings of the National Academy of Sciences of the United States of America, 108*(44), 18120–18125.

Hotaling, J. M. (2013). *Decision field theory-planning: A cognitive model of planning and dynamic decision making* (Unpublished doctoral dissertation). Indiana University.

Hotaling, J. M., & Busemeyer, J. R. (2012). DFT-D: A cognitive-dynamical model of dynamic decision making. *Synthese, 189*(1), 67–80.

Huys, Q. J. M., Eshel, N., O'Nions, E., Sheridan, L., Dayan, P., & Roiser, J. P. (2012). Bonsai trees in your head: How the Pavlovian system sculpts goal-directed choices by pruning decision trees. *PLoS Computational Biology, 8*(3), e1002410.

Huys, Q. J. M., Lally, N., Faulkner, P., Eshel, N., Seifritz, E., Gershman, S. J., … Roiser, J. P. (2015). Interplay of approximate planning strategies. *Proceedings of the National Academy of Sciences of the United States of America, 112*(10), 3098–3103.

Kersten, D., Mamassian, P., & Yuille, A. (2004). Object perception as Bayesian inference. *Annual Review of Psychology, 55*, 271–304.

Krajbich, I., Armel, C., & Rangel, A. (2010). Visual fixations and the computation and comparison of value in simple choice. *Nature Neuroscience, 13*(10), 1292–1298.

Krajbich, I., & Rangel, A. (2011). Multialternative drift-diffusion model predicts the relationship between visual fixations and choice in value-based decisions. *Proceedings of the National Academy of Sciences of the United States of America, 108*(33), 13852–13857.

Luce, R. D. (1986). *Response times: Their role in inferring elementary mental organization.* New York, NY: Oxford University Press.

Milosavljevic, M., Malmaud, J., Huth, A., Koch, C., & Rangel, A. (2010). The drift diffusion model can account for the accuracy and reaction time of value-based choices under high and low time pressure. *Judgment and Decision Making, 5*(6), 437–449.

Morey, R. D. (2008). Confidence intervals from normalized data: A correction to Cousineau (2005). *Tutorial in Quantitative Methods for Psychology, 4*(2), 61–64.

Newsome, W. T., Britten, K. H., & Movshon, J. A. (1989). Neuronal correlates of a perceptual decision. *Nature, 341*, 52–54.

Oud, B., Krajbich, I., Miller, K., Cheong, J. H., Botvinick, M., & Fehr, E. (2016). Irrational time allocation in decision-making. *Proceedings of the Royal Society B: Biological Sciences, 283*(1822), 20151439.

Pearl, J. (1988). *Probabilistic reasoning in intelligent systems: Networks of plausible inference.* San Francisco, CA: Morgan Kaufmann.

Ratcliff, R. (1978). A theory of memory retrieval. *Psychological Review, 85*(2), 59–108.

Ratcliff, R. (2002). A diffusion model account of response time and accuracy in a brightness discrimination task: Fitting real data and failing to fit fake but plausible data. *Psychonomic Bulletin & Review, 9*(2), 278–291.

Ratcliff, R., & McKoon, G. (2008). The diffusion decision model: Theory and data for two-choice decision tasks. *Neural Computation, 20*(4), 873–922.

Shadlen, M. N., & Shohamy, D. (2016). Decision making and sequential sampling from memory. *Neuron, 90*(5), 927–939.

Simen, P., Cohen, J. D., & Holmes, P. (2006). Rapid decision threshold modulation by reward rate in a neural network. *Neural Networks, 19*(8), 1013–1026.

Simen, P., Contreras, D., Buck, C., Hu, P., Holmes, P., & Cohen, J. D. (2009). Reward rate optimization in two-alternative decision making: Empirical tests of theoretical predictions. *Journal of Experimental Psychology: Human Perception and Performance, 35*(6), 1865–1897.

Solway, A., & Botvinick, M. M. (2012). Goal-directed decision making as probabilistic inference: A computational framework and potential neural correlates. *Psychological Review, 119*(1), 120–154.

Solway, A., & Botvinick, M. M. (2015). Evidence integration in model-based tree search. *Proceedings of the National Academy of Sciences of the United States of America, 112*(37), 11708–11713.

Sutton, R. S., & Barto, A. G. (1998). *Reinforcement learning: An introduction.* The MIT Press.

Tajima, S., Drugowitsch, J., & Pouget, A. (2016). Optimal policy for value-based decision-making. *Nature Communications, 7*(12400).

Wagenmakers, E.-J., Ratcliff, R., Gomez, P., & McKoon, G. (2008). A diffusion model account of criterion shifts in the lexical decision task. *Journal of Memory and Language, 58*(1), 140−159.

Wunderlich, K., Rangel, A., & O'Doherty, J. P. (2010). Economic choices can be made using only stimulus values. *Proceedings of the National Academy of Sciences of the United States of America, 107*(34), 15005−15010.

CHAPTER 4

Case-Based Decision Neuroscience: Economic Judgment by Similarity

Rahul Bhui
Departments of Psychology and Economics & Center for Brain Science, Harvard University, Cambridge, MA, United States

It is said that the only constant in life is change. We are routinely faced with different situations, no two exactly alike. We visit new places, try new foods, meet new people, find new jobs, and invent new products. You have probably never read this very sentence before. The pervasiveness of novelty can be paralyzing if one is not prepared for it. By their nature, unfamiliar situations challenge our ability to draw on past experience. And by our nature, humans make do.

In general, how do we appraise courses of action in various contexts? We could form projections of what's likely to happen as a result of each action in a context and combine that with an evaluation of how desirable those outcomes are. Or we could lean on the automatic attitudes drilled into us by extensive experience. These are the two prevailing theories in neuroeconomics, expressed mathematically in terms of expected utility and reinforcement learning. The biggest success story of decision neuroscience to date has been in uncovering neural instantiations of these decision-making rules.

How might these two systems go awry when there is little direct experience to work from? The former relies heavily on a cognitive map or mental structure but does not have much to hang its structure on. The latter depends vitally on preexisting experience, but this direct experience is unavailable. The capacity to cope in new circumstances is an important but tricky skill.

A plausible alternative is to recall how well or poorly similar actions turned out in similar contexts in the past. Such an approach to decision-making enables us to draw on the variety of disparate experiences we acquire over time and respond gracefully to the novelty and complexity that pervades real life. It imposes fewer assumptions about the structure of the world compared with sophisticated probabilistic judgments, while squeezing more information out of background knowledge than simple value caching.

I lay out a "case-based" system combining theory and empirical evidence from economics, psychology, neuroscience, statistics, and computer science. Value judgment by similarity corresponds to an economic model called case-based decision theory (CBDT), inspired in part by a computational problem-solving process known as case-based reasoning (CBR). This theory has links to nonparametric statistics, suggesting why and when the system works well. Recent evidence from neuroscience indicates that we use this kind of

Goal-Directed Decision Making
ISBN 978-0-12-812098-9, https://doi.org/10.1016/B978-0-12-812098-9.00004-8

system and implicates the hippocampus and related medial temporal lobe (MTL) regions as neural loci. This can be thought of as a complementary narrative to what has been described as episodic control (Dayan, 2008; Lengyel & Dayan, 2008).

JUDGMENT AND DECISION-MAKING FROM SIMILARITY

Judgments based on similarity are ubiquitous. Consciously and unconsciously, we map athletes onto predecessors to forecast performance ("LeBron James is the next Michael Jordan"—ESPN), we react to people based on group stereotypes, we talk about new businesses in terms of existing analogues (describing various start-ups as the Ubers of food delivery, flowers, laundry, lawn care, marijuana, and mortgage lending), we evaluate products based on brand lines, we search historic economic events for relations to modern ones ("The Great Recession is just like the Great Depression"—Forbes), we hold to legal precedent as a guide for future cases, and we pitch new TV shows or movies as mixtures of old concepts ("The pitch [for Hollywood movie *Man's Best Friend*] was 'Jaws with Paws' ... Investors were told that if the movie *Jaws* was a huge success, a similar plot but on land with a dog could also be a huge success."—Reid Hoffman).

People are psychologically attuned to similarity. This is for good reason. In a sense, all learning is premised on finding similarity. Heraclitus said that "you cannot step twice into the same river," which is not only a deep philosophical truth but also an evolutionary problem. If every instant is unique, how can we learn and make decisions from experience? We are thus tasked with recognizing useful parallels that allow us to generalize from the past.

At its best, similarity-based judgment constitutes an ecologically valid heuristic for summarizing a vast landscape of information in service of decision-making. A neural network trained to classify handwritten digits holds the potential to perform well on digits it has never seen before, provided it has access to data on similarity between digits (as implicitly evaluated by the classification probabilities from another neural network; Hinton, Vinyals, & Dean, 2015). It has never encountered a "3," but knowing that certain "2's" are visually similar while "1's" are quite different implicitly contains a fair amount of information about what exactly a "3" looks like. This is precisely the kind of quality required for good transfer learning. This aspect of similarity is intimately tied to our propensity to associate and connect and categorize. We may not be wired to easily navigate probabilistic state spaces, but we are able to effortlessly form comparisons and associations between concepts in our memory.

This is not to say that similarities are always well founded. The movie *Man's Best Friend* turned out to be terrible, for instance. Classic examples of irrationality can be explained by indiscriminate similarity judgment. When asked how likely it is that the outspoken socially involved philosophy major Linda is a bank teller or a feminist bank teller, people respond that she is more likely to be the latter than the former (Tversky & Kahneman, 1983). Although this belief violates the laws of probability, Linda better resembles our idea of a feminist bank teller and so we judge that possibility to be more likely (Bar-Hillel, 1974; Kahneman & Tversky, 1972). We can be unconsciously misled by superficial connections, even when we are experts. Prominent venture capitalist Paul Graham was quoted as saying "I can be tricked by anyone who looks like Mark Zuckerberg. There was a guy once who we funded who was terrible. I said: 'How could he be bad? He looks like Zuckerberg!'" While this was said in jest, such biases are plausible. Gilovich (1981) asked professional sportswriters and varsity football coaches to predict the success of fictitious young players based on written profiles. In one manipulation, a player won an award named after a famous pro who either played in the same position or a different position. Success ratings turned out to be higher when the pro played in the same position.

For better or worse, we often form evaluations based on examples considered similar to our present situation. A body of research in economics explores the theme of valuation based on similarity. This work centers on a theoretical framework that reflects the mental contagion of value.

Case-based decision theory

CBDT is a model of decision-making, which takes past experiences as its primitives and weights those experiences based on their similarity to the current choice situation. It was developed and originally axiomatized by Gilboa and Schmeidler (1995a) as a psychologically plausible complement to expected utility theory. In order to apply classical expected utility theory, the agent must hold subjective probabilities over all pertinent states of the world. In many situations, this state space and its associated probabilities can be extremely complicated, intricate, or unnatural to construct. When deciding on a new restaurant to visit for dinner, one might not naturally estimate probability distributions over the quality of food and service for each place. Instead, one might simply call to mind their experiences at places thought to be roughly similar. From the start, Bayesian decision theory was primarily considered appropriate inside what Savage (1954) called a *small world*, where knowledge is plentiful. CBDT was meant to tackle decision-making in large worlds.

The primitive concepts of CBDT are a set of past cases and subjective similarity assessments between each case and the current situation. The agent's memory M is a set of cases formally described as triples (q, a, r), where q represents the problem situation, a is the action taken, and r is the result. The agent evaluates an action by combining the utilities of outcomes that occurred when that action was taken in the past. These utilities are weighted by the similarity between the current situation (with description p) and each past case (with description q), $s(p, q)$:

$$U(a) = \sum_{(q,a,r)\in M} s(p,q)u(r).$$

Table 4.1 Case-based decision theory calculation example

City	s(city, Paris)	s(city, Sydney)	Utility
Montreal	0.8	0.1	5
Los Angeles	0	0.5	6
Vancouver	0.5	0.5	10

Gilboa and Schmeidler (1997a) generalized this to allow similarity between cases to depend on acts as well as descriptions, so $U(a) = \sum_{(q,b,r) \in M} s((p, a), (q, b))u(r)$, and Gilboa, Schmeidler, and Wakker (2002) provided two additional axiomatic derivations clarifying its empirical content. A variant also formulated in Gilboa and Schmeidler (1995a) uses averaged similarity:

$$V(a) = \sum_{(q,a,r) \in M} \frac{s(p, q)}{\sum_{(q',a,r) \in M} s(p, q')} u(r).$$

To illustrate, suppose you are deciding which city to visit for a vacation and have narrowed the options down to Paris and Sydney. Though you have been to neither, you recall your past trips to Montreal, Los Angeles, and Vancouver, as laid out in Table 4.1. Though French-speaking urban Montreal was chilly, you had a decent time there (utility 5). You feel the city is quite similar to Paris (similarity 0.8) but hardly at all like Sydney (similarity 0.1). LA was hot, which you like, and occasionally smoggy, which you do not, but it was pleasant overall (utility 6). You consider LA to moderately resemble Sydney (similarity 0.5) but not Paris (similarity 0). The metropolis of Vancouver was special with its beautiful mountains, oceans, and fresh air—your favorite trip by far (utility 10). The city seems to you halfway between Paris and Sydney (similarity 0.5 each). As a standard case-based decision-maker, the projected utility of visiting Paris is $0.8 \times 5 + 0 \times 6 + 0.5 \times 10 = 9$, while the projected utility of visiting Sydney is $0.1 \times 5 + 0.5 \times 6 + 0.5 \times 10 = 8.5$, a calculation about which you reminisce on your flight to Paris. (If you were using the averaged variant, this decision would be reversed.)

Similarity functions

What might the similarity function look like? Goldstone and Son (2005) organize psychological models of similarity into four types: geometric, feature-based, alignment-based, and transformational. (Research in machine learning has developed more computationally sophisticated takes on these styles; see for example, Chen, Garcia, Gupta, Rahimi, & Cazzanti, 2009.)

Geometric models represent objects as multidimensional points in a metric space. The similarity between objects is calculated as inversely related to the distance between them in this space. A basic form may be found in models of generalization gradients originating from experiments on behavioral responses to stimuli varying in simple physical dimensions like wavelength of light (Ghirlanda & Enquist, 2003; Spence, 1937). Similarity of behavioral response is usually described as decreasing in the distance of stimulus qualities with exponential decay, $s(x, y) = \alpha \exp(-|x - y|/\beta)$, or Gaussian decay, $s(x, y) = \alpha \exp(-(x - y)^2/\beta^2)$, where α and β are scaling parameters. While having the appeal of parsimony, these models are typically applied to low-level stimuli and imply properties such as symmetry that are at odds with experimental results in other circumstances.

Feature-based models represent objects as collections of features. Similarity is based on a linear combination of the common and distinctive features of each object and is not in general symmetric. In Tversky's (1977) contrast model, $s(x, y) = \theta f(X \cap Y) - \alpha f(X - Y) - \beta f(Y - X)$, where X and Y are the feature sets of stimuli x and y, f is a monotonically increasing function, and θ, α, and β are nonnegative weights. With additional restrictions, stimulus similarity based on the contrast model forms a natural category structure that can be compactly represented in a hierarchical tree. In the ratio model, $s(x, y) = f(X \cap Y)/(f(X \cap Y) + \alpha f(X - Y) + \beta f(Y - X))$, normalizing similarity between 0 and 1.

Alignment-based models involve more complex mappings of features based on higher-order structure mapping. Similarity depends on the degree to which object features can be structurally aligned. For example, Goldstone's (1994) model of "similarity, interactive activation, and mapping" comprises a neural network that learns about the correspondences between stimulus features. Each node reflects the hypothesis that given features map onto each other across stimuli, with excitatory and inhibitory activation encouraging an exclusive one-to-one correspondence. Similarity is based on the weighted mean of feature proximity weighted by activation of the node representing the feature pair.

Transformational models are based on topological warping operations such as rotation, scaling, and translation. Similarity is computed from transformational distance, the degree of warping required to transform one stimulus into another. This may be defined in simple ways such as the minimum number of transformations needed (Imai, 1977), or in more complicated ways like Kolmogorov complexity, the length of the shortest computer program that describes the necessary transformations (Hahn, Chater, & Richardson, 2003). This style of model is typically applied to perceptual stimuli.

Empirical studies of case-based decision theory

CBDT has been applied to study consumer behavior (Gilboa & Schmeidler, 1993, 1997b, 2003; Gilboa, Postlewaite, & Schmeidler, 2015), brand choice (Gilboa & Pazgal, 1995), individual learning (Gilboa & Schmeidler, 1996), social learning (Blonski, 1999; Heinrich, 2013; Krause, 2009), sequential planning (Gilboa & Schmeidler, 1995b), asset pricing (Guerdjikova, 2006), real estate (Gayer, Gilboa, & Lieberman, 2007), portfolio choice (Golosnoy & Okhrin, 2008), technology adoption (Eichberger & Guerdjikova, 2012), manufacturing capacity (Jahnke, Chwolka, & Simons, 2005), macroeconomic expectations (Pape & Xiao, 2014), and Japanese TV drama watching (Kinjo & Sugawara, 2016). The idea of similarity between strategic games on both structural and perceptual levels has been used to analyze learning, transfer, and spillover across different games and institutional setups (Bednar, Chen, Liu, & Page, 2012; Cason, Savikhin, & Sheremeta, 2012; Cooper & Kagel, 2003; Cownden, Eriksson, & Strimling, 2015; Di Guida & Devetag, 2013; Guilfoos & Pape, 2016; LiCalzi, 1995; Mengel & Sciubba, 2014; Rankin, Van Huyck, & Battalio, 2000; Samuelson, 2001; Sarin & Vahid, 2004; Spiliopoulos, 2013; Steiner & Stewart, 2008).

Experimental tests specifically conducted on CBDT have yielded encouraging results. Ossadnik, Wilmsmann, and Niemann (2013) ran a ball and urn experiment with a twist. Every ball had three separate payoffs on it, identified by colors (which were the same across balls). On each trial, participants had to choose a color. A ball was drawn from the urn (which contained a known number of balls) and only the payoff associated with the chosen color was revealed. After a number of trials, a second round began in which a few balls were removed from the urn without being revealed. Later on, a third round began in which several balls were similarly added to the urn. Given the limited information available and the high number of possible ball–color–value combinations, full Bayesian updating would be difficult. The experimenters found that, as compared with maximin-type criteria and simple model-free reinforcement learning, the data conformed best to CBDT supposing that similarity across trials was proportional to the number of balls in common.

Participants in the study of Grosskopf, Sarin, and Watson (2015) were in the role of a company having to choose production levels for an economic good. The amount of profit for a given production level depended on "market conditions," which were represented by a list of five symbols. In each round, participants had access to only a few past cases, which were combinations of market conditions (case descriptions), production choices (actions), and profit levels (outcomes). Similarity between the vectors of past and present market conditions was taken to be the number of symbols in common, a special case of Tversky's (1977) contrast model. CBDT described participant behavior better than a heuristic, which ignored market conditions and chose the production level yielding the highest past value.

Bleichrodt, Filko, Kothiyal, and Wakker (2017) used a special design to test the core of CBDT without making any structural assumptions about similarity. Participants made choices on the basis of hypothetical case banks, one of which consisted of true values and would be used for payment. Cases dealt with the monthly value appreciation of real estate investments in various parts of the Netherlands. Participants had to choose between gambles with payoffs based on the appreciation percentage of a new piece of real estate. This experimental design allows certain functions of similarity weights to be estimated, which can be used to test the implications of CBDT from binary choices alone. CBDT predicts that people choose by combining, in a specific way, the hypothetical memory with their personal assessment of similarity across types and locations of real estate. The theory's axioms impose behavioral restrictions reflecting the consistency of similarity weights across decisions. These restrictions were generally satisfied by the data.

Pape and Kurtz (2013) combined CBDT with the ALCOVE neural network model to analyze classification learning. In this model, the relative importance of each feature dimension is updated from feedback, with overall learning rate, aspiration level, and degree of imperfect recall estimated as model parameters. A simulated case-based agent predicted the speed of learning well across categorization schemes of various difficulty levels (Nosofsky & Palmeri, 1996; Nosofsky, Gluck, Palmeri, McKinley, & Glauthier, 1994). Moreover, additive similarity was found to fit the data better than averaged similarity.

CBDT is attractive because it forces us to link choice to the set of cases in our memory in a way that offers a platform for the impact of memory and associations in economic modeling. Particularly in the most complex of situations, all cases may not be immediately recalled. Rather, we have to engage in mental search. Evaluation may derive from finite samples drawn from memory, as some theories posit. At the extreme, people often retrieve only a single case to work from. If the probability of retrieval is proportional to the similarity between cases, then the averaged case-based assessment constitutes the expectation of retrieved value. When we take into account that people draw small samples from similar cases in memory (Qian & Brown, 2005), regularly observed biases affecting judgment and decision-making can be parsimoniously explained (Gayer, 2010; Hertwig, Barron, Weber, & Erev, 2004; Marchiori, Di Guida, & Erev, 2015; Stewart, Chater, & Brown, 2006).

Computational models of association can be integrated with CBDT to produce a unified model for studying the effects of framing on economic decisions. After all, such phenomena are about altered patterns of mental association stemming from the way a problem is presented. In novel conditions, the case-based estimate represents a kind of half-educated guess. It is stitched together, Frankenstein-like, from whatever comes to mind. It is not an exceptionally consistent estimate and is prone to being jostled by the vagaries of memory. Preferences are therefore unstable, cobbled-together assessments of value that shift as different memories are emphasized. In this vein, Gonzalez and colleagues have developed case-based (aka instance-based) models that incorporate similarity and selective retrieval in the ACT-R architecture to predict and explain a variety of economic choices (Dutt & Gonzalez, 2012; Dutt, Arló-Costa, Helzner, & Gonzalez, 2014; Gonzalez, 2013; Gonzalez & Dutt, 2011; Gonzalez, Lerch, & Lebiere, 2003; Harman & Gonzalez, 2015; Lebiere, Gonzalez, & Martin, 2007; Lejarraga et al., 2012, 2014). Some of this work focuses explicitly on framing, accounting for variation in preferences based on differences in the retrieval process (Gonzalez & Mehlhorn, 2016).

COMPUTATIONAL CHARACTERIZATIONS
Case-based reasoning

The idea of computational connections should not be entirely surprising as CBDT was conceived of with a certain computational backdrop in mind—a problem-solving process known as CBR that stores training data and waits to make judgments until a new problem is posed (Stanfill & Waltz, 1986; Riesbeck & Schank, 1989; Aha, Kibler, & Albert, 1991; Kolodner, 1992, 1993). The heart of CBR lies in solving new problems by reusing and adapting solutions to similar old problems. It is captured by the "CBR cycle" consisting of the 4 *R*'s: Retrieve, Reuse, Revise, and Retain (Richter & Weber, 2013). When a new problem is encountered, similar past cases are *retrieved* from the case base, their information is *reused* to construct solutions, their solutions are *revised* to fit current needs, and the new experience is *retained* for future use.

CBR has been fruitfully applied to commercial tasks as diverse as customer service, vehicle fault diagnosis and repair, and aircraft part construction (Watson & Marir, 1994; Leake, 1994, 1996; Montani & Jain, 2010, 2014). For example, a critical task in the aerospace industry is to precisely bond together composite materials using extreme heat and pressure in an autoclave. However, the right way to arrange these materials in the autoclave is complicated because its heating properties are not perfectly understood, and identical examples are unavailable because product designs are always changing. The company Lockheed successfully tackled this problem with a software system called Clavier, which recommended new layouts by adapting previous similar layouts. Clavier proved useful even with a small case base, and with more experience its "performance 'grew' to approach that of the most experienced autoclave operator in the shop" (Hennessy & Hinkle, 1992).

Why has CBR proven so successful? It can be flexibly applied to a wide range of problems, even difficult ones encountered for the first time. CBR is a type of lazy learning, meaning that the answer is only generated when a new query arises. This just-in-time approach is helpful when faced with an infinite number of unencountered and unforeseen possibilities. We are commonly forced to perform in novel circumstances where causal relationships are not well understood but background knowledge can still prove useful, and CBR can support transfer learning here (Aha, Molineaux, & Sukthankar, 2009; Klenk, Aha, & Molineaux, 2011). We can further understand case-based decision-making by comparing control systems from a statistical standpoint.

Bias—variance trade-off

CBDT shares properties with nonparametric estimation. The case-based estimate is a similarity-weighted sum of case values. It takes the same kind of form as a nonparametric kernel estimate, which is a kernel-weighted sum of data points. The similarity function plays the role of the kernel, assessing how close the new input value is to each of the old input values in psychological space, and then blending the old output values accordingly. The case-based estimate with averaged similarity especially mimics the Nadaraya—Watson kernel regression estimator, a locally weighted average of data points. If only a single case is retrieved due to cognitive limitations, the model coincides with nearest-neighbor interpolation. Case-based estimation may not be as agnostic as statistical technique about the domain of application since background information is contained in the shape of the similarity function. It may also exhibit properties such as asymmetry (Tversky, 1977) that are atypical in statistical applications. Nonetheless, formal links have been established between case-based and kernel-based methods (Gilboa, Lieberman, & Schmeidler, 2011; see also Hüllermeier, 2007). We may thus view CBDT from one angle as a nonparametric estimate of value. This link helps us see why and when the case-based estimate is useful.

A case-based controller exhibits a different statistical trade-off than model-free and model-based controllers. This entails a distinct pattern of advantages and disadvantages. A case-based system stakes out an intermediate position between model-free and model-based systems on the bias—variance spectrum.

Case-based control employs knowledge derived from unsupervised or other subtler forms of learning to a greater degree than a model-free system. It better leverages experience by casting a wider net in the sea of memory. In other words, it engages in greater generalization from other circumstances to its present condition. Simplistic reinforcement learning models relinquish this power and neglect background relationships between acts or contexts. Continuous state or action spaces provide extreme examples of the need to generalize. Continuity has been a classic issue in reinforcement learning partly because it implies that an agent never encounters the exact same action or state more than once. Incorporating the values of similar actions in similar contexts sharpens predictions. A kernel approach turns out to be robust to convergence problems that other solutions suffer from in continuous state spaces (Gershman & Daw, 2017; Ormoneit & Sen, 2002). Generalizing does come with the cost of statistical bias as the extra data reflect circumstances that may only be marginally relevant and can significantly degrade performance when poorly selected. In line with this, nonparametric estimators carry an intrinsic smoothing bias, which results from using data far from the focal point to reduce the estimator variance. But when one has almost no direct experience, using imperfectly relevant knowledge is worthwhile. For this reason, statisticians regulate smoothing bias via choice of bandwidth and find that the optimal window is larger when the sample is small. The benefit of even limited or noisy additional information is high when facing new stimuli.

However, a case-based controller is not as bold as a model-based controller. Model-based estimates impose strong assumptions in order to hone their predictions and reduce the portion of generalization error stemming from variance. This is the benefit of a cognitive map. But it comes at the cost of bias from two sources. First is the coarsening inherent in the construction of any mental model. All practical models must be simplifications, otherwise they would be far too complicated to represent. Second is the more egregious misspecification resulting from a mistaken understanding of the world. This issue is made worse by conditions of limited experience, when little data are available to constrain the map. Like nonparametric objects, case-based estimates avoid structure in order to mitigate bias but yield to the error from variance. A model-based system goes out on a limb in an attempt to make sharp predictions across new circumstances. In this sense, the model-based controller is the staunch one that sticks to its guns, while the case-based controller exhibits a more flexible and graceful judgment. A drawback is that the latter will learn more slowly—it hesitates to draw inferences even when those inferences may be justifiable—but as the maxim goes, it is better to be approximately right than definitely wrong. When traveling through new and complex surroundings, where the risk of a misstep can be high, clinging stubbornly to potentially outmoded conclusions is especially maladaptive.

Gilboa, Samuelson, and Schmeidler (2013) construct a unified model containing multiple classes of reasoning. They show that an agent may exhibit cycles where Bayesian reasoning is used until an unexpected event occurs, at which point case-based and rule-based reasoning take the lead until more data are collected and a new probabilistic model is formed. CBR can thus be inductively rational in the face of the unexpected. Lengyel and Dayan (2008) show that a kind of episodic memory-based control can outperform model-based control when the world is novel and complex. Erroneous or misspecified aspects of the model-based belief structure, represented as inferential noise, produce costly mistakes particularly when problems are multistage and experience is limited. Researchers at Google DeepMind recently demonstrated that in the low-data regime, such episodic control prevails over other state-of-the-art algorithms in complicated sequential decision-making tasks like video games (Blundell et al., 2016), especially when the feature mapping can also be trained (Pritzel et al., 2017).

Despite these useful characterizations of case-based control, our understanding of how similarity is realistically learned and processed in the brain has more to say. This understanding could inspire further hybrid models that draw out the economic implications of lifelike neural architectures. The idea of a case-based system dovetails with recent interest in the role of the MTL, and specifically the hippocampus, in decision-making. This region might be considered a primary neural locus for the processes of learning and memory that instantiate a case-based system.

NEURAL PATHWAYS

Generalization and the hippocampus

Hippocampal function is traditionally conceived in terms of spatial knowledge and episodic memory. However, growing attention is being paid to how its associational processing flexibly subserves the learning and construction of value, especially in novel and complex situations (Seger & Peterson, 2013; Shohamy & Turk-Browne, 2013; Wimmer & Shohamy, 2011). Though the canonical view of feedback learning focuses on the basal ganglia, recent work suggests expanding the previously overlooked role of the MTL. Dopamine-driven striatal learning turns out to be limited, for example, when feedback is delayed or withheld. The hippocampal region, which appears to play a central role in generalization, is then required to bind information about cues and outcomes across time and space. While the striatum is responsible for encoding stimulus—response links, the hippocampus is responsible for encoding stimulus—stimulus links. The MTL supports generalization by this process of bundling stimulus representations into associative networks, within which items are considered similar neurally and psychologically based on shared connections.

We tend to view memory as dealing with the past, but it actually exists to help us predict the future. The process of association carried out by the hippocampus has two purposes from a decision-making perspective: First is to retrieve relevant memories, particularly those elements corresponding to value, in service of present decisions; second is to construct, modify, and consolidate memory in service of future decisions.

The hippocampus tugs the mental strings connected to an encountered configuration of stimuli in an attempt to anticipate forthcoming stimuli and rewards. The ingredients needed for decision-making and value learning appear to be represented in the hippocampus. Human neuroimaging has revealed concurrent value and choice signals in area CA1 of the hippocampus shortly before choices are made, as well as outcome signals following choice (Lee, Ghim, Kim, Lee, & Jung, 2012). Striking evidence for a control system distinct from standard dopaminergic and striatal mechanisms comes from feedback learning experiments, which involve comparisons and dissociations with Parkinson's disease (PD) patients and MTL amnesics (Reber, Knowlton, & Squire, 1996; Moody, Bookheimer, Vanek, & Knowlton, 2004; Shohamy, Myers, Onlaor, & Gluck, 2004, 2009). Foerde, Race, Verfaellie, and Shohamy (2013) documented a double dissociation on a standard probabilistic learning task with either immediate or delayed feedback. When faced with immediate feedback, PD patients were impaired while amnesics performed as well as controls, whereas with delayed feedback, PD patients performed as well as controls while amnesics were impaired. Remarkably, the delay difference producing the effect was not long (1 s vs. 7 s).

Consistent with the idea that a case-based system is most advantageous under novelty, Poldrack et al. (2001) showed that control appears to be transferred from MTL to the striatum as classification learning proceeds. Moreover, several studies demonstrate involvement of the hippocampus in spillover of value to stimuli and actions that are new but similar to those observed or taken in the past (Kahnt, Park, Burke, & Tobler, 2012; Wimmer, Daw, & Shohamy, 2012). Barron, Dolan, and Behrens (2013) created especially novel stimuli, which were new combinations of familiar foods, such as an avocado and raspberry smoothie. This forced participants to construct assessments of the novel goods via combination of past experiences. Activity in the hippocampus was found to be related to this construction process.

Though the exact mechanisms by which the MTL comes to generalize value are as yet unknown, any theories must respect the fundamental associative nature of hippocampal function (Horner & Burgess, 2013). As a multimodal convergence zone, it takes in signals from many regions. By the manner in which the hippocampus recognizes stimulus bundles, it links lower level stimuli to higher level associations and concepts, illustrated by sparse coding cells, extreme versions of which are popularly known as "grandmother cells" or "Jennifer Aniston neurons" (Kreiman, Koch, & Fried, 2000; Quiroga, Reddy, Kreiman, Koch, & Fried, 2005, 2008, 2014). It thus exerts some control over one's degree of conceptual granularity, thereby impacting degrees of generalization. One proposed mechanism of generalization that fits this picture is integrative encoding, wherein episodes with overlapping elements are integrated into a linked network of mnemonic associations (Shohamy & Wagner, 2008). Retrieval under novel circumstances then activates this network and can indirectly draw upon associations between concepts or stimuli that were never directly experienced together (Walther, 2002), mechanically similar to the creation of false memories (Roediger & McDermott, 1995). Even bumblebees may similarly merge memories after feedback learning (Hunt & Chittka, 2015). Indeed, when a new memory is formed, older memories with overlapping events are reactivated (Schlichting, Zeithamova, & Preston, 2014), alongside the rewards tied to those older memories (Kuhl, Shah, DuBrow, & Wagner, 2010; Wimmer & Büchel, 2016).

Stimulus associations and the hippocampus

A large body of human and animal studies reveals that whenever stimuli are separated in time and space, the hippocampus is central to connecting them to each other as well as their spatial and temporal context (Staresina & Davachi, 2009). This is especially the case when the configurations are stable and consistent (Mattfeld & Stark, 2015). The hippocampus is engaged during sequence learning (Schendan, Searl, Melrose, & Stern, 2003), and lesions impair the ability to learn and remember temporal regularities (Curran, 1997; Farovik, Dupont, & Eichenbaum, 2010; Schapiro, Gregory, Landau, McCloskey, & Turk-Browne, 2014). It is usually crucial for "trace conditioning" in which there is a significant interval between the end of the conditioned stimulus and beginning of the unconditioned stimulus presentation (Bangasser, Waxler, Santollo, & Shors, 2006; Cheng, Disterhoft, Power, Ellis, & Desmond, 2008) and also seems involved in "delay conditioning" when there is a long delay between conditioned and unconditioned stimulus onset even if they overlap (Berger, Alger, & Thompson, 1976; Christian & Thompson, 2003; Green & Arenos, 2007; Tam & Bonardi, 2012). Computational models are able to predict hippocampal learning in such paradigms by focusing on how its stimulus representations change over the course of a trial (Ludvig, Sutton, Verbeek, & Kehoe, 2009, 2008; Moustafa et al., 2013).

When new stimulus configurations are encountered, the hippocampus binds the components together and associates them with past bundles of stimuli, whether learning is explicit or implicit (Degonda et al., 2005; Rose, Haider, Weiller, & Büchel, 2002). The ultimate goal is to make better predictions through generalization. Accordingly, when there is reason to believe that different stimuli will foreshadow similar prospects, these stimuli actually become represented more similarly by neural activity patterns, so that they will be treated similarly in further processing. The stimuli come to activate similar networks and also become embedded and integrated more strongly within these networks, leading them to be better remembered (Kuhl et al., 2010; LaRocque et al., 2013; Staresina, Gray, & Davachi, 2009). Intriguingly, the degree of this representational overlap for a given memory is negatively related to the strength of its unique episodic reinstatement, suggesting a trade-off between integration of the memory into the network and retrieval of its specific details (Tompary & Davachi, 2017).

This enhanced pattern similarity can be triggered in multiple ways. Most directly, cues that are associated with the same outcome are mentally bundled together, and information learned about one is generalized to the others. This phenomenon of acquired equivalence relies on the hippocampal formation (Bódi, Csibri, Myers, Gluck, & Kéri, 2009; Coutureau et al., 2002; Myers et al., 2003; Preston, Shrager, Dudukovic, & Gabrieli, 2004). The stimuli come to be coded more similarly in the hippocampus (McKenzie et al., 2014) and become easier to confuse with each other (Meeter, Shohamy, & Myers, 2009). Stimuli that merely appear close together in time and context, absent outcomes, are like-wise informationally linked. This sensory preconditioning also depends on the hippocampal formation (Port & Patterson, 1984; Wimmer & Shohamy, 2012), and so might higher-order conditioning, when the original cue is conditioned before cues are paired (Gilboa, Sekeres, Moscovitch, & Winocur, 2014). Such stimuli become represented more similarly by MTL activity patterns (Hsieh, Gruber, Jenkins, & Ranganath, 2014; Schapiro, Kustner, & Turk-Browne, 2012), and pattern similarity at the time of retrieval is related to one's subjective sense of temporal and contextual proximity between the objects (Ezzyat & Davachi, 2014), as well as successful memory for their order (DuBrow & Davachi, 2014).

Neural pattern similarity in the temporal lobe appears representative of psychological similarity (Charest, Kievit, Schmitz, Deca, & Kriegeskorte, 2014; Davis & Poldrack, 2014; Davis, Xue, Love, Preston, & Poldrack, 2014), perhaps because psychological category structure may be represented in such a dimension-reduced and hierarchical manner that it can be smoothly mapped onto a two-dimensional neural substrate (Huth, Nishimoto, Vu, & Gallant, 2012; Kriegeskorte et al., 2008). There is some evidence that hippocampal coding for nonsemantic item-context bundles also follows a hierarchical structure. McKenzie et al. (2014) recorded activity from neuronal ensembles of rats in a learning task and found context to be of primary importance to coding similarity, followed by position of items within the environment, followed by the item valence (reward status), and lastly the item identity itself. These results can help us understand how similarity is constructed on deep levels.

Neural computations of the hippocampus

Some of the mechanisms contributing to such high-level patterns are reasonably well understood. Computational theories describe the associative retrieval and encoding functions of the hippocampal region in terms of information processing by each of its anatomical substructures in turn (Gluck & Myers, 2001; Hasselmo & Eichenbaum, 2005; Marr, 1971; McNaughton & Nadel, 1990; Treves & Rolls, 1994). Input from the neocortex is first processed through hippocampal afferents in a specialized manner, with the perirhinal and lateral entorhinal cortices supporting item memory and the parahippocampal and medial entorhinal cortices supporting context memory (Diana, Yonelinas, & Ranganath, 2013, 2007; Kragel, Morton, & Polyn, 2015; Libby, Hannula, & Ranganath, 2014; Reagh & Yassa, 2014). The entorhinal cortex (EC) acts as a primary gateway between the hippocampus and the rest of the brain. Information travels through a loop with recurrence and multiple paths (Andersen, Bliss, & Skrede, 1971): The EC projects to the dentate gyrus (DG), area CA3 (through the perforant pathway), and area CA1; DG projects sparsely to CA3 via mossy fibers; CA3 exhibits a relatively large amount of recurrent collaterals feeding back onto itself, and projects to CA1 via Schaffer collaterals; and CA1 projects out of hippocampus via subiculum and EC back out to neocortex, and via fornix to other regions in cortex.

A computational linchpin is area CA3, thought to form a recurrent autoassociative network that reconstructs complete memories from partial inputs (Gluck & Myers, 1997). In this process of pattern completion, the presentation of cues reinstates networks of activity based on the nexus of associated places, times, histories, concepts, and outcomes, particularly those that are most pivotally and centrally connected. Pattern completion by CA3 is integral to both memory retrieval and encoding, though they invoke different neural paths. Mice and rats with lesions to CA3 are impaired on spatial learning tasks especially when given a smaller number of cues with which to retrieve the full memory (Gold & Kesner, 2005; Nakazawa et al., 2002) and single-unit recording shows CA3 output as being closer to stored representations than to degraded input patterns (Neunuebel & Knierim, 2014). This sort of retrieval is predominantly initiated by direct input from the EC and is accordingly disrupted by lesions of the perforant pathway (Lee & Kesner, 2004), though some evidence suggests that dentate granule cells also help with pattern completion (Gu et al., 2012; Nakashiba et al., 2012). However, the projection from the EC is too weak to handle the encoding of new memories.

Autoassociative encoding can be powerful enough to yield one-shot learning, in which a single trial alone is enough to firmly store a memory (Day, Langston, & Morris, 2003; Nakazawa et al., 2003; Rutishauser, Mamelak, & Schuman, 2006). Incoming patterns must be separated if they are to be stored distinctively, which is considered a function of the DG (Bakker, Kirwan, Miller, & Stark, 2008; Leutgeb, Leutgeb, Moser, & Moser, 2007; McHugh et al., 2007; Schmidt, Marrone, & Markus, 2012). Encoding is indeed driven by the mossy fibers from the DG, and new learning is disturbed if these are inactivated, although retrieval is usually spared (Lassalle, Bataille, & Halley, 2000; Lee & Kesner, 2004). The mossy fiber synapses come close to the bodies of CA3 pyramidal neurons and are sometimes called "detonator synapses" because they hold the ability to forcefully induce associative plasticity among CA3 neurons and their afferents (Brandalise & Gerber, 2014; Chierzi, Stachniak, Trudel, Bourque, & Murai, 2012; Lee et al., 2013; Lysetskiy, Földy, & Soltesz, 2005; Rebola, Carta, Lanore, Blanchet, & Mulle, 2011).

Reencoding must normally happen when the stimuli anticipated by cued associations fail to match the stimuli actually encountered—that is, when there is a prediction error. Signals of expectancy violation have been detected in the hippocampus with a range of methods (Fyhn, Molden, Hollup, Moser, & Moser, 2002; Hannula & Ranganath, 2008; Honey, Watt, & Good, 1998; Knight, 1996; Kumaran & Maguire, 2006, 2007) and appear to be associative in that they are based on unexpected combinations of stimuli rather than merely novelty of stimuli alone (Kafkas & Montaldi, 2015; Shohamy & Wagner, 2008). These signals have been localized to area CA1, which is ideally placed to act as a comparator or match—mismatch detector, as it receives sensory information about the environment from the EC along with the associative predictions formed by CA3 (Chen, Olsen, Preston, Glover, & Wagner, 2011; Duncan, Ketz, Inati, & Davachi, 2012). Such signals are likely needed to switch between the retrieval and encoding modes of CA3 autoassociation. When expectations are not met, encoding is triggered and memories are updated, by either strengthening or weakening connections and representations as needed. The mnemonic representations of items that fail to materialize when expected become weaker, making them easier to forget (Kim, Lewis-Peacock, Norman, & Turk-Browne, 2014). These associative prediction errors guide learning in many circumstances, of which novelty is an important class (Kumaran & Maguire, 2007, 2009). In this way, the prediction error induces plasticity to adaptively enhance learning under novelty, complementing the adaptive properties of choice under novelty discussed earlier.

INTERACTIONS BETWEEN SYSTEMS

Many doors are open for interaction between control systems. The hippocampus is anatomically embedded in multiple dopaminergic pathways. The neurophysiological record shows direct connections between the hippocampal formation and the ventral striatum (Floresco, Todd, & Grace, 2001; Legault, Rompré, & Wise, 2000; Lisman & Grace, 2005) and possibly dorsal striatum (Finch, 1996; Finch, Gigg, Tan, & Kosoyan, 1995; Jung, Hong, & Haber, 2003; La Grutta & Sabatino, 1988; Sabatino, Ferraro, Liberti, Vella, & La Grutta, 1985; Sørensen & Witter, 1983) and Scimeca and Badre (2012) discuss several ways the striatum could support retrieval. Direct dopaminergic projections from the ventral tegmental area (VTA) have been shown to enhance long-term potentiation in the hippocampus to support plasticity and encoding (Duncan, Tompary, & Davachi, 2014; Lisman & Grace, 2005; Shohamy & Adcock, 2010; Wittmann et al., 2005). Recent evidence indicates that dopamine release from the locus coeruleus also plays a pivotal role in hippocampal signaling (Kempadoo, Mosharov, Choi, Sulzer, & Kandel, 2016; Takeuchi et al., 2016), especially for one-shot learning in highly novel contexts (Wagatsuma et al., 2018). Dopamine modulates hippocampal plasticity on timescales from minutes to hours (Bethus, Tse, & Morris, 2010; Frey et al., 1990; Lisman, Grace, & Duzel, 2011; O'Carroll, Martin, Sandin, Frenguelli, & Morris, 2006), improving memory encoding and consolidation (Apitz & Bunzeck, 2013; Axmacher et al., 2010; Imai, Kim, Sasaki, & Watanabe, 2014; Kafkas & Montaldi, 2015; McNamara, Tejero-Cantero, Trouche, Campo-Urriza, & Dupret, 2014; Murayama & Kitagami, 2014; Rosen, Cheung, & Siegelbaum, 2015; Rossato, Bevilaqua, Izquierdo, Medina, & Cammarota, 2009). Memory strength can thus be easily enhanced by reward, and information acquisition itself can provide pseudorewards or bonuses (Kakade & Dayan, 2002), strengthening memory via similar neural pathways (Bunzeck, Doeller, Dolan, & Duzel, 2012; Gruber, Gelman, & Ranganath, 2014; Kang et al., 2009; Wittmann, Daw, Seymour, & Dolan, 2008, 2007).

In the other direction, the hippocampus can activate dopaminergic neurons in the VTA by sending CA1 novelty signals through the subiculum, nucleus accumbens, and ventral pallidum (Bunzeck & Düzel, 2006; Lisman & Grace, 2005). Contextual information straight from CA3 also travels through lateral septum to the VTA (Luo, Tahsili-Fahadan, Wise, Lupica, & Aston-Jones, 2011). Hippocampal pattern completion, replay of experience, and autobiographical recollection evoke or reinstate representations of value in the striatum to help accurately consolidate memories and associations relating to stimuli (Han, Huettel, Raposo, Adcock, & Dobbins, 2010; Schwarze, Bingel, Badre, & Sommer, 2013) or rewards (Kuhl et al., 2010; Lansink, Goltstein, Lankelma, McNaughton, & Pennartz, 2009; Speer, Bhanji, & Delgado, 2014).

Growing evidence reveals that episodic memory can guide value-based decision-making and is starting to shed light on how the hippocampus and striatum interact in the process (Pennartz, Ito, Verschure, Battaglia, & Robbins, 2011). In a simple value learning paradigm, Duncan and Shohamy (2016) documented behaviorally that contextual familiarity encouraged the retrieval and use of past episodes in decision-making. Murty, FeldmanHall, Hunter, Phelps, and Davachi (2016) showed that cues were used to adaptively guide lottery choice when learned cue—outcome associations were strong.

Wimmer and Büchel (2016) cued the retrieval of single past episodes in which stimuli were associated with rewards and found that risk preferences were biased by reactivation of the reward values, which were represented in the striatum. Gluth, Sommer, Rieskamp, and Büchel (2015) found that evaluation of snack food was biased toward items that were better remembered, and they observed corresponding value signals in the striatum, hippocampus, and ventromedial prefrontal cortex (vmPFC). They further found that this bias was mediated by hippocampal–vmPFC functional connectivity. Several other studies have observed that the strength of hippocampal–striatal connectivity during reward learning and at rest is related to value generalization (Gerraty, Davidow, Wimmer, Kahn, & Shohamy, 2014; Kumaran, Summerfield, Hassabis, & Maguire, 2009; Wimmer & Shohamy, 2012; Wimmer et al., 2012). Thus, the distributed neural representation of stimuli, values, and their associations depends crucially on what type of information must be retrieved and applied.

Interactions between case-based and model-free systems

Both competitive and cooperative links have been observed between case-based and model-free behaviors, as well as their presumed neural substrates.

Several experiments indicate competitive links between MTL-dependent declarative learning and striatum-dependent procedural learning (Moody et al., 2004; Poldrack & Packard, 2003; Poldrack, Prabhakaran, Seger, & Gabrieli, 1999), which may be mediated by PFC (Poldrack & Rodriguez, 2004). Rats with hippocampal lesions actually perform better on procedural learning tasks (Eckart, Huelse-Matia, & Schwarting, 2012). It may be that hippocampal context–outcome associations interfere with striatal action–outcome contingencies that could be more important in such circumstances over the long run (Cheung & Cardinal, 2005). Collins, Ciullo, Frank, and Badre (2017) imposed working memory load by increasing the number of stimuli to be learned and found that this strengthened model-free reward prediction errors. Wimmer, Braun, Daw, and Shohamy (2014) used a drifting probabilistic reward learning task in which a unique incidental picture accompanied each trial. Better episodic memory for the pictures on a surprise memory test the following day was negatively correlated with reward and reinforcement learning rate during the task. For individual trials on which the picture was successfully remembered, reward had a weaker influence on the subsequent choice, and reward prediction error signals in the putamen were negligible.

At the same time, cooperative links have been demonstrated in similar paradigms (Ferbinteanu, 2016). Bornstein, Khaw, Shohamy, and Daw (2017) showed that decision-making in a multiarmed bandit task was biased by incidental reminders of past trials, consistent with a version of model-free reinforcement learning that incorporates episodic sampling. Aberg, Müller, and Schwartz (2017) found that delivered and anticipated rewards were positively related to associative memory encoding, and valence-dependent asymmetries in these effects were modulated by individual differences in sensitivity to reward versus punishment. Dickerson, Li, and Delgado (2011) observed that prediction errors in feedback learning correlated positively with activity in both the putamen and the hippocampus. In some experiments centered on either episodic memory encoding or probabilistic reward learning, activity in the hippocampus appears to positively correlate with activity in the putamen on feedback trials when stimuli are successfully remembered later (Sadeh, Shohamy, Levy, Reggev, & Maril, 2011; Wimmer et al., 2014). In a probabilistic learning task with feedback accompanied by incidental trial-unique images, Davidow, Foerde, Galván, and Shohamy (2016) found that stronger episodic memory encoding was correlated with enhanced reinforcement learning among adolescents but not adults. Moreover, functional hippocampal–striatal connectivity was positive only for adolescents. Thus the process of development may play an important role in how these systems interact. Kahnt et al. (2012) looked at value updating in a perceptual association paradigm, augmenting a standard reinforcement learning model with a similarity-based generalization gradient. They found that hippocampal–striatal connectivity was negatively correlated with the width of the generalization window, suggesting a discriminative mechanism.

The exact nature of such interactions thus remains an open question. Computational theory may help suggest possible mechanisms, especially cooperative ones. Various strands of the artificial intelligence literature synergistically combine CBR with model-free reinforcement learning to enhance transfer learning. This is particularly valuable when state and action spaces are large or continuous (Santamaría, Sutton, & Ram, 1997). Similar past cases can accelerate learning by contributing to initial guesses of the value function, which can then be revised according to temporal difference learning, retaining its promises of long-run convergence (Drummond, 2002; Gabel & Riedmiller, 2005; Sharma et al., 2007; Bianchi, Ribeiro, & Costa, 2008, 2009; Celiberto, Matsuurade Ma'ntaras, & Bianchi, 2010, 2011). Once learning has converged in a task, the optimal policy can be abstracted for transfer to future tasks (Von Hessling & Goel, 2005). In return, reinforcement learning is able to influence the retrieval of cases by helping with online assessment of the best similarity metrics for CBR (Juell & Paulson, 2003). Cases may be stored preferentially when the agent is attaining high rewards (Auslander, Lee-Urban, Hogg, & Munoz-Avila, 2008) and selectively pruned when they hinder prediction or exceed storage (Gabel & Riedmiller, 2005; Wilson & Martinez, 2000).

Such computational models suggest further possibilities for neural interplay between case-based and model-free systems. Along related lines, an influential body of work has been inspired by the well-known Dyna architecture (Sutton, 1990), which is used to explain hippocampal replay of recent memories during rest (Johnson & Redish, 2007; Kurth-Nelson, Economides, Dolan, & Dayan, 2016). Modified temporal difference algorithms with offline replay of previously experienced sequences allow extra practice, substantially speeding up early learning (Johnson & Redish, 2005; Johnson & Venditto, 2015).

Another important form of generalization relies on eligibility traces, which are computational accessories to temporal difference learning attached to states or actions that facilitate their value updates across temporal gaps (Barto, Sutton, & Brouwer, 1981; Sutton & Barto, 1998). Given its responsibilities in connecting stimuli across delays, the hippocampus may be involved in instantiating eligibility traces (Ludvig et al., 2009). Implementation could happen by means of synaptic tagging, in which recently active synapses are tagged for increased susceptibility to long-term potentiation or depression over longer periods of time (Frey & Morris, 1997; Izhikevich, 2007). Dopamine firing patterns do appear to reflect eligibility traces (Pan, Schmidt, Wickens, & Hyland, 2005), and we have seen how strongly entangled the hippocampus is with various dopamine circuits. However, such hypotheses remain to be empirically verified. Notably, eligibility traces are most beneficial in non-Markovian environments. Among other reasons, this could occur when agents are unsure of what to attend to in an unfamiliar setting, rendering the state space only partially observable.

Interactions between case-based and model-based systems

Model-based decision-making relies on sophisticated forecasting, typically involving the estimation of state transition probabilities. One source of these subjective probabilities may be a case-based system (Blok, Medin, & Osherson, 2003; Taylor, Jong, & Stone, 2008).

Some evidence supports the existence of a hippocampal process for learning transition probabilities that operates in parallel with the striatum and is linked to model-based decision-making (Bornstein & Daw, 2012, 2013). Hippocampal similarity-based learning is also thought to be one mechanism for learning word transition probabilities of artificial grammars (Opitz & Friederici, 2004). Such belief updating may be premised on sequential association learning (Amso, Davidson, Johnson, Glover, & Casey, 2005) and the binding of regularities across time and space as discussed earlier. Consistent with a key role for association, Doll, Shohamy, and Daw (2015) found that generalization in an acquired equivalence task was correlated with use of a model-based strategy in a separate sequential learning task. Theoretically, probabilities constructed from stimulus associations might reflect the successor representation (Dayan, 1993), which assesses the expected future visitations of states based on their sequential cooccurrence. This can be done latently prior to the introduction of reward and sheds light on how cognitive maps may be neurally instantiated in the hippocampus (Stachenfeld, Botvinick, & Gershman, 2014, 2017). The successor representation could explain why sensitivity to contingency degradation is impaired in rats with lesions of the hippocampal region but sensitivity to outcome devaluation is spared (Corbit & Balleine, 2000; Corbit, Ostlund,

& Balleine, 2002). A mild disparity in these sensitivities occurs even under normal circumstances in humans, which may be explained by a hybrid successor representation/model-based mechanism (Momennejad et al., 2017). Moreover, the temporal context model of episodic memory can be viewed as estimating the successor representation (Gershman, Moore, Todd, Norman, & Sederberg, 2012), revealing a deep connection between episodic memory and reinforcement learning.

Outcome projection based on similar cases is common in the world at large and has proven successful when facing complex problems. As John Locke said, "in things which sense cannot discover, analogy is the great rule of probability." Similarity-based approaches can help accurately predict college admissions (Klahr, 1969), movie revenue (Lovallo, Clarke, & Camerer, 2012), and legal case outcomes (Teitelbaum, 2014). Historically, weather forecasting was done by seeing how conditions evolved on similar recorded days (Kruizinga & Murphy, 1983). In general, this method of "reference class forecasting" suggested by Kahneman and Tversky (1982) has been found helpful in project management to the point where it is officially endorsed by the American Planning Association, particularly for "nonroutine projects … and other local one-off projects"—in other words, novel problems with limited past data. One branch of decision theoretic models formalizes the idea by constructing probabilities from similarity-weighted frequencies of past outcomes—a kernel estimate of event occurrence. Billot, Gilboa, Samet, and Schmeidler (2005) provide an axiomatized representation of probabilities as similarity-weighted frequencies. Others have relaxed their assumptions in various ways, such as by allowing beliefs to depend on the database size, having multiple beliefs to reflect ambiguity (Eichberger & Guerdjikova, 2010), and combining similarity-weighted frequencies with a prior in a nested Bayesian framework (Bordley, 2011). Theoretical predictions from these models await empirical testing.

Another line of research focuses on a more flexible form of forecasting based on imagination. Imagining potential outcomes in detail can help agents evaluate options, and the hippocampus plays a significant role in this mental simulation (Buckner & Carroll, 2007; Gilbert & Wilson, 2007; Suddendorf & Corballis, 2007). Just as the hippocampus enables us to reconstruct vivid scenes from past episodes, it also helps us to conjure up potential future scenarios from reconstituted episodes (Schacter, Addis, & Buckner, 2007, 2008, 2012). In the process, it may interact with vmPFC to integrate related events in a flexible and prospectively useful form (Benoit, Szpunar, & Schacter, 2014; Weilbächer & Gluth, 2017; Zeithamova & Preston, 2010; Zeithamova, Dominick, & Preston, 2012). Future events are imagined in more detail when they would occur in familiar or recently experienced settings, revealing their origins in past episodes (Szpunar & McDermott, 2008). Envisioning future events recruits similar temporal and prefrontal regions as envisioning the past (Addis, Wong, & Schacter, 2007; Okuda et al., 2003; Schacter & Addis, 2007; Szpunar, Watson, & McDermott, 2007), and hippocampal amnesics typically exhibit impaired episodic prospection (Klein, Loftus, & Kihlstrom, 2002; Hassabis, Kumaran,

& Maguire, 2007; Addis, Sacchetti, Ally, Budson, & Schacter, 2009; Andelman, Hoofien, Goldberg, Aizenstein, & Neufeld, 2010; Kwan, Carson, Addis, & Rosenbaum, 2010; Race, Keane, & Verfaellie, 2011, 2013). Such imaginative prospection may be goal-relevant and enhanced by reward (Bulganin & Wittmann, 2015), though not always adaptively (Gershman, Zhou, & Kommers, 2017). Animals in choice experiments exhibit a phenomenon known as vicarious trial and error, in which they pause at choice points and orient themselves toward potential options, as if they were envisioning the future implications of taking a given path (Johnson, van der Meer, & Redish, 2007; Muenzinger, 1938; Tolman, 1938). This behavior appears to rely on the hippocampus (Hu & Amsel, 1995; Hu, Xu, & Gonzalez-Lima, 2006), and hippocampal activity represents positions traveling down each path ahead of the animal (Johnson & Redish, 2007).

This type of goal-relevant simulation sometimes plays a role in intertemporal choice, as the constructed representation of future reward may feed into previously identified frontoparietal control regions associated with a preference for longer-term options (McClure, Laibson, Loewenstein, & Cohen, 2004). Rats with hippocampal lesions tend to pick smaller, immediate rewards (Abela & Chudasama, 2013; Cheung & Cardinal, 2005; Mariano et al., 2009; McHugh, Campbell, Taylor, Rawlins, & Bannerman, 2008; Rawlins, Feldon, & Butt, 1985). People who are prompted to consciously imagine spending a delayed reward in the future tend to choose the delayed option more often, and the strength of this bias is correlated with simulation richness (Benoit, Gilbert, & Burgess, 2011; Daniel, Stanton, & Epstein, 2013; Lebreton et al., 2013; Lin & Epstein, 2014; Liu, Feng, Chen, & Li, 2013; Peters & Büchel, 2010). Hippocampal amnesics do not display this effect, although their intertemporal choices appear to be comparable to controls who are not prompted to use imagination (Palombo, Keane, & Verfaellie, 2014; though see Kwan et al., 2015), in accordance with multiple process hypotheses. From a theoretical standpoint, associative neural network models of region CA3 naturally generate standard reward discounting curves derived from the predicted similarity representations they produce with respect to future states (Laurent, 2013).

A final intriguing angle centers on analogical reasoning, which depends on higher-order structural similarity and enables powerful generalization (Gentner, James Holyoak, & Kokinov, 2001; Holyoak, 2012; Kolodner, 1997). Analogizing appears to be a problem-solving ability near the peak of cognition and decision-making. Raven's Matrices, which test abstract relational reasoning, rank highly among mental tests in their g-loading (Jensen, 1998). The flexible application and recombination of past cases invokes more conscious processing involving our evolutionarily well-developed PFC (Krawczyk, 2012; Zeithamova & Preston, 2010). Analogical thinking has been tested in other species as well, and only chimpanzees have succeeded at a level modestly comparable to humans (Zentall, Wasserman, Lazareva, Thompson, & Rattermann, 2008). Notably, successful chimps were those with prior training in symbolic representations like language or tokens. Thus, high-level relational comparisons may be key to both generalization and intelligence. These skills make a difference even at the frontier of human ability. The brokerage firm Merrill Lynch was styled after cofounder Charlie Merrill's experience in the supermarket industry (Gavetti, Levinthal, & Rivkin, 2005), and mathematician Stefan Banach often said that "good mathematicians see analogies between theorems or theories; the very best ones see analogies between analogies."

CONCLUSION

Decision neuroscience has been guided by the formal characterization of habitual and goal-directed control in terms of model-free and model-based systems. Research emerging from multiple fields points to the importance of alternative memory-based mechanisms in learning and valuation, straining the boundaries of the traditional dichotomy. I have reviewed the behavioral and neural evidence characterizing these "case-based" mechanisms from several angles.

Empirical research in psychology and economics shows that evaluation often occurs on the basis of similarity judgments (Gilovich, 1981). Theoretical work from economics, psychology, and computer science describes how decisions can be made by drawing on similar past cases (Gilboa & Schmeidler, 1995a; Kolodner, 1992). Computational and statistical perspectives reveal that such methods have different properties than typical model-free and model-based rules, analogous to nonparametric techniques (Gilboa et al., 2011). In particular, case-based evaluation makes fewer assumptions about problem structure than model-based evaluation, while still generalizing beyond the circumstances of past observations more than model-free evaluation. As a result, case-based approaches can be adaptive compared to other systems in novel and complex settings (Gilboa et al., 2013). This provides a normative justification for such alternative mechanisms and suggests under which conditions we might expect them to be mobilized.

The hippocampus and broader MTL structures are natural candidates to subserve a case-based system. Recent work in neuroscience indicates that these regions are involved in value-based judgment to a previously unrecognized extent. The hippocampus can reinstate memories of stimulus and reward associations when triggered by task-relevant or external cues (Wimmer & Büchel, 2016), and it sometimes even represents value signals directly (Gluth et al., 2015). Hippocampal involvement occurs especially with stimuli that are novel or natural objects (Barron et al., 2013) or when learning occurs over relatively long timescales (Foerde et al., 2013). Case-based computations could also support or compete with those of model-free or model-based systems, and both types of interaction have been observed (Bornstein & Daw, 2013; Wimmer et al., 2014).

A number of open questions follow from this perspective:

- Can tighter correspondences be found between brain activity and computational theory for a case-based system? Representations of value derived from reinforcement learning and expected utility have been observed in the striatum and prefrontal cortex

(Daw & O'Doherty, 2013; Knutson, Taylor, Kaufman, Peterson, & Glover, 2005), as have more exotic quantities such as regret (Lohrenz, McCabe, Camerer, & Montague, 2007). Case-based decision theory provides another quantitative account of value that may help explain neural activity.

- How does the interaction between hippocampal and striatal functions depend on the properties of the decision problem? In some tasks, the hippocampus reinstates contextual features to support striatal value representations (Wimmer & Büchel, 2016), while in other tasks, value signals appear to be represented in the hippocampus itself (Gluth et al., 2015). The distributed representation of value likely depends on properties of the stimulus and environment, such as familiarity and complexity, but clear principles are still to be laid out.

- What other adaptive properties might normatively justify contributions of episodic memory to decision-making? Computational noise in a model-based system can stem from stringent memory demands, so episodic control may exhibit more robust performance due to lower cognitive costs (Lengyel & Dayan, 2008). Such arguments may suggest new predictions about how factors like cognitive load affect learning and behavior.

To address these questions will require moving beyond traditional neuroeconomic paradigms in which artificial stimuli are presented repeatedly and value is learned incrementally. Neuroimaging techniques with high spatial resolution must also be used to measure brain activity in humans because the size, shape, and cytoarchitecture of the hippocampus make it difficult to image.

Evaluation based on similarity has arisen time and again across the behavioral and computational sciences. I have attempted to synthesize a wide range of relevant theoretical and empirical findings into a cohesive foundation for neuroeconomics to build on. These ideas reveal the need for studies that reflect the novel, unstructured, non-Markovian, discontiguous—in short, *messy*—nature of the world at large. Decision-making under such conditions may call upon different sets of mechanisms than those traditionally considered. The transparency and simplicity of most neuroeconomic experiments may obstruct our view of what happens when matters are not so tidy—and we live in an untidy world.

REFERENCES

Abela, A. R., & Chudasama, Y. (2013). Dissociable contributions of the ventral hippocampus and orbitofrontal cortex to decision-making with a delayed or uncertain outcome. *European Journal of Neuroscience, 37*(4), 640–647.

Aberg, K. C., Müller, J., & Schwartz, S. (2017). Trial-by-trial modulation of associative memory formation by reward prediction error and reward anticipation as revealed by a biologically plausible computational model. *Frontiers in Human Neuroscience, 11*(56).

Addis, D. R., Sacchetti, D. C., Ally, B. A., Budson, A. E., & Schacter, D. L. (2009). Episodic simulation of future events is impaired in mild Alzheimer's disease. *Neuropsychologia, 47*(12), 2660–2671.

Addis, D. R., Wong, A. T., & Schacter, D. L. (2007). Remembering the past and imagining the future: Common and distinct neural substrates during event construction and elaboration. *Neuropsychologia, 45*(7), 1363–1377.

Aha, D. W., Kibler, D., & Albert, M. K. (1991). Instance-based learning algorithms. *Machine Learning, 6*(1), 37–66.

Aha, D. W., Molineaux, M., & Sukthankar, G. (2009). Case-based reasoning in transfer learning. In *Case-based reasoning research and development* (pp. 29–44). Springer.

Amso, D., Davidson, M. C., Johnson, S. P., Glover, G., & Casey, B. J. (2005). Contributions of the hippocampus and the striatum to simple association and frequency-based learning. *NeuroImage, 27*(2), 291–298.

Andelman, F., Hoofien, D., Goldberg, I., Aizenstein, O., & Neufeld, M. Y. (2010). Bilateral hippocampal lesion and a selective impairment of the ability for mental time travel. *Neurocase, 16*(5), 426–435.

Andersen, P., Bliss, T. V. P., & Skrede, K. K. (1971). Lamellar organization of hippocampal excitatory pathways. *Experimental Brain Research, 13*(2), 222–238.

Apitz, T., & Bunzeck, N. (2013). Dopamine controls the neural dynamics of memory signals and retrieval accuracy. *Neuropsychopharmacology, 38*(12), 2409–2417.

Auslander, B., Lee-Urban, S., Hogg, C., & Munoz-Avila, H. (2008). Recognizing the enemy: Combining reinforcement learning with strategy selection using case-based reasoning. In *Advances in case-based reasoning* (pp. 59–73). Springer.

Axmacher, N., Cohen, M. X., Fell, J., Haupt, S., Dümpelmann, M., Elger, C. E., ... Ranganath, C. (2010). Intracranial EEG correlates of expectancy and memory formation in the human hippocampus and nucleus accumbens. *Neuron, 65*(4), 541–549.

Bakker, A., Kirwan, C. B., Miller, M., & Stark, C. E. L. (2008). Pattern separation in the human hippocampal CA3 and dentate gyrus. *Science, 319*(5870), 1640–1642.

Bangasser, D. A., Waxler, D. E., Santollo, J., & Shors, T. J. (2006). Trace conditioning and the hippocampus: The importance of contiguity. *Journal of Neuroscience, 26*(34), 8702–8706.

Bar-Hillel, M. (1974). Similarity and probability. *Organizational Behavior and Human Performance, 11*(2), 277–282.

Barron, H. C., Dolan, R. J., & Behrens, T. E. J. (2013). Online evaluation of novel choices by simultaneous representation of multiple memories. *Nature Neuroscience, 16*(10), 1492–1498.

Barto, A. G., Sutton, R. S., & Brouwer, P. S. (1981). Associative search network: A reinforcement learning associative memory. *Biological Cybernetics, 40*(3), 201–211.

Bednar, J., Chen, Y., Liu, T. X., & Page, S. (2012). Behavioral spillovers and cognitive load in multiple games: An experimental study. *Games and Economic Behavior, 74*(1), 12–31.

Benoit, R. G., Gilbert, S. J., & Burgess, P. W. (2011). A neural mechanism mediating the impact of episodic prospection on farsighted decisions. *Journal of Neuroscience, 31*(18), 6771–6779.

Benoit, R. G., Szpunar, K. K., & Schacter, D. L. (2014). Ventromedial prefrontal cortex supports affective future simulation by integrating distributed knowledge. *Proceedings of the National Academy of Sciences of the United States of America, 111*(46), 16550–16555.

Berger, T. W., Alger, B., & Thompson, R. F. (1976). Neuronal substrate of classical conditioning in the hippocampus. *Science, 192*(4238), 483–485.

Bethus, I., Tse, D., & Morris, R. G. M. (2010). Dopamine and memory: Modulation of the persistence of memory for novel hippocampal NMDA receptor-dependent paired associates. *Journal of Neuroscience, 30*(5), 1610–1618.

Bianchi, R. A. C., Ribeiro, C. H. C., & Costa, A. H. R. (2008). Accelerating autonomous learning by using heuristic selection of actions. *Journal of Heuristics, 14*(2), 135–168.

Bianchi, R. A. C., Ros, R., & Mantaras, R. L. De (2009). Improving reinforcement learning by using case based heuristics. In *Case-based reasoning research and development* (pp. 75–89). Springer.

Billot, A., Gilboa, I., Samet, D., & Schmeidler, D. (2005). Probabilities as similarity-weighted frequencies. *Econometrica, 73*(4), 1125–1136.

Bleichrodt, H., Filko, M., Kothiyal, A., & Wakker, P. P. (2017). Making case-based decision theory directly observable. *American Economic Journal: Microeconomics, 9*(1), 123–151.

Blok, S., Medin, D., & Osherson, D. (2003). Probability from similarity. In *AAAI conference on commonsense reasoning*.

Blonski, M. (1999). Social learning with case-based decisions. *Journal of Economic Behavior & Organization, 38*(1), 59–77.

Blundell, C., Uria, B., Pritzel, A., Li, Y., Ruderman, A., Leibo, J. Z., … Hassabis, D. (2016). *Model-free episodic control*. arXiv preprint arXiv:1606.04460.

Bódi, N., Csibri, É., Myers, C. E., Gluck, M. A., & Kéri, S. (2009). Associative learning, acquired equivalence, and flexible generalization of knowledge in mild Alzheimer disease. *Cognitive and Behavioral Neurology, 22*(2), 89–94.

Bordley, R. F. (2011). Using Bayes' rule to update an event's probabilities based on the outcomes of partially similar events. *Decision Analysis, 8*(2), 117–127.

Bornstein, A. M., & Daw, N. D. (2012). Dissociating hippocampal and striatal contributions to sequential prediction learning. *European Journal of Neuroscience, 35*(7), 1011–1023.

Bornstein, A. M., & Daw, N. D. (2013). Cortical and hippocampal correlates of deliberation during model-based decisions for rewards in humans. *PLoS Computational Biology, 9*(12), e1003387.

Bornstein, A. M., Khaw, M. W., Shohamy, D., & Daw, N. D. (2017). Reminders of past choices bias decisions for reward in humans. *Nature Communications, 8*, 15958.

Brandalise, F., & Gerber, U. (2014). Mossy fiber-evoked subthreshold responses induce timing-dependent plasticity at hippocampal CA3 recurrent synapses. *Proceedings of the National Academy of Sciences of the United States of America, 111*(11), 4303–4308.

Buckner, R. L., & Carroll, D. C. (2007). Self-projection and the brain. *Trends in Cognitive Sciences, 11*(2), 49–57.

Bulganin, L., & Wittmann, B. C. (2015). Reward and novelty enhance imagination of future events in a motivational-episodic network. *PLoS One, 10*(11), e0143477.

Bunzeck, N., Doeller, C. F., Dolan, R. J., & Duzel, E. (2012). Contextual interaction between novelty and reward processing within the mesolimbic system. *Human Brain Mapping, 33*(6), 1309–1324.

Bunzeck, N., & Düzel, E. (2006). Absolute coding of stimulus novelty in the human substantia nigra/VTA. *Neuron, 51*(3), 369–379.

Cason, T. N., Savikhin, A. C., & Sheremeta, R. M. (2012). Behavioral spillovers in coordination games. *European Economic Review, 56*(2), 233–245.

Celiberto, L. A., Matsuura, J. P., de Mantaras, R. L., Bianchi, R., et al. (2011). Using cases as heuristics in reinforcement learning: A transfer learning application. In *Proceedings of the twenty-second international joint conference on artificial intelligence* (pp. 1211–1217).

Celiberto, L. A., Matsuura, J. P., de Màntaras, R. L., & Bianchi, R. A. C. (2010). Using transfer learning to speed-up reinforcement learning: A cased-based approach. In *Proceedings of the latin american robotics symposium and intelligent robotics meeting* (pp. 55–60). Latin American: IEEE.

Charest, I., Kievit, R. A., Schmitz, T. W., Deca, D., & Kriegeskorte, N. (2014). Unique semantic space in the brain of each beholder predicts perceived similarity. *Proceedings of the National Academy of Sciences of the United States of America, 111*(40), 14565–14570.

Chen, Y., Garcia, E. K., Gupta, M. R., Rahimi, A., & Cazzanti, L. (2009). Similarity-based classification: Concepts and algorithms. *Journal of Machine Learning Research, 10*(3), 747–776.

Cheng, D. T., Disterhoft, J. F., Power, J. M., Ellis, D. A., & Desmond, J. E. (2008). Neural substrates underlying human delay and trace eyeblink conditioning. *Proceedings of the National Academy of Sciences of the United States of America, 105*(23), 8108–8113.

Chen, J., Olsen, R. K., Preston, A. R., Glover, G. H., & Wagner, A. D. (2011). Associative retrieval processes in the human medial temporal lobe: Hippocampal retrieval success and CA1 mismatch detection. *Learning & Memory, 18*(8), 523–528.

Cheung, T. H. C., & Cardinal, R. N. (2005). Hippocampal lesions facilitate instrumental learning with delayed reinforcement but induce impulsive choice in rats. *BMC Neuroscience, 6*(1), 36.

Chierzi, S., Stachniak, T. J., Trudel, E., Bourque, C. W., & Murai, K. K. (2012). Activity maintains structural plasticity of mossy fiber terminals in the hippocampus. *Molecular and Cellular Neuroscience, 50*(3), 260–271.

Christian, K. M., & Thompson, R. F. (2003). Neural substrates of eyeblink conditioning: Acquisition and retention. *Learning & Memory, 10*(6), 427–455.

Collins, A. G. E., Ciullo, B., Frank, M. J., & Badre, D. (2017). Working memory load strengthens reward prediction errors. *Journal of Neuroscience, 37*(16), 4332–4342.

Cooper, D., & Kagel, J. H. (2003). Lessons learned: Generalizing learning across games. *American Economic Review, 93*(2), 202–207.

Corbit, L. H., & Balleine, B. W. (2000). The role of the hippocampus in instrumental conditioning. *Journal of Neuroscience, 20*(11), 4233–4239.

Corbit, L. H., Ostlund, S. B., & Balleine, B. W. (2002). Sensitivity to instrumental contingency degradation is mediated by the entorhinal cortex and its efferents via the dorsal hippocampus. *Journal of Neuroscience, 22*(24), 10976–10984.

Coutureau, E., Simon Killcross, A., Good, M., Marshall, V. J., Ward-Robinson, J., & Honey, R. C. (2002). Acquired equivalence and distinctiveness of cues: II. Neural manipulations and their implications. *Journal of Experimental Psychology: Animal Behavior Processes, 28*(4), 388.

Cownden, D., Eriksson, K., & Strimling, P. (2015). The implications of learning across perceptually and strategically distinct situations. *Synthese*, 1–18.

Curran, T. (1997). Higher-order associative learning in amnesia: Evidence from the serial reaction time task. *Journal of Cognitive Neuroscience, 9*(4), 522–533.

Daniel, T. O., Stanton, C. M., & Epstein, L. H. (2013). The future is now: Comparing the effect of episodic future thinking on impulsivity in lean and obese individuals. *Appetite, 71*, 120–125.

Davidow, J. Y., Foerde, K., Galván, A., & Shohamy, D. (2016). An upside to reward sensitivity: The hippocampus supports enhanced reinforcement learning in adolescence. *Neuron, 92*(1), 93–99.

Davis, T., & Poldrack, R. A. (2014). Quantifying the internal structure of categories using a neural typicality measure. *Cerebral Cortex, 24*(7), 1720–1737.

Davis, T., Xue, G., Love, B. C., Preston, A. R., & Poldrack, R. A. (2014). Global neural pattern similarity as a common basis for categorization and recognition memory. *Journal of Neuroscience, 34*(22), 7472–7484.

Daw, N. D., & O'Doherty, J. P. (2013). Multiple systems for value learning. In P. W. Glimcher, & E. Fehr (Eds.), *Neuroeconomics: Decision making, and the brain*.

Dayan, P. (1993). Improving generalization for temporal difference learning: The successor representation. *Neural Computation, 5*(4), 613–624.

Dayan, P. (2008). The role of value systems in decision making. In C. Engel, & W. Singer (Eds.), *Better than conscious? Decision making, the human mind, and implications for institutions* (pp. 51–70). MIT Press.

Day, M., Langston, R., & Morris, R. G. M. (2003). Glutamate-receptor-mediated encoding and retrieval of paired-associate learning. *Nature, 424*(6945), 205–209.

Degonda, N., Mondadori, C. R. A., Bosshardt, S., Schmidt, C. F., Boesiger, P., Nitsch, R. M., … Henke, K. (2005). Implicit associative learning engages the hippocampus and interacts with explicit associative learning. *Neuron, 46*(3), 505–520.

Di Guida, S., & Devetag, G. (2013). Feature-based choice and similarity perception in normal-form games: An experimental study. *Games, 4*(4), 776–794.

Diana, R. A., Yonelinas, A. P., & Ranganath, C. (2007). Imaging recollection and familiarity in the medial temporal lobe: A three-component model. *Trends in Cognitive Sciences, 11*(9), 379–386.

Diana, R. A., Yonelinas, A. P., & Ranganath, C. (2013). Parahippocampal cortex activation during context reinstatement predicts item recollection. *Journal of Experimental Psychology: General, 142*(4), 1287.

Dickerson, K. C., Li, J., & Delgado, M. R. (2011). Parallel contributions of distinct human memory systems during probabilistic learning. *NeuroImage, 55*(1), 266–276.

Doll, B. B., Shohamy, D., & Daw, N. D. (2015). Multiple memory systems as substrates for multiple decision systems. *Neurobiology of Learning and Memory, 117*, 4–13.

Drummond, C. (2002). Accelerating reinforcement learning by composing solutions of automatically identified subtasks. *Journal of Artificial Intelligence Research, 16*(1), 59–104.

DuBrow, S., & Davachi, L. (2014). Temporal memory is shaped by encoding stability and intervening item reactivation. *Journal of Neuroscience, 34*(42), 13998–14005.

Duncan, K., Ketz, N., Inati, S. J., & Davachi, L. (2012). Evidence for area CA1 as a match/mismatch detector: A high-resolution fMRI study of the human hippocampus. *Hippocampus, 22*(3), 389–398.

Duncan, K. D., & Shohamy, D. (2016). Memory states influence value-based decisions. *Journal of Experimental Psychology: General, 145*(11), 1420.

Duncan, K., Tompary, A., & Davachi, L. (2014). Associative encoding and retrieval are predicted by functional connectivity in distinct hippocampal area CA1 pathways. *Journal of Neuroscience, 34*(34), 11188–11198.

Dutt, V., Arló-Costa, H., Helzner, J., & Gonzalez, C. (2014). The description–experience gap in risky and ambiguous gambles. *Journal of Behavioral Decision Making, 27*(4), 316–327.

Dutt, V., & Gonzalez, C. (2012). The role of inertia in modeling decisions from experience with instance-based learning. *Frontiers in Psychology, 3*(177).

Eckart, M. T., Huelse-Matia, M. C., & Schwarting, R. K. W. (2012). Dorsal hippocampal lesions boost performance in the rat sequential reaction time task. *Hippocampus, 22*(5), 1202–1214.

Eichberger, J., & Guerdjikova, A. (2010). Case-based belief formation under ambiguity. *Mathematical Social Sciences, 60*(3), 161–177.

Eichberger, J., & Guerdjikova, A. (2012). Technology adoption and adaptation to climate change?A case-based approach. *Climate Change Economics, 3*(02), 1–41.

Ezzyat, Y., & Davachi, L. (2014). Similarity breeds proximity: Pattern similarity within and across contexts is related to later mnemonic judgments of temporal proximity. *Neuron, 81*(5), 1179–1189.

Farovik, A., Dupont, L. M., & Eichenbaum, H. (2010). Distinct roles for dorsal CA3 and CA1 in memory for sequential nonspatial events. *Learning & Memory, 17*(1), 12–17.

Ferbinteanu, J. (2016). Contributions of hippocampus and striatum to memory-guided behavior depend on past experience. *Journal of Neuroscience, 36*(24), 6459–6470.

Finch, D. M. (1996). Neurophysiology of converging synaptic inputs from the rat prefrontal cortex, amygdala, midline thalamus, and hippocampal formation onto single neurons of the caudate/putamen and nucleus accumbens. *Hippocampus, 6*(5), 495–512.

Finch, D. M., Gigg, J., Tan, A. M., & Kosoyan, O. P. (1995). Neurophysiology and neuropharmacology of projections from entorhinal cortex to striatum in the rat. *Brain Research, 670*(2), 233–247.

Floresco, S. B., Todd, C. L., & Grace, A. A. (2001). Glutamatergic afferents from the hippocampus to the nucleus accumbens regulate activity of ventral tegmental area dopamine neurons. *Journal of Neuroscience, 21*(13), 4915–4922.

Foerde, K., Race, E., Verfaellie, M., & Shohamy, D. (2013). A role for the medial temporal lobe in feedback-driven learning: Evidence from amnesia. *Journal of Neuroscience, 33*(13), 5698–5704.

Frey, U., & Morris, R. G. M. (1997). Synaptic tagging and long-term potentiation. *Nature, 385*(6616), 533–536.

Frey, U., Schroeder, H., et al. (1990). Dopaminergic antagonists prevent long-term maintenance of posttetanic LTP in the CA1 region of rat hippocampal slices. *Brain Research, 522*(1), 69–75.

Fyhn, M., Molden, S., Hollup, S., Moser, M.-B., & Moser, E. I. (2002). Hippocampal neurons responding to first-time dislocation of a target object. *Neuron, 35*(3), 555–566.

Gabel, T., & Riedmiller, M. (2005). CBR for state value function approximation in reinforcement learning. In *Case-based reasoning research and development* (pp. 206–221). Springer.

Gavetti, G., Levinthal, D. A., & Rivkin, J. W. (2005). Strategy making in novel and complex worlds: The power of analogy. *Strategic Management Journal, 26*(8), 691–712.

Gayer, G. (2010). Perception of probabilities in situations of risk: A case based approach. *Games and Economic Behavior, 68*(1), 130–143.

Gayer, G., Gilboa, I., & Lieberman, O. (2007). Rule-based and case-based reasoning in housing prices. *BE Journal of Theoretical Economics, 7*(1).

Gentner, D., James Holyoak, K., & Kokinov, B. N. (2001). *The analogical mind: Perspectives from cognitive science.* MIT Press.

Gerraty, R. T., Davidow, J. Y., Wimmer, G. E., Kahn, I., & Shohamy, D. (2014). Transfer of learning relates to intrinsic connectivity between hippocampus, ventromedial prefrontal cortex, and large-scale networks. *Journal of Neuroscience, 34*(34), 11297–11303.

Gershman, S. J., & Daw, N. D. (2017). Reinforcement learning and episodic memory in humans and animals: An integrative framework. *Annual Review of Psychology, 68*, 101–128.

Gershman, S. J., Moore, C. D., Todd, M. T., Norman, K. A., & Sederberg, P. B. (2012). The successor representation and temporal context. *Neural Computation, 24*(6), 1553–1568.

Gershman, S. J., Zhou, J., & Kommers, C. (2017). Imaginative reinforcement learning: Computational principles and neural mechanisms. *Journal of Cognitive Neuroscience, 29*(12), 2103–2113.

Ghirlanda, S., & Enquist, M. (2003). A century of generalization. *Animal Behaviour, 66*(1), 15–36.

Gilbert, D. T., & Wilson, T. D. (2007). Prospection: Experiencing the future. *Science, 317*(5843), 1351–1354.

Gilboa, I., Lieberman, O., & Schmeidler, D. (2011). A similarity-based approach to prediction. *Journal of Econometrics, 162*(1), 124–131.

Gilboa, I., & Pazgal, A. (1995). *History dependent brand switching: Theory and evidence.* Northwestern University Center for Mathematical Studies in Economics and Management Science Discussion Paper 1146.

Gilboa, I., Postlewaite, A., & Schmeidler, D. (2015). *Consumer choice as constrained imitation.* Penn Institute for Economic Research Working Paper 15–013.

Gilboa, I., Samuelson, L., & Schmeidler, D. (2013). Dynamics of inductive inference in a unified framework. *Journal of Economic Theory, 148*(4), 1399–1432.

Gilboa, I., & Schmeidler, D. (1993). *Case-based consumer theory.* Northwestern University Center for Mathematical Studies in Economics and Management Science Discussion Paper 1025.

Gilboa, I., & Schmeidler, D. (1995a). Case-based decision theory. *Quarterly Journal of Economics, 110*(3), 605–639.

Gilboa, I., & Schmeidler, D. (1995b). *Case-based knowledge and planning.* Northwestern University Center for Mathematical Studies in Economics and Management Science Discussion Paper 1127.

Gilboa, I., & Schmeidler, D. (1996). Case-based optimization. *Games and Economic Behavior, 15*(1), 1–26.

Gilboa, I., & Schmeidler, D. (1997a). Act similarity in case-based decision theory. *Economic Theory, 9*(1), 47–61.

Gilboa, I., & Schmeidler, D. (1997b). Cumulative utility consumer theory. *International Economic Review, 38*(4), 737–761.

Gilboa, I., & Schmeidler, D. (2003). Reaction to price changes and aspiration level adjustments. In *Markets, games, and organizations* (pp. 89–97). Springer.

Gilboa, I., Schmeidler, D., & Wakker, P. P. (2002). Utility in case-based decision theory. *Journal of Economic Theory, 105*(2), 483–502.

Gilboa, A., Sekeres, M., Moscovitch, M., & Winocur, G. (2014). Higher-order conditioning is impaired by hippocampal lesions. *Current Biology, 24*(18), 2202–2207.

Gilovich, T. (1981). Seeing the past in the present: The effect of associations to familiar events on judgments and decisions. *Journal of Personality and Social Psychology, 40*(5), 797–808.

Gluck, M. A., & Myers, C. E. (1997). Psychobiological models of hippocampal function in learning and memory. *Annual Review of Psychology, 48*(1), 481–514.

Gluck, M. A., & Myers, C. E. (2001). *Gateway to memory: An introduction to neural network modeling of the hippocampus and learning.* MIT Press.

Gluth, S., Sommer, T., Rieskamp, J., & Büchel, C. (2015). Effective connectivity between hippocampus and ventromedial prefrontal cortex controls preferential choices from memory. *Neuron, 86*(4), 1078–1090.

Gold, A. E., & Kesner, R. P. (2005). The role of the CA3 subregion of the dorsal hippocampus in spatial pattern completion in the rat. *Hippocampus, 15*(6), 808–814.

Goldstone, R. L. (1994). Similarity, interactive activation, and mapping. *Journal of Experimental Psychology: Learning, Memory, and Cognition, 20*(1), 3–28.

Goldstone, R. L., & Son, Ji Y. (2005). Similarity. In K. J. Holyoak, & R. G. Morrison (Eds.), *Cambridge handbook of thinking and reasoning.* Cambridge University Press.

Golosnoy, V., & Okhrin, Y. (2008). General uncertainty in portfolio selection: A case-based decision approach. *Journal of Economic Behavior & Organization, 67*(3), 718–734.

Gonzalez, C. (2013). The boundaries of Instance-Based Learning Theory for explaining decisions from experience. In V. S. Chandrasekhar Pammi, & N. Srinivasan (Eds.), *Vol. 202. Decision making: Neural and behavioural approaches* (pp. 73–98).

Gonzalez, C., & Dutt, V. (2011). Instance-based learning: Integrating sampling and repeated decisions from experience. *Psychological Review, 118*(4), 523–551.

Gonzalez, C., Lerch, J. F., & Lebiere, C. (2003). Instance-based learning in dynamic decision making. *Cognitive Science, 27*(4), 591–635.

Gonzalez, C., & Mehlhorn, K. (2016). Framing from experience: Cognitive processes and predictions of risky choice. *Cognitive Science, 40*(5), 1163–1191.

Green, J. T., & Arenos, J. D. (2007). Hippocampal and cerebellar single-unit activity during delay and trace eyeblink conditioning in the rat. *Neurobiology of Learning and Memory, 87*(2), 269–284.

Grosskopf, B., Sarin, R., & Watson, E. (2015). An experiment on case-based decision making. *Theory and Decision, 79*(4), 639–666.

Gruber, M. J., Gelman, B. D., & Ranganath, C. (2014). States of curiosity modulate hippocampus-dependent learning via the dopaminergic circuit. *Neuron, 84*(2), 486–496.

Gu, Y., Arruda-Carvalho, M., Wang, J., Janoschka, S. R., Josselyn, S. A., Frankland, P. W., & Ge, S. (2012). Optical controlling reveals time-dependent roles for adult-born dentate granule cells. *Nature Neuroscience, 15*(12), 1700–1706.

Guerdjikova, A. (2006). *Evolution of wealth and asset prices in markets with case-based investors.* Working Paper.

Guilfoos, T., & Pape, A. D. (2016). Predicting human cooperation in the prisoner's dilemma using case-based decision theory. *Theory and Decision, 80*(1), 1–32.

Hahn, U., Chater, N., & Richardson, L. B. (2003). Similarity as transformation. *Cognition, 87*(1), 1–32.

Han, S., Huettel, S. A., Raposo, A., Adcock, R. A., & Dobbins, I. G. (2010). Functional significance of striatal responses during episodic decisions: Recovery or goal attainment? *Journal of Neuroscience, 30*(13), 4767–4775.

Hannula, D. E., & Ranganath, C. (2008). Medial temporal lobe activity predicts successful relational memory binding. *Journal of Neuroscience, 28*(1), 116–124.

Harman, J. L., & Gonzalez, C. (2015). Allais from experience: Choice consistency, rare events, and common consequences in repeated decisions. *Journal of Behavioral Decision Making, 28*(4), 369–381.

Hassabis, D., Kumaran, D., & Maguire, E. A. (2007). Using imagination to understand the neural basis of episodic memory. *Journal of Neuroscience, 27*(52), 14365–14374.

Hasselmo, M. E., & Eichenbaum, H. (2005). Hippocampal mechanisms for the context-dependent retrieval of episodes. *Neural Networks, 18*(9), 1172–1190.

Heinrich, T. (2013). Endogenous negative stereotypes: A similarity-based approach. *Journal of Economic Behavior & Organization, 92*, 45–54.

Hennessy, D., & Hinkle, D. (1992). Applying case-based reasoning to autoclave loading. *IEEE Expert, 7*(5), 21–26.

Hertwig, R., Barron, G., Weber, E. U., & Erev, I. (2004). Decisions from experience and the effect of rare events in risky choice. *Psychological Science, 15*(8), 534–539.

Hinton, G., Vinyals, O., & Dean, J. (2015). *Distilling the knowledge in a neural network.* arXiv preprint arXiv: 1503.02531.

Holyoak, K. J. (2012). Analogy and relational reasoning. In *Oxford handbook of thinking and reasoning* (pp. 234–259).

Honey, R. C., Watt, A., & Good, M. (1998). Hippocampal lesions disrupt an associative mismatch process. *Journal of Neuroscience, 18*(6), 2226–2230.

Horner, A. J., & Burgess, N. (2013). The associative structure of memory for multielement events. *Journal of Experimental Psychology: General, 142*(4), 1370.

Hsieh, L.-T., Gruber, M. J., Jenkins, L. J., & Ranganath, C. (2014). Hippocampal activity patterns carry information about objects in temporal context. *Neuron, 81*(5), 1165–1178.

Hu, D., & Amsel, A. (1995). A simple test of the vicarious trial-and-error hypothesis of hippocampal function. *Proceedings of the National Academy of Sciences of the United States of America, 92*(12), 5506–5509.

Hüllermeier, E. (2007). *Case-based approximate reasoning* (Vol. 44). Springer Science & Business Media.

Hunt, K. L., & Chittka, L. (2015). Merging of long-term memories in an insect. *Current Biology, 25*(6), 741–745.

Huth, A. G., Nishimoto, S., Vu, An T., & Gallant, J. L. (2012). A continuous semantic space describes the representation of thousands of object and action categories across the human brain. *Neuron, 76*(6), 1210–1224.

Hu, D., Xu, X., & Gonzalez-Lima, F. (2006). Vicarious trial-and-error behavior and hippocampal cytochrome oxidase activity during Y-maze discrimination learning in the rat. *International Journal of Neuroscience, 116*(3), 265–280.

Imai, S. (1977). Pattern similarity and cognitive transformations. *Acta Psychologica, 41*(6), 433–447.

Imai, H., Kim, D., Sasaki, Y., & Watanabe, T. (2014). Reward eliminates retrieval-induced forgetting. *Proceedings of the National Academy of Sciences of the United States of America, 111*(48), 17326–17329.

Izhikevich, E. M. (2007). Solving the distal reward problem through linkage of STDP and dopamine signaling. *Cerebral Cortex, 17*(10), 2443–2452.

Jahnke, H., Chwolka, A., & Simons, D. (2005). Coordinating service-sensitive demand and capacity by adaptive decision making: An application of case-based decision theory. *Decision Sciences, 36*(1), 1–32.

Jensen, A. R. (1998). *The g factor: The science of mental ability*. Westport: Praeger.

Johnson, A., & Redish, A. D. (2005). Hippocampal replay contributes to within session learning in a temporal difference reinforcement learning model. *Neural Networks, 18*(9), 1163–1171.

Johnson, A., & Redish, A. D. (2007). Neural ensembles in CA3 transiently encode paths forward of the animal at a decision point. *Journal of Neuroscience, 27*(45), 12176–12189.

Johnson, A., van der Meer, M. A. A., & Redish, A. D. (2007). Integrating hippocampus and striatum in decision-making. *Current Opinion in Neurobiology, 17*(6), 692–697.

Johnson, A., & Venditto, S. (2015). Reinforcement learning and hippocampal dynamics. In *Analysis and modeling of coordinated multi-neuronal activity* (pp. 299–312). Springer.

Juell, P., & Paulson, P. (2003). Using reinforcement learning for similarity assessment in case-based systems. *IEEE Intelligent Systems, 18*(4), 60–67.

Jung, Y., Hong, S., & Haber, S. N. (2003). Organization of direct hippocampal projections to the different regions of the ventral striatum in primate. *Korean Journal of Anatomy, 36*(1).

Kafkas, A., & Montaldi, D. (2015). Striatal and midbrain connectivity with the hippocampus selectively boosts memory for contextual novelty. *Hippocampus, 25*(11), 1262–1273.

Kahneman, D., & Tversky, A. (1972). Subjective probability: A judgment of representativeness. *Cognitive Psychology, 3*(3), 430–454.

Kahneman, D., & Tversky, A. (1982). Intuitive prediction: Biases and corrective procedures. In D. Kahneman, P. Slovic, & A. Tversky (Eds.), *Judgment under uncertainty: Heuristics and biases*. London: Cambridge University Press.

Kahnt, T., Park, S. Q., Burke, C. J., & Tobler, P. N. (2012). How glitter relates to gold: Similarity-dependent reward prediction errors in the human striatum. *Journal of Neuroscience, 32*(46), 16521–16529.

Kakade, S., & Dayan, P. (2002). Dopamine: Generalization and bonuses. *Neural Networks, 15*(4), 549–559.

Kang, M. J., Hsu, M., Krajbich, I. M., Loewenstein, G., McClure, S. M., Wang, J. T., & Camerer, C. F. (2009). The wick in the candle of learning: Epistemic curiosity activates reward circuitry and enhances memory. *Psychological Science, 20*(8), 963–973.

Kempadoo, K. A., Mosharov, E. V., Choi, Se J., Sulzer, D., & Kandel, E. R. (2016). Dopamine release from the locus coeruleus to the dorsal hippocampus promotes spatial learning and memory. *Proceedings of the National Academy of Sciences of the United States of America, 113*(51), 14835–14840.

Kim, G., Lewis-Peacock, J. A., Norman, K. A., & Turk-Browne, N. B. (2014). Pruning of memories by context-based prediction error. *Proceedings of the National Academy of Sciences of the United States of America, 111*(24), 8997–9002.

Kinjo, K., & Sugawara, S. (2016). Predicting empirical patterns in viewing Japanese TV dramas using case-based decision theory. *BE Journal of Theoretical Economics, 16*(2), 679–709.

Klahr, D. (1969). Decision making in a complex environment: The use of similarity judgements to predict preferences. *Management Science, 15*(11), 595–618.

Klein, S. B., Loftus, J., & Kihlstrom, J. F. (2002). Memory and temporal experience: The effects of episodic memory loss on an amnesic patient's ability to remember the past and imagine the future. *Social Cognition, 20*(5), 353–379.

Klenk, M., Aha, D. W., & Molineaux, M. (2011). The case for case-based transfer learning. *AI Magazine, 32*(1), 54–69.

Knight, R. T. (1996). Contribution of human hippocampal region to novelty detection. *Nature, 383*(6597), 256–259.

Knutson, B., Taylor, J., Kaufman, M., Peterson, R., & Glover, G. (2005). Distributed neural representation of expected value. *Journal of Neuroscience, 25*(19), 4806–4812.

Kolodner, J. L. (1992). An introduction to case-based reasoning. *Artificial Intelligence Review, 6*(1), 3–34.

Kolodner, J. (1993). *Case-based reasoning*. Morgan Kaufmann.

Kolodner, J. L. (1997). Educational implications of analogy: A view from case-based reasoning. *American Psychologist, 52*(1), 57–66.

Kragel, J. E., Morton, N. W., & Polyn, S. M. (2015). Neural activity in the medial temporal lobe reveals the fidelity of mental time travel. *Journal of Neuroscience, 35*(7), 2914–2926.

Krause, A. (2009). Learning and herding using case-based decisions with local interactions. *IEEE Transactions on Systems, Man, and Cybernetics—Part A: Systems and Humans, 39*(3), 662–669.

Krawczyk, D. C. (2012). The cognition and neuroscience of relational reasoning. *Brain Research, 1428*, 13–23.

Kreiman, G., Koch, C., & Fried, I. (2000). Category-specific visual responses of single neurons in the human medial temporal lobe. *Nature Neuroscience, 3*(9), 946–953.

Kriegeskorte, N., Marieke, M., Ruff, D. A., Kiani, R., Bodurka, J., Esteky, H., ... Bandettini, P. A. (2008). Matching categorical object representations in inferior temporal cortex of man and monkey. *Neuron, 60*(6), 1126–1141.

Kruizinga, S., & Murphy, A. H. (1983). Use of an analogue procedure to formulate objective probabilistic temperature forecasts in The Netherlands. *Monthly Weather Review, 111*(11), 2244–2254.

Kuhl, B. A., Shah, A. T., DuBrow, S., & Wagner, A. D. (2010). Resistance to forgetting associated with hippocampus-mediated reactivation during new learning. *Nature Neuroscience, 13*(4), 501–506.

Kumaran, D., & Maguire, E. A. (2006). An unexpected sequence of events: Mismatch detection in the human hippocampus. *PLoS Biology, 4*(12), e424.

Kumaran, D., & Maguire, E. A. (2007). Match-mismatch processes underlie human hippocampal responses to associative novelty. *Journal of Neuroscience, 27*(32), 8517–8524.

Kumaran, D., & Maguire, E. A. (2009). Novelty signals: A window into hippocampal information processing. *Trends in Cognitive Sciences, 13*(2), 47–54.

Kumaran, D., Summerfield, J. J., Hassabis, D., & Maguire, E. A. (2009). Tracking the emergence of conceptual knowledge during human decision making. *Neuron, 63*(6), 889–901.

Kurth-Nelson, Z., Economides, M., Dolan, R. J., & Dayan, P. (2016). Fast sequences of non-spatial state representations in humans. *Neuron, 91*(1), 194–204.

Kwan, D., Carson, N., Addis, D. R., & Rosenbaum, R. S. (2010). Deficits in past remembering extend to future imagining in a case of developmental amnesia. *Neuropsychologia, 48*(11), 3179–3186.

Kwan, D., Craver, C. F., Green, L., Myerson, J., Gao, F., Black, S. E., & Rosenbaum, R. S. (2015). Cueing the personal future to reduce discounting in intertemporal choice: Is episodic prospection necessary? *Hippocampus, 25*(4), 432–443.

La Grutta, V., & Sabatino, M. (1988). Focal hippocampal epilepsy: Effect of caudate stimulation. *Experimental Neurology, 99*(1), 38–49.

Lansink, C. S., Goltstein, P. M., Lankelma, J. V., McNaughton, B. L., & Pennartz, C. M. A. (2009). Hippocampus leads ventral striatum in replay of place-reward information. *PLoS Biology, 7*(8), e1000173.

LaRocque, K. F., Smith, M. E., Carr, V. A., Witthoft, N., Grill-Spector, K., & Wagner, A. D. (2013). Global similarity and pattern separation in the human medial temporal lobe predict subsequent memory. *Journal of Neuroscience, 33*(13), 5466–5474.

Lassalle, J.-M., Bataille, T., & Halley, H. (2000). Reversible inactivation of the hippocampal mossy fiber synapses in mice impairs spatial learning, but neither consolidation nor memory retrieval, in the Morris navigation task. *Neurobiology of Learning and Memory, 73*(3), 243–257.

Laurent, P. A. (2013). A neural mechanism for reward discounting: Insights from modeling hippocampal–striatal interactions. *Cognitive Computation, 5*(1), 152–160.

Leake, D. B. (1994). Case-based reasoning. *Knowledge Engineering Review, 9*(01), 61–64.

Leake, D. B. (1996). CBR in context: The present and future. In *Case-based reasoning, experiences, lessons & future directions* (pp. 1–30).

Lebiere, C., Gonzalez, C., & Martin, M. (2007). Instance-based decision making model of repeated binary choice. In *Proceedings of the 8th international conference on cognitive modeling*.

Lebreton, M., Bertoux, M., Boutet, C., Lehericy, S., Dubois, B., Fossati, P., & Pessiglione, M. (2013). A critical role for the hippocampus in the valuation of imagined outcomes. *PLoS Biology, 11*(10).

Lee, H., Ghim, J.-W., Kim, H., Lee, D., & Jung, M. W. (2012). Hippocampal neural correlates for values of experienced events. *Journal of Neuroscience, 32*(43), 15053–15065.

Lee, I., & Kesner, R. P. (2004). Encoding versus retrieval of spatial memory: Double dissociation between the dentate gyrus and the perforant path inputs into CA3 in the dorsal hippocampus. *Hippocampus, 14*(1), 66–76.

Lee, K. J., Queenan, B. N., Rozeboom, A. M., Bellmore, R., Lim, S. T., Vicini, S., & Pak, D. T. S. (2013). Mossy fiber-CA3 synapses mediate homeostatic plasticity in mature hippocampal neurons. *Neuron, 77*(1), 99–114.

Legault, M., Rompré, P.-P., & Wise, R. A. (2000). Chemical stimulation of the ventral hippocampus elevates nucleus accumbens dopamine by activating dopaminergic neurons of the ventral tegmental area. *Journal of Neuroscience, 20*(4), 1635–1642.

Lejarraga, T., Dutt, V., & Gonzalez, C. (2012). Instance-based learning: A general model of repeated binary choice. *Journal of Behavioral Decision Making, 25*(2), 143–153.

Lejarraga, T., Lejarraga, J., & Gonzalez, C. (2014). Decisions from experience: How groups and individuals adapt to change. *Memory & Cognition, 42*(8), 1384–1397.

Lengyel, M., & Dayan, P. (2008). Hippocampal contributions to control: The third way. In *Advances in neural information processing systems* (pp. 889–896).

Leutgeb, J. K., Leutgeb, S., Moser, M.-B., & Moser, E. I. (2007). Pattern separation in the dentate gyrus and CA3 of the hippocampus. *Science, 315*(5814), 961–966.

Libby, L. A., Hannula, D. E., & Ranganath, C. (2014). Medial temporal lobe coding of item and spatial information during relational binding in working memory. *Journal of Neuroscience, 34*(43), 14233–14242.

LiCalzi, M. (1995). Fictitious play by cases. *Games and Economic Behavior, 11*(1), 64–89.

Lin, H., & Epstein, L. H. (2014). Living in the moment: Effects of time perspective and emotional valence of episodic thinking on delay discounting. *Behavioral Neuroscience, 128*(1), 12–19.

Lisman, J. E., & Grace, A. A. (2005). The hippocampal-VTA loop: Controlling the entry of information into long-term memory. *Neuron, 46*(5), 703–713.

Lisman, J., Grace, A. A., & Duzel, E. (2011). A neoHebbian framework for episodic memory; role of dopamine-dependent late LTP. *Trends in Neurosciences, 34*(10), 536–547.

Liu, L., Feng, T., Chen, J., & Li, H. (2013). The value of emotion: How does episodic prospection modulate delay discounting? *PLoS One, 8*(11), e81717.

Lohrenz, T., McCabe, K., Camerer, C. F., & Montague, P. R. (2007). Neural signature of fictive learning signals in a sequential investment task. *Proceedings of the National Academy of Sciences of the United States of America, 104*(22), 9493–9498.

Lovallo, D., Clarke, C., & Camerer, C. (2012). Robust analogizing and the outside view: Two empirical tests of case-based decision making. *Strategic Management Journal, 33*(5), 496–512.

Ludvig, E. A., Sutton, R. S., & Kehoe, E. J. (2008). Stimulus representation and the timing of reward-prediction errors in models of the dopamine system. *Neural Computation, 20*(12), 3034–3054.

Ludvig, E. A., Sutton, R. S., Verbeek, E., & Kehoe, E. J. (2009). A computational model of hippocampal function in trace conditioning. In *Advances in neural information processing systems* (pp. 993–1000).

Luo, A. H., Tahsili-Fahadan, P., Wise, R. A., Lupica, C. R., & Aston-Jones, G. (2011). Linking context with reward: A functional circuit from hippocampal CA3 to ventral tegmental area. *Science, 333*(6040), 353–357.

Lysetskiy, M., Földy, C., & Soltesz, I. (2005). Long-and short-term plasticity at mossy fiber synapses on mossy cells in the rat dentate gyrus. *Hippocampus, 15*(6), 691–696.

Marchiori, D., Di Guida, S., & Erev, I. (2015). Noisy retrieval models of over- and undersensitivity to rare events. *Decision, 2*(2), 82–106.

Mariano, T. Y., Bannerman, D. M., McHugh, S. B., Preston, T. J., Rudebeck, P. H., Rudebeck, S. R., … Campbell, T. G. (2009). Impulsive choice in hippocampal but not orbitofrontal cortex-lesioned rats on a nonspatial decision-making maze task. *European Journal of Neuroscience, 30*(3), 472–484.

Marr, D. (1971). Simple memory: A theory for archicortex. *Philosophical Transactions of the Royal Society B: Biological Sciences, 262*(841), 23–81.

Mattfeld, A. T., & Stark, C. E. L. (2015). Functional contributions and interactions between the human hippocampus and subregions of the striatum during arbitrary associative learning and memory. *Hippocampus, 25*(8), 900–911.

McClure, S. M., Laibson, D. I., Loewenstein, G., & Cohen, J. D. (2004). Separate neural systems value immediate and delayed monetary rewards. *Science, 306*(5695), 503—507.

McHugh, S. B., Campbell, T. G., Taylor, A. M., Rawlins, J. N. P., & Bannerman, D. M. (2008). A role for dorsal and ventral hippocampus in inter-temporal choice cost-benefit decision making. *Behavioral Neuroscience, 122*(1), 1—8.

McHugh, T. J., Jones, M. W., Quinn, J. J., Balthasar, N., Coppari, R., Elmquist, J. K., … Tonegawa, S. (2007). Dentate gyrus NMDA receptors mediate rapid pattern separation in the hippocampal network. *Science, 317*(5834), 94—99.

McKenzie, S., Frank, A. J., Kinsky, N. R., Porter, B., Rivière, P. D., & Eichenbaum, H. (2014). Hippocampal representation of related and opposing memories develop within distinct, hierarchically organized neural schemas. *Neuron, 83*(1), 202—215.

McNamara, C. G., Tejero-Cantero, Á., Trouche, S., Campo-Urriza, N., & Dupret, D. (2014). Dopaminergic neurons promote hippocampal reactivation and spatial memory persistence. *Nature Neuroscience, 17*(12), 1658—1660.

McNaughton, B. L., & Nadel, L. (1990). Hebb-Marr networks and the neurobiological representation of action in space. In M. A. Gluck, & D. E. Rumelhart (Eds.), *Developments in connectionist theory. Neuroscience and connectionist theory* (pp. 1—63). Hillsdale, NJ, US: Lawrence Erlbaum Associates, Inc.

Meeter, M., Shohamy, D., & Myers, C. E. (2009). Acquired equivalence changes stimulus representations. *Journal of the Experimental Analysis of Behavior, 91*(1), 127—141.

Mengel, F., & Sciubba, E. (2014). Extrapolation and structural similarity in games. *Economics Letters, 125*(3), 381—385.

Momennejad, I., Russek, E. M., Cheong, J. H., Botvinick, M. M., Daw, N., & Gershman, S. J. (2017). The successor representation in human reinforcement learning. *Nature Human Behaviour, 1,* 680—692.

Montani, S., & Jain, L. C. (2010). *Successful case-based reasoning applications.* Springer.

Montani, S., & Jain, L. C. (2014). *Successful case-based reasoning applications-2.* Springer.

Moody, T. D., Bookheimer, S. Y., Vanek, Z., & Knowlton, B. J. (2004). An implicit learning task activates medial temporal lobe in patients with Parkinson's disease. *Behavioral Neuroscience, 118*(2), 438—442.

Moustafa, A. A., Wufong, E., Servatius, R. J., Pang, K. C. H., Gluck, M. A., & Myers, C. E. (2013). Why trace and delay conditioning are sometimes (but not always) hippocampal dependent: A computational model. *Brain Research, 1493,* 48—67.

Muenzinger, K. F. (1938). Vicarious trial and error at a point of choice: I. A general survey of its relation to learning efficiency. *The Pedagogical Seminary and Journal of Genetic Psychology, 53*(1), 75—86.

Murayama, K., & Kitagami, S. (2014). Consolidation power of extrinsic rewards: Reward cues enhance long-term memory for irrelevant past events. *Journal of Experimental Psychology: General, 143*(1), 15—20.

Murty, V. P., FeldmanHall, O., Hunter, L. E., Phelps, E. A., & Davachi, L. (2016). Episodic memories predict adaptive value-based decision-making. *Journal of Experimental Psychology: General, 145*(5), 548—558.

Myers, C. E., Shohamy, D., Gluck, M. A., Grossman, S., Kluger, A., Ferris, S., … Schwartz, R. (2003). Dissociating hippocampal versus basal ganglia contributions to learning and transfer. *Journal of Cognitive Neuroscience, 15*(2), 185—193.

Nakashiba, T., Cushman, J. D., Pelkey, K. A., Renaudineau, S., Buhl, D. L., McHugh, T. J., … Tonegawa, S. (2012). Young dentate granule cells mediate pattern separation, whereas old granule cells facilitate pattern completion. *Cell, 149*(1), 188—201.

Nakazawa, K., Quirk, M. C., Chitwood, R. A., Watanabe, M., Yeckel, M. F., Sun, L. D., … Tonegawa, S. (2002). Requirement for hippocampal CA3 NMDA receptors in associative memory recall. *Science, 297*(5579), 211—218.

Nakazawa, K., Sun, L. D., Quirk, M. C., Rondi-Reig, L., Wilson, M. A., & Tonegawa, S. (2003). Hippocampal CA3 NMDA receptors are crucial for memory acquisition of one-time experience. *Neuron, 38*(2), 305—315.

Neunuebel, J. P., & Knierim, J. J. (2014). CA3 retrieves coherent representations from degraded input: Direct evidence for CA3 pattern completion and dentate gyrus pattern separation. *Neuron, 81*(2), 416—427.

Nosofsky, R. M., Gluck, M. A., Palmeri, T. J., McKinley, S. C., & Glauthier, P. (1994). Comparing modes of rule-based classification learning: A replication and extension of Shepard, Hovland, and Jenkins (1961). *Memory & Cognition, 22*(3), 352—369.

Nosofsky, R. M., & Palmeri, T. J. (1996). Learning to classify integral-dimension stimuli. *Psychonomic Bulletin & Review, 3*(2), 222–226.

Okuda, J., Fujii, T., Ohtake, H., Tsukiura, T., Tanji, K., Suzuki, K., … Yamadori, A. (2003). Thinking of the future and past: The roles of the frontal pole and the medial temporal lobes. *NeuroImage, 19*(4), 1369–1380.

Opitz, B., & Friederici, A. D. (2004). Brain correlates of language learning: The neuronal dissociation of rule-based versus similarity-based learning. *Journal of Neuroscience, 24*(39), 8436–8440.

Ormoneit, D., & Sen, Ś. (2002). Kernel-based reinforcement learning. *Machine Learning, 49*(2–3), 161–178.

Ossadnik, W., Wilmsmann, D., & Niemann, B. (2013). Experimental evidence on case-based decision theory. *Theory and Decision, 75*(2), 211–232.

O'Carroll, C. M., Martin, S. J., Sandin, J., Frenguelli, B., & Morris, R. G. M. (2006). Dopaminergic modulation of the persistence of one-trial hippocampus-dependent memory. *Learning & Memory, 13*(6), 760–769.

Palombo, D. J., Keane, M. M., & Verfaellie, M. (2014). The medial temporal lobes are critical for reward-based decision making under conditions that promote episodic future thinking. *Hippocampus, 25*(3), 345–353.

Pan, W.-X., Schmidt, R., Wickens, J. R., & Hyland, B. I. (2005). Dopamine cells respond to predicted events during classical conditioning: Evidence for eligibility traces in the reward-learning network. *Journal of Neuroscience, 25*(26), 6235–6242.

Pape, A. D., & Kurtz, K. J. (2013). Evaluating case-based decision theory: Predicting empirical patterns of human classification learning. *Games and Economic Behavior, 82*, 52–65.

Pape, A., & Xiao, W. (2014). *Case-based learning in the Cobweb model.* Working Paper.

Pennartz, C. M. A., Ito, R., Verschure, P. F. M. J., Battaglia, F. P., & Robbins, T. W. (2011). The hippocampal–striatal axis in learning, prediction and goal-directed behavior. *Trends in Neurosciences, 34*(10), 548–559.

Peters, J., & Büchel, C. (2010). Episodic future thinking reduces reward delay discounting through an enhancement of prefrontal-mediotemporal interactions. *Neuron, 66*(1), 138–148.

Poldrack, R. A., Clark, J., Pare-Blagoev, E. J., Shohamy, D., Moyano, J. C., Myers, C., & Gluck, M. A. (2001). Interactive memory systems in the human brain. *Nature, 414*(6863), 546–550.

Poldrack, R. A., & Packard, M. G. (2003). Competition among multiple memory systems: Converging evidence from animal and human brain studies. *Neuropsychologia, 41*(3), 245–251.

Poldrack, R. A., Prabhakaran, V., Seger, C. A., & Gabrieli, J. D. E. (1999). Striatal activation during acquisition of a cognitive skill. *Neuropsychology, 13*(4), 564–574.

Poldrack, R. A., & Rodriguez, P. (2004). How do memory systems interact? Evidence from human classification learning. *Neurobiology of Learning and Memory, 82*(3), 324–332.

Port, R. L., & Patterson, M. M. (1984). Fimbrial lesions and sensory preconditioning. *Behavioral Neuroscience, 98*(4), 584–589.

Preston, A. R., Shrager, Y., Dudukovic, N. M., & Gabrieli, J. D. E. (2004). Hippocampal contribution to the novel use of relational information in declarative memory. *Hippocampus, 14*(2), 148–152.

Pritzel, A., Uria, B., Srinivasan, S., Puigdomènech, A., Vinyals, O., Hassabis, D., … Blundell, C. (2017). *Neural episodic control.* arXiv preprint arXiv:1703.01988.

Qian, J., & Brown, G. D. A. (2005). Similarity-based sampling: Testing a model of price psychophysics. In *Proceedings of the 27th annual conference of the Cognitive Science Society* (pp. 1785–1790).

Quiroga, R. Q., Kraskov, A., Mormann, F., Fried, I., & Koch, C. (2014). Single-cell responses to face adaptation in the human medial temporal lobe. *Neuron, 84*(2), 363–369.

Quiroga, R. Q., Kreiman, G., Koch, C., & Fried, I. (2008). Sparse but not 'grandmother-cell' coding in the medial temporal lobe. *Trends in Cognitive Sciences, 12*(3), 87–91.

Quiroga, R. Q., Reddy, L., Kreiman, G., Koch, C., & Fried, I. (2005). Invariant visual representation by single neurons in the human brain. *Nature, 435*(7045), 1102–1107.

Race, E., Keane, M. M., & Verfaellie, M. (2011). Medial temporal lobe damage causes deficits in episodic memory and episodic future thinking not attributable to deficits in narrative construction. *Journal of Neuroscience, 31*(28), 10262–10269.

Race, E., Keane, M. M., & Verfaellie, M. (2013). Losing sight of the future: Impaired semantic prospection following medial temporal lobe lesions. *Hippocampus, 23*(4), 268–277.

Rankin, F. W., Van Huyck, J. B., & Battalio, R. C. (2000). Strategic similarity and emergent conventions: Evidence from similar stag hunt games. *Games and Economic Behavior, 32*(2), 315–337.

Rawlins, J. N. P., Feldon, J., & Butt, S. (1985). The effects of delaying reward on choice preference in rats with hippocampal or selective septal lesions. *Behavioural Brain Research, 15*(3), 191–203.

Reagh, Z. M., & Yassa, M. A. (2014). Object and spatial mnemonic interference differentially engage lateral and medial entorhinal cortex in humans. *Proceedings of the National Academy of Sciences of the United States of America, 111*(40), E4264–E4273.

Reber, P. J., Knowlton, B. J., & Squire, L. R. (1996). Dissociable properties of memory systems: Differences in the flexibility of declarative and nondeclarative knowledge. *Behavioral Neuroscience, 110*(5), 861.

Rebola, N., Carta, M., Lanore, F., Blanchet, C., & Mulle, C. (2011). NMDA receptor-dependent metaplasticity at hippocampal mossy fiber synapses. *Nature Neuroscience, 14*(6), 691–693.

Richter, M. M., & Weber, R. (2013). *Case-based reasoning: A textbook.* Springer Science & Business Media.

Riesbeck, C. K., & Schank, R. C. (1989). *Inside case-based reasoning.* Psychology Press.

Roediger, H. L., & McDermott, K. B. (1995). Creating false memories: Remembering words not presented in lists. *Journal of Experimental Psychology: Learning, Memory, and Cognition, 21*(4), 803–814.

Rose, M., Haider, H., Weiller, C., & Büchel, C. (2002). The role of medial temporal lobe structures in implicit learning: An event-related FMRI study. *Neuron, 36*(6), 1221–1231.

Rosen, Z. B., Cheung, S., & Siegelbaum, S. A. (2015). Midbrain dopamine neurons bidirectionally regulate CA3-CA1 synaptic drive. *Nature Neuroscience, 18*(12), 1763–1771.

Rossato, J. I., Bevilaqua, L. R. M., Izquierdo, I., Medina, J. H., & Cammarota, M. (2009). Dopamine controls persistence of long-term memory storage. *Science, 325*(5943), 1017–1020.

Rutishauser, U., Mamelak, A. N., & Schuman, E. M. (2006). Single-trial learning of novel stimuli by individual neurons of the human hippocampus-amygdala complex. *Neuron, 49*(6), 805–813.

Sabatino, M., Ferraro, G., Liberti, G., Vella, N., & La Grutta, V. (1985). Striatal and septal influence on hippocampal theta and spikes in the cat. *Neuroscience Letters, 61*(1), 55–59.

Sadeh, T., Shohamy, D., Levy, D. R., Reggev, N., & Maril, A. (2011). Cooperation between the hippocampus and the striatum during episodic encoding. *Journal of Cognitive Neuroscience, 23*(7), 1597–1608.

Samuelson, L. (2001). Analogies, adaptation, and anomalies. *Journal of Economic Theory, 97*(2), 320–366.

Santamaría, J. C., Sutton, R. S., & Ram, A. (1997). Experiments with reinforcement learning in problems with continuous state and action spaces. *Adaptive Behavior, 6*(2), 163–217.

Sarin, R., & Vahid, F. (2004). Strategy similarity and coordination. *Economic Journal, 114*(497), 506–527.

Savage, L. (1954). *The foundations of statistics.*

Schacter, D. L., & Addis, D. R. (2007). The cognitive neuroscience of constructive memory: Remembering the past and imagining the future. *Philosophical Transactions of the Royal Society B: Biological Sciences, 362*(1481), 773–786.

Schacter, D. L., Addis, D. R., & Buckner, R. L. (2007). Remembering the past to imagine the future: The prospective brain. *Nature Reviews Neuroscience, 8*(9), 657–661.

Schacter, D. L., Addis, D. R., & Buckner, R. L. (2008). Episodic simulation of future events. *Annals of the New York Academy of Sciences, 1124*(1), 39–60.

Schacter, D. L., Addis, D. R., Hassabis, D., Martin, V. C., Spreng, R. N., & Szpunar, K. K. (2012). The future of memory: Remembering, imagining, and the brain. *Neuron, 76*(4), 677–694.

Schapiro, A. C., Gregory, E., Landau, B., McCloskey, M., & Turk-Browne, N. B. (2014). The necessity of the medial temporal lobe for statistical learning. *Journal of Cognitive Neuroscience, 26*(8), 1736–1747.

Schapiro, A. C., Kustner, L. V., & Turk-Browne, N. B. (2012). Shaping of object representations in the human medial temporal lobe based on temporal regularities. *Current Biology, 22*(17), 1622–1627.

Schendan, H. E., Searl, M. M., Melrose, R. J., & Stern, C. E. (2003). An FMRI study of the role of the medial temporal lobe in implicit and explicit sequence learning. *Neuron, 37*(6), 1013–1025.

Schlichting, M. L., Zeithamova, D., & Preston, A. R. (2014). CA1 subfield contributions to memory integration and inference. *Hippocampus, 24*(10), 1248–1260.

Schmidt, B., Marrone, D. F., & Markus, E. J. (2012). Disambiguating the similar: The dentate gyrus and pattern separation. *Behavioural Brain Research, 226*(1), 56–65.

Schwarze, U., Bingel, U., Badre, D., & Sommer, T. (2013). Ventral striatal activity correlates with memory confidence for old-and new-responses in a difficult recognition test. *PLoS One, 8*(3), e54324.

Scimeca, J. M., & Badre, D. (2012). Striatal contributions to declarative memory retrieval. *Neuron, 75*(3), 380–392.

Seger, C. A., & Peterson, E. J. (2013). Categorization = decision making + generalization. *Neuroscience and Biobehavioral Reviews, 37*(7), 1187–1200.

Sharma, M., Holmes, M. P., Santamaría, J. C., Irani, A., Isbell, C. L., Jr., & Ram, A. (2007). Transfer learning in real-time strategy games using hybrid CBR/RL. In *Proceedings of the 20th international joint conference on artifical intelligence* (pp. 1041–1046).

Shohamy, D., & Adcock, R. A. (2010). Dopamine and adaptive memory. *Trends in Cognitive Sciences, 14*(10), 464–472.

Shohamy, D., Myers, C. E., Hopkins, R. O., Sage, J., & Gluck, M. A. (2009). Distinct hippocampal and basal ganglia contributions to probabilistic learning and reversal. *Journal of Cognitive Neuroscience, 21*(9), 1820–1832.

Shohamy, D., Myers, C. E., Onlaor, S., & Gluck, M. A. (2004). Role of the basal ganglia in category learning: How do patients with Parkinson's disease learn? *Behavioral Neuroscience, 118*(4), 676–686.

Shohamy, D., & Turk-Browne, N. B. (2013). Mechanisms for widespread hippocampal involvement in cognition. *Journal of Experimental Psychology: General, 142*(4), 1159–1170.

Shohamy, D., & Wagner, A. D. (2008). Integrating memories in the human brain: Hippocampal-midbrain encoding of overlapping events. *Neuron, 60*(2), 378–389.

Sørensen, K. E., & Witter, M. P. (1983). Entorhinal efferents reach the caudato-putamen. *Neuroscience Letters, 35*(3), 259–264.

Speer, M. E., Bhanji, J. P., & Delgado, M. R. (2014). Savoring the past: Positive memories evoke value representations in the striatum. *Neuron, 84*(4), 847–856.

Spence, K. W. (1937). The differential response in animals to stimuli varying within a single dimension. *Psychological Review, 44*(5), 430–444.

Spiliopoulos, L. (2013). Beyond fictitious play beliefs: Incorporating pattern recognition and similarity matching. *Games and Economic Behavior, 81*, 69–85.

Stachenfeld, K. L., Botvinick, M., & Gershman, S. J. (2014). Design principles of the hippocampal cognitive map. In *Advances in neural information processing systems* (pp. 2528–2536).

Stachenfeld, K. L., Botvinick, M. M., & Gershman, S. J. (2017). The hippocampus as a predictive map. *Nature Neuroscience, 20*, 1643–1653.

Stanfill, C., & Waltz, D. (1986). Toward memory-based reasoning. *Communications of the ACM, 29*(12), 1213–1228.

Staresina, B. P., & Davachi, L. (2009). Mind the gap: Binding experiences across space and time in the human hippocampus. *Neuron, 63*(2), 267–276.

Staresina, B. P., Gray, J. C., & Davachi, L. (2009). Event congruency enhances episodic memory encoding through semantic elaboration and relational binding. *Cerebral Cortex, 19*(5), 1198–1207.

Steiner, J., & Stewart, C. (2008). Contagion through learning. *Theoretical Economics, 3*(4), 431–458.

Stewart, N., Chater, N., & Brown, G. D. A. (2006). Decision by sampling. *Cognitive Psychology, 53*(1), 1–26.

Suddendorf, T., & Corballis, M. C. (2007). The evolution of foresight: What is mental time travel, and is it unique to humans? *Behavioral and Brain Sciences, 30*(03), 299–313.

Sutton, R. S. (1990). Integrated architectures for learning, planning, and reacting based on approximating dynamic programming. In *Proceedings of the 7th international conference on machine learning* (pp. 216–224).

Sutton, R. S., & Barto, A. G. (1998). *Introduction to reinforcement learning*. MIT Press.

Szpunar, K. K., & McDermott, K. B. (2008). Episodic future thought and its relation to remembering: Evidence from ratings of subjective experience. *Consciousness and Cognition, 17*(1), 330–334.

Szpunar, K. K., Watson, J. M., & McDermott, K. B. (2007). Neural substrates of envisioning the future. *Proceedings of the National Academy of Sciences of the United States of America, 104*(2), 642–647.

Takeuchi, T., Duszkiewicz, A. J., Sonneborn, A., Spooner, P. A., Yamasaki, M., Watanabe, M., … Morris, R. G. (2016). Locus coeruleus and dopaminergic consolidation of everyday memory. *Nature, 537*(7620), 357–362.

Tam, S. K. E., & Bonardi, C. (2012). Dorsal hippocampal involvement in appetitive trace conditioning and interval timing. *Behavioral Neuroscience, 126*(2), 258.

Taylor, M. E., Jong, N. K., & Stone, P. (2008). Transferring instances for model-based reinforcement learning. In *Machine learning and knowledge discovery in databases* (pp. 488–505). Springer.

Teitelbaum, J. C. (2014). Analogical legal reasoning: Theory and evidence. *American Law and Economics Review, 17*(1), 160–191.

Tolman, E. C. (1938). The determiners of behavior at a choice point. *Psychological Review, 45*(1), 1.

Tompary, A., & Davachi, L. (2017). Consolidation promotes the emergence of representational overlap in the hippocampus and medial prefrontal cortex. *Neuron, 96*(1), 228–241.

Treves, A., & Rolls, E. T. (1994). Computational analysis of the role of the hippocampus in memory. *Hippocampus, 4*(3), 374–391.

Tversky, A. (1977). Features of similarity. *Psychological Review, 84*(4), 327–352.

Tversky, A., & Kahneman, D. (1983). Extensional versus intuitive reasoning: The conjunction fallacy in probability judgment. *Psychological Review, 90*(4), 293–315.

Von Hessling, A., & Goel, A. K. (2005). Abstracting reusable cases from reinforcement learning. In *International conference on case-based reasoning workshops* (pp. 227–236).

Wagatsuma, A., Okuyama, T., Sun, C., Smith, L. M., Abe, K., & Tonegawa, S. (2018). Locus coeruleus input to hippocampal CA3 drives single-trial learning of a novel context. *Proceedings of the National Academy of Sciences of the United States of America, 115*(2), E310–E316.

Walther, E. (2002). Guilty by mere association: Evaluative conditioning and the spreading attitude effect. *Journal of Personality and Social Psychology, 82*(6), 919–934.

Watson, I., & Marir, F. (1994). Case-based reasoning: A review. *Knowledge Engineering Review, 9*(04), 327–354.

Weilbächer, R. A., & Gluth, S. (2017). The interplay of hippocampus and ventromedial prefrontal cortex in memory-based decision making. *Brain Sciences, 7*(1), 4.

Wilson, D. R., & Martinez, T. R. (2000). Reduction techniques for instance-based learning algorithms. *Machine Learning, 38*(3), 257–286.

Wimmer, G. E., Braun, E. K., Daw, N. D., & Shohamy, D. (2014). Episodic memory encoding interferes with reward learning and decreases striatal prediction errors. *Journal of Neuroscience, 34*(45), 14901–14912.

Wimmer, G. E., & Büchel, C. (2016). Reactivation of reward-related patterns from single past episodes supports memory-based decision making. *Journal of Neuroscience, 36*(10), 2868–2880.

Wimmer, G. E., Daw, N. D., & Shohamy, D. (2012). Generalization of value in reinforcement learning by humans. *European Journal of Neuroscience, 35*(7), 1092–1104.

Wimmer, G. E., & Shohamy, D. (2011). The striatum and beyond: Contributions of the hippocampus to decision making. In M. Delgado, E. A. Phelps, & T. W. Robbins (Eds.), *Decision Making, Affect, and Learning: Attention and Performance XXII* (pp. 281–310). Oxford University Press.

Wimmer, G. E., & Shohamy, D. (2012). Preference by association: How memory mechanisms in the hippocampus bias decisions. *Science, 338*(6104), 270–273.

Wittmann, B. C., Bunzeck, N., Dolan, R. J., & Düzel, E. (2007). Anticipation of novelty recruits reward system and hippocampus while promoting recollection. *NeuroImage, 38*(1), 194–202.

Wittmann, B. C., Daw, N. D., Seymour, B., & Dolan, R. J. (2008). Striatal activity underlies novelty-based choice in humans. *Neuron, 58*(6), 967–973.

Wittmann, B. C., Schott, B. H., Guderian, S., Frey, J. U., Heinze, H.-J., & Düzel, E. (2005). Reward-related fMRI activation of dopaminergic midbrain is associated with enhanced hippocampus-dependent long-term memory formation. *Neuron, 45*(3), 459–467.

Zeithamova, D., Dominick, A. L., & Preston, A. R. (2012). Hippocampal and ventral medial prefrontal activation during retrieval-mediated learning supports novel inference. *Neuron, 75*(1), 168–179.

Zeithamova, D., & Preston, A. R. (2010). Flexible memories: Differential roles for medial temporal lobe and prefrontal cortex in cross-episode binding. *Journal of Neuroscience, 30*(44), 14676–14684.

Zentall, T. R., Wasserman, E. A., Lazareva, O. F., Thompson, R. K. R., & Rattermann, M. J. (2008). Concept learning in animals. *Comparative Cognition & Behavior Reviews, 3*, 13–45.

CHAPTER 5

Learning Structures Through Reinforcement

Anne G.E. Collins
Department of Psychology, University of California, Berkeley, Berkeley, CA, United States

INTRODUCTION

The flexible and efficient decision-making that characterizes human behavior requires quick adaptation to changes in the environment and good use of gathered information. Thus, investigating the mechanisms by which humans learn complex behaviors is critical to understanding goal-directed decision-making. In the past 20 years, cognitive neuroscience has progressed immensely in understanding how humans learn from rewards and punishment, particularly for simpler behaviors shared in common with other mammals, such as learning simple associations between stimuli and actions. Reinforcement learning (RL) theory (Sutton & Barto, 1998) has provided a crucial theoretical framework explaining how humans learn to represent the value of choices and/or make decisions that are more likely to lead to rewards than to punishments. However, both cognitive neuroscience and artificial intelligence fields struggle with explaining more complex, and more characteristically human, learning behaviors, such as rapid learning in completely new and complex environments.

This chapter discusses the use of the RL framework to understand many complex learning behaviors, focusing specifically on model-free RL algorithms for learning values of or policies over states and actions, since we have a good understanding of how cortico–basal ganglia loops use dopaminergic input to implement an approximate form of this computation. We will show that many forms of complex human RL can be framed by applying this RL computation, provided that we model the inputs and outputs of the algorithm appropriately. Specifically, we argue that by better defining the state and action spaces for which humans learn values or policies, we can broadly widen the types of behaviors for which RL can account. We support this statement with examples from the literature showing how the brain may be performing the same computations for different types of inputs/outputs and how this can account for complex behavior, such as hierarchical RL (HRL), structure learning, generalization, and transfer.

We will first provide a short introduction to RL, both from a computational point of view, highlighting the limitations and difficulties encountered by this algorithm, and from a cognitive neuroscience point of view, mapping these computations to neural mechanisms. We will then attempt to unify multiple frameworks from the human learning literature, such as representation learning (Wilson & Niv, 2012), HRL (Botvinick, Niv, & Barto, 2009), rule learning (Collins & Koechlin, 2012), and structure learning (Collins & Frank, 2013), into a single framework, whereby the brain uses a single mechanistic computation—defined by a model-free RL mechanism—and applies it to different input and output spaces, notably, state and action spaces. We will first focus on how we can mitigate the curse of dimensionality by altering how we define state spaces, leading to more complex and efficient learning. We will then show that assuming different action spaces, in particular, by introducing temporal abstraction or rule abstraction, leads to faster learning and to an ability to generalize information. Last, we will show that humans sometimes create latent state or action spaces, which seemingly makes learning problems more complicated but comes with a number of behavioral advantages. Finally, we will conclude by broadening to other open questions in flexible learning: the role of the reward function in RL, the various algorithms other than model-free RL that may also contribute to efficient learning, and the roles of models of the environment in learning.

REINFORCEMENT LEARNING

Reinforcement learning algorithms

RL models are a class of algorithms designed to solve specific kinds of learning problems for an agent interacting with an environment that provides rewards and/or punishments (Fig. 5.1A). The following type of "grid world" problem exemplifies an archetypal RL problem (Fig. 5.2A). The agent (black square) sits in one of the cells of a grid environment and can navigate through the grid by choosing one of four actions (up, down,

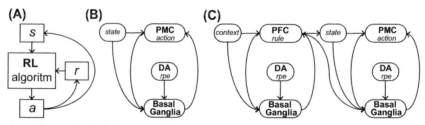

Figure 5.1 *Schematic of reinforcement learning (RL) systems.* (A) RL algorithms observe a state *s* as input and select an action *a* as output. The environment provides reinforcement *r*, which is used to update the RL algorithm and transitions to the next state. (B) An approximation of these computations is performed in the cortico–basal ganglia loop (Frank et al., 2004). For example, a sensory observation leads to preactivation of possible actions in the premotor cortex (PMC); the PMC–basal ganglia loops allow gating of one action; dopamine (DA) signals a reward prediction error (RPE) signal that reinforces corticostriatal synapses, allowing the gating mechanism to select the actions most likely to lead to reward. (C) This learning process occurs at multiple hierarchical levels in the brain in parallel (Collins & Frank, 2013). For example, loops involving the prefrontal cortex allow learning to occur between abstract contexts and high-level rules, which then constrains the lower-level learning loop.

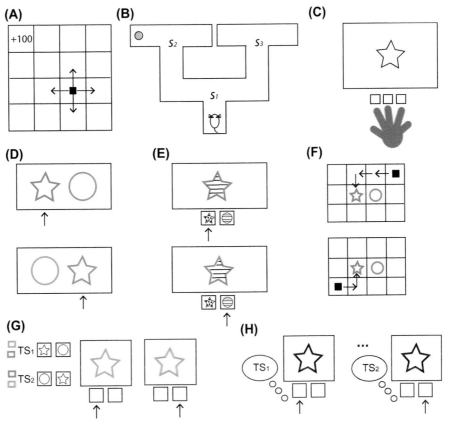

Figure 5.2 *Examples of reinforcement learning (RL) problems.* (A) Grid world: The artificial agent navigates between cells using one of four directions. (B) Animals navigate in a maze to obtain reward; the states s_i are physical locations. (C) Instrumental learning task: Participants use reward feedback to learn to select the correct button for each possible stimulus (e.g., shapes). (D) Representation learning task: Participants need to select one of two patterns; only one dimension matters (here, shape matters, with the star being the most rewarding of the two shapes). (E) Hierarchical learning: Participants learn that for one color (red—top), the shape of the input determines the correct action, but for the other color (blue—bottom), the texture determines the correct action. (F) Options framework or hierarchical RL (HRL): In both cases, participants select the same high-level action (or option): go to the star. This constrains a different sequence of low-level actions. (G) Structure learning: Participants learn to select one high-level abstract action (a rule, or task set, TS1) for some colors and another (TS2) for other colors; in parallel, they learn to associate low-level actions (button presses) to stimuli (here, shapes) for each of the high-level abstract rules. (H) Latent rule learning: Participants learn high-level rules as in (G) but do not observe the contexts. Instead, they infer the latent context from their observations of the outcomes to their choices.

left, or right). The agent can collect points by selecting some actions or by entering some cells (e.g., the top–left corner in Fig. 5.2A). The goal of the agent is to maximize points earned. Defined more technically, an RL problem is characterized by a state space S (here, the cells in the grid world), an action space A (here, the four available actions), a transition function $T(s,a,s') = p(s'|s,a)$ that controls the probability of the next state s' given that the agent chose action a in a state s and a reward function $R(s,a,s')$. The goal of the agent is to optimize the expected sum of future discounted rewards and, specifically, to find a policy $\pi(s,a) = p(a|s)$ that maximizes this sum. One way to achieve this goal is to estimate the expected value of each state or of each state and action under the optimal policy (where value is the expected sum of discounted future rewards). If one can do this, the optimal policy falls out by selecting the action with the highest value.

There are many different algorithms that propose solutions to this problem and offer guarantees of convergence. We focus on a simple class of algorithms, called model-free because they do not require a model of the environment (i.e., knowledge of the transition function and the reward function). We focus on model-free RL algorithms, such as temporal difference learning, Q-learning, SARSA, and actor-critic algorithms (Sutton & Barto, 1998), because they have been extremely helpful in understanding animal behavior and neural correlates of learning. Model-free RL algorithms use a key quantity, called the reward prediction error, to learn to estimate values of states or of state—action pairs. At each trial t, the reward prediction error is defined as the difference between what is expected for future discounted reward after taking an action (the sum of reward r and discounted value of next step $\gamma V(s_{t+1})$, where γ is the discount factor) and what was expected prior to taking that action ($V(s_t)$). Using the reward prediction error $rpe = r + \gamma V(s_{t+1}) - V(s_t)$ to update the previous estimate of $V(s_t)$ by a small increment of the error $\alpha . rpe$ (where α is the learning rate) is a good algorithm under certain assumptions and constraints (Sutton & Barto, 1998).

Reinforcement learning in the brain

Through this model-free RL algorithm, an artificial agent can learn the optimal way to attain a reward in the simple grid world of the example in Fig. 5.2A, after many attempts to solve this problem (Sutton & Barto, 1998). It is a good model of behavior for an animal learning to find its way toward a reward in a maze (Fig. 5.2B). Further, this algorithm has been a critical source of progress in the cognitive neuroscience of learning because it provides a useful model of the neural correlates of RL. Specifically, researchers have discovered that dopaminergic neurons fire in a pattern that is consistent with a reward prediction error signal: Their firing increases phasically with unexpected reward, decreases phasically with missed expected reward or with unexpected punishment, and stays at the tonic level for expected rewards (Montague, Dayan, & Sejnowski, 1996). Dopamine release in the striatum follows parametrically what would be expected for a bidirectional reward prediction error signal (Hart, Rutledge, Glimcher, & Phillips, 2014). Furthermore, dopamine signaling in the striatum modulates plasticity of corticostriatal synapses, with increased dopamine strengthening associations in the pathway facilitating action selection and decreasing them in the pathway blocking it; decreased dopamine has the opposite effects in these pathways (Adamantidis et al., 2011; Hamid et al., 2015; Kravitz, Tye, & Kreitzer, 2012; Tai, Lee, Benavidez, Bonci, & Wilbrecht, 2012). Cortico—basal ganglia loops act as a gate for action selection that is dependent on the strength of these two corticostriatal pathways (Collins & Frank, 2014; Frank, Seeberger, & O'Reilly, 2004). Thus, there is strong evidence that cortico—basal ganglia loops implement a model-free RL computation, with dopamine reward prediction errors training corticostriatal associations to help select choices that lead to reward and avoid those that lead to punishment (Fig. 5.1B).

Limitations

Model-free RL algorithms are thus very successful at explaining animal learning because they capture many behaviors well, including for example, probabilistic reward learning (Frank, Moustafa, Haughey, Curran, & Hutchison, 2007), and they have a plausible mechanistic implementation in the brain. Using these computational models to link between brain and behavior has increased understanding of individual differences in RL, of learning deficit in some pathologies (e.g., Parkinson) and of the effect of dopaminergic drugs on learning (Frank, 2005). However, model-free RL also has a number of limitations that have led cognitive neuroscience and artificial intelligence researchers to look at other algorithms to better model human learning and enhance artificial agents, respectively. One major limitation of RL is that it suffers from the curse of dimensionality: While RL can be relatively efficient in small problem spaces, learning with this algorithm in relatively bigger problem spaces would take an enormous amount of practice, making it extremely inefficient. In contrast, humans can often learn new behaviors very quickly (e.g., how to drive a car). Another limitation is that model-free RL is inflexible: When the environment changes (e.g., the position of the reward in the grid world), model-free RL algorithms need to slowly unlearn. By contrast, humans (and animals) are sensitive to changes in the environment and can quickly alter their behavior toward their goal. To solve these and other limitations of model-free RL algorithms, researchers in artificial intelligence and cognitive neuroscience have proposed new algorithms. For example, model-based RL algorithms offer some solutions to the inflexibility problem by proposing a different way of computing expected values that integrates knowledge about the model of the world. However, we will show here that we can understand many complex human behaviors in the framework of the same simple model-free algorithm, with its grounding in a well-understood neural implementation, by carefully considering the state and action spaces over which model-free computations of estimated values or policies are performed.

Framing the problem

What are state and action spaces when modeling human behavior? This modeling choice is often dictated by the experimental design and is assumed away as obvious. We give some examples in Fig. 5.2B and C. The most direct translation from original RL algorithms, such as grid worlds, is the modeling of spatial learning tasks in which animals need to learn to find a reward in a maze. States are modeled as discrete places in the maze at which a decision is needed, and actions are modeled as choices of direction (e.g., left or right; Fig. 5.2B). Note that making different choices for the state/action space could lead to a very different model (e.g., with more discrete places in the maze, actions could include stop, groom, etc.). For human behavior, state spaces are often replaced with sets of stimuli, and actions are replaced with simple choices, such as key presses (Fig. 5.2C); this modeling choice retains a fairly unambiguous interpretation of the environment. Probabilistic reward learning tasks offer a good example of the ambiguity of defining state/action spaces. In these tasks (e.g., Davidow, Foerde, Galvan, & Shohamy, 2016; Frank et al., 2004), subjects may be asked to choose between two shapes (e.g., Fig. 5.2D). There is some ambiguity in how this task should be modeled. Is the state the current pair of stimuli? Is this pair dependent or independent of their left/right position? More generally, this task tends to be modeled as a single state and two actions: "picking the star" or "picking the circle." It is important to note that (1) this action state is much more abstract than "press the left/right button," as it does not map to a single set of motor commands, and (2) a different choice, for example, "pick left" versus "pick right," would be unable to capture behavior in this task, since left and right are not informative about reward. Despite the abstraction of this action space, the model-free RL algorithm excels at capturing the behavior and neural effects in this task (Davidow et al., 2016). We show here that we can capture many behaviors of higher complexity in the model-free RL framework by carefully considering the state and action spaces over which the computations occur. Table 5.1 shows in pseudocode how this can be done in the examples of Fig. 5.2. We will show that developing appropriate states and action spaces overcomes many issues thought of as classic limitations of model-free RL.

STATE SPACES

Simplifying the state space

Figuring out an appropriate state space over which RL operates can dramatically improve RL performance by reducing the curse of dimensionality. Learning to drive is a task, which teenagers may accomplish in a few hours but which many top artificial intelligence researchers and companies have been unable to get an artificial agent to perform without major issues. How do we use our experience from 15 years of life to accomplish such fast learning? Taking all visual inputs into account would be overwhelming to a learning agent, as we essentially never see the same scene twice when driving. However, if one can discern that the relevant information for making a decision whether to stop or to go at an intersection is the color of the light (red, yellow, or green) then one part of the problem is suddenly reduced to a one-dimensional, two-feature state space. Niv and colleagues investigated such state space learning in a series of studies (Leong et al., 2017;

Table 5.1 Pseudocode for learning examples in Fig. 5.2

A) Grid world	$RL(S = \{all\ (x_i, y_j)\}, A = \{up, down, left, right\})$
B) Maze	$RL(S = \{all\ (x_i, y_j)\}, A = \{forward, left, right\})$
C) Instrumental learning	$RL(S = \{star, circle\}, A = \{button1, button2, button3\})$
D) Representation learning	$RL(S = \{all\ (shape, color, texture)\}, A = \{left, right\})$
	$RL(S = \{star, circle\}, A = \{left, right\})$
E) Hierarchical reinforcement learning	$RL(S = \{all\ (shape, color, texture)\}, A = \{left, right\})$
	$RL(S_1 = \{colors\},$
	$A_1 = \{attend(texture) = RL(S_2 = \{texture\},$
	$A_2 = \{left, right\}),$
	$attend(shape) = RL(S_3 = \{shape\}, A_2 = \{left, right\})\})$
F) Options— hierarchical reinforcement learning	$RL(S = \{all\ (x_i, y_j)\}, A = \{up, down, left, right\})$
	$RL(S = \{all\ (x_i, y_j)\},$
	$A_1 = \{go\ to\ circle = RL(S = \{all\ (x_i, y_j)\}, A_2 = A),$
	$go\ to\ star = RL(S = \{all\ (x_i, y_j)\}, A_2 = A)\})\})$
G) Structure learning	$RL(S = \{all\ (color, shape)\}, A = \{button1, button2\})$
	$RL(S_1 = \{colors\},$
	$A_1 = \{policy1 = RL(S_2 = \{shapes\}, A_2 = \{button1, button2\}),$
	$policy2 = RL(S_2 = \{shapes\}, A_2 = \{button1, button2\})\})$
H) Latent rule learning	$RL(S = \{shapes\}, A = \{button1, button2\})$
	$RL(S_1 = \{context1, context2, ...\},$
	$A_1 = \{policy1 = RL(S_2 = \{shapes\}, A_2 = \{button1, button2\}),$
	$policy2 = RL(S_2 = \{shapes\}, A_2 = \{button1, button2\})\})$

RL represents a single learning algorithm producing a policy over given state or action spaces **S, A**. Light blue is the "naïve" or flat modeling of a problem, using the simplest state spaces for inputs and action spaces for outputs. Black models structure learning, as observed in participants.

Niv et al., 2015; Wilson & Niv, 2012; see also Chapter 12 by Shuck, Wilson, and Niv), and a simplified example is schematized in Fig. 5.2D. At each trial, participants were shown three items and needed to choose one item to try to win points. Each item had three dimensions (shape, color, and texture), and each dimension had three features (e.g., red, blue, and green). In a learning problem, only one feature from one dimension (e.g., the star) had a high likelihood of leading to reward; thus, if participants were able to learn that the other two dimensions did not matter and that they should learn to represent the problem as an RL problem concerned only with shapes, they could significantly simplify the dimensionality of the problem and thus improve their performance (Wilson & Niv, 2012). Results showed that behavior was best explained by a process where participants learned to focus their attention on a single dimension and applied simple RL to features of this dimension. Thus, they effectively created a relevant, smaller state space, over which an RL algorithm was run (Table 5.1D); indeed, reward prediction error signals in the striatum were better explained by assuming RL happened over the state space defined by the focus of attention than by other models. This is one of the most direct examples of how humans define nonobvious state spaces over which to learn values or policies with RL. An important question is how we create the state space itself; in the example given here, how do we learn the feature on which we should focus our attention? A study by Leong et al. (2017) showed that creation of state space can be performed using reward feedback, such that there is a bidirectional interaction: Attention told subjects over which dimensions they should perform RL, and reward prediction errors helped participants direct their attention to the correct dimension and thus create the state space over which to operate RL.

Multiple state spaces

A state space that is appropriate for one goal may not be appropriate for another. Consider our driving example with the traffic light: If you are in the lane to go straight, the main round lights are relevant to your decision to stop or go, but if you are in the lane to turn left, you should ignore these lights and instead pay attention to the left arrow lights. Said differently, your state space should be conditioned on an additional aspect of the environment: which lane you are in. Being able to create multiple state spaces and knowing the one to which you should apply RL would allow significantly more complex learning behavior. Indeed, it would allow a hierarchical contextualization of learning by context. A series of studies (Badre & Frank, 2011; Badre, Kayser, & Esposito, 2010; Frank & Badre, 2011) has shown that healthy young adults are able to hierarchically contextualize the learning space and that it strongly improves their learning. Participants saw a single three-dimensional item on the screen and had to learn which of three actions to pick to receive points. In a flat condition, all the three dimensions were needed to figure out the correct action for an item, leading to three-dimensional state spaces with an overwhelming 18 items. In a hierarchical condition, one of the dimensions (color) controlled which of the other two dimensions was relevant for learning (e.g., if the item was red, only the shape mattered, but if it was blue, only the texture mattered; Fig. 5.2E). Thus, participants could essentially build two small state spaces (one corresponding to three textures and another to three shapes) and at each trial determine which state space to use based on the color of the item (Table 5.1E). Badre et al. (2010) showed that participants did learn this way, as evidenced by much more efficient learning in the hierarchical condition than in the flat condition. Further, studies (Badre & Frank, 2011; Frank & Badre, 2011) have shown that this method of learning could be computationally understood as RL computations happening over two hierarchical loops and different state (and action—see below) spaces (Fig. 5.1B): The top loop learned through RL which of the two state spaces to select for a given color, while the bottom loop learned which key press to select for either of the two simpler state spaces.

This example highlights a number of important points. First, RL computations may happen over multiple state spaces in the same learning problem, with other signals serving as a contextualizing factor. Second, they may happen simultaneously over multiple state spaces (in the previous example, learning which state space to select for a color state and which key to press for a given shape or texture). This latter point implies two further important features: (1) a notion of hierarchy, whereby the choice from one of the RL loops has an influence over the learning and decision of a "lower-level" loop and (2) the choice in the higher hierarchical loop is more abstract than the one at the lower level—indeed, in this example, RL in the top loop happens not only on a subpart of the original state space (the color dimension) but also on a new abstract action space, indicating the dimension on which a subject must focus attention. Below, we will come back to hierarchical representations in RL and to the importance of learning action spaces, in addition to state spaces.

ACTION SPACES

Abstract hierarchical action spaces

The study by Frank and Badre (2011), discussed above, showed that learning the hierarchical structure of the environment, which simplifies a large unstructured state space into two smaller state spaces selected conditionally on a context, can facilitate learning. It introduced the need to operate RL not only over multiple state spaces but also over an abstract action space, where the action is the decision of which lower-level state space to use. More generally, other complex learning behavior can be obtained by this combination of two characteristics: (1) RL at multiple hierarchical levels simultaneously and (2) RL over abstract higher-level action spaces that control lower-level decisions. In that sense, the higher-level actions are themselves policies mapping lower-level stimuli to lower-level actions. A body of work extended the previous notion of HRL by showing that such abstract actions could be more than just attentional filters (i.e., the dimension of the input to which I should focus my attention for making my decision), and could instead be abstract policies, also called "rules" or task sets (Collins & Frank, 2016a,b, 2013; Collins & Koechlin, 2012). Specifically, similarly to the studies of Badre and colleagues, these studies showed that participants learned to make a choice at a higher level in response to a feature of the environment (e.g., a color) and that the higher-level choice constrained answers to other features of the environment. However, in this case, the higher-level choice was not that one should focus on one dimension and neglect another dimension—indeed stimuli were only two-dimensional. Rather, the higher-level choice constrained the correct set of choices for the features of the second dimension (Fig. 5.2G).

Going back to the driving example, whether you are in France or in the United Kingdom, you need to pay attention to all the same visual signals to drive correctly. However, the actions you take in answer to these signals depend on the context: arriving at a circle in France requires you to turn right, but the same in the United Kingdom requires you to turn left. Thus, more complex behavior sometimes requires us not only to use context to determine where to focus our attention but also to determine how to respond to the focus of our attention. We showed that participants create such high-level abstract choice spaces, where choices correspond to this high-level policy choice; we call them rules or task sets (Collins & Frank, 2016a,b, 2013; Collins & Koechlin, 2012, Table 5.1G). Creating rules that one selects in response to a context, but that are not bound or equated to that context, is a critical factor in flexible, efficient learning. Indeed, because participants created these choice spaces, they were also able to try these choices in new contexts; this means that they were able to generalize a high-level policy to a new context (for example, the rules of driving in France apply mostly as a whole to driving in Germany). Furthermore, the new associations were stored by the policy learned at the lower level, constrained by the higher-level choice, without being tied to the context in which it was learned. Thus, participants were able to transfer knowledge learned in one context to other contexts that required selecting the same rule (for example, after having observed that driving is similar in Boston and Berkeley, learning how to handle a four-way stop in one location would immediately transfer to the other).

Creating an abstract action space (where actions are rules or task sets and can be viewed as a policy over another state action space) greatly increases the flexibility and efficiency of learning because it allows generalization and transfer. It also provides some form of *divide and conquer*, whereby a complicated decision over a large state space (all possible input features) is transformed into a series of simpler, hierarchical decisions: first selecting a rule in response to a context; then, given that rule, selecting an action in response to a stimulus. We showed with computational modeling and electroencephalography (Collins, Cavanagh, & Frank, 2014; Collins & Frank, 2016a,b, 2013) that this process can be performed in a model that applies RL computations in hierarchical cortico–basal ganglia loops (Fig. 5.1C). Thus, such hierarchical structure learning can also be understood as RL over appropriate state (at multiple hierarchical levels) and action (at multiple abstraction levels) spaces.

Temporally abstract actions

The previously described form of RL is clearly hierarchical: It consists of selecting a higher-level rule, which is really a policy in that it constrains selection of actions at the lower level. This feature allows us to draw a parallel to a specific class of algorithms that are known in the literature as "HRL," also called the "options framework." The options framework also seeks to improve on simple RL mechanisms by building a more complex action space and, specifically, by introducing options. Options can be seen as local policies or hierarchical actions (Table 5.1F). In the simplest case, options correspond to a class of sequences of simple actions that lead to a subgoal. For example, reaching the door of a room in a grid world is a high-level option and may define a local policy (how to reach a door from any point in the room or the star in the example of Fig. 5.2F). In the driving example, an example of a high-level option is shifting gears. You may learn at the high level when to shift or not to shift gears, but then once you select that option, it requires a series of lower-level actions (engage the clutch, shift the gear, then release the clutch) over which you can also learn.

Using options can partially solve the curse of dimensionality by facilitating exploration (Botvinick et al., 2009). Indeed, a single higher-level choice may lead an agent to explore further and more efficiently. Options also capture an important feature of human sequential behavior, which often includes hierarchical sequences of actions. A few studies have shown evidence of human learning and neural computations being well explained by the options framework, whereby learning happens hierarchically, both for the option itself and for the actions within the option (Diuk, Tsai, Wallis, Botvinick, & Niv, 2013; Ribas-Fernandes et al., 2011; Solway et al., 2014). In these studies, participants made choices in sequential environments that provided a possibility for HRL. Further, these studies showed evidence in the brain for reward prediction errors corresponding to learning over both action spaces (within the option policy and at the higher hierarchical level).

LATENT STATE AND ACTION SPACES

We have shown that many complex learning behaviors can be explained as applying a simple model-free RL algorithm to the correct state and action space, or sometimes as applying more than one RL computation to multiple appropriate state and action spaces in parallel. An interesting feature is that in hierarchical forms of RL (structure learning, options framework, and hierarchical rule learning), the higher-level action space is abstract in the form of a policy. In particular, it cannot be described as a concrete motor action. Here, we show that abstraction in the state space can also help understand more complex learning behaviors. In particular, assuming unobservable, or latent, states can greatly enhance the flexibility of the learning agent (Gershman, Norman, & Niv, 2015). For example, if you are driving in winter, you might not be able to see that the road is icy, but if you observe that your usual actions lead to undesirable consequences (slipping), you might deduce that the latent cause in the environment is the weather and adapt your behavior based on this latent cause. This example captures some of the important features for which RL over latent states or causes can better explain human learning: when the contingencies of the environment change suddenly but not in an observable way (e.g., in reversal learning experiments (Hampton, Bossaerts, & O'Doherty, 2006)), an RL agent operating over the observable state space needs to unlearn previous associations before being able to learn new associations. By contrast, humans may identify a change point, infer a new unobservable context or latent cause, and learn over this state. Several studies (Gershman, Blei, & Niv, 2010; Gershman et al., 2015; Soto, Gershman, & Niv, 2014) have shown how this assumption can explain a number of learning phenomena, such as extinction and compound generalization.

Latent spaces enrich the state representations over which RL operates. In combination with other previously described mechanisms, such as abstract action spaces (rules) that hierarchically constrain simultaneous learning over other state and action spaces, the mechanism of creating latent spaces provides an explanation for additional aspects of human fast and flexible learning. One behavioral study (Collins & Koechlin, 2012) had participants learn associations between one-dimensional stimuli and actions (task sets) using probabilistic reward feedback (Fig. 5.2H). The task sets changed periodically without warning and, unbeknownst to participants, could be reused as a whole later in the experiment.

Results showed that participants were able to create both an abstract action space of task sets and an abstract state space of latent temporal contexts (Table 5.1H); they identified the current temporal context as a state in which a given task set was to be selected, constraining RL over association between an observable state space (stimuli) and actions (key presses). Furthermore, when they identified a new temporal context (after an inferred switch in the environmental contingencies), they explored in the abstract action space of task sets, reselecting previously learned strategies as a whole, rather than exploring only in the low-level state space (Collins & Koechlin, 2012; Donoso, Collins, & Koechlin, 2014). This strategy allowed participants to transfer task sets to new contexts and thus to adapt more quickly than they would have otherwise.

The examples given above show that much of complex human learning does not require any learning algorithm more complex than model-free RL, provided that the latter algorithm is applied to the right inputs and outputs (state and action spaces). This process may require (1) running this algorithm over more than one set of spaces in parallel, a task for which the cortico—basal ganglia loops are well configured (Alexander & DeLong, 1986), and (2) using hierarchical influence of one output over another input, for which the prefrontal cortex is well organized (Badre, 2008; Koechlin, Ody, & Kouneiher, 2003; Koechlin & Summerfield, 2007; Nee & D'Esposito, 2016). These features enable much more efficient and flexible learning than was originally thought possible with a simple model-free algorithm for RL value estimation. Specifically, they allow for fast and efficient exploration, improvement of performance by massive simplification of problems, and fast learning in new environments by generalization and transfer of information.

HOW DO WE CREATE THE STATE/ACTION SPACES?

Efficiently modeling complex human learning with model-free RL crucially relies on operating over the right state and action spaces. Using inappropriate spaces instead strongly impairs learning, as shown, for example, by Botvinick et al. (2009) in simulations where using incorrect options lead to slowed exploration. The question of how we acquire the appropriate state and action spaces for our current environments remains largely open, although the previous examples do suggest some potential mechanisms.

For learning state spaces when the optimal state space is a subspace of the full sensory space, some studies (Leong et al., 2017; Niv et al., 2015) suggest that we use a frontoparietal mechanism to focus attention specifically on that subspace and that we learn to do so using reinforcement. Frank and Badre (2011) suggest that the gating mechanisms of the prefrontal cortex—basal ganglia loops may learn which aspects of the environment to keep in working memory, as well as which items should be allowed to influence other loops, thus also using the simple RL mechanism to create the ad hoc state spaces required for HRL. Collins and Frank (2013) also showed that such mechanisms enabled the creation of abstract action spaces. Furthermore, there seems to be a strong bias toward learning occurring hierarchically. Specifically, some studies (Badre & Frank, 2011; Badre et al., 2010) have shown that participants engaged anterior portions of the prefrontal cortex a priori initially, even in problems that could not be simplified. Further, other studies (Collins & Frank, 2013; Collins et al., 2014) have shown that participants built a hierarchical abstract rule structure even in environments that did not immediately benefit from it, highlighting a more general drive toward this kind of organization. This bias toward hierarchical learning could be due to a prior belief that hierarchical structures are useful (Collins & Frank, 2016a,b) or to constraints that result from the way our hierarchical cortico—basal ganglia loops evolved from motor cortex—originating loops (Collins & Frank, 2016a,b), or, more likely, it could be due to both.

A series of models from Alexander, Brown, and colleagues (Alexander & Brown, 2011, 2014, 2015) also point out the potential importance of medial prefrontal cortex in learning rules for cognitive control. Their models assume that the medial prefrontal cortex learns to represent errors of prediction at various hierarchical levels, thus teaching the lateral prefrontal cortex to represent useful state and action spaces to minimize such errors of prediction. These models also resonate with work by Holroyd and colleagues (Holroyd & McClure, 2015; Holroyd & Yeung, 2012), which points out the importance of the anterior cingulate cortex (ACC) in extended motivated behavior. Specifically, they argue that the ACC enables HRL (in the sense of the options framework), whereby the hierarchy is in the choice of higher-level actions that constrain sequences of lower-level actions.

This HRL/options framework also raises the question of how the action space is created, or the "options discovery" problem: How do we create options that take us to the doors of the room rather than to the windows? Theoretical work suggests that using pseudoreward when reaching a subgoal and using RL with this pseudoreward to learn the option may help option creation (Botvinick et al., 2009), and there is some evidence that such a mechanism may occur in the brain (Diuk et al., 2013; Ribas-Fernandes et al., 2011; Solway et al., 2014). However, how do we determine useful subgoals? Work by Schapiro and colleagues (Schapiro, Rogers, Cordova, Turk-Browne, & Botvinick, 2013; Schapiro, Turk-Browne, Botvinick, & Norman, 2016) has shown that humans are able to identify bottlenecks in the environments we navigate and, if given a chance, create options with these bottlenecks as subgoals, which might be one mechanism for creating a useful action space in the framework of options.

Interestingly, some methods for learning useful state and action spaces require a model of the environment. For example, creating useful options may require identifying bottlenecks in a mental map of the environment. In the case of latent state spaces, in particular, a model of the environment consists of a likelihood function, defining expected outcomes (for example rewards) in response to interactions with the environment under a given latent space (Collins & Koechlin, 2012; Gershman et al., 2015). Using this likelihood function allows both an inference about the current hidden state and the online creation of what the latent state space is (Collins & Frank, 2013; Gershman et al., 2010). It is important to note that these models are used to create a state space and to infer a state but that, despite this use of a model, the learning algorithm in operation may still be a model-free RL algorithm. This highlights the blurry line between what we should label as model-free and model-based learning (see also Chapter 18 by Miller, Ludvig, Pezzulo, and Shenhav); most learning may use a model of the environment, even in the absence of a mechanism of forward planning, as is usually defined in formal model-based RL algorithms (Daw, Gershman, Seymour, Dayan, & Dolan, 2011). RL with a model can reach many more types of behaviors than those usually understood by model-based RL.

OPEN QUESTIONS

We have shown that thinking of human learning as a simple computation occurring over well-tailored state and action spaces can explain many feats of flexible and efficient decision-making. However, many open questions remain, one of which we have already discussed: how these state and action spaces are built. Two other classes of questions also merit further research to better understand human learning. Learning from reinforcement requires four elements: a state and action space, a reward function, and an algorithm to learn a policy. We have focused here on the role of the state and action spaces and have just assumed a simple model-free RL algorithm and reward function. However, both learning algorithms and reward functions should be further investigated.

Reward function

Most RL experiments use primary or secondary rewards or punishments, such as food, pain, points, and money (gains or losses), as reinforcers. However, other features might also contribute to the reward function. For instance, theoretical and experimental results have suggested various "bonuses" to the reward function, related for example to novelty (Kakade & Dayan, 2002) and information (Bromberg-Martin & Hikosaka, 2009); these and other influences may be reflected in the dopamine reward prediction error signal (see also , Chapter 11 by Sharpe and Schoenbaum). Other results have shown costs in the form of mental effort and conflict (Cavanagh, Masters, Bath, & Frank, 2014; Kool & Botvinick, 2014; Westbrook & Braver, 2015; see also Chapter 7 by Kool, Cushman, and Gershman). Furthermore, the movement of gamification relies on the notion that learners are motivated by nonrewarding outcomes (e.g., stars) that mark the attainment of subgoals (Deterding, Dixon, Khaled, & Nacke, 2011; Hamari, Koivisto, & Sarsa, 2014). This notion relates to pseudoreward, which may be useful for learning options in the HRL framework: Maintaining motivation over extended behaviors when real reward is infrequent might require us to consider intermediary, symbolic subgoals as rewarding (Diuk et al., 2013; Ribas-Fernandes et al., 2011; Lieder & Griffiths, n.d.). Theoretical work has shown that this notion could tremendously improve learning in complex situations (Lieder & Griffiths, n.d.). Thus, future research in human learning should aim to better understand what outcomes contribute to the reward function used by the RL algorithm for learning and to determine whether humans manipulate this reward function beyond normal reward to create better representations of the learning problem.

Algorithms

Separating the algorithm of learning from its inputs and outputs—the state and action spaces—enables us to better understand how a rich collection of human learning behaviors can be explained with this framework. However, this argument should not be taken to mean that we propose the brain uses only the learning algorithm presented and exactly this algorithm to learn to make decisions from reward information. In fact, much remains poorly understood about the computations performed by the brain to learn policies. For the model-free RL algorithm, we understand that the cortico—basal ganglia loops with dopamine reward prediction errors approximate it, but many precise aspects of this computation remain under debate. For example, the direct and indirect pathways apparently have redundant roles in learning (Collins & Frank, 2014; Dunovan & Verstynen, 2016); more research is needed to better understand their distinct contributions to model-free RL.

Furthermore, it is very likely that the brain also uses, in parallel, other algorithms to learn policies from reward. One method is simple memorization of associations in working memory, which accounts for part of learning from rewards in simple associative learning tasks (Collins & Frank, 2012; Collins, Ciullo, Frank, & Badre, 2017). Similarly, by allowing us to sample from past events, episodic memory may play an important role in policies learned from reward (Bornstein & Norman, 2017; Bornstein, Khaw, Shohamy, & Daw, 2017). Furthermore, there is also ample evidence that humans also perform model-based planning RL in parallel to model-free RL (Daw et al., 2011; Doll, Duncan, Simon, Shohamy, & Daw, 2015). Exactly how this prospective planning occurs, especially many steps ahead, is not well understood—it may depend on the use of heuristics to simplify the forward search (Huys et al., 2015) or inferential processes (Chapter 3 by Solway and Botvinick). Thus, much remains unknown about the algorithms themselves.

CONCLUSION

Human learning is incredibly efficient and flexible and does much to promote human intelligence and goal-directed behavior. In this chapter, we explored how a very simple family of algorithms—that we know are approximately implemented by a precise neural circuitry in the brain—can explain a surprisingly wide array of complex learning, unifying literature on HRL, the options framework, structure learning, and representation learning. Specifically, we show that this simple computation of expected value (or policy weight), obtained by incremental updates with reward prediction errors, can lead to very efficient learning, exploring, transfer, and generalization when applied to useful state and action spaces. Understanding how we construct these useful spaces and how we interlock multiple computational loops in parallel to learn at multiple levels simultaneously is a future challenge. One important point is that finding useful spaces is not simply a matter of simplifying the sensory and motor space by factoring it into lower-dimensional or discrete subspaces but can rather also involve making the spaces more complex—creating new states that are not a subspace of sensory and motor space but are abstract states and actions carrying more information about the structure of the problem. These state and action spaces of higher complexity can counter-intuitively lead to an eventual improvement in behavior by rendering decision-making more flexible and by providing useful subpolicies that achieve subgoals or other generalizable chunks of behavior.

REFERENCES

Adamantidis, A. R., Tsai, H.-C., Boutrel, B., Zhang, F., Stuber, G. D., Budygin, E. A., … de Lecea, L. (2011). Optogenetic interrogation of dopaminergic modulation of the multiple phases of reward-seeking behavior. *The Journal of Neuroscience: the Official Journal of the Society for Neuroscience, 31*(30), 10829–10835. http://doi.org/10.1523/JNEUROSCI.2246-11.2011.

Alexander, W. H., & Brown, J. W. (2011). Medial prefrontal cortex as an action-outcome predictor. *Nature Neuroscience, 14*(10), 1338–1344. http://doi.org/10.1038/nn.2921.

Alexander, W. H., & Brown, J. W. (2014). A general role for medial prefrontal cortex in event prediction. *Frontiers in Computational Neuroscience, 8*(69). http://doi.org/10.3389/fncom.2014.00069.

Alexander, W. H., & Brown, J. W. (2015). Hierarchical error representation: A computational model of anterior cingulate and dorsolateral prefrontal cortex. *Neural Computation, 27*(11), 2354–2410. http://doi.org/10.1162/NECO_a_00779.

Alexander, G., & DeLong, M. (1986). Parallel organization of functionally segregated circuits linking basal ganglia and cortex. *Annual Review of Neuroscience*.

Badre, D. (2008). Cognitive control, hierarchy, and the rostro-caudal organization of the frontal lobes. *Trends in Cognitive Sciences, 12*(5), 193–200. http://doi.org/10.1016/j.tics.2008.02.004.

Badre, D., & Frank, M. J. (2011). Mechanisms of hierarchical reinforcement learning in cortico-striatal circuits 2: Evidence from fMRI. *Cerebral Cortex (New York, N.Y.: 1991)*, 1–10. http://doi.org/10.1093/cercor/bhr117.

Badre, D., Kayser, A. S., & Esposito, M. D. (2010). Article frontal Cortex and the Discovery of abstract action rules. *Neuron, 66*(2), 315–326. http://doi.org/10.1016/j.neuron.2010.03.025.

Bornstein, A. M., Khaw, M. W., Shohamy, D., & Daw, N. D. (2017). What's past is present: Reminders of past choices bias decisions for reward in humans. *bioRxiv*.

Bornstein, A. M., & Norman, K. A. (2017). Putting value in context: A role for context memory in decisions for reward. *bioRxiv*.

Botvinick, M. M., Niv, Y., & Barto, A. C. (2009). Hierarchically organized behavior and its neural foundations: A reinforcement-learning perspective. *Cognition, 113*(3), 262–280.

Bromberg-Martin, E. S., & Hikosaka, O. (2009). Midbrain dopamine neurons signal preference for advance information about upcoming rewards. *Neuron, 63*(1), 119–126. http://doi.org/10.1016/j.neuron.2009.06.009.

Cavanagh, J. F., Masters, S. E., Bath, K., & Frank, M. J. (2014). Conflict acts as an implicit cost in reinforcement learning. *Nature Communications, 5*(5394). http://doi.org/10.1038/ncomms6394.

Collins, A. G. E., Cavanagh, J. F., & Frank, M. J. (2014). Human EEG uncovers latent generalizable rule structure during learning. *The Journal of Neuroscience, 34*(13), 4677–4685. http://doi.org/10.1523/JNEUROSCI.3900-13.2014.

Collins, A. G. E., Ciullo, B., Frank, M. J., & Badre, D. (2017). Working memory load strengthens reward prediction errors. *The Journal of Neuroscience, 37*(16), 2700–2716. http://doi.org/10.1523/JNEUROSCI.2700-16.2017.

Collins, A. G. E., & Frank, M. J. (2012). How much of reinforcement learning is working memory, not reinforcement learning? A behavioral, computational, and neurogenetic analysis. *The European Journal of Neuroscience, 35*(7), 1024–1035. http://doi.org/10.1111/j.1460-9568.2011.07980.x.

Collins, A. G. E., & Frank, M. J. M. J. (2013). Cognitive control over learning: Creating, clustering, and generalizing task-set structure. *Psychological Review, 120*(1), 190–229. http://doi.org/10.1037/a0030852.

Collins, A. G. E., & Frank, M. J. (2014). Opponent actor learning (OpAL): Modeling interactive effects of striatal dopamine on reinforcement learning and choice incentive. *Psychological Review, 121*(3), 337–366. http://doi.org/10.1037/a0037015.

Collins, A. G. E., & Frank, M. J. (2016a). Neural signature of hierarchically structured expectations predicts clustering and transfer of rule sets in reinforcement learning. *Cognition, 152*, 160–169. http://doi.org/10.1016/j.cognition.2016.04.002.

Collins, A. G. E., & Frank, M. J. (2016b). Motor demands constrain cognitive rule structures. *PLoS Computational Biology, 12*(3), e1004785. http://doi.org/10.1371/journal.pcbi.1004785.

Collins, A. G. E., & Koechlin, E. (2012). Reasoning, learning, and creativity: Frontal lobe function and human decision-making. *Plos Biology, 10*(3), e1001293. http://doi.org/10.1371/journal.pbio.1001293.

Davidow, J. Y., Foerde, K., Galvan, A., & Shohamy, D. (2016). An upside to reward Sensitivity: The Hippocampus supports enhanced reinforcement learning in adolescence. *Neuron, 92*(1), 93–99. http://doi.org/10.1016/j.neuron.2016.08.031.

Daw, N. D., Gershman, S. J., Seymour, B., Dayan, P., & Dolan, R. J. (2011). Model-based influences on humans' choices and striatal prediction errors. *Neuron, 69*(6), 1204–1215. http://doi.org/10.1016/j.neuron.2011.02.027.

Deterding, S., Dixon, D., Khaled, R., & Nacke, L. (2011). From game design elements to gamefulness. In *Proceedings of the 15th International Academic MindTrek Conference on Envisioning future Media environments - MindTrek '11 (p. 9)*. New York, New York, USA: ACM Press. http://doi.org/10.1145/2181037.2181040.

Diuk, C., Tsai, K., Wallis, J., Botvinick, M., & Niv, Y. (2013). Hierarchical learning induces two simultaneous, but separable, prediction errors in human basal ganglia. *The Journal of Neuroscience : The Official Journal of the Society for Neuroscience, 33*(13), 5797–5805. http://doi.org/10.1523/JNEUROSCI.5445-12.2013.

Doll, B. B., Duncan, K. D., Simon, D. A., Shohamy, D., & Daw, N. D. (2015). Model-based choices involve prospective neural activity. *Nature Neuroscience*, (February), 1–9. http://doi.org/10.1038/nn.3981.

Donoso, M., Collins, A. G. E., & Koechlin, E. (2014). Foundations of human reasoning in the prefrontal cortex. *Science, 344*(6191), 1481–1486. http://doi.org/10.1126/science.1252254.

Dunovan, K., & Verstynen, T. (2016). Believer-skeptic meets actor-critic: Rethinking the role of basal ganglia pathways during decision-making and reinforcement learning. *Frontiers in Neuroscience, 10*(106). http://doi.org/10.3389/fnins.2016.00106.

Frank, M. J. (2005). Dynamic dopamine modulation in the basal ganglia: A neurocomputational account of cognitive deficits in medicated and nonmedicated parkinsonism. *Journal of Cognitive Neuroscience, 17*(1), 51–72. http://doi.org/10.1162/0898929052880093.

Frank, M. J., & Badre, D. (2011). Mechanisms of hierarchical reinforcement learning in corticostriatal circuits 1: Computational analysis. *Cerebral Cortex (New York, N.Y.: 1991), 2010*, 1–18. http://doi.org/10.1093/cercor/bhr114.

Frank, M. J., Moustafa, A. A., Haughey, H. M., Curran, T., & Hutchison, K. E. (2007). Genetic triple dissociation reveals multiple roles for dopamine in reinforcement learning. *Proceedings of the National Academy of Sciences of the United States of America, 104*(41), 16311–16316. http://doi.org/10.1073/pnas.0706111104.

Frank, M. J., Seeberger, L. C., & O'Reilly, R. C. (2004). By carrot or by stick: Cognitive reinforcement learning in parkinsonism. *Science (New York, N.Y.), 306*(5703), 1940–1943. http://doi.org/10.1126/science.1102941.

Gershman, S. J., Blei, D. M., & Niv, Y. (2010). Context, learning, and extinction. *Psychological Review, 117*(1), 197–209. http://doi.org/10.1037/a0017808.

Gershman, S. J., Norman, K. A., & Niv, Y. (2015). Discovering latent causes in reinforcement learning. *Current Opinion in Behavioral Sciences, 5*, 43–50. http://doi.org/10.1016/j.cobeha.2015.07.007.

Hamari, J., Koivisto, J., & Sarsa, H. (2014). Does gamification Work? – a literature review of empirical studies on gamification. In *2014 47th Hawaii International Conference on system Sciences* (pp. 3025–3034). IEEE. http://doi.org/10.1109/HICSS.2014.377.

Hamid, A. A., Pettibone, J. R., Mabrouk, O. S., Hetrick, V. L., Schmidt, R., Vander Weele, C. M., … Berke, J. D. (2015). Mesolimbic dopamine signals the value of work. *Nature Neuroscience, 19*(1), 117–126. http://doi.org/10.1038/nn.4173.

Hampton, A. N., Bossaerts, P., & O'Doherty, J. P. (2006). The role of the ventromedial prefrontal cortex in abstract state-based inference during decision making in humans. *The Journal of Neuroscience: the Official Journal of the Society for Neuroscience, 26*(32), 8360–8367. http://doi.org/10.1523/JNEUROSCI.1010-06.2006.

Hart, A. S., Rutledge, R. B., Glimcher, P. W., & Phillips, P. E. M. (2014). Phasic dopamine release in the rat nucleus accumbens symmetrically encodes a reward prediction error term. *Journal of Neuroscience, 34*(3), 698–704. http://doi.org/10.1523/JNEUROSCI.2489-13.2014.

Holroyd, C. B., & McClure, S. S. M. (2015). Hierarchical control over effortful behavior by rodent medial frontal cortex: A computational model. *Psychological Review, 122*(1), 54–83. http://doi.org/10.1037/a0038339.

Holroyd, C. B., & Yeung, N. (2012). Motivation of extended behaviors by anterior cingulate cortex. *Trends in Cognitive Sciences, 16*(2), 122–128. http://doi.org/10.1016/j.tics.2011.12.008.

Huys, Q. J. M., Lally, N., Faulkner, P., Eshel, N., Seifritz, E., Gershman, S. J., … Roiser, J. P. (2015). Interplay of approximate planning strategies. *Proceedings of the National Academy of Sciences of the United States of America, 112*(10), 3098–3103. http://doi.org/10.1073/pnas.1414219112.

Kakade, S., & Dayan, P. (2002). Dopamine: Generalization and bonuses. *Neural Networks, 15*(4), 549–559. http://doi.org/10.1016/S0893-6080(02)00048-5.

Koechlin, E., Ody, C., & Kouneiher, F. (2003). The architecture of cognitive control in the human prefrontal cortex. *Science (New York, N.Y.),, 302*(5648), 1181–1185. http://doi.org/10.1126/science.1088545.

Koechlin, E., & Summerfield, C. (2007). An information theoretical approach to prefrontal executive function. *Trends in Cognitive Sciences, 11*(6), 229–235. http://doi.org/10.1016/j.tics.2007.04.005.

Kool, W., & Botvinick, M. (2014). A labor/leisure tradeoff in cognitive control. *Journal of Experimental Psychology: General, 143*(1), 131–141. http://doi.org/10.1037/a0031048.

Kravitz, A. V., Tye, L. D., & Kreitzer, A. C. (2012). Distinct roles for direct and indirect pathway striatal neurons in reinforcement. *Nature Neuroscience, 15*(6), 816–818. http://doi.org/10.1038/nn.3100.

Leong, Y. C., Radulescu, A., Daniel, R., Dewoskin, V., Niv, Y., & Partners, T. (2017). Dynamic interaction between reinforcement learning and attention in multidimensional environments. *Neuron, 93*(2), 451–463. http://doi.org/10.1016/j.neuron.2016.12.040.

Lieder, F., Griffiths, T.L. (n.d.). Helping people make better decisions using optimal gamification.

Montague, P. R., Dayan, P., & Sejnowski, T. J. (1996). A framework for mesencephalic dopamine systems based on predictive Hebbian learning. *The Journal of Neuroscience: the Official Journal of the Society for Neuroscience, 16*(5), 1936–1947.

Nee, D. E., & D'Esposito, M. (2016). The hierarchical organization of the lateral prefrontal cortex. *eLife, 5*(March 2016), 1–26. http://doi.org/10.7554/eLife.12112.

Niv, Y., Daniel, R., Geana, A., Gershman, S. J., Leong, Y. C., Radulescu, A., & Wilson, R. C. (2015). Reinforcement learning in multidimensional environments relies on attention mechanisms. *The Journal of Neuroscience : The Official Journal of the Society for Neuroscience, 35*(21), 8145–8157. http://doi.org/10.1523/JNEUROSCI.2978-14.2015.

Ribas-Fernandes, J. J. F., Solway, A., Diuk, C., McGuire, J. T., Barto, A. G., Niv, Y., & Botvinick, M. M. (2011). A neural signature of hierarchical reinforcement learning. *Neuron, 71*(2), 370–379. http://doi.org/10.1016/j.neuron.2011.05.042.

Schapiro, A. C., Rogers, T. T., Cordova, N. I., Turk-Browne, N. B., & Botvinick, M. M. (2013). Neural representations of events arise from temporal community structure. *Nature Neuroscience, 16*(4), 486–492. http://doi.org/10.1038/nn.3331.

Schapiro, A. C., Turk-Browne, N. B., Botvinick, M. M., & Norman, K. A. (2016). Complementary learning systems within the hippocampus: A neural network modeling approach to reconciling episodic memory with statistical learning. *bioRxiv, 51870.* http://doi.org/10.1101/051870.

Solway, A., Diuk, C., Córdova, N., Yee, D., Barto, A. G., Niv, Y., & Botvinick, M. M. (2014). Optimal behavioral hierarchy. *PLoS Computational Biology, 10*(8). http://doi.org/10.1371/journal.pcbi.1003779.

Soto, F. A., Gershman, S. J., & Niv, Y. (2014). Explaining compound generalization in associative and causal learning through rational principles of dimensional generalization. *Psychological Review, 121*(3), 526–558. http://doi.org/10.1037/a0037018.

Sutton, R. S., & Barto, A. G. (1998). *Reinforcement learning* (Vol. 9). MIT Press.

Tai, L.-H., Lee, A. M., Benavidez, N., Bonci, A., & Wilbrecht, L. (2012). Transient stimulation of distinct subpopulations of striatal neurons mimics changes in action value. *Nature Neuroscience, 15*(9), 1281–1289. http://doi.org/10.1038/nn.3188.

Westbrook, A., & Braver, T. S. (2015). Cognitive effort: A neuroeconomic approach. *Cognitive, Affective, & Behavioral Neuroscience, 15*(2), 395–415. http://doi.org/10.3758/s13415-015-0334-y.

Wilson, R. C., & Niv, Y. (2012). Inferring relevance in a changing world. *Frontiers in Human Neuroscience, 5*(January), 1–14. http://doi.org/10.3389/fnhum.2011.00189.

CHAPTER 6

Goal-Directed Sequences in the Hippocampus

Brandy Schmidt[1], Andrew M. Wikenheiser[2], A. David Redish[1]

[1]Department of Neuroscience, University of Minnesota, Minneapolis, MN, United States; [2]National Institute on Drug Abuse Intramural Research Program, Baltimore, MD, United States

Humans make goal-directed decisions every day. New data suggest that other mammals also make goal-directed decisions. Current theories hypothesize that goal-directed decisions arise from search processes through imagined forward models by which we work out the consequences of specific actions then choose from among those actions based on the utility of the outcomes (Niv, Joel, & Dayan, 2006). In this chapter, we will review the processes that underlie goal-directed decision-making in mammalian brains and make the case that the hippocampus is a key component of the imagination process. First, however, we will need to address the question of imagination because if you need imagination for goal-directed decision-making and nonhuman animals make goal-directed decisions, then we need to determine what imagination is, neurally, so that we can measure it in nonhuman animals.

In humans, the term *episodic future thinking* refers to the capacity to imagine an auto-biographical experience that happens in the future (Buckner & Carroll, 2007). Episodic future thinking engages the same neural mechanisms as remembering past experiences (Addis, Wong, & Schacter, 2007; Hassabis, Kumaran, Vann, & Maguire, 2007; Schacter et al., 2012). The fact that recall of past events is fragile (Talarico & Rubin, 2003) and varies depending on the presently available cues (Loftus & Palmer, 1974) suggests that remembering past experiences, like imagining future outcomes, entails flexibly retrieving previously stored information and recombining that information into an imagined situation. Studies involving aging populations (Schacter, Gaesser, & Addis, 2013), amnesiacs (Cole, Morrison, Barak, Pauly-Takacs, & Conway, 2016; Hassabis, Kumaran, & Maguire, 2007; Kurzcek et al., 2015; Race, Keane, & Verfaellie, 2011; Tulving, 1985; Zeman, Butler, Muhlert, & Milton, 2013), patients with Alzheimer's disease (Haj, Antoine, & Kapogiannis, 2015; Irish & Piolino, 2016) and prefrontal lesions (Ramussen & Bersten, 2016) all show a reduction in both remembering the past and imagining the future. Imaging studies have shown that a similar neural network is activated during episodic future thinking and remembering past experiences, including the medial temporal lobe, retrosplenial cortex, medial prefrontal cortex (mPFC), and lateral temporal and parietal regions (Addis et al., 2007; Hassabis, Kumaran, Vann, et al., 2007;

Goal-Directed Decision Making
ISBN 978-0-12-812098-9, https://doi.org/10.1016/B978-0-12-812098-9.00006-1

Schacter, Addis, & Buckner, 2007). Additionally, the ventromedial prefrontal cortex may facilitate access to the conceptual knowledge of a scenario necessary to simulate an episodic event, as well as the valuation of these events (Bonnici et al., 2012; Kumaran, Summerfield, Hassabis, & Maguire, 2009; Lin, Horner, Bisby, & Burgess, 2015; Peters & Buchel, 2010).

Theoretically, planning requires the ability to predict consequences of actions and outcomes, and thus requires a model of the world, including both a categorization of the states of the world and the transitions between those states. In reinforcement learning models, determining action policies through planning is termed "model-based decision-making" because of its dependence on a model of the world (Niv et al., 2006; Sutton & Barto, 1998).

Although they cannot demonstrate it linguistically, behavioral observations and neural recordings suggest that rodents are capable of developing these models of the world. Tolman (1948) termed this a "cognitive map." Tolman was led to this conclusion through the observation of latent learning: In an early study by Tolman and Honzik (1930), rats were trained in a complex maze full of turns and dead ends. The end of the maze contained food reward that one group of rats received after reaching the end of the maze; the second group of rats had a barrier between them and the reward and were taken out of the maze once they reached the end. The rats that had access to the food reward learned the maze quickly; however, the rats that did not have access to the food reward failed to run the maze reliably. Interestingly, after 10 trials, these rats then had access to the food reward and their performance on the maze immediately improved, even outperforming the original group of rats. The data show that the rats had learned the maze, even if they lacked the motivation to run it.

Tolman's "cognitive map" concept was that the rat had an internal representation of the structure of the environment. From this internal representation of the structure of the environment, it is theoretically possible to simulate the possible actions and to imagine the consequences of your actions. Computationally, this allows the discovery of shortcuts (O'Keefe & Nadel, 1978; Redish, 1999; Samsonovich & Ascoli, 2005) and the evaluation of the consequences of one's actions in the light of one's current needs (Niv et al., 2006). Importantly, planning using the cognitive map could be contrasted with situation—action decisions, in which one learns to take an action in response to the current situation, with no explicit representation of the consequences of the action (Daw, Niv, & Dayan, 2005; Hull, 1943; Niv et al., 2006; van der Meer, Kurth-Nelson, & Redish, 2012). Tolman hypothesized that rats (and people) were learning the structure of the world so that they could later plan action paths through it, while Hull hypothesized that rats (and presumably people) were learning what actions to take in given situations.

This dichotomy between Tolman's cognitive map and Hull's stimulus-response can be most easily seen in the T-choice plus maze (Barnes, Nadel, & Honig, 1980; Packard & McGaugh, 1996; Tolman, 1948; Fig. 6.1). In this task, rats are first allowed to explore a plus-shaped maze, presumably allowing them to derive the structure of that maze. They

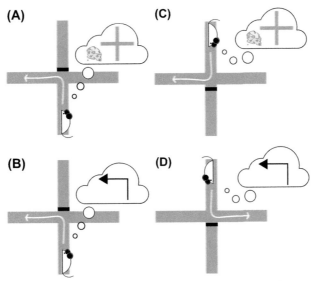

Figure 6.1 The plus maze task can dissociate which navigational strategy the rat is using (Packard & McGaugh, 1996). (A) In this plus maze task, rats are trained to turn left from the South arm to the West arm. The rat can either use a planning-based (Tolmanian) algorithm, in which it knows where it is (on the South arm) and knows where it wants to go (to the West arm), (B) or the rat can use a situation— action association (Hullian) algorithm (bottom mazes), in which it knows to turn left when placed on the maze. (C and D) When a rat is placed on the North arm, it is possible to determine which naviga- tion strategy the rat is using. A Tolmanian rat uses spatial cues to make this decision and goes to the place (where the food reward is located). In contrast, a Hullian rat will continue to turn left, this time ending up on the East arm. Rats with limited training show Tolmanian choices, turning left to the West arm, but rats with extended training show Hullian responses, turning right to the East arm.

are then trained to turn left from the South arm to the West arm. The rat can learn this task either through a planning-based (Tolmanian) algorithm, in which it knows where it is (on the South arm) and knows where it wants to go (to the West arm), or through a situation—action association (Hullian) algorithm, in which it knows to turn left when placed on the maze. Although these two algorithms are not dissociable from the South arm, when a rat is placed on the North arm, these algorithms produce different behaviors. A Tolmanian rat will turn right to the West arm, taking a different action to achieve the same result, while a Hullian rat will turn left to the East arm, taking the same action but achieving a different result. Of course, it is not that one of these options is correct and the other wrong, but they are different generalizations of the changed situation. Rats with limited training show Tolmanian choices, turning right to the West arm, but rats with extended training show Hullian responses, turning left to the East arm (Packard & McGaugh, 1996).

This task has been extensively studied. Manipulations that make the cognitive map easier to learn (more cues, rats with better vision) shift rats toward Tolmanian mapping processes (Chang & Gold, 2004), as do manipulations that make learning the situation—action associations less useful (Gardner et al., 2013; Schmidt, Papale, Redish, & Markus, 2013). Importantly, anterior dorsolateral striatum is a key structure in the development of the Hullian situation—action process (Chang & Gold, 2003; Kesner, Bolland, & Dakis, 1993; Packard, 1999; Packard & McGaugh, 1992, 1996; Yin, Knowlton, & Balleine, 2004), while the hippocampus, mPFC, and the posterior dorsomedial striatum are critical to behavioral flexibility and the use of the cognitive map in Tolmanian decisions (Bissonette & Roesch, 2017; Chang & Gold, 2003; Packard, 1999; Packard & McGaugh, 1992, 1996; Ragozzino, Detrick, & Kesner, 1999; Rich & Shapiro, 2007, 2009; Yin et al., 2004). As can be seen in the plus maze example, the cognitive map is easiest to study in the light of navigation, where the map can be directly observed and map-based navigation can be contrasted with learning specific routes (i.e., action sequences). In this navigation framework, a map places external information onto a coordinate system, allowing one to infer novel relationships between them (Gallistel, 1990; O'Keefe & Nadel, 1978; Redish, 1999). Importantly, a map is more than a coordinate system. While a map requires a coordinate system as input, the map is the relationship between the external information and the coordinate system and is unlikely to include the coordinate system internally (Redish & Touretzky, 1997). Extensive evidence suggests that the hippocampus maintains these relationships of objects in the environment in regard to each other and to the animal by relating them to this extrahippocampal coordinate system. This cognitive map would then allow an animal to have awareness of its environment irrespective of any particular sensory input and to mentally combine different parts of the environment even if they have never been experienced at that same time (O'Keefe & Nadel, 1978; Redish, 1999; Worden, 1992).

When rats reach a choice point, they often pause, orienting and reorienting toward their potential routes—a behavior termed vicarious trial and error (VTE; Gardner et al., 2013; Hu & Amsel, 1995; Muenzinger, 1938; Muenzinger & Gentry, 1931; Redish, 2016; Tolman, 1938). VTE is seen during early learning and decreases with task proficiency (Tolman, 1939). VTE increases with changes in task demands (Blumenthal, Steiner, Seeland, & Redish, 2011; Steiner & Redish, 2012) or by increasing the number of choices/options (Bett et al., 2012). We have found that VTE increases in rats when learning and/or using a hippocampal place strategy, during strategy conflicts, and immediately after error trials, again suggesting that rats are engaged in deliberation during VTE (Schmidt et al., 2013). VTE is most likely a behavioral reflection of indecision in deliberative decision-making (Amemiya & Redish, 2016; Gardner et al., 2013; Papale, Stott, Powell, Regier, & Redish, 2012; van der Meer et al., 2012; see Redish, 2016 for a review).

In humans and rodents, the hippocampus and prefrontal cortex are both engaged during spatial navigation and planning (O'Keefe & Nadel, 1978; Redish, 1999; Spiers & Maguire, 2007). For example, in a recent fMRI study by Kaplan et al. (2017), participants were trained on novel spatial navigation paradigm where they needed to plan paths of varying difficulty on novel mazes. The authors found that the prefrontal cortex and the hippocampus were both engaged during navigation planning. Interestingly, the functional connectivity between these two structures was higher when planning required more deliberation and preceding correct choices. Similarly, the rodent hippocampus and mPFC are functionally engaged during deliberative decision-making, showing increased coherence in the theta frequency specifically at choice points and phase locking of mPFC neurons to hippocampal theta oscillations (Benchenane et al., 2010; Hyman, Zilli, Paley, & Hasselmo, 2005; Jones & Wilson, 2005; Siapas & Wilson, 1998). Lesions to the hippocampus impair VTE behavior (Bett et al., 2012; Hu & Amsel, 1995); however, disrupting normal hippocampal functions can actually lead to an increase in VTE behavior (Papale, 2015; Robbe et al., 2007). This leads us to hypothesize that the hippocampus is not the driving force for VTE behavior but that VTE is engaged by another neural system. Wang et al. (2015) proposed that during decision-making, the lateral prefrontal cortex generates numerous potential action plans (i.e., take this choice, skip this choice) and that this information is sent to the hippocampus, which retrieves the stored representations related to these specific actions. The hippocampus then iteratively engages the mPFC as it sorts through different hippocampal-generated behavioral simulations, in order to determine the best choice of action. Lesion studies have found that the mPFC facilitates behavioral flexibility during new learning (Ragozzino et al., 1999), the same time period when VTE behavior is prevalent. In further support of this hypothesis, our lab has recently found that disrupting the mPFC with Designer Receptors Exclusively Activated by Designer Drugs reduces VTE behavior (Schmidt & Redish, 2016, *Society for Neuroscience Abstract*). Recent studies have found that mPFC is engaged during strategy changes, particularly during times when VTE is increased (Benchenane et al., 2010; Bissonette & Roesch, 2017; Powell & Redish, 2016).

HOW CAN WE EXAMINE EPISODIC FUTURE THINKING/MENTAL TIME TRAVEL?

Exactly how can we measure episodic future thinking or mental time travel in rodents? Try as we might, we have so far been unable to get our rats to fill out any of our post-behavioral training questionnaires. Instead, one must infer cognition through behavioral observation, which historically engendered much debate about the reliability of such inferences (Hull, 1943; MacCorquodale & Meehl, 1948; Skinner, 1948; Watson, 1913). However, the recognition that imagination entails activation of the same neural systems as during active perception and action suggests that it may be possible to observe episodic future thinking (mental time travel), even in nonlinguistic animals such as rodents (Johnson, Fenton, Kentros, & Redish, 2009). Imagination of sensory objects activates the same sensory areas as when those objects are perceived (Haxby, Connolly, & Guntupalli, 2014; Kosslyn, 1994; O'Craven & Kanwisher, 2000; Pearson, Naselaris, Holmes, & Kosslyn, 2015). Similarly, imagination of motor actions activates the motor areas (Jeannerod, 1994; Rizzolatti & Craigero, 2004). It has even been possible to use these imagination processes to directly observe planning in humans (Abram, 2017; Doll, Duncan, Simon, Shohamy, & Daw, 2015).

Doll et al. (2015) trained subjects on the two-step decision task (Daw, Gershman, Seymour, Dayan, & Dolan, 2011; Fig. 6.2). In this task, subjects are given two choices (C1 = A or B). This choice leads to a second layer of two possible choices (C2 = C or D, or C3 = E or F). Choosing A in C1 leads to C2 (C vs. D) 80% of the time and C3 (E vs. F) 20% of the time, while choosing B in C1 leads to C2 20% of the time and C3 80% of the time. Choosing C, D, E, or F leads to a probabilistically delivered reward. The key to this task is that the probability of reward delivery changes slowly over time, so the goal of the task is to return to a winning outcome. Because planning systems take the structure of the world into account, after a rare transition (A → C3 or B → C2), a planning-based (Tolmanian) algorithm would choose the other choice (A → C3 → E/F → reward → B; B → C2 C/D → reward → A), while a habit/ procedural/do-it-again situation–action association (Hullian) algorithm would repeat the original choice (A →…reward → A; B →…reward → B). Thus, this task is able

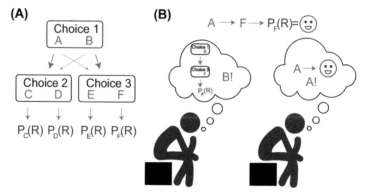

Figure 6.2 Two-step decision task (Daw et al., 2011). (A) In this task, subjects are initially given a choice (Choice 1 = A or B). This leads to a second layer of two possible choices (Choice 2 = C or D or Choice 3 = E or F). Choosing A in Choice 1 leads to Choice 2 (C or D) 80% of the time and Choice 3 (E or F) 20% of the time, while choosing B in Choice 1 leads to Choice 2 20% of the time and Choice 3 80% of the time. Choosing C, D, E, or F leads to a probabilistically delivered reward. The probability of reward delivery changes slowly over time, and the goal is to return to a winning outcome. (B) Because planning systems take the structure of the world into account, after a rare transition (A → Choice 3 or B → Choice 2), a planning-based (Tolmanian) algorithm would choose the other choice (A → Choice 3 → E/F → reward → B; B → Choice 2C/D → reward → A), while a habit/procedural/do-it-again situation–action association (Hullian) algorithm would repeat the original choice (A → …reward → A; B → …reward → B). Thus, this task is able to differentiate Tolmanian planning processes from Hullian situation–action processes.

to differentiate Tolmanian planning processes from Hullian situation–action processes, much like the plus maze described earlier. Doll et al. (2015) designed this task using cues that could be differentiated in fMRI (faces, tools, body parts, landscapes) and found that when subjects showed planning behaviors, the fMRI signals indicated imagination of the upcoming cues.

A similar process can be used in neurophysiological recordings from awake, behaving nonhuman animals (such as rats) (Johnson et al., 2009). Pyramidal cells in the hippocampus, aka "place cells," show spatially specific firing properties (O'Keefe & Dostrovsky, 1971; O'Keefe & Nadel, 1978; Redish, 1999), typically showing a peak firing in a small location in the environment and remaining mostly quiet in the rest of the environment. The area of maximal firing is referred to as the "place field." The place fields of different cells are distributed throughout the environment (Muller, 1996), creating a maplike representation of the environment (O'Keefe & Nadel, 1978; Redish, 1999). In addition to firing at the rat's current location, place cells also show rare extrafield firing, i.e., firing occasionally in locations separate from their place field. This nonlocal firing is typically seen at feeder/reward sites (see Redish, 1999 for review) and decision points (Johnson & Redish, 2007). With large enough neural ensembles, it is possible to decode the information represented within the ensemble (Wilson & McNaughton, 1993; Zhang, Ginzburg, McNaughton, & Sejnowski, 1998). During these extrafield firing events, decoding reveals nonlocal representations of space (Jensen & Lisman, 2000; Johnson & Redish, 2007; Pfeiffer & Foster, 2013).

More recent studies have determined that during this nonlocal firing, the place cells are activated in behaviorally relevant sequences that can represent trajectories the rat previously traversed or could traverse (Davidson, Kloosterman, & Wilson, 2009; Foster & Wilson, 2006; Gupta, van der Meer, Touretzky, & Redish, 2010, 2012; Pfeiffer & Foster, 2013; Skaggs & McNaughton, 1996). What was once believed to be noise is now hypothesized to reflect the rodent "thinking" about another location. The answer to how the hippocampus engages in episodic future thinking thus lies in the firing sequences of hippocampal place cells and their relation to local field potentials.

Place cell ensemble firing sequences are typically seen during two oscillatory events (Fig. 6.3): sharp-wave ripple complexes (SWR; 150 ms 150–220 Hz burst events), which occur during sleep and awake quiescence (Buzsaki, Leung, & Vanderwolf, 1983; O'Keefe & Nadel, 1978), and theta oscillations (more continuous 6–10 Hz processes), which occur during movement and attentive states (Buzsaki, 2002; O'Keefe & Nadel, 1978; Vanderwolf, 1969). During sleep (Kudrimoti, Barnes, & McNaughton, 1999; Lee & Wilson, 2002; Skaggs & McNaughton, 1996; Wilson & McNaughton, 1994) and quiet wakefulness (Csicsvari, O'Neill, Allen, & Senior, 2007; Diba & Buzsaki, 2007; Foster & Wilson, 2006; Gupta, van der Meer, Touretzky, & Redish, 2012; Jackson, Johnson, & Redish, 2006; Jadhav, Kemere, German, & Frank, 2012; O'Neill, Senior, & Csicsvari, 2006; Pfeiffer & Foster, 2013; Singer, Carr, Karlsson, & Frank,

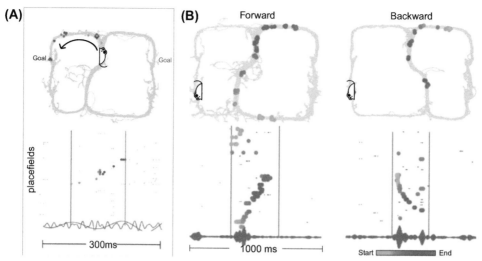

Figure 6.3 Examples of sequences. (A) Top: example of theta sequence while the rat is located at the choice point. Each place field center is represented by a *colored dot* (place in sequences corresponds to color bar in bottom right panel, blue is early, pink is later). Bottom: place cells sorted relative to the rat's location over a single theta cycle. Local Field Potential filtered for theta (6—10 Hz) and gamma (40—100 Hz). (B) Example of sharp-wave ripple sequences for forward (top left) and backward (top right) sequences. Bottom: place cells sorted relative to the rat's location over a sharp-wave ripple. *(Adapted with permission from Gupta et al. (2010, 2012).)*

2013), brief episodes of high-amplitude, fast-frequency SWR dominate the local field potential in CA1 and CA3 as a result of synchronous CA3 and CA1 activity (Buzsaki, 2015; Buzsaki et al., 1983; Csicsvari, Hirase, Mamiya, & Buzsaki, 2000). During SWR, place cell assemblies "replay" spatial trajectories previously traversed in a temporally condensed manner. These reactivation and replay sequences were first observed during sleep after behaviors (Buzsaki, 2015; Pavlides & Winson, 1989; Wilson & McNaughton, 1994). Note that reactivation and replay during sleep are examples of imagination and mental time travel—representations of other places and other times, such as reactivation of recently experienced behaviors on a track, while the rat rests on a separate platform.

From their first discovery, SWR sequences were hypothesized to facilitate memory consolidation, by continually recapitulating previous experiences during sleep (Alvarez & Squire, 1994; Buzsaki et al., 1983; Gais & Born, 2004; Marr, 1971; Sutherland & McNaughton, 2000). During sleep, pyramidal cell firing sequences are generally replayed in the original order of firing (forward replays) supporting their theorized role in memory consolidation. However, when SWR sequences were discovered during awake quiescence, not only did they fire in the original order of the trajectory traversed but also in the reverse order (backward replay; Csicsvari et al., 2007; Foster & Wilson, 2006;

Gupta et al., 2010). They also traversed novel trajectories never before experienced by the rat (Gupta et al., 2010), which suggests that they likely play a role in exploring the cognitive map (Derdikman & Moser, 2010; Samsonovich & Ascoli, 2005), much like mind-wandering in humans (Christoff, Irving, Fox, Spreng, & Andrews-Hanna, 2016).

Other studies, however, have found that firing during wake SWRs can predict the subsequent path of the animal. Pfeiffer and Foster (2013) trained rats on a goal-directed navigation task to forage for food reward between randomly distributed locations and a stationary "home" location. During events with large multiunit cellular activity, though not specifically during SWR, but usually coinciding with, sequences represented trajectories to behaviorally relevant locations; for example, when the rat was away from the home location, sequences predicted trajectories going home; however, this was not seen during random foraging (Fig. 6.4). Interestingly, these trajectory events were not simply paths in front of the rat; sequences represented future paths regardless of the head direction of the rat. Similar to Gupta et al. (2010), sequences even represented novel trajectories back to the home location.

The specific roles played by reactivations during SWR events remain unclear. There is some evidence that sequences during awake quiescence are more variable than sequences during sleep (Wikenheiser & Redish, 2013), including both forward and backward sequences, and seem to be related to attended areas of the maze, such as recent and future paths (Davidson et al., 2009; Foster & Wilson, 2006; Pfeiffer & Foster, 2013; Silva, Feng, & Foster, 2015), as well as novel and important, but not recently experienced, paths (Gupta et al., 2010). One possibility is that the sequences seen during quiet waking states are akin to imagination in the human default mode network (Raichle et al., 2001), allowing the novel connection of new concepts (Samsonovich & Ascoli, 2005). Another possibility is that it is a potential substrate for memory retrieval to be used in planning processes (Carr, Jadhav, & Frank, 2011; Pfeiffer & Foster, 2013; Schmidt & Redish, 2013). Disrupting SWRs in waking states impairs working memory and learning (Jadhav et al., 2012) and increases VTE behavior (Papale, Zielinski, Frank, Jadhav, & Redish, 2016).

Jadhav et al. (2012) selectively disrupted awake SWR events in rats trained on a hippocampal-dependent spatial alternation task. In the W maze, the rats were rewarded for alternating between the three arms of the maze. When the rats were on the outside arm, they were rewarded for entering the center arm of the maze. When the rats were on the center arm, however, they were only rewarded for visiting the outermost arm that was not previously visited (i.e., left—center—right—center—left). This allowed the comparison between two arm trajectories, one with a memory component and the other without. Awake SWR ripples were disrupted through a stimulation electrode targeting the ventral hippocampal commissure. Electrical stimulation within 25 ms of SWR detection disrupted SWR events and multiunit activity. Interestingly, SWR disruption impaired spatial working memory on the W maze by selectively impairing outbound

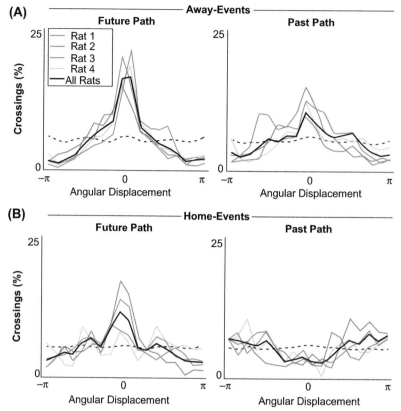

Figure 6.4 Sequences depict future trajectories to home location. (A) In order to determine whether sequences predicted future paths, the angular displacement between the future projected path and the actual future and previous paths taken were measured. The angular displacement was measured between the projected and actual trajectories at progressively increasing radii from the rat's location. Angular displacements at zero represent trajectories taken that matched with the predicted trajectory. (A) Differences between future paths and projected paths to goal locations were concentrated around zero angular displacement and more uniformly distributed when compared to the past path. (B) Differences between future paths and projected paths to home locations showed weaker relationships. *(Adapted with permission from Pfeiffer and Foster (2013).)*

trials while sparing inbound trials. These deficits were found despite no overall change to place cell firing characteristics or fields as well as intact sleep SWR sequences. These data suggest that disrupting awake SWR impaired spatial memory performance by disrupting the link between recent and remote experiences that SWR are believed to provide.

In contrast, sequences during sleep seem to be more veridical (i.e., forward) (Skaggs & McNaughton, 1996; Wikenheiser & Redish, 2013) and include both the hippocampal sequence and the consequence of those sequences (seen as activation of reward-related information in downstream nucleus accumbens, Lansink, Goltstein, Lankelma,

McNaughton, & Pennartz, 2009; Pennartz et al., 2004). Reactivation during sleep is generally hypothesized to facilitate the consolidation of contextual information by strengthening synaptic connections and transferring information from the hippocampus to the cortex (Sutherland & McNaughton, 2000). Supporting a role for replay as goal-directed exploration, Lansink et al. (2008) found that ventral striatal reward-related information appeared time-locked to hippocampal replays—cells representing the appropriate reward site fired at the end of SWR sequences replaying approaching that reward site. Disruption of SWRs during postbehavior sleep disrupts learning and consolidation effects (Ego-Stengel & Wilson, 2010; Girardeau, Benchenane, Wiener, Buzsaki, & Zugaro, 2009), and activation of dopaminergic signals during sleep-based reactivation leads to learning of that reactivated site as a goal (de Lavilleon, Lacroix, Rondi-Reig, & Benchenane, 2015). Recently, de Lavilleon et al. (2015) stimulated dopamine neurons every time a specific place cell was active during sleep SWRs and found that rats preferred to approach that goal the next day.

Sequences seen during theta oscillations, in contrast, represent time-compressed spatial trajectories that could facilitate spatial navigation and planning (Foster & Wilson, 2007; Wikenheiser & Redish, 2015). Johnson and Redish (2007) found that theta sequences serially traverse potential routes. Subsequent studies suggest theta sequences run to the potential goal locations (Gupta et al., 2012; Wikenheiser & Redish, 2015). Therefore, the activation of these sequences may support different behavioral processes whether they are active during SWR or theta oscillations.

In 1993, O'Keefe and Recce reported that the relationship between hippocampal cell firing and the theta rhythm changed as an animal passed through the place field—with spiking beginning at the end of each theta cycle on entry and precessing earlier and earlier as the animal passed through the field. This phenomenon, termed *phase precession*, because the phase of firing precesses as the animal runs through the field, has been robustly replicated by numerous labs (Dragoi & Buzsaki, 2006; Foster & Wilson, 2007; Gupta et al., 2012). Several labs quickly noted that this phenomenon meant that there was a sequence within each theta cycle, progressing along the path of the animal (Jensen & Lisman, 1996; Skaggs & McNaughton, 1996; Tsodyks, Skaggs, Sejnowski, & McNaughton, 1996). Two important questions remained: (1) Were the sequences a consequence of phase precession or vice versa? (2) Were the sequences running from behind the animal to the location of the animal, from the animal forward, or from behind to in front?

Studies attempting to answer the first question found that in well-learned environments, sequences better described the data than phase precession. Dragoi and Buzsaki (2006) found that the timing between pairs of cells better explained the data than the phase of firing of each of those cells. Other labs looking at learned environments have found that the sequences can occur without phase precession—Johnson and Redish (2007) found that during VTE, sequences alternated between options, even though no phase precession was occurring. Comparing place field firing on the running wheel with and without a goal, phase precession occurred when there was a goal (Pastalkova, Itskov, Amarasingham, & Buzsaki, 2008), but when there was no goal, the phase of firing remained constant (Hirase, Czurko, Csicsvari, & Buzsaki, 1999), suggesting that without a goal, the rat was running the same sequence over and over again (Lisman & Redish, 2009).

Although it would seem that phase precession and theta sequences are two ways of looking at the same phenomenon, Feng, Silva, and Foster (2015) recently found that one could get phase precession without sequences. On the first pass through a place field, cells phase precessed but did not line up into sequences until the second pass, because while individual cells phase precessed on the first lap, the starting phase shifted from cell to cell, so they did not start line up to create sequences. Recently, Wang, Roth, and Pastalkova (2016) examined whether theta sequences are dependent upon internally generated neural activity or if sensory input is sufficient. Silencing the medial septum, which provides theta input to the hippocampus, disrupted theta sequences while preserving firing fields. These data suggest that while phase precession could arise from sensory input, theta sequences are integrally generated by hippocampal network dynamics and not sensory input. So far, theta sequences have always been observed to follow the rat's direction of motion, even when animals move backward. Cei, Girardeau, Drieu, Kanbi, and Zugaro (2014) geared a car so that when the rat ran forward, the car ran backward. Similarly, Maurer, Lester, Burke, Ferng, and Barnes (2014) trained a rat to actually walk backward. In both of these cases, both phase precession and theta sequences ran along the trajectory of the rat (i.e., not the direction the rat was facing), implying that these sequences are encoding the path of the rat.

An important question about theta sequences is whether they are about the future path of the rat or about the past path already run. As Skaggs and McNaughton (1996) noted, this could be determined by where these phase precessions crossed in different approaches to a place field: If sequences were about the past, place cell firing on two paths that crossed would cooccur at the start of the place field, while if sequences were about the future, the two paths that converged would converge at the end of the place field. Later data definitively proved that multidirectional place fields in open environments and bidirectional fields in linear tracks converged at the end of the field, implying that these sequences were running from the animal forward, predicting future paths of the animal (Battaglia, Sutherland, & McNaughton, 2004; Huxter et al., 2008; see Lisman & Redish, 2009 for review). Further studies have consistently shown that place fields align from lap to lap at the end of their place fields, even if the starting point can change (Wikenheiser & Redish, 2015; Zheng et al., 2016). Newly formed place fields emerge from back to front, with firing first locked to the end of the place field and later expanding backward to earlier positions with subsequent experience (Bittner et al., 2015; Mehta, Barnes, & McNaughton, 1997; Monaco, Rao, Roth, & Knierim, 2014).

However, in more complex mazes, these results were more complicated, with sequences appearing behind the animal (running from behind to where the animal was) when the animal approached a goal and sequences appearing in front of the animal as it left the goal (running from the location of the animal forward to future positions) (Gupta et al., 2012). Direct examination of these sequences suggested that the sequences ran to the actual goal of the animal, bypassing earlier potential goals that the rat planned to skip (Wikenheiser & Redish, 2015).

As mentioned, VTE is believed to behaviorally reflect the neurophysiological generation and evaluation of future actions. During VTE, place cells transiently "sweep" forward, in a serial manner, spatially representing specific routes to goal locations (Amemiya & Redish, 2016; Johnson & Redish, 2007; Papale et al., 2016). These sequences are consistent with the results of Gupta et al. (2012) who found that theta sequences appear to segment the maze in a task-related manner, representing areas ahead of the animal as it left maze locations. In contrast to the nonlocal representations seen during SWR, the sequences seen during VTE occur during strong theta oscillations (Johnson & Redish, 2007; Papale et al., 2016).

Theta sequences are believed to only represent in a forward direction, unlike SWR sequences that show representations in both the forward and backward directions. This suggests that theta sequences may support planning, but their exact role in goal-directed decision-making is not yet clear. In order to elucidate the role of theta sequences in planning, Wikenheiser and Redish (2015) trained rats in a foraging task on a circular maze for food reward. Rats ran in a circle with three evenly dispersed reward sites, each site with a different fixed-length delay required in order to receive the food reward. The rats encountered a series of stay/go decisions where the rat could wait out the delay for the food reward or skip the current reward site and travel to the next reward site. The rat's choices could be qualified into three behaviors: one-segment, in which the rat ran to the next reward site and waited out the delay; two-segment, in which the rat skipped the next reward site but stopped at the second, subsequent reward site to wait out the delay; and three-segment, in which the rat skipped the next two reward sites, returning to the original reward site (i.e., running a full lap around the circle) before waiting out the delay (see Fig. 6.6). This task permitted the authors to examine how theta sequences are connected to goal-directed decision-making by examining how far theta sequences "looked ahead" during these one-, two-, and three-segment trials. Theta sequences were compared on the first segment of all the three trajectory types, which held the behavior constant and only varied in the goal destination. The distance traveled for the theta sequences were commensurate with the trajectory length, shortest for one-segment, longer for two-segment, and longest for three-segment trajectories. In contrast, when approaching their goal locations, theta sequences were comparable for all three trial types. Taken together these data suggest that hippocampal theta sequences do facilitate planning mechanisms for goal-directed decision-making.

Theta sequences are also necessary for correct performance on hippocampal-dependent behavioral paradigms. A study by Robbe et al. (2007) measured the effects of cannabinoids on theta and SWR oscillations, as well as theta sequences. Cannabinoids impair memory in hippocampal-dependent tasks in humans and rodents alike (Litchtman, Diemen, & Martin, 1995; Litchman & Martin, 1996; Robbe et al., 2007). In the Robbe et al. (2007) study, place cells were recorded from CA1 in rats under the influence of a cannabinoid receptor (CB1) agonist on a hippocampal-dependent spatial alternation task (Ainge, van der Meer, Langston, & Wood, 2007). In addition to the decreasing power in the theta and SWR frequencies, CB1 agonists severely impaired the temporal synchrony of hippocampal pyramidal cells without affecting the overall population firing rates. In a subsequent study, Robbe and Buzsaki (2009) replicated the behavioral deficits on the hippocampal-dependent spatial memory task and temporal organization of cell firing. Interestingly, the rodents showed more VTE and likely increased indecision. Despite the preserved place field firing characteristics, coordinated place cell firing and likely theta sequences were disrupted. This study demonstrated that disrupting the organization of theta sequences increased VTE and impairs behavioral performance on hippocampal-dependent tasks.

On the flip side, clonidine is an α-adrenergic autoreceptor agonist that decreases tonic levels of noradrenaline pharmacologically; behaviorally it decreases indecision in humans (Coull, Middleton, Robbins, & Sahakian, 1995; Jakala et al., 1999), potentially by limiting mental exploration. Similarly, clonidine in rodents also suppresses VTE behavior and, therefore, increases decisiveness (Amemiya, Noji, Kubota, Nishijima, & Kita, 2014). In a subsequent study, Amemiya and Redish (2016) examined whether the reduced VTE behavior seen in rats given clonidine also resulted in reduced mental exploration. Consistent with other results (Johnson & Redish, 2007; Papale et al., 2016), theta sequences represented both the chosen and unchosen paths during VTE under saline but more often represented the chosen path during non-VTE behavior. Interestingly, clonidine suppressed theta sequences that represented the unchosen path during VTE, suggesting that clonidine induced decisiveness resulted from a reduction in mental exploration of options.

Anatomical and physiological studies support the hypothesis of an inverse relationship between SWR and theta oscillations. Subcortical inputs to the hippocampus have suppressing effects on CA3 recurrent excitation, thereby suppressing SWR events (Buzsaki, 2015; Buzsaki et al., 1983; Vandecastelle et al., 2014). Numerous studies have shown that during theta oscillations, SWR are suppressed via presynaptic cholinergic muscarinic receptors (Hasselmo, 1995, 1999, 2006), cannabinoid CB1 receptors (Robbe et al., 2007), as well as cholinergic inputs from the medial septum (Vandecastelle et al., 2014). Lesions that reduce theta oscillations, including lesions to the medial septum, fimbria fornix, and entorhinal cortex, all increase SWR events (Buzsaki, 2015; Buzsaki et al., 1983).

OPEN QUESTIONS

There are still many unknowns regarding sequences. Despite the decades of research, we are still unclear about how sequences are generated. What is the mechanistic relationship between sequences and phase precession? Are they controlled by the same mechanism? Do we need sequences for planning? When engaging in episodic future thinking, humans may mentally travel serially along all the required steps to reach a goal but in other cases may only mentally travel to the final outcome (Schacter, Benoit, & Szpunar, 2017; Suddendorff, 2013). How much do sequences help the rodent plan their future paths? Are there differences between dorsal and ventral hippocampal sequences given that place field size can vary along the septotemporal axis (Jung, Wiener, & McNaughton, 1994; Kjelstrup et al., 2008; Royer, Sirota, Patel, & Buzsaki, 2010), potentially reflecting a gradient of contextual representation along the dorsal–ventral axis (Schmidt, Satvat, Argraves, Markus, & Marrone, 2012).

How are sequences generated?

Early models suggested that sequences were a passive product of theta phase precession or at least a product of the same mechanism that generates phase precession (Lisman & Redish, 2009; Maurer & McNaughton, 2007; O'Keefe & Recce, 1993; Skaggs, McNaughton, Wilson, & Barnes, 1996). However, sequences can still be seen within each theta cycle, even when the rat is paused. For example, Johnson and Redish (2007) found theta sequences occurring while the rat was paused during VTE; although there were sequences proceeding ahead of the rat within each theta cycle, the cells themselves did not phase precess. As mentioned above, Feng et al. (2015) found that theta sequences and phase precession can be dissociated, at least upon first exposure to an environment. Without experience of the maze, place cells did show phase precession, but the ensemble failed to show sequences; however, one traversal of the track was sufficient to organize the place cell assembly, so that sequences appeared on the second traversal.

Recently, it has been suggested that theta sequences/phase precession could be generated by the entorhinal cortex. The entorhinal cortex sends spatial and sensory information to the hippocampus. The medial entorhinal cortex has a plethora of spatially firing cells, including grid cells, border cells, and head direction cells (Hafting, Fyhn, Molden, Moser, & Moser, 2005; Sargolini et al., 2006; Solstad et al., 2008; Quirk, Muller, Kubie, & Ranck, 1992). Unlike place cells, which fire in a specific location in the environment, grid cells in the entorhinal cortex fire in a triangular grid that spans the length of the environment (Hafting et al., 2005; Sargolini et al., 2006). A computational model by Jaramillo, Schmidt, and Kempter (2014) suggests that phase precession is generated by grid cells and then driven onto downstream structures like the hippocampus. This model is supported by data showing that interfering with grid cells in the entorhinal cortex impairs phase precession and theta sequences in the hippocampus (Schlesiger et al., 2015).

Sanders, Renno-Costa, Idiart, and Lisman (2015) suggest that the phase precession, generated in the entorhinal cortex, is imposed upon downstream place cells to produce sequences that can travel linearly ahead of the animal. In this model, sequences only go forward, yet sweeps have been found to go around corners (Gupta et al., 2012; Johnson & Redish, 2007) even in enclosed mazes (Amemiya & Redish, 2016). One possibility is that sequences going around corners may depend on the cognitive map and the hippocampus itself.

How much do sequences improve/increase/predict planning?

Because theta sequences usually represent trajectories in front of the rat, they are believed to be necessary for planning future paths, instead of replaying the past. Redish et al. have found that theta sequences occurring during behavioral tasks where the rat is engaged in more deliberative/planning behaviors subsequently decrease when the behavior automates (Amemiya & Redish, 2016; Johnson & Redish, 2007; Papale et al., 2016; Regier, Amemiya, & Redish, 2015). On the multiple T-maze, for example, Johnson and Redish (2007) found that theta sequences initially go down both arms of the maze, but then, as the rat proceeds to know its target, the sequences go down only one direction. Furthermore, as the rat starts to automate its behavior, the length of the sequences decreases with experience. Redish et al. have suggested that this entails three stages: deliberation, planning, and automation (Redish, 2016; van der Meer, Johnson, Schmitzer-Torbert, & Redish, 2010).

Though the cumulative data suggest that SWR and theta sequences facilitate planning and spatial navigation, exactly how much do they improve or predict behavior? Pfeiffer and Foster (2013) suggest that there is an increase in SWR sequences toward the goal of the rat just before movement; however, while highly significant, this is a very small increase of only 3%. Nevertheless, studies do suggest that increased coordination between cells during SWR predicts improved performance.

Singer et al. (2013) have found increased place cell firing coordination during SWR on correct trials on a hippocampal-dependent spatial navigation task. As previously described, the hippocampal-dependent W maze (Kim & Frank, 2009) requires the rat to alternate between outbound trials (i.e., left arm—center arm—right arm—center arm). The proportion of cells that had coordinated activity during SWR was measured across learning. During early learning, when the behavioral performance was close to chance, coordinated activity during SWR failed to predict whether the next trial would be correct or incorrect (Fig. 6.5). However, when performance was greater than 65% place cell coordinated firing during SWR was greater preceding correct trials. Further analyses revealed that coordinated firing could predict correct or incorrect performance on a trial-by-trial basis, during early learning. Because this was a binary choice, the SWR activity increased the ability to predict the path of the animal by 10% (60% compared to

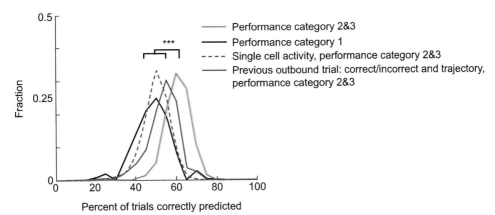

Figure 6.5 Pairwise spiking activity during SWRs accurately predict subsequent trial outcome. The proportion of coactive cell pairs was predictive of trial-by-trial performance for performance categories 2 and 3 (65%–85% and >85%; *green line*). In contrast, coactive cell pairs for performance category 1 (<65%; *black solid line*) were closer to chance. Predications based on single-cell activity were slightly better than chance (*gray dashed line*), as was the model based on prior outbound trial trajectory. *SWR*, sharp-wave ripple. *(Adapted with permission from Singer et al. (2013).)*

chance of 50%). Once at asymptotic performance firing coordination failed to predict correct or incorrect trials, thereby suggesting that coordinated firing during SWRs were no longer necessary once the task was well learned (i.e., the task was potentially automated).

In a similar unpublished analysis, Wikenheiser and Redish (2015) used linear discriminant analysis on decoded SWR representations to predict which feeder the rat would run to next on their three-step goal task described earlier. They decoded 200 ms windows centered on SWR events using a standard one-step Bayesian decoding operation with a uniform prior (Zhang et al., 1998) and then averaged the representation across space (thereby ignoring any temporal information in the representation). Thus, each SWR event produced an averaged decoded probability distribution over 100 spatial bins. For prediction analyses, only SWRs that occurred when the animal's speed was <5 cm/s were included. Each decoded distribution was categorized using linear discriminant analysis. A unique training set was constructed for each event by randomly drawing a subset of probability distributions, with equal numbers of one-, two-, and three-step cases. The distribution to be categorized was never included in the training set. Analysis was performed within each session, with statistics across sessions. To generate shuffled distributions, they followed the same classification procedure, as described above, except the identity of the training set that was randomized. They found an increase in prediction of the outbound target (where the rat was going to go on the next trial) of approximately 12% (45% relative to chance of 33%); shuffled data came out as chance (Fig. 6.6). Interestingly, it was also possible to predict the previous goal

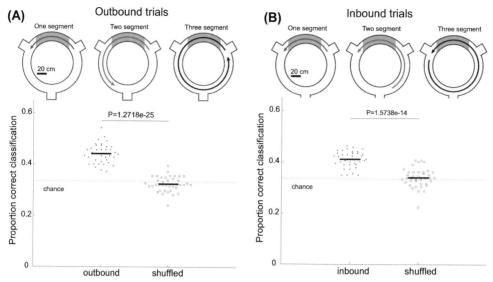

Figure 6.6 SWR sequences predict future paths. (A) We used linear discriminant analysis on decoded SWR representations to predict which feeder the rat would run to next (outbound prediction) and (B) which feeder the rat had arrived from (inbound prediction). We decoded 200 ms windows centered on ripple events (one-step, uniform prior) and then averaged the representation across space (thereby ignoring any temporal information in the representation). Thus, each ripple event produced an averaged decoded probability distribution over 100 spatial bins. For prediction analyses, only ripples that occurred when the animal's speed was <5 cm/s were included. Each decoded distribution was categorized using linear discriminant analysis. A unique training set was constructed for each event by randomly drawing a subset of probability distributions, with equal numbers of one-, two-, and three-step cases. The distribution to be categorized was never included in the training set. This analysis was performed within each session, so each dot on the plots represents one behavioral session. To generate shuffled distributions, we followed the same classification procedure, as described earlier, except the identity of the training set was randomized. All P values indicated on the plots come from paired t tests comparing the fraction of correctly predicted trials in each session to the fraction correctly achieved by the shuffles. *SWR*, sharp-wave ripple.

(where the rat had just come from) with similar proportions (41% relative to chance of 33%, for an increase of 8%).

Can theta sequences go backward?

Though SWR sequences have been observed to proceed in both forward (along the experienced path of the rat) and backward (against that experienced path) directions (Davidson et al., 2009; Foster & Wilson, 2007; Gupta et al., 2010), theta sequences seem to only go forward, consistent with a role in planning. Are theta sequences capable of going backward? As noted above, both Cei et al. (2014) and Maurer et al. (2014) found that theta sequences proceeded along the trajectory of the rat, even when that trajectory was opposite to the head direction of the rat. That is, both of these studies found that when the rat ran backward, the theta sequences proceeded along the experienced trajectory of the rat. Taken together these studies imply that theta sequences reflect the future path of the rat, not the direction the rat is facing.

CONCLUDING THOUGHTS

Deliberative decision-making encompasses understanding and exploring the environment, imagining and predicting possible outcomes, evaluating the outcomes, and then taking action. During the imagining and planning stage, humans engage in episodic future thinking, where they project themselves into the future situations (Buckner & Carroll, 2007). Sequences seen in rodents could facilitate a rodent analogue of episodic future thinking. Though research suggests that SWR sequences support planning (Pfeiffer & Foster, 2013), that disrupting them impairs future planning (Jadhav et al., 2012), and that SWR sequences provide information about the future goal such that it is possible to improve one's prediction of that goal from coactivation within SWRs (Singer et al., 2013; Wikenheiser & Redish, unpublished data in Fig. 6.6), what role SWRs play in goal-directed decision-making remains unclear. Similarly, although research suggests that theta sequences run along the trajectory of the rat (Cei et al., 2014; Foster & Wilson, 2007; Maurer et al., 2014) to the goal (Amemiya & Redish, 2016; Gupta et al., 2012; Papale et al., 2016; Wikenheiser & Redish, 2015), when the goal is clear, the specific role of theta sequences is unclear. During VTE (which is essentially an indecision between goals, Redish, 2013, 2016), theta sequences run to alternate goals, but so far it has not been possible to predict which goal an animal will take during those indecisive trials (Amemiya & Redish, 2016; Johnson & Redish, 2007; Papale et al., 2016; Redish, 2016).

Moreover, the distinction between episodic future thinking in humans and sequences in rodents should not go unnoticed. Though sequences appear to traverse a traditional series of events (Gupta et al., 2012), episodic future thinking in humans rarely progresses through the entire series of events to reach the end goal. Humans typically project directly to the end goal and after evaluating their respective outcomes, then consider the series of steps required to accomplish that outcome (Newell, Shaw, & Simon, 1959; Kurth-Nelson et al., 2012; Suddendorft, 2013).

REFERENCES

Abram, S. V. (2017). *Towards a translational model of decision-making: Findings from the web-surf task.* Thesis. Univ. Minnesota.

Addis, D. R., Wong, A. T., & Schacter, D. L. (2007). Remembering the past and imagining the future: Common and distinct neural substrates during event construction and elaboration. *Neuropsychologia, 45*(7), 1363−1377.

Ainge, J. A., van der Meer, M. A., Langston, R. F., & Wood, E. R. (2007). Exploring the role of context-dependent hippocampal activity in spatial alternation behavior. *Hippocampus, 17*(10), 988−1002.

Alvarez, P., & Squire, L. R. (1994). Memory consolidation and the medial temporal lobe: A simple network model. *Proceedings of the National Academy of Sciences of the United States of America, 91*, 7041−7045.

Amemiya, S., Noji, T., Kubota, N., Nishijima, T., & Kita, I. (2014). Noradrenergic modulation of vicarious trial-and-error behavior during a spatial decision-making task in rats. *Neuroscience, 265*, 291−301.

Amemiya, S., & Redish, A. D. (2016). Manipulating decisiveness in decision making: Effects of clonidine on hippocampal search strategies. *The Journal of Neuroscience, 36*(3), 814−827.

Barnes, C. A., Nadel, L., & Honig, W. K. (1980). Spatial memory deficit in senescent rats. *Canadian Journal of Psychology, 34*(1), 29−39.

Battaglia, F. P., Sutherland, G. R., & McNaughton, B. L. (2004). Local sensory cues and place cell directionality: Additional evidence of prospective coding in the hippocampus. *The Journal of Neuroscience, 24*(19), 4541−4550.

Benchenane, K., Peyrache, A., Khamassi, M., Tierney, P. L., Gioanni, Y., Battaglia, F. P., & Wiener, S. I. (2010). Coherent theta oscillations and reorganization of spike timing in the hippocampal-prefrontal network upon learning. *Neuron, 66*, 921−936.

Bett, D., Allison, E., Murdoch, L. H., Kaefer, K., Wood, E. R., & Dudchenko, P. A. (2012). The neural substrates of deliberative decision making: Contrasting effects of hippocampus lesions on performance and vicarious trial-and-error behavior in a spatial memory task and a visual discrimination task. *Frontiers in Behavioral Neuroscience, 6*(70). https://doi.org/10.3389/fnbeh.2012.00070.

Bissonette, G. B., & Roesch, M. R. (2017). Neurophysiology of rule switching in the corticostriatal circuit. *Neuroscience, 345*, 64−76.

Bittner, K. C., Grienberger, C., Vaidya, S. P., Milstein, A. D., Macklin, J. J., Suh, J., ... Magee, J. C. (2015). Conjunctive input processing drives feature selectivity in hippocampal CA1 neurons. *Nature Neuroscience, 18*(8), 1133−1142.

Blumenthal, A., Steiner, A., Seeland, K., & Redish, A. D. (2011). Effects of pharmacological manipulations of NMDA receptors on deliberation in the multiple T task. *Neurobiology of Learning and Memory, 95*, 376−384.

Bonnici, H. M., Chadwick, M. J., Lutti, A., Hassabis, D., Weiskopf, N., & Maguire, E. A. (2012). Detecting representations of recent and remote autobiographical memories in vmPFC and hippocampus. *The Journal of Neuroscience, 32*(47), 16982−16991.

Buckner, R. L., & Carroll, D. C. (2007). Self projection and the brain. *Trends in Cognitive Sciences, 11*(2), 49−57.

Buzsaki, G. (2002). Theta oscillations in the hippocampus. *Neuron, 33*(3), 325−340.

Buzsaki, G. (2015). Hippocampal sharp wave-ripple: A cognitive biomarker for episodic memory and planning. *Hippocampus, 25*(10), 1073−1188.

Buzsaki, G., Leung, L. W., & Vanderwolf, C. H. (1983). Cellular bases of hippocampal EEG in the behaving rat. *Brain Research, 287*(2), 139−171.

Carr, M. F., Jadhav, S. P., & Frank, L. M. (2011). Hippocampal replay in the awake state: A potential substrate for memory consolidation and retrieval. *Nature Neuroscience, 14*(2), 147−153.

Cei, A., Girardeau, G., Drieu, C., Kanbi, K. E., & Zugaro, M. (2014). Reversed theta sequences of hippocampal cell assemblies during backward travel. *Nature Neuroscience, 17*(5), 719−724.

Chang, Q., & Gold, P. E. (2003). Switching memory systems during learning: Changes in patterns of brain acetylcholine release in the hippocampus and striatum in rats. *The Journal of Neuroscience, 23*(7), 3001−3005.

Chang, Q., & Gold, P. E. (2004). Inactivation of dorsolateral striatum impairs acquisition of response learning in cue-deficient, but not cue-available, conditions. *Behavioral Neuroscience, 118*(2), 383−388.

Christoff, K., Irving, Z. C., Fox, K. C., Spreng, R. N., & Andrews-Hanna, J. R. (2016). Mind-wandering as spontaneous thought: A dynamic framework. *Nature Reviews Neuroscience, 17*(11), 718−731.

Cole, S. N., Morrison, C. M., Barak, O., Pauly-Takacs, K., & Conway, M. A. (2016). Amnesia and future thinking: Exploring the role of memory in the quantity and quality of episodic future thoughts. *British Journal of Clinical Psychology, 55*(2), 206−224.

Coull, J. T., Middleton, H. C., Robbins, T. W., & Sahakian, B. J. (1995). Contrasting effects of clonidine and diazepam on tests of working memory and planning. *Psychopharmacology (Berlin)*, *120*, 311–321.

Csicsvari, J., Hirase, H., Mamiya, A., & Buzsaki, G. (2000). Ensemble patterns of hippocampal CA3-CA1 neurons during sharp wave-associated population events. *Neuron*, *28*, 585–594.

Csicsvari, J., O'Neill, J., Allen, K., & Senior, T. (2007). Place-selective firing contributes to the reverse-order reactivation of CA1 pyramidal cells during sharp waves in open-field exploration. *The European Journal of Neuroscience*, *26*(3), 704–716.

Davidson, T. J., Kloosterman, F., & Wilson, M. A. (2009). Hippocampal replay of extended experience. *Neuron*, *63*(4), 497–507.

Daw, N. D., Gershman, S. J., Seymour, B., Dayan, P., & Dolan, R. J. (2011). Model-based influences on human's choices and striatal prediction errors. *Neuron*, *69*, 1204–1215.

Daw, N. D., Niv, Y., & Dayan, P. (2005). Uncertainty-based competition between prefrontal and dorsolateral striatal systems for behavioral control. *Nature Neuroscience*, *8*, 1704–1711.

Derdikman, D., & Moser, M. B. (2010). A dual role for hippocampal replay. *Neuron*, *65*(5), 582–584.

Diba, K., & Buzsaki, G. (2007). Forward and reverse hippocampal place-cell sequences during ripples. *Nature Neuroscience*, *19*(10), 1241–1242.

Doll, B. B., Duncan, K. D., Simon, D. A., Shohamy, D., & Daw, N. D. (2015). Model-based choices involve prospective neural activity. *Nature Neuroscience*, *18*(5), 767–772.

Dragoi, G., & Buzsaki, G. (2006). Temporal encoding of place sequences by hippocampal cell assemblies. *Neuron*, *50*(1), 145–157.

Ego-Stengel, V., & Wilson, M. A. (2010). Disruption of ripple-associated hippocampal activity during rest impairs spatial learning in the rat. *Hippocampus*, *20*(1), 1–10.

Feng, T., Silva, D., & Foster, D. J. (2015). Dissociation between the experience-dependent development of hippocampal theta sequences and single-trial phase precession. *The Journal of Neuroscience*, *35*(12), 4890–4902.

Foster, D. J., & Wilson, M. A. (2006). Reverse replay of behavioral sequences in hippocampal place cells during the awake state. *Nature*, *440*, 680–683.

Foster, D. J., & Wilson, M. A. (2007). Hippocampal theta sequences. *Hippocampus*, *17*(11), 1093–1099.

Gais, S., & Born, J. (2004). Declarative memory consolidation: Mechanisms acting during human sleep. *Learning & Memory*, *11*, 679–685.

Gallistel, C. R. (1990). Representations in animal cognition: An introduction. *Cognition*, *37*(1–2), 1–22.

Gardner, R. S., Uttaro, M. R., Fleming, S. E., Suarez, D. F., Ascoli, G. A., & Dumas, T. C. (2013). A secondary working memory challenge preserves primary place strategies despite overtraining. *Learning & Memory*, *20*(11), 648–656.

Girardeau, G., Benchenane, K., Wiener, S. I., Buzsaki, G., & Zugaro, M. B. (2009). Selective suppression of hippocampal ripples impair spatial memory. *Nature Neuroscience*, *12*(10), 1222–1223.

Gupta, A. S., van der Meer, M. A. A., Touretzky, D. S., & Redish, A. D. (2010). Hippocampal replay is not a simple function of experience. *Neuron*, *65*, 695–705.

Gupta, A. S., van der Meer, M. A. A., Touretzky, D. S., & Redish, A. D. (2012). Segmentation of spatial experience by hippocampal these sequences. *Nature Neuroscience*, *15*(7), 1032–1041.

Hafting, T., Fyhn, M., Molden, S., Moser, M. B., & Moser, E. I. (2005). Microstructure of a spatial map in the entorhinal cortex. *Nature*, *436*(7052), 801–806.

Haj, M. E., Antoine, P., & Kapogiannis, D. (2015). Flexibility decline contributes to similarity of past and future thinking in Alzheimer's disease. *Hippocampus*, *25*, 1447–1455.

Hassabis, D., Kumaran, D., & Maguire, E. A. (2007). Using imagination to understand the neural basis of episodic memory. *The Journal of Neuroscience*, *27*(52), 14365–14374.

Hassabis, D., Kumaran, D., Vann, S. D., & Maguire, E. A. (2007). Patients with hippocampal amnesia cannot imagine experiences. *Proceedings of the National Academy of Sciences of the United States of America*, *104*, 1726–1731.

Hasselmo, M. E. (1995). Neuromodulation and cortical function: Modeling the physiological basis of behavior. *Behavioral Brain Research*, *67*, 1–27.

Hasselmo, M. E. (1999). Neuromodulation: Acetylcholine and memory consolidation. *Trends in Cognitive Sciences*, *3*, 351–359.

Hasselmo, M. E. (2006). The role of acetylcholine in learning and memory. *Current Opinion in Neurobiology,* *16,* 710–715.

Haxby, J. V., Connolly, A. C., & Guntupalli, J. S. (2014). Decoding neural representational spaces using multivariate pattern analysis. *Annual Review of Neuroscience, 37,* 435–456.

Hirase, H., Czurko, A., Csicsvari, J., & Buzsaki, G. (1999). Firing rate and theta-phase coding by hippocampal pyramidal neurons during 'space clamping'. *The European Journal of Neuroscience, 11*(12), 4373–4380.

Hu, D., & Amsel, A. (1995). A simple test of the vicarious trial-and-error hypothesis of hippocampal function. *Proceedings of the National Academy of Sciences of the United States of America, 92*(12), 5506–5509.

Hull, C. L. (1943). *Principles of behavior.* Appleton-Century-Crofts.

Huxter, J. R., Senior, T. J., Allen, K., & Csicsvari, J. (2008). Theta phase-specific code for two-dimensional position, trajectory and heading in the hippocampus. *Nature Neuroscience, 11*(5), 587–594.

Hyman, J. M., Zilli, E. A., Paley, A. M., & Hasselmo, M. E. (2005). Medial prefrontal cortex cells show dynamic modulation with the hippocampal theta rhythm dependent on behavior. *Hippocampus, 15,* 739–749.

Irish, M., & Piolino, P. (2016). Impaired capacity for prospection in the dementias—Theoretical and clinical implications. *British Journal of Clinical Psychology, 55*(1), 49–68.

Jackson, J. C., Johnson, A., & Redish, A. D. (2006). Hippocampal sharp waves and reactivation during awake states depend on repeated sequential experience. *The Journal of Neuroscience, 26*(48), 12415–12426.

Jadhav, S. P., Kemere, C., German, P. W., & Frank, L. M. (2012). Awake hippocampal sharp-wave ripples support spatial memory. *Science, 336*(6087), 1454–1458.

Jakala, P., Riekkinen, M., Sirvio, J., Koivisto, E., Kejonen, K., Vanhanen, M., & Riekkinen, P., Jr. (1999). Guanfacine, but not clonidine, improves planning and working memory performance in humans. *Neuropsychopharmacology, 20,* 460–470.

Jaramillo, J., Schmidt, R., & Kempter, R. (2014). Modeling inheritance of phase precession in the hippocampal formation. *The Journal of Neuroscience, 34*(22), 7715–7731.

Jeannerod, M. (1994). The hand and the object: The role of posterior parietal cortex in forming motor representations. *Canadian Journal of Physiology and Pharmacology, 72*(5), 535–541.

Jensen, O., & Lisman, J. E. (1996). Hippocampal CA3 region predicts memory sequences: Accounting for phase precession of place cells. *Learning & Memory, 3*(2–3), 279–287.

Jensen, O., & Lisman, J. E. (2000). Position reconstruction from and ensemble of hippocampal place cells: Contribution of theta phase coding. *Journal of Neurophysiology, 83*(5), 2602–2609.

Johnson, A., Fenton, A. A., Kentros, C., & Redish, A. D. (2009). Looking for cognition in the structure within the noise. *Trends in Cognitive Sciences, 13,* 55–64.

Johnson, A., & Redish, A. D. (2007). Neural ensembles in CA3 transiently encode paths forward of the animal at a decision point. *The Journal of Neuroscience, 27,* 12176–12189.

Jones, M. W., & Wilson, M. A. (2005). Theta rhythms coordinate hippocampal-prefrontal interactions in a spatial memory task. *PLoS Biology, 3,* e402.

Jung, M. W., Wiener, S. I., & McNaughton, B. L. (1994). Comparison of spatial firing characteristics of units in dorsal and ventral hippocampus of the rat. *The Journal of Neuroscience, 14*(12), 7347–7356.

Kaplan, R., King, J., Koster, R., Penny, W. D., Burgess, N., & Friston, K. J. (2017). The neural representation of prospective choice during spatial planning and decisions. *PLoS Biology, 15*(1), e1002588. https://doi.org/10.1371/journal.pbio.1002588.

Kesner, R. P., Bolland, B. L., & Dakis, M. (1993). Memory for spatial locations, motor responses, and objects: Triple dissociation among the hippocampus, caudate nucleus, and extrastriate visual cortex. *Experimental Brain Research, 93*(3), 462–470.

Kim, S. M., & Frank, L. M. (2009). Hippocampal lesions impair rapid learning of a continuous spatial alternation task. *PLoS One, 4*(5), 5494. https://doi.org/10.1371/journal.pone.0005494.

Kjelstrup, K. B., Solstad, T., Brun, V. H., Hafting, T., Leutgeb, S., Witter, M. P., … Moser, M. B. (2008). Finite scale of representation in the hippocampus. *Science, 321*(5885), 140–143.

Kosslyn, S. M. (1994). *Image and the brain.* Cambridge, MA: MIT Press.

Kudrimoti, H. S., Barnes, C. A., & McNaughton, B. L. (1999). Reactivation of hippocampal cell assemblies: Effects of behavioral state, experience, and EEG dynamics. *The Journal of Neuroscience, 19*(10), 4090–4101.

Kumaran, D., Summerfield, J. J., Hassabis, D., & Maguire, E. A. (2009). Tracking the emergence of conceptual knowledge during human decision making. *Neuron, 63*(6), 889−901.

Kurth-Nelson, Z., Bickel, W. K., & Redish, A. D. (2012). A theoretical account of cognitive effects in delay discounting. *European Journal of Neuroscience, 35*, 1052−1064.

Kurzcek, J., Welcher, E., Abuja, S., Jensen, U., Cohen, N. J., Tranel, D., & Duff, M. C. (2015). Differential contributions of the hippocampus and medial prefrontal cortex to self-projection and self-referential processing. *Neuropsychologia, 73*, 116−126.

Lansink, C. S., Goltstein, P. M., Lankelma, J. V., Jossten, R. N., McNaughton, B. L., & Pennartz, C. M. A. (2008). Preferential reactivation of motivationally relevant information in the ventral striatum. *The Journal of Neuroscience, 28*, 6372−6382.

Lansink, C. S., Goltstein, P. M., Lankelma, J. V., McNaughton, B. L., & Pennartz, C. M. A. (2009). Hippocampus leads ventral striatum in replay of place-reward information. *PLoS Biology, 7*(8), e1000173. https://doi.org/10.1371/journal.pbio.1000173.

de Lavilleon, G., Lacroix, M. M., Rondi-Reig, L., & Benchenane, K. (2015). Explicit memory creation during sleep demonstrates a causal roles of pace cells in navigation. *Nature Neuroscience, 18*(4), 493−495.

Lee, A. K., & Wilson, M. A. (2002). Memory of sequential experience in the hippocampus during slow wave sleep. *Neuron, 36*(6), 1183−1194.

Lin, W.-J., Horner, A. J., Bisby, J. A., & Burgess, N. (2015). Medial prefrontal cortex: Adding value to imagined scenarios. *Journal of Cognitive Neuroscience, 27*, 1957−1967.

Lisman, J., & Redish, A. D. (2009). Prediction, sequences and the hippocampus. *Philosophical Transactions of the Royal Society of London, Biological Sciences, 364*(1521), 1193−1201.

Litchtman, A. H., Diemen, K. R., & Martin, B. R. (1995). Systemic or intrahippocampal cannabinoid administration impairs spatial memory in rats. *Psychopharmacology (Berlin), 119*, 282−290.

Litchman, A. H., & Martin, B. R. (1996). Delta 9-tetrahydrocannabinol impairs spatial memory through a cannabinoid receptor mechanism. *Psychopharmacology (Berlin), 126*, 125−131.

Loftus, E. F., & Palmer, F. C. (1974). Reconstruction of automobile destruction: An example of the interactions between language and memory. *Journal of Verbal Learning and Verbal Behavior, 13*, 585−589.

MacCorquodale, K., & Meehl, P. E. (1948). On a distinction between hypothetical constructs and intervening variables. *Psychological Review, 55*, 95−107.

Marr, D. (1971). Simple memory: A theory for archicortex. *Philosophical Transactions of the Royal Society of London. Series B, Biological Sciences, 262*, 23−81.

Maurer, A. P., Lester, A. W., Burke, S. N., Ferng, J. J., & Barnes, C. A. (2014). Back to the future: Preserved hippocampal network activity during reverse ambulation. *The Journal of Neuroscience, 34*(45), 15022−15031.

Maurer, A. P., & McNaughton, B. L. (2007). Network and intrinsic cellular mechanisms underlying theta phase precession of hippocampal neurons. *Trends in Neurosciences, 30*(7), 325−333.

Mehta, M. R., Barnes, C. A., & McNaughton, B. L. (1997). Experience-dependent, asymmetric expansion of hippocampal place fields. *Proceedings of the National Academy of Sciences of the United States of America, 94*(16), 8918−8921.

van der Meer, M. A. A., Johnson, A., Schmitzer-Torbert, N. C., & Redish, A. D. (2010). Triple dissociation of information processing in dorsal striatum, ventral striatum, and hippocampus on a learned spatial decision task. *Neuron, 67*(1), 25−32.

van der Meer, M. A. A., Kurth-Nelson, Z., & Redish, A. D. (2012). Information processing in decision-making systems. *The Neuroscientist, 18*(4), 342−359.

Monaco, J. D., Rao, G., Roth, E. D., & Knierim, J. J. (2014). Attentive scanning behavior drives one-trial potentiation of hippocampal place fields. *Nature Neuroscience, 17*(5), 725−731.

Muenzinger, K. F. (1938). Vicarious trial and error at a point of choice. I. A general survey of its relation to learning efficiency. *The Journal of Genetic Psychology, 53*, 75−86.

Muenzinger, K. F., & Gentry, E. (1931). Tone discrimination in white rats. *Journal of Comparative Psychology, 12*, 195−206.

Muller, R. (1996). A quarter of a century of place cells. *Neuron, 17*(5), 813−822.

Newell, A., Shaw, J. C., & Simon, H. A. (1959). Report on a general problem-solving program. In *Proceedings of the international conference on information processing* (pp. 256−264).

Niv, Y., Joel, D., & Dayan, P. (2006). A normative perspective on motivation. *Trends in Cognitive Sciences, 10*(8), 375−381.

O'Craven, K. M., & Kanwisher, N. (2000). Mental imagery of faces and places activates corresponding stimulus-specific brain regions. *Journal of Cognitive Neuroscience, 12*(6), 1013−1023.

O'Keefe, J., & Dostrovsky, J. (1971). The hippocampus as a spatial map. Preliminary evidence from unit activity in the freely-moving rat. *Brain Research, 34*, 171−175.

O'Keefe, J., & Nadel, L. (1978). *The hippocampus as a cognitive map.* Oxford, New York: Clarendon Press; Oxford University Press.

O'Keefe, J., & Recce, M. L. (1993). Phase relationship between hippocampal place units and the EEG theta rhythm. *Hippocampus, 3*(3), 317−330.

O'Neill, J., Senior, T., & Csicsvari, J. (2006). Place-selective firing of CA1 pyramidal cells during sharp wave/ripple network patterns in exploratory behavior. *Neuron, 49*(1), 143−155.

Packard, M. G. (1999). Glutamate infused posttraining into the hippocampus or caudate-putamen differentially strengthens place and response learning. *Proceedings of the National Academy of Sciences of the United States of America, 96*(22), 12881−12886.

Packard, M. G., & McGaugh, J. L. (1992). Double dissociation of fornix and caudate nucleus lesions on acquisition of two-water maze tasks: Further evidence for multiple memory systems. *Behavioral Neuroscience, 106*(3), 439−446.

Packard, M. G., & McGaugh, J. L. (1996). Inactivation of hippocampus or caudate nucleus with lidocaine differentially affects expression of place and response learning. *Neurobiology of Learning and Memory, 65*(1), 65−72.

Papale, A. E. (2015). *Hippocampal representations on the spatial delay discounting task.* Thesis. Univ. Minnesota.

Papale, A. E., Stott, J. J., Powell, N. J., Regier, P. S., & Redish, A. D. (2012). Interactions between deliberation and delay-discounting in rats. *Cognitive, Affective & Behavioral Neuroscience, 12*(3), 513−526.

Papale, A. E., Zielinski, M. C., Frank, L. M., Jadhav, S. P., & Redish, A. D. (2016). Interplay between hippocampal sharp-wave-ripple events and vicarious trial and error behaviors in decision making. *Neuron, 92*(5), 975−982.

Pastalkova, E., Itskov, V., Amarasingham, A., & Buzsaki, G. (2008). Internally generated cell assembly sequences in the rat hippocampus. *Science, 321*(5894), 1322−1327.

Pavlides, C., & Winson, J. (1989). Influences of hippocampal place cell firing in the awake sate on the activity of these cells during subsequent sleep episodes. *The Journal of Neuroscience, 9*, 2907−2918.

Pearson, J., Naselaris, T., Holmes, E. A., & Kosslyn, S. M. (2015). Mental imagery: Functional mechanisms and clinical applications. *Trends in Cognitive Sciences, 19*(10), 590−602.

Pennartz, C. M., Lee, E., Verheul, J., Lipa, P., Barnes, C. A., & McNaughton, B. L. (2004). The ventral striatum in off-line processing: Ensemble reactivation during sleep and modulation by hippocampal ripples. *The Journal of Neuroscience, 24*, 6446−6456.

Peters, J., & Buchel, C. (2010). Episodic future thinking reduces reward delay discounting through and enhancement of prefrontal-mediotemporal interactions. *Neuron, 66*, 138−148.

Pfeiffer, B. E., & Foster, D. J. (2013). Hippocampal place-cell sequences depict future paths to remembered goals. *Nature, 497*(7447), 74−79.

Powell, N. J., & Redish, A. D. (2016). Representational changes of latent strategies in rat medial prefrontal cortex precede changes in behavior. *Nature Communications, 7*, 12830. https://doi.org/10.1038/ncomms12830.

Quirk, G. J., Muller, R. U., Kubie, J. L., & Ranck, J. B., Jr. (1992). The positional firing properties of medial entorhinal neurons: Description and comparison with hippocampal place cells. *The Journal of Neuroscience, 12*(5), 1945−1963.

Race, E., Keane, M. M., & Verfaellie, M. (2011). Medial temporal lobe damage causes deficits in episodic memory and episodic future thinking not attributable to deficits in narrative construction. *The Journal of Neuroscience, 31*(28), 10262−10269.

Ragozzino, M. E., Detrick, S., & Kesner, R. P. (1999). Involvement of the prelimbic-infralimbic areas of the rodent prefrontal cortex in behavioral flexibility for place and response learning. *The Journal of Neuroscience, 19*(11), 4585−4594.

Raichle, M. E., MacLeod, A. M., Snyder, A. Z., Powers, W. J., Gusnard, D. A., & Shulman, G. L. (2001). A default mode of brain function. *Proceedings of the National Academy of Sciences of the United States of America, 98*(2), 676–682.

Ramussen, K. W., & Bersten, D. (2016). Deficits in remembering the past and imagining the future in patients with prefrontal lesions. *Journal of Neuropsychology.* https://doi.org/10.1111/jnp.12108.

Redish, A. D. (1999). *Beyond the cognitive map: From place cells to episodic memory.* MIT Press.

Redish, A. D. (2013). *The mind within the brain: How we make decisions and how those decision go wrong.* Oxford University Press.

Redish, A. D. (2016). Vicarious trial and error. *Nature Reviews Neuroscience, 17*(3), 147–153.

Redish, A. D., & Touretzky, D. S. (1997). Cognitive maps beyond the hippocampus. *Hippocampus, 7*(1), 15–35.

Regier, P. S., Amemiya, S., & Redish, A. D. (2015). Hippocampus and subregions of the dorsal striatum respond differently to a behavioral strategy change on a spatial navigation task. *Journal of Neurophysiology, 114*(3), 1399–1416.

Rich, E. L., & Shapiro, M. (2007). Prelimbic/infralimbic inactivation impairs memory for multiple task switches, but not flexible selection of familiar tasks. *The Journal of Neuroscience, 27*(17), 4747–4755.

Rich, E. L., & Shapiro, M. (2009). Rat prefrontal cortical neurons selectively code strategy switches. *The Journal of Neuroscience, 29*(22), 7208–7219.

Rizzolatti, G., & Craigero, L. (2004). The mirror-neuron system. *Annual Review of Neuroscience, 27*, 169–192.

Robbe, D., & Buzsaki, G. (2009). Alternation of theta timescale dynamics of hippocampal place cells by a cannabinoid is associated with memory impairment. *The Journal of Neuroscience, 29*(4), 12597–12605.

Robbe, D., Montgomery, S. M., Thome, A., Rudea-Orozco, P. E., McNaughton, B. L., & Buzsaki, G. (2007). Cannabinoids reveal importance of spike timing coordination in hippocampal function. *Nature Neuroscience, 9*(12), 1526–1533.

Royer, S., Sirota, A., Patel, J., & Buzsaki, G. (2010). Distinct representations and theta dynamics in dorsal and ventral hippocampus. *The Journal of Neuroscience, 30*(5), 1777–1787.

Samsonovich, A. V., & Ascoli, G. A. (2005). A simple neural network model of the hippocampus suggesting it's pathfinding role in episodic memory retrieval. *Learning & Memory, 12*(2), 193–208.

Sanders, H., Renno-Costa, C., Idiart, M., & Lisman, J. (2015). Grid cells and place cells: An integrated view of their navigational and memory function. *Trends in Neurosciences, 38*(12), 763–775.

Sargolini, F., Fyhn, M., Hafting, T., McNaughton, B. L., Witter, M. P., Moser, M. B., & Moser, E. I. (2006). Conjunctive representation of position, direction, and velocity in entorhinal cortex. *Science, 312*(5774), 758–762.

Schacter, D. L., Addis, D. R., & Buckner, R. L. (2007). Remembering the past to imagine the future: The prospective brain. *Nature Reviews Neuroscience, 8*, 657–661.

Schacter, D. L., Addis, D. R., Hassabis, D., Martin, V. C., Spreng, R. N., & Szpunar, K. K. (2012). The future of memory: Remembering, imagining, and the brain. *Neuron, 76*(4), 677–694.

Schacter, D. L., Benoit, R. G., & Szpunar, K. K. (2017). Episodic future thinking: Mechanisms and functions. *Current Opinion in Behavioral Sciences, 17*, 41–50.

Schacter, D. L., Gaesser, B., & Addis, D. R. (2013). Remembering the past and imagining the future in the elderly. *Gerontology, 59*(2), 143–151.

Schlesiger, M. I., Cannova, C. C., Boubil, B. L., Hales, J. B., Mankin, E. A., Brandon, M. P., … Leutgeb, S. (2015). The medial entorhinal cortex is necessary for temporal organization of hippocampal neuronal activity. *Nature Neuroscience, 18*(8), 1123–1132.

Schmidt, B., Papale, A., Redish, A. D., & Markus, E. J. (2013). Conflict between place and response navigation: Effects on vicarious trial and error (VTE) behaviors. *Learning & Memory, 20*(3), 130–138.

Schmidt, B., & Redish, A. D. (November 2016). *DREADDs disruption of prelimbic cortex alters hippocampal SWR dynamics during rest after a foraging task in rats.* San Diego, CA: Society for Neuroscience.

Schmidt, B., & Redish, A. D. (2013). Neuroscience: Navigation with a cognitive map. *Nature, 497*(7447), 42–43.

Schmidt, B., Satvat, E., Argraves, M., Markus, E. J., & Marrone, D. F. (2012). Cognitive demands induce selective hippocampal reorganization: Arc expression in a place and response task. *Hippocampus*, *22*(11), 2114−2126.

Siapas, A. G., & Wilson, M. A. (1998). Coordinated interactions between hippocampal ripples and cortical spindles during slow-wave sleep. *Neuron*, *21*, 1123−1128.

Silva, D., Feng, T., & Foster, D. J. (2015). Trajectory events across hippocampal place cells require previous experience. *Nature Neuroscience*, *18*(12), 1772−1779.

Singer, A. C., Carr, M. F., Karlsson, M. P., & Frank, L. M. (2013). Hippocampal SWR activity predicts correct decisions during the initial learning of alternation task. *Neuron*, *77*(6), 1163−1173.

Skaggs, W. E., & McNaughton, B. L. (1996). Replay of neuronal firing sequences in rat hippocampus during sleep following spatial experience. *Science*, *271*(5257), 1870−1873.

Skaggs, W. E., McNaughton, B. L., Wilson, M. A., & Barnes, C. A. (1996). Theta phase precession in hippocampal neuronal populations and the compression of temporal sequences. *Hippocampus*, *6*(2), 149−172.

Skinner, B. F. (1948). Superstition in the pigeon. *Journal of Experimental Psychology*, *38*, 168−172.

Solstad, T., Boccara, C. N., Kropff, E., Moser, M. B., & Moser, E. I. (2008). Rpresentation of geometric borders in the entorhinal cortex. *Science*, *322*(5909), 1865−1868.

Spiers, H. J., & Maguire, E. A. (2007). A navigational guidance system in the human brain. *Hippocampus*, *17*(8), 618−626.

Steiner, A. P., & Redish, A. D. (2012). The road not taken: Neural correlates of decision making in orbitofrontal cortex. *Frontiers in Behavioral Neuroscience*, *6*(131). https://doi.org/10.3389/fnins.2012.00131.

Suddendorf, T. (2013). Mental time travel: Continuities and discontinuities. *Trends in Cognitive Sciences*, *17*(4), 151−152.

Sutherland, G. R., & McNaughton, B. (2000). Memory trace reactivation in hippocampal and neocortical neuronal ensembles. *Current Opinion in Neurobiology*, *10*, 180−186.

Sutton, R. S., & Barto, A. G. (1998). *Reinforcement learning: An introduction*. Cambridge, MA: MIT Press.

Talarico, J. M., & Rubin, D. C. (2003). Confidence, not consistency, characterizes flashbulb memories. *Psychological Science*, *14*(5), 455−461.

Tolman, E. C. (1938). The determiners of behavior at a choice point. *Psychological Review*, *45*(1), 1−41.

Tolman, E. C. (1939). Prediction of vicarious trial and error by means of the schematic sowbug. *Psychological Review*, *39*, 318−336.

Tolman, E. C. (1948). Cognitive maps in rats and men. *Psychological Review*, *55*, 189−208.

Tolman, E. C., & Honzik, C. H. (1930). Introduction and removal of reward, and maze performance in rats. *University of California Publications in Psychology*, *4*, 257−275.

Tsodyks, M. V., Skaggs, W. E., Sejnowski, T. J., & McNaughton, B. L. (1996). Population dynamics and theta rhythm phase precession of hippocampal place cell firing: A spiking neuron model. *Hippocampus*, *6*(3), 271−280.

Tulving, E. (1985). Memory and consciousness. *Canadian Psychology*, *26*, 1−12.

Vandecastelle, M., Varga, V., Berenyi, A., Papp, E., Bartho, P., Venance, L., ... Buzsaki, G. (2014). Optogenetic activation of septal cholinergic neurons suppresses sharp wave ripples and enhances theta oscillations in the hippocampus. *Proceedings of the National Academy of Sciences of the United States of America*, *111*, 13535−13540.

Vanderwolf, C. H. (1969). Hippocampal electrical activity and voluntary movement in the rat. *Electroencephalography and Clinical Neurophysiology*, *26*, 207−418.

Wang, J. X., Cohen, N. J., & Voss, J. L. (2015). Covert rapid-action memory simulation (CREAMS): A hypothesis of hippocampal-prefrontal interactions for adaptive behavior. *Neurobiology of Learning and Memory*, *117*, 22−33.

Wang, Y., Roth, Z., & Pastalkova, E. (2016). Synchronized excitability in a network enables generation of internal neuronal sequences. *eLife*, *5*, e20697. https://doi.org/10.7554/eLife.20697.

Watson, J. (1913). Psychology as the behaviorist views it. *Psychological Review*, *20*, 158−177.

Wikenheiser, A. M., & Redish, A. D. (2013). The balance of forward and backward hippocampal sequences shifts across behavioral states. *Hippocampus*, *23*(1), 22−29.

Wikenheiser, A. M., & Redish, A. D. (2015). Hippocampal theta sequences reflect current goals. *Nature Neuroscience, 18*(2), 289—294.

Wilson, M., & McNaughton, B. (1993). Dynamics of the hippocampal ensemble code for space. *Science, 261,* 1055—1058.

Wilson, M. A., & McNaughton, B. L. (1994). Reactivation of hippocampal ensemble memories during sleep. *Science, 265,* 676—679.

Worden, R. (1992). Navigation by fragment fitting: A theory of hippocampal function. *Hippocampus, 2*(2), 165—187.

Yin, H. H., Knowlton, B., & Balleine, B. W. (2004). Lesions of dorsolateral striatum preserve outcome expectancy but disrupt habit formation in instrumental learning. *The European Journal of Neuroscience, 19,* 181—189.

Zeman, A., Butler, C., Muhlert, N., & Milton, F. (2013). Novel forms of forgetting in temporal lobe epilepsy. *Epilepsy & Behavior, 26*(3), 335—342.

Zhang, K., Ginzburg, I., McNaughton, B. L., & Sejnowski, T. J. (1998). Interpreting neuronal population activity by reconstruction: Unified framework with application to hippocampal place cells. *Journal of Neurophysiology, 79,* 1017—1044.

Zheng, C., Bieri, K. W., Hsiao, Y.-T., & Colgin, L. L. (2016). Spatial sequence coding differs during slow and fast gamma rhythms in the hippocampus. *Neuron, 89*(2), 398—408.

CHAPTER 7

Competition and Cooperation Between Multiple Reinforcement Learning Systems

Wouter Kool[1], Fiery A. Cushman[1], Samuel J. Gershman[1,2]
[1]Department of Psychology, Harvard University, Cambridge, MA, United States; [2]Center for Brain Science, Harvard University, Cambridge, MA, United States

INTRODUCTION

As you leave work each day, how do you choose a route home? Prominent dual-system accounts posit two distinct cognitive systems that solve this task in different ways (Balleine & O'Doherty, 2009; Dickinson, 1985; Fudenberg & Levine, 2006; Kahneman, 1973; Sloman, 1996). On the one hand, you could decide your route home by relying on *habit*. Since you have successfully taken one particular route to your house many times, this route has been ingrained into your motor system and can be executed quickly and automatically. Habits are useful because they make often-repeated behavior efficient and automatized; however, they are also inflexible and therefore more likely to produce errors. For example, consider the case where your significant other asked you to buy some toilet paper on your way back home. In this case, it would be better to suppress the habitual route and engage in *goal-directed* control. This involves the recall of the alternate goal (picking up toilet paper), and planning a new route that goes past the convenience store, using an internal model ("cognitive map") of the environment. Goal-directed planning is useful because it is more flexible and consequently more accurate than relying on habit. However, it also carries significant computational costs (Gershman & Daw, 2012).

These two systems are typically theorized as *competitors*, vying for control of behavior. A major goal of modern decision research is understanding how control is allocated between the two systems. We will attempt to summarize and extend this line of research.

Yet, the two systems may also interact *cooperatively*. For example, you might learn a habit to check traffic reports before you leave work because this facilitates planning an optimal route. Moreover, the act of "checking" could involve elements of goal-directed planning—for instance, searching for radio stations—even if initiated out of habit. These illustrate just two forms of cooperation: habitual actions can support effective goal pursuit and even drive the selection of goals themselves.

Until recently, the computational principles underlying the competition and cooperation between habitual and goal-directed systems were poorly understood. Armed with a new set of sequential decision tasks, researchers are now able to track habitual

Goal-Directed Decision Making
ISBN 978-0-12-812098-9, https://doi.org/10.1016/B978-0-12-812098-9.00007-3

and goal-directed influences on behavior across an experimental session (Daw, Gershman, Seymour, Dayan, & Dolan, 2011; Doll, Duncan, Simon, Shohamy, & Daw, 2015; Keramati, Smittenaar, Dolan, & Dayan, 2016; Kool, Cushman, & Gershman, 2016). This work has spurred new computational approaches to multisystem reinforcement learning (RL) and control architectures.

In this chapter, we review recent work on both competition and cooperation. First, we will provide a short, nontechnical exposition of the computational framework underlying this research (see Gershman, 2017 for a technical review). Next, we will discuss recent work that suggests how competition between habit and planning can be understood as a cost–benefit trade-off. Finally, we describe several studies that detail how the complementary strengths of habitual and goal-directed systems can be combined cooperatively to achieve both efficiency and accuracy.

MODEL-FREE AND MODEL-BASED CONTROL IN REINFORCEMENT LEARNING

The core problem in RL is estimating the *value* (expected discounted return) of state–action pairs in order to guide action selection. Broadly speaking, there are two strategies for solving this problem: a model-free strategy that estimates values incrementally from experience and a model-based strategy that learns a world model (reward and transition functions), which can then be used to plan an optimal policy. A central tenet of modern RL theory posits that the model-free strategy is implemented by the habitual system and the model-based strategy is implemented by the goal-directed system (Daw, Niv, & Dayan, 2005; Dolan & Dayan, 2013).

Roughly speaking, the model-free strategy is a form of Thorndike's law of effect, which states that actions that led to a reward become more likely to be repeated (Thorndike, 1911). This strategy is referred to as "model-free" because it does not rely on an internal model of the environment. Instead, values are stored in a cached format (a look-up table or function approximator), which allows them to be quickly retrieved. These values can be updated incrementally using simple error-driven learning rules like the temporal difference learning algorithm (Sutton & Barto, 1998). The main downside of the model-free strategy is its inflexibility: when a change in the environment or task occurs, the entire set of cached values needs to be relearned through experience. This inflexibility, ingrained by repetition, is what makes the model-free strategy habitual. In summary, the model-free strategy achieves efficiency of learning and control at the expense of flexibility in the face of change.

The model-based strategy, by contrast, represents its knowledge in the form of an internal model that can be modified locally when changes occur (e.g., if a particular route is blocked, only that part of the model is modified). These local changes can then induce global effects on the value function, which is computed on the fly using planning or dynamic programming algorithms. Thus, the model-based strategy, unlike the model-free strategy, need not cache values. As a consequence, the model-based strategy can flexibly modify its policy in pursuit of a goal without relearning the entire model. This flexibility is only available at a computational cost; however, since model-based algorithms are inevitably more time- and resource-intensive than querying a look-up table of cached values or function approximator (Daw et al., 2005; Keramati, Dezfouli, & Piray, 2011).

PRINCIPLES OF COMPETITION

Distinguishing habit from planning in humans

A long line of research in psychology and neuroscience has sought empirical evidence for the distinction between these model-free and model-based RL systems. Early studies tended to focus on animal models, and this literature has been reviewed extensively elsewhere (Dolan & Dayan, 2013; Gershman, 2017), so we will not cover it here. Instead, we focus on more recent studies with human subjects. We will describe how one particular experimental paradigm, a sequential decision task, which we will refer to as the "Daw two-step task" (Daw et al., 2011), has been pivotal in revealing the competition between model-free and model-based control in humans. We then turn to our main topic of interest in this section: Given that the systems can compete for control, how is this competition arbitrated?

Many recent studies of model-free and model-based control in humans have used the Daw two-step (Daw et al., 2011), summarized in Fig. 7.1 (following Decker, Otto, Daw, & Hartley, 2016, the depicted version of this task features a space travel

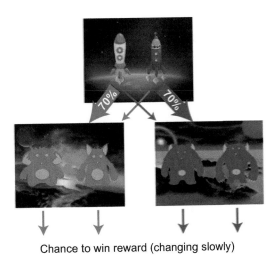

Chance to win reward (changing slowly)

Figure 7.1 Design and state transition structure of Daw two-step task (Daw et al., 2011; Decker et al., 2016). Each first-stage choice has a high probability (70%) of transitioning to one of two second-stage states and a low probability of transitioning to the other. Each second-stage choice is associated with a probability of obtaining a binary reward (between 0.25 and 0.75) that slowly changes across the duration of the experiment according to a Gaussian random walk with $\sigma = 0.025$.

cover story to make it more engaging for participants). The key appeal of this task is that it can be used to quantitatively distinguish the influence of model-free and model-based control on choices (see Akam, Costa, & Dayan, 2015). Each trial of the Daw two-step task starts with a choice between two stimuli (spaceships), which lead probabilistically to one of two second-stage states (planets). At these second-stage states, the participant then makes a second choice between two stimuli (aliens) that both offer a chance of obtaining a monetary reward (space treasure). The reward probabilities for these second-stage stimuli change slowly and independently throughout the task in order to encourage continuous learning. The most important feature of the Daw two-step task is its transition structure from the first-stage stimuli to the second-stage states. Specifically, each first-stage option leads to one of the second-stage states with a high probability (a "common" transition), whereas on a minority of the trials they lead to the other state (a "rare" transition).

Through these low-probability transitions between actions and rewards, the Daw two-step task can behaviorally distinguish between model-free and model-based choice. Because the model-free strategy does not have access to the task structure, it will increase the probability of taking the previous action if it led to reward, regardless of whether this was obtained through a common or a rare transition. Therefore, choice dictated by a purely model-free agent looks like a main effect of reward, with increased probability of repeating the previous action after a reward and with no effect of the previous transition (Fig. 7.2A). The model-based strategy, on the other hand, computes the first-stage action values through planning, using the transition structure to compute the

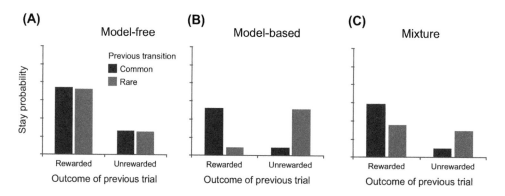

Figure 7.2 Probability of repeating the first-stage choice for three agents. (A) For model-free agents, the probability of repeating the previous choice is dependent only on whether a reward was obtained and not on transition structure. (B) Model-based behavior is reflected in an interaction between previous transition and outcome, increasing the probability of transitioning to the state where the reward was obtained. (C) Behavioral performance on this task reflects features of both model-based and model-free decision-making, the main effect of previous reward and its interaction with the previous transition. *(Reprinted from Kool et al. (2016).)*

expected value at the second stage for either action. Therefore, this system will reduce the likelihood of repeating the first-stage action after a reward obtained through a rare transition, since the other first-stage action has a higher likelihood to lead to the previously rewarded second-stage state. This behavior is reflected as a crossover interaction between the previous transition type and previous reward on the probability of staying: after rare transitions, wins predict a switch and losses predict a stay (Fig. 7.2B).

Interestingly, behavioral performance on the Daw two-step task reflects a mixture of these strategies (Fig. 7.2C). The probability of repeating the previous actions shows both the model-free main effect of previous reward and the model-based crossover interaction between previous transition type and previous reward. The relative influence of the model-based and model-free systems on this task can be estimated by fitting a reinforcement learning model to participants' behavior. Here, both strategies compute first-stage action values, which are then combined according to a weight parameter that determines the relative balance between model-free and model-based control.

The relative balance between model-based and model-free control indexed by this task has been linked to a broad range of other cognitive, neural, and clinical phenomena. For example, Decker et al. (2016) showed that children show virtually no signs of model-based control and that our ability for model-based planning develops through adolescence into adulthood (see Chapter 13 by Hartley). Gillan, Kosinski, Whelan, Phelps, and Daw (2016) have reported that the degree of model-based control in this task positively predicts psychiatric symptoms related to compulsive behavior (see Chapter 15 by de Wit and Chapter 17 by Morris), and others have shown that it also negatively predicts personality traits such as alcohol dependence (Sebold et al., 2014; see Chapter 16 by Corbit) and extraversion (Skatova, Chan, & Daw, 2015).

In addition to these findings that bolster the applicability of the two-step task to the broader field of psychology, it can also account for important phenomena in the RL literature, such as the finding that overtraining of an action—reward association induces insensitivity to subsequent outcome devaluation (a hallmark feature of habitual control; Gillan, Otto, Phelps, & Daw, 2015).

Arbitration between habit and planning as a cost—benefit trade-off

The finding that people show a balance between model-based and model-free control on the Daw two-step task raises the question of whether and how people decide, from moment to moment, which strategy to use. Although there are several theoretical proposals on this topic (Boureau, Sokol-Hessner, & Daw, 2015; Gershman, Horvitz, & Tenenbaum, 2015; Griffiths, Lieder, & Goodman, 2015; Keramati et al., 2011; Pezzulo, Rigoli, & Chersi, 2013), it has received surprisingly little empirical focus (but see Daw et al., 2005; Lee, Shimojo, & O'Doherty, 2014).

Several experimental manipulations have been discovered to alter the balance between model-free and model-based control, and these provide key clues about the form and function of arbitration between RL systems. As we review, many of these implicate some form of executive function or working memory in model-based control.

In one such case (Otto, Gershman, Markman, & Daw, 2013), participants performed the Daw two-step task while they were sometimes required to perform a numerical Stroop task that taxed their working memory and therefore reduced the amount of available cognitive resources. At the start of those trials, participants kept two numbers of different value and physical size in working memory. After the reward outcome of the two-step trial was presented, participants were then prompted to indicate on what side of the screen the number with larger size or value had appeared. Interestingly, on trials with this "load" condition, subjects showed a strong reliance on the model-free strategy and virtually no influence of a model-based strategy (Otto, Gershman, et al., 2013). This study suggests that the exertion of model-based control relies, at least in part, on executive functioning or cognitive control. This set of cognitive processes, which are dependent on computations in the frontal cortex, allow us to reconfigure information processing in order to execute novel and effortful tasks (Miller & Cohen, 2001).

Another clue for the involvement of executive functioning in model-based planning comes from a study by Smittenaar, FitzGerald, Romei, Wright, and Dolan (2013). In this experiment, participants performed the Daw two-step task while activity in their right dorsolateral prefrontal cortex (dlPFC), a region that is critical for the functioning of cognitive control, was sometimes disrupted using transcranial magnetic stimulation. Interestingly, performance on the task showed increased reliance on habitual control during those trials, indicating a crucial role for the dlPFC and executive functioning in model-based planning (see also Gläscher, Daw, Dayan, & O'Doherty, 2010; Lee et al., 2014).

Several other reports have yielded consistent evidence, in the form of robust correlations between individual differences in the degree of model-based control used in the Daw two-step task and measures of cognitive control ability. For example, Otto, Skatova, Madlon-Kay, and Daw (2015) showed that people with reduced performance in a response conflict task (such as the Stroop task; Stroop, 1935) also showed reduced employment of model-based control. In another study, participants with increased working memory capacity showed a reduced shift toward model-free control under stress (Otto, Raio, Chiang, Phelps, & Daw, 2013). In addition, Schad et al. (2014) showed that measures of general intelligence predicted reliance on model-based control. Their participants completed both the Daw two-step task and also the trail-making task (Army Individual Test Battery, 1944), in which participants use a pencil to connect numbers and letters, randomly distributed on a sheet of paper, in ascending order, while also alternating between numbers and letters (i.e., 1-A-2-B-3-C, etc.). Interestingly, individuals with increased processing speed on this task, indicating increased ability for cognitive control in the form of task switching, also showed a greater reliance on model-based control in the Daw two-step task.

We now address the question of whether, and how, the brain arbitrates between model-based and model-free control. One potential metacontrol strategy would simply be to always use the more accurate model-based system when the necessary cognitive resources are available and only use the habitual system when they are occupied or otherwise inoperative. Note that, although this would lead to increased average accuracy, such a model does not describe how its resources should be allocated when they could be devoted to multiple tasks. In other words, this model does not predict how people allocate control resources when the available tasks together demand more resources than available.

When aiming to describe such a trade-off, it would be sensible for a model to be sensitive to the elevated computational costs that are associated with model-based control, since those cognitive resources could be applied to other rewarding tasks. Consequently, we propose that allocation of control is based on the costs and benefits associated with each system in a given task. In this case, model-based control would be deployed when it generates enough of a reward advantage over model-free control to offset its costs.

Consistent with this possibility, recent experimental evidence suggests that demands for cognitive control register as intrinsically costly (Kool, McGuire, Rosen, & Botvinick, 2010; Schouppe, Ridderinkhof, Verguts, & Notebaert, 2014; Westbrook, Kester, & Braver, 2013). For example, in the demand selection task (Kool et al., 2010), participants freely choose between task options that require different amounts of cognitive control and subsequently show a strong bias toward the lines of action with the smallest control demands. The intrinsic cost account predicts, in addition, that this avoidance bias should be offset by incentives. Indeed, several studies provide evidence for this hypothesis by showing increased willingness to perform demanding tasks when appropriately rewarded (Westbrook et al., 2013), even if this commits them to increased time toward goal attainment (Kool et al., 2010).

Based on these, and other, findings (for a review, see Botvinick & Braver, 2015), recent accounts of executive functioning propose that the exertion of cognitive control can best be understood as a form of cost–benefit decision-making. For example, Shenhav, Botvinick, and Cohen (2013) have proposed that the brain computes an "expected value of control" for each action—the expected rewarded discounted by the cost of associated control demands—and then chooses the action with highest value. Other researchers have proposed similar accounts (Gershman et al., 2015; Griffiths et al., 2015), whereby metacontrol between different systems is determined by the "value of computation," the expected reward for a given action subtracted by the costs of computation and time.

An older, but related, account was developed by Payne, Bettman, and Johnson (1988; 1993), who proposed that humans are "adaptive decision-makers," choosing among strategies by balancing accuracy against cognitive effort. Finally, a recent model from Kurzban, Duckworth, Kable, and Myers (2013) addresses the cost–benefit trade-off from a slightly different angle. They argue that the cost of effort, and therefore the subsequent implementation of control for a certain action, is dependent on the opportunity costs of the alternatively available actions. This model predicts that subjective experiences of effort, and subsequent reductions in control, depend on the value of the next-best line of action. In summary, even though these proposals differ in terms of how costs influence decision-making, they all center on the idea that the mobilization of control can best be understood as a form of cost–benefit trade-off.

Below we sketch our own recent efforts to combine these insights from RL theory in general—and the Daw two-step task, in particular—with the emerging view of cognitive control as value-based decision-making. We then review several other related approaches in the contemporary literature.

Control–reward trade-off in the two-step task

We propose that arbitration between model-based and model-free control is achieved by integrating the costs and benefits of each system. The rewards obtained by each system can be calculated by observing the average returns obtained by each control system, independently, and conditioned on the present task. Next, the brain uses these resulting "controller values" to select actions that maximize future cumulative reward. In doing so, it imposes an intrinsic, subject "cost" on the model-based controller. This cost represents the foregone reward due to model-based control, for instance due to the potentially longer decision time and due to the foregone opportunity to deploy limited cognitive control resources on other, concurrent tasks.

A core prediction of this model is that manipulating the rewards available during a decision-making task should alter the balance between model-free and model-based control. A natural candidate task to test this prediction is the Daw two-step task. Indeed, the model-based strategy in this task has been described as "optimal" (e.g., Sebold et al., 2014). Thus, one would predict that the more money at stake on any given trial of the task, the more willing the participant should be to pay the intrinsic cost of cognitive control in order to obtain the benefits of accurate performance.

In practice, however, recent research on this task shows that increased reliance on the model-based system does not predict increased performance accuracy on the Daw two-step task (Akam et al., 2015; Kool et al., 2016). To show this, Kool et al. (2016) recorded the average reward rate of many RL agents that varied across a range from pure model-free control to pure model-based control (see Fig. 7.3A). These simulations showed no systematic relationship between reward rate and model-based control for the original Daw two-step task, or for several related variants of this task (Dezfouli & Balleine, 2013; Doll et al., 2015) across a wide range of RL parameters. Consistent with this simulation result, they also found no correlation between model-based control and average reward in a subsequent experiment (Fig. 7.3B). The absence of this relation is produced by the interaction of at least five factors, several of which appear to prevent the model-

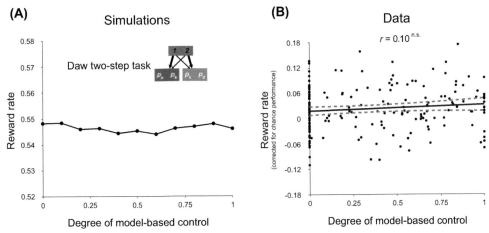

Figure 7.3 Control–reward trade-off in the Daw two-step task. (A) The relationship between the degree of model-based control and reward rate across 1000 simulations (with reinforcement learning parameters mirroring the median fits reported by Daw et al. (2011)). Importantly, these simulation results show that the task does not embody a trade-off between model-based control and reward. (B) Relationship between the estimated degree of model-based control and reward rate in the Daw two-step task (Daw et al., 2011). Consistent with simulation results, there is no correlation between these variables ($n = 197$). *Dashed lines* indicate the 95% confidence interval. *(Adapted from Kool et al. (2016).)*

based system from obtaining sufficiently reliable reward estimates (Kool et al., 2016). In short, the effectiveness of the model-based strategy is weakened on the Daw two-step task, because the first-stage choices carry relatively decreased importance and because this strategy does not have access to accurate representations of the second-stage reward outcomes. The fact that there is no control–reward trade-off in the Daw two-step task makes it ill-suited to test the cost–benefit hypothesis of RL arbitration, for example, by testing the effect of increased "stakes" on controller selection.

A novel two-step paradigm

In order to gain more experimental and computational traction on a control–reward trade-off in RL, Kool et al. (2016) developed a novel two-step task that theoretically and empirically achieves a trade-off between control and reward. The changes in this new task are based on the factors that were identified to produce the absence of this relationship in the Daw two-step task. One of the more notable changes to this paradigm is that it adopts a different task structure (Fig. 7.4; Doll et al., 2015). This task uses two first-stage states (randomly selected at the start of each trial) that both offer deterministic choices to one of two second-stage states. In both these second-stage states, the choices again are associated with a reward outcome that randomly changes across the experimental session. Specifically, the drifting reward probabilities at the second stage are

Opportunity to obtain reward (changing over time)

Figure 7.4 Design and state transition structure of the novel two-step task. Each first-stage choice deterministically transitions to one of two second-stage states. Each second-stage choice is associated with a scalar reward (between 0 and 9), which changes over the duration of the experiment according to a random Gaussian walk with $\sigma = 2$.

replaced with drifting scalar rewards (ranging from a negative to a positive number), so that the payoff of each action is identical to its value. This change was implemented to increase the informativeness of each reward outcome and thus to increase model-based accuracy.

The dissociation between model-free and model-based control in this task follows a different logic than the Daw two-step task. Since the model-free system only learns state—action—reward outcomes, it will not be able to transfer information learned in one starting state to the other starting state. In other words, rewards that are obtained in one starting state only increase the likelihood of revisiting that second-stage when the next trial starts in the same starting state but should not affect subsequent choices from the other starting state. The model-based system, on the other hand, treats the two starting states as functionally equivalent because it realizes the implicit equivalence of their action outcomes. Therefore, it will be able to generalize knowledge across them. So, reward outcomes at the second-stage should equally affect first-stage choices in the next trial, independent of whether this trial starts with the same state as the previous one.

This novel version of the two-step task incorporates many changes that increase the importance of the first-stage state and the ability of the system to learn the second-stage action values. Because of this, it achieves a trade-off between control and reward. This was first demonstrated through the simulation of RL agents performing this novel task (Kool et al., 2016). These simulations showed that the degree of model-based control was positively associated with average reward rate on the novel two-step paradigm (see Fig. 7.5A). A subsequent experiment provided convergent evidence for this theoretical result. Kool et al. (2016) found that, across participants, the degree of model-based control positively predicted the average reward rate (Fig. 7.5B), and this correlation was significantly stronger than that in the Daw two-step task.

Interestingly, Kool et al. (2016) also observed that participants spontaneously increased their rates of model-based control on the novel two-step task compared to the Daw two-step task. This suggests that the existence of the control−demand trade-off in the novel paradigm may have triggered a shift toward model-based control. Note that this result is consistent with the cost−benefit hypothesis of arbitration between habit and planning. However, alternative explanations are possible. For example, it may be the case that the introduction of negative reward in the novel paradigm triggered a shift toward model-based control, due to loss aversion. Such a shift would be the result of a decision heuristic signaling that certain features of the task should lead to increased model-based control, regardless of whether it actually yield larger overall reward than model-free control.

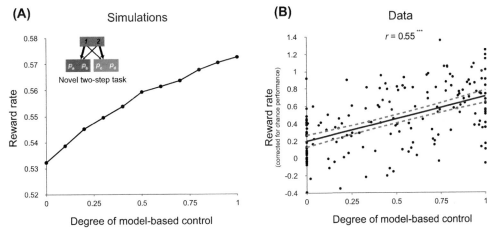

Figure 7.5 Control−reward trade-off in the novel two-step task. (A) The relationship between the degree of model-based control and reward rate across 1000 simulations. In contrast with the Daw two-step task, these simulation results show that the novel two-step task successfully achieves a trade-off between model-based control and reward. (B) Relationship between the estimated degree of model-based control and reward rate in the novel two-step task. Consistent with simulation results, there was a strong correlation between these variables ($n = 184$). *Dashed lines* indicate the 95% confidence interval. ***$P < .001$. (*Adapted from Kool et al. (2016).*)

Testing the cost—benefit model of arbitration

To distinguish between these two accounts, Kool, Gershman, and Cushman (2017) adapted the novel two-step paradigm so that the size of potential reward (the "stakes") changes randomly from trial to trial. In this new task, participants are cued about the size of the stakes at the outset of each trial. The size of the stakes is randomly selected on each trial, with high stakes calculated as a quintupling of baseline rewards. If behavior on this task is determined by a cost—benefit analysis, then people should employ more model-based control in the face of increased incentives, since on those trials the cost—benefit trade-off would be most beneficial. The results from this experiment were consistent with this hypothesis. Participants showed increased reliance on model-based control on high-stakes trials, indicating an increased willingness to engage in effortful planning (Kool et al., 2017).

Even though this result is consistent with the trade-off hypothesis, it is also consistent with an account that does not rely on the flexible and adaptive integration of costs and benefits. Specifically, participants may have simply acted on a decision heuristic, which reflexively increases model-based control in high-stake situations, regardless of whether this provides a reward advantage. To test this possibility, Kool et al. (2017) also implemented the stakes manipulation in the Daw two-step paradigm, since in this task there exists no trade-off between control and reward. If the stakes effect is driven by an incentive heuristic, high stakes should trigger increased model-based control in both tasks. However, under a cost—benefit account, where the brain estimates task-specific controller values for both systems, model-based control should not increase on high-stakes trials in the stakes version of the Daw two-step task. The results supported the latter hypothesis. Participants who completed the original Daw two-step task were insensitive to reward amplification through the stakes manipulation, in contrast with the increase in model-based control to reward amplification in the novel paradigm (Kool et al., 2017).

These results provide the first evidence that the brain attaches a cost to the exertion of model-based control. Furthermore, they provide insight into the way humans arbitrate between control mechanisms. Rather than relying on a heuristic of increasing model-based control when presented with larger incentives or other task features, participants seemed to engage in an adaptive integration of costs and benefits for either strategy in the current environment. Participants flexibly estimated the expected rewards for each system and then weighed this against the increased costs of model-based control.

Alternative models of arbitration

The cost—benefit account of competition between RL systems is broadly consistent with two bodies of research. First, the assumption that model-based control carries an intrinsic effort cost finds resonance in a large literature on the aversive nature of cognitive control (Botvinick & Braver, 2015; Gershman et al., 2015; Griffiths et al., 2015; Payne et al.,

1993; Rieskamp & Otto, 2006; Shenhav et al., 2013). This work suggests that the exertion of cognitive control can best be understood as the output of cost—benefit analysis. The comparison of behavior between the novel and Daw two-step tasks described above indicates that a similar trade-off guides the allocation of model-based control, presumably because this also requires the exertion of cognitive control (Otto, Gershman, et al., 2013; Smittenaar et al., 2013).

Second, there are now several other models of arbitration between competing RL system that are, to varying degrees, compatible with the cost—benefit trade-off account, but which differ in their details (Daw et al., 2005; Keramati et al., 2011; Lee et al., 2014; Pezzulo et al., 2013). Below, we will describe how these models implement the competition between model-free and model-based control and contrast them with our cost—benefit account.

According to Daw et al. (2005), arbitration between conflicting systems for behavioral control is primarily determined on the basis of uncertainty. Specifically, this model estimates each system's value uncertainty for each state—action pair. The model-based system has uncertainty due to bounded computational resources, whereas the model-free system has uncertainty due to limited experience in the environment. These measures of uncertainty are computed through Bayesian implementations of both RL systems as the posterior variance of the action values. After estimating these two different forms of uncertainty, the competition is then resolved by choosing the action value of the system with lower uncertainty.

A related metacontrol model uses signals of the systems' reliability as a means of arbitration (Lee et al., 2014). Here, the measure of reliability for a system is proportional to the absolute size of their prediction errors, the degree to which the systems predicted future states or rewards accurately. Similar to the Daw et al. (2005) model, Bayesian estimation of reliability still occurs for the model-based system, while a Pearce—Hall associability-like rule is used to estimate the reliability of the model-free system. In addition, this model also incorporates a "model bias" term, which favors the model-free system all else being equal, so as to account for differences in cognitive effort. The resulting arbitration process transforms these variables into a weighting parameter, which is then used to compute a weighted combination of action values to guide decision-making. Note that, in contrast to the Daw et al. (2005) model, the competition is resolved as a function of the average reliability of the model-based and model-free systems, and not separately for each action.

These closely related models of metacontrol account for many experimental findings, such as the finding that as the model-free system becomes more accurate, agents become increasingly insensitive toward outcome devaluation (since the model-free system needs to incrementally relearn its action—outcome contingencies). Furthermore, the reliability signals in the Lee et al. (2014) model have been shown to have a neural correlate in the inferior lateral prefrontal cortex. They cannot, however, explain the observation of increased model-based control on high-stakes trials (Kool et al., 2017), since the accuracy of either system's prediction does not change as a result of the amplification of reward. Therefore, these models do not predict an increase in proactive model-based control in the face of increased reward potential.

Instead, our cost–benefit hypothesis and the data described above align more strongly with metacontrol models that balance accuracy against control costs. One such model is proposed by Keramati et al. (2011). According to this account, the choice between model-based and model-free control is essentially about maximizing total reward. At each time point, the decision-maker estimates the expected gain in reward from running a model-based estimation of action values. This measure, also known as the value of information (Howard, 1966), was originally developed as a way to negotiate the exploration–exploitation trade-off in RL. Next, the agent also estimates the cost of running those simulations. This cost is explicitly formalized as the amount of potential reward that the model-free system could have accrued while the model-based system is engaged in these prospective simulations. In other words, the cost of model-based control is explicitly an opportunity cost directly proportional to the required processing time. Finally, the costs and gains are compared against each other, and their relative size determines whether the model-based system is invoked. If the costs outweigh the gains, the faster the habitual system is employed, otherwise the agent engages in slower model-based planning.

Pezzulo et al. (2013) have developed a related value-based account of arbitration between habit and planning. Similar to the proposal of Keramati et al. (2011), the agent assesses each available action in the current state by first computing the value of information (Howard, 1966) associated with model-based planning. This variable encompasses both the uncertainty about the action's value and also the difference in value between each action and the best available alternative action. The value of information increases when the uncertainty about the current action is high and also if the difference between possible action values is small (that is, if the decision is more difficult). Next, this measure of the expected gains of model-based control is compared against a fixed threshold that represents the effort cost (Gershman & Daw, 2012) or time cost associated with planning. Again, if the cost exceeds the value of information, the agent relies on cached values; otherwise it will employ model-based simulations over an internal representation of the environment to reduce the uncertainty about the current action values (Solway & Botvinick, 2012).

Both the Keramati et al. (2011) and Pezzulo et al. (2013) models account for a range of behavioral findings. The time-based account of Keramati et al. (2011) model accounts for the increasing insensitivity to outcome devaluation over periods of training. It can also naturally incorporate the finding that response times increase with the number of options, especially early in training, since at those moments the model will engage in time-consuming model-based simulations across the decision tree. Relatedly, Pezzulo et al. (2013) showed that, in a multistage RL task, their model switches from a large number of model-based simulations in earlier stages toward more reliance on model-free control in later stages. In other words, when the model-free system has generated a sufficiently accurate representation of the world, the agent then prefers to avoid the cost of model-based control. The Pezzulo et al. (2013) model is also able to flexibly shift between systems. For example, it shows a rebalancing toward model-based control in response to a change in reward structure of the environment, i.e., an increase in uncertainty of action outcomes.

However, these models still arbitrate between habit and planning as a function of the amount of uncertainty about value estimates in the model-free action values: both models assume an advantage for model-based control when uncertainty about model-free estimates is high (Keramati et al., 2011; Pezzulo et al., 2013). In doing so, they are not immediately able to explain the effect of increased stakes on model-based control (Kool et al., 2017). Those data instead favor a mechanism that directly contrasts the rewards obtained by model-based and model-free control, discounted by their respective cost. Furthermore, the fact that these models require the explicit computation of the expected gains from model-based simulations (the value of information; Howard, 1966) creates the problem of infinite regress (Boureau et al., 2015). If the purpose of metacontrol is to avoid unnecessary deployment of cognitive control, then this purpose is undermined by engaging in an explicit and demanding computation to determine whether cognitive demands are worthwhile.

Based on the evidence described here, we make two suggestions for new formal models of arbitration between RL systems. First, they should incorporate a direct contrast between the costs and benefits of both model-free and model-based learning strategies in their current environment, perhaps in addition to a drive to increase reliability of controller predictions. This property should afford flexible adaptive control in response to the changing potential for reward, such as in the stake size experiment described above. Second, in order to avoid the issue of infinite regress, the arbitration between habit and planning should be guided by a process that does not involve control-demanding computations of reward advantage, such as the value of information (Howard, 1966). Instead, new models of metacontrol should focus on more heuristic forms of arbitration. Notably, a system that attaches an intrinsic cost to model-based planning might guide metacontrol with enhanced efficiency, by circumventing the need for an explicit computation of those costs in terms of effort, missed opportunities, and time. In sum, these properties motivate our proposal that a form of model-free RL integrates the reward history and control costs associated with different control mechanisms. The resulting "controller values" dictate controller arbitration.

PRINCIPLES OF COOPERATION

While the evidence reviewed in the previous section supports competitive architectures, recent evidence also suggests a variety of cooperative interactions between model-free and model-based RL. In this section, we review three different flavors of cooperation.

Model-based simulation as a source of training data for model-free learning

One way to think about the trade-off between model-free and model-based algorithms is in terms of *sample complexity* and *time complexity*. Sample complexity refers to the number of training examples a learning algorithm needs to achieve some level of accuracy. Time complexity refers to how long an algorithm takes to execute. Intuitively, these correspond to "learning time" and "decision time."

Model-free algorithms have high sample complexity but low time complexity—in other words, learning is slow but deciding is fast. Model-based algorithms have the opposite property: relatively low sample complexity, assuming that the model can be learned efficiently, but high time complexity. Since the amount of data that an agent has access to is typically fixed (by the world or by the experimenter) and thus beyond algorithmic improvement, it might seem that this trade-off is inevitable. However, it is possible to create additional examples simply by *simulating* from the model and allowing model-free algorithms to learn from these simulated examples. In this way, the model-based system can manufacture an arbitrarily large number of examples. As a consequence, the model-free system's sample complexity is no longer tied to its real experience in the world; model-based simulations, provided they are accurate, are a perfectly good substitute.

Sutton (1990) proposed a cooperative architecture called *Dyna* that exploits this idea. A model-free agent, by imbibing model-based simulations, can become arbitrarily proficient without increasing either sample complexity or time complexity. The only requirement is that the agent has sufficient spare time to process these simulations. Humans and many other animals have long periods of sleep or quiet wakefulness during which such simulation could plausibly occur. Notably, neurons in the hippocampus tuned to spatial location ("place cells") replay sequences of firing patterns during rest and sleep (see, Carr, Jadhav, & Frank, 2011 for a review), suggesting they might act as a neural substrate for a Dyna-like simulator (Johnson & Redish, 2005). Furthermore, it is well known that motor skills can improve following a rest period without additional training (Korman, Raz, Flash, & Karni, 2003; Walker, Brakefield, Morgan, Hobson, & Stickgold, 2002) and reactivating memories during sleep can enhance subsequent task performance (Oudiette & Paller, 2013). Ludvig, Mirian, Kehoe, and Sutton (2017) have argued that simulation may underlie a number of animal learning phenomena (e.g., spontaneous recovery, latent inhibition) that are vexing for classical learning theories (which are essentially variants of model-free algorithms).

A series of experiments reported by Gershman, Markman, and Otto (2014) attempted to more directly test Dyna as a theory of human RL. The experimental design is summarized in Fig. 7.6A. In Phase 1, subjects learn the structure of a simple two-step sequential decision problem. In Phase 2, they learn that taking action A in state 1 is superior to

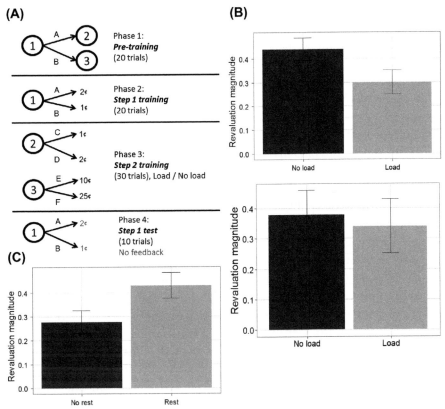

Figure 7.6 (A) The sequential decision problem consists of three states (indicated by *numbered circles*) and two mutually exclusive actions in each state (indicated by *letters*). Deterministic transitions between states conditional upon the chosen action are indicated by *arrows*. Rewards for each state–action pair are indicated by amounts (in cents). In Phase 4, reward feedback is delayed until the end of the phase. (B) Revaluation in load and no load conditions. Revaluation magnitude is measured as $P_4(\text{action} = B|\text{state} = 1) - P_2(\text{action} = B|\text{state} = 1)$, where $P_i(\text{action} = a|\text{state} = s)$ is the probability of choosing action a in state s during Phase i. Top: load applied during Phase 3; Bottom: load applied during Phase 4. (C) A brief rest phase prior to Phase 4 ameliorates the effects of load.

taking action B. They then learn in Phase 3 that state 3 is superior to state 2. This sets up a conflict with what they learned in Phase 2 because taking the preferred action A in state 1 will lead them to state 2 (the inferior state). In Phase 4, Gershman et al. (2014) tested whether they switch their preference for action A following their experience in the second-step states.

Standard model-free learning algorithms like temporal difference learning do not predict any revaluation because they rely on unbroken trajectories through the state space in order to chain together reward predictions. These trajectories were deliberately broken in the experimental structure so as to handicap model-free learning. Less obviously, standard model-based learning algorithms *also* predict no revaluation because subjects are explicitly instructed in Phase 4 that they are only being rewarded for their actions in the first state. Thus, the optimal model-based policy should completely ignore information about the second step. Crucially, Dyna predicts a positive revaluation effect because model-based simulation can effectively stitch together the state sequences, which were not explicitly presented to subjects, allowing model-free algorithms to revise the value estimate in state 1 following experience in states 2 and 3.

The experimental results showed clear evidence for a revaluation effect (Fig. 7.6B), supporting the predictions of Dyna. Additional support came from several other findings. First, cognitive load during Phase 3 reduced the revaluation effect. This is consistent with the idea that model-based simulation, like other model-based processes, is computationally intensive and thus susceptible to disruption by competition for resources. Second, the load effect could be mitigated by increasing the number of trials (i.e., opportunities for revaluation) during Phase 3. Third, a brief rest (quiet wakefulness) prior to Phase 4 increased revaluation, consistent with the hypothesis of offline simulation driving model-free learning (Fig. 7.6C). Finally, applying cognitive load during Phase 4 had no effects on the results, supporting our proposal that performance is driven by model-free control (recall that cognitive load has a selective, deleterious effect on model-based control; Otto, Gershman, et al., 2013).

Taken together, these results provide some of the first behavioral evidence for cooperative interaction between model-based and model-free RL. The same framework may explain the observation that model-based control on the Daw two-step task becomes resistant to disruption by cognitive load over the course of training (Economides, Kurth-Nelson, Lübbert, Guitart-Masip, & Dolan, 2015). If one effect of training is to inject model-based knowledge into the model-free value function, then the model-free system will be able to exhibit model-based behavior autonomously. Dyna may also shed light on the recent observation that dopamine neurons signal prediction errors based on *inferred* (i.e., simulated) values (Doll & Daw, 2016; Sadacca, Jones, & Schoenbaum, 2016).

Partial evaluation

Keramati et al. (2016) have investigated an alternative way to combine model-based and model-free systems, which they refer to as "planning-until-habit," a strategy closely related to "partial evaluation" in the computer science literature (see Daw & Dayan, 2014). The basic idea, illustrated in Fig. 7.7, is to do limited-depth model-based planning and then insert cached model-free values at the leaves of the decision tree. The sum of these two components will equal the full value at the root node. This model nests pure model-based (infinite depth) and pure model-free (depth 0) algorithms as special cases. The primary computational virtue of partial evaluation is that it can efficiently

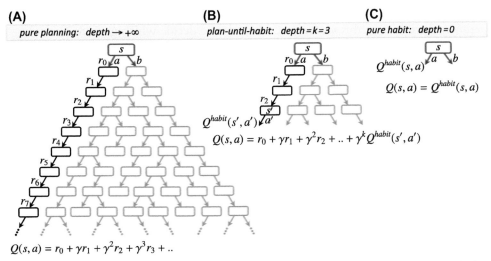

(A) pure planning: depth → +∞

(B) plan-until-habit: depth = k = 3

(C) pure habit: depth = 0

$$Q(s,a) = r_0 + \gamma r_1 + \gamma^2 r_2 + \gamma^3 r_3 + ..$$

$$Q(s,a) = r_0 + \gamma r_1 + \gamma^2 r_2 + .. + \gamma^k Q^{habit}(s',a')$$

$$Q(s,a) = Q^{habit}(s,a)$$

Figure 7.7 (A) Pure planning: rewards are mentally accumulated over an infinite horizon. (B) Plan-until-habit: rewards are partially accumulated and then combined with a cached value function. (C) Pure habit: actions are evaluated using only cached values, no reward accumulation. *(Reprinted from Keramati et al. (2016).)*

exploit cached values to augment model-based planning. This will work well when cached values are accurate in some states but not others (where planning is required).

Keramati et al. (2016) provided behavioral evidence for this proposal using a novel three-step extension of the Daw two-step task. Using the same logic of analyzing the interaction between reward outcome and transition probability on subsequent choices, they found differences in the mixture of model-based and model-free behavior at different steps of the task. In particular, subjects appeared model-based with respect to the second step but model-free with respect to the third step, precisely what was predicted by the partial evaluation strategy. Moreover, putting people under time pressure shifted them to a pure model-free strategy at both steps, consistent with the idea that the depth of model-based planning is adaptive and depends on resource availability.

Habitual goal selection

An advantage of model-based control is its capacity to plan toward goals. That is, a model-based agent can specify any particular state of the world that she wishes to attain (e.g., being at the dentist's office at 2 p.m. with a bottle of ibuprofen) and then evaluate candidate policies against their likelihood of attaining that goal state. In many laboratory tasks, the number of possible goal states may be very small, or they may be explicitly stated by the experimenter. For instance, in the classic "two-step" task presented in Fig. 7.1, there are only six states toward which the agent might plan (two intermediate states and four terminal states). In the real world, however, the number of possible goal states that we might select at any given moment is very large. Usually, there are no ex-perimenters restricting this set for you. How do we decide which goals to pursue?

One possibility is exhaustive search, but this is computationally prohibitive. Consider, for instance, evaluating candidate goals alphabetically: You could set the goal of *absconding with an aardvark*, or *absconding with an abacus*, and so on, until eventually considering selecting the goal of *X-raying with a Xerox*. For the same reason—i.e., the large set of possible goals in most real-world settings—it is not practical to employ model-based eval-uation of the rewards of candidate goals in order to decide which goal to select. Is there a more efficient way to decide which particular goal to pursue from moment to moment?

An obvious alternative is to select goals by model-free methods—in other words, to store a state-specific cached value of the likely value of pursuing different goals. Put simply, an agent might ask himself/herself, "when I've been in this situation in the past, what have been rewarding goals for me to select?" Of course, once a goal is selected, it falls to model-based processes to plan toward that goal. This entails a cooperative relationship between the two control mechanisms: Cached, model-free values may be used to decide *which goal to pursue*, while model-based planning is used in order to determine *how to attain it*.

The utility of this approach is best appreciated through a specific example (Cushman & Morris, 2015). Consider an experienced journalist who sets out to report on different news events each day. At a high level of abstraction, his/her job is structured around a regular series of goals to pursue: "Find out what has happened this morning"; "Consult with my editor"; "Obtain interviews"; "Write a draft," and so forth. Thus, selecting goals may be efficiently accomplished by considering their cached value: "Obtaining inter-views" was a valuable goal yesterday, and it will remain so today. Yet, pursuing any one of these goals would require flexible model-based planning—for instance, the ac-tions necessary to interview the president one day will be different than the actions neces-sary to interview a political dissident the next day. In sum, then, a favorable architecture for many tasks would select goals according to model-free value but then attains goals by model-based planning.

Cushman and Morris (2015) found empirical support for this architecture using several modified versions of the classic Daw two-step task. An example is illustrated in Fig. 7.8. The essence of the design is to prompt people to choose an action that reveals their goal but then occasionally transition them to a nongoal state. If this reinforcement history affects their subsequent choice despite its low probability, then it can be attributed to a model-free value update process. Subsequently, participants are tested on different actions that are associated with common goal states. Influence of reinforcement history even upon these different actions implies a model-free value assignment not to the action itself, but rather to the goal state with which it is associated.

Beyond the particular case of goal selection, this research points toward a more gen-eral form of cooperative interaction between model-free and model-based systems. For typical real-world problems, full model-based evaluation of all possible action sequences

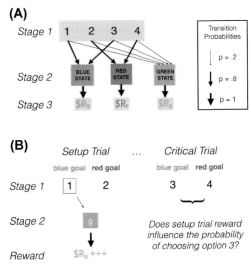

Figure 7.8 A modified version of the two-step task designed to test a model of habitual goal selection. (A) At stage 1, participants are presented with two available actions drawn from a set of four (1, 2, 3, and 4). These transitions are with high probability to either a blue or red intermediate state, and with equal low probability to a green state. (B) On critical trials, the low-probability green-state transition occurs. The key question is whether the reward obtained following the green state influences subsequent choice of different actions that share the same goal (e.g., whether a reward following the sequence 1, green influences the probability of subsequently choosing action 3, which shares the blue-state goal with action 1). Across several experiments, participants exhibited precisely this effect. *(Reprinted from Cushman and Morris (2015).)*

will always pose prohibitive computational demands. One solution to this problem is to use cached, model-free values to weight the probability with which all possible actions are introduced into the subset of actions that receive further model-based evaluation. (This subset might be described as the "choice set.") All else being equal, the higher the cached value of an action, the more likely that the benefits of a more precise model-based estimate of its current value will outweigh the computational demands involved. Investigating this general process of "choice set construction" is an important direction for future research.

CONCLUSION

Over the last century, the idea that human behavior is controlled by two systems, one habitual and one goal-directed, has become a cornerstone of psychological and behavioral theories of cognition and decision-making (Dickinson, 1985; Dolan & Dayan, 2013; Fudenberg & Levine, 2006; Kahneman, 1973; Sloman, 1996). Recent RL theory has brought mathematical precision to this area of research by formalizing this distinction in terms of model-based and model-free control (Daw et al., 2005, 2011; Gläscher et al., 2010). We have reviewed the surge of empirical and theoretical research emanating from this formalism.

First, we reviewed work that addresses how the habitual and goal-directed systems are engaged in a competition for control of behavior. We proposed that this competition is arbitrated as a trade-off between the costs and benefits of employing each system. At the core of this proposal is the idea that the exertion of model-based control carries an intrinsic effort cost associated with the exertion of cognitive control. This account is supported by the findings that model-based planning is dependent on cognitive resources (Otto, Gershman, et al., 2013; Otto, Raio, et al., 2013; Otto et al., 2015; Schad et al., 2014) and that humans attach intrinsic disutility to the exertion of cognitive control (Kool et al., 2010; Westbrook et al., 2013). Current research indicates that model-based control is spontaneously increased in response to reward amplification, but only when the model-based system is associated with increased accuracy (Kool et al., 2016, 2017). Together, these findings suggest that the brain estimates values for each system, integrating their costs and benefits into a single metacontrol value that it uses to guide controller arbitration.

Second, we reviewed a new line of research that focuses on the ways in which habit and planning act in a cooperative fashion to achieve both efficiency and accuracy. Evidence suggests a plethora of cooperative strategies: the model-free system can learn from data simulated from the model-based system (Gershman et al., 2014), can truncate model-based planning (Keramati et al., 2016), or can facilitate the selection of rewarding goals (Cushman & Morris, 2015). At present, it is unclear whether these different strategies occur simultaneously or are adaptively invoked much like in the controller arbitration problem.

In the work described here, the idea of an intrinsic effort cost for model-based control has only come to the fore in the research on the competitive interaction between habit and planning. However, given the ubiquitous nature of the cost for cognitive control (Botvinick & Braver, 2015; Westbrook & Braver, 2015), such a cost is likely to also play a role in the collaborative interactions between these two systems. From this perspective, several intriguing questions arise.

Some of these questions concern the basic algorithmic approach that the brain takes to decision-making. For instance, is habitual goal selection (Cushman & Morris, 2015) more prevalent for people who attach a higher intrinsic cost to model-based planning? Does the intrinsic cost of cognitive control establish the threshold at which estimation of action values switches from planning to habit in the situations described by Keramati et al. (2016)? In light of our cost—benefit theory of controller arbitration, one may view the cooperative interaction between habit and planning as a case of bounded rationality (Gigerenzer & Goldstein, 1996). From this perspective, costly cognitive resources would be deployed to maximize accuracy among a restricted region of the action space while preserving a net gain in value, and habit would provide complementary assistance for those actions not analyzed through model-based control. Note that this framework predicts that increased potential incentives (as used in Kool et al., 2017) will lead to deeper planning in the Keramati et al. (2016) task and a reduced reliance on habitual goal selection in the Cushman and Morris (2015) task.

Other questions involve neural implementation. Ever since the recent resurgence of RL theory in modern psychological research, the neuromodulator dopamine has come to the fore as playing a key role. Most famously, Schultz, Dayan, and Montague (1997) showed that reward prediction errors, the signals that drive learning of action—outcome contingencies, are encoded by the phasic firing of dopamine neurons that project to the ventral striatum in the basal ganglia. More important for the current purpose, it has been suggested that tonic levels of dopamine encode an average reward signal that determines response vigor in operant conditioning tasks (Hamid et al., 2016; Niv, Daw, & Dayan, 2006), so higher dopamine levels yield increased responding on free-operant conditioning tasks. Based on these and related results, Salamone and colleagues (Salamone & Correa, 2012; Salamone, Correa, Farrar, Nunes, & Pardo, 2009) have proposed that baseline levels of dopamine in the basal ganglia may actually serve to discount the perceived costs of physical effort. For example, rats in an effort-based decision-making task show reduced willingness to climb over barriers to obtain rewards after depletion of dopamine in the nucleus accumbens (Cousins, Atherton, Turner, & Salamone, 1996). Westbrook and Braver (2016) have proposed a very similar view for the case of mental effort. According to this account, increases in baseline dopamine levels in response to high-reward situations facilitate subsequent cognitive processing by enhancing stability of working memory representations in the prefrontal cortex. Intriguingly, recent experiments indicate that baseline dopamine levels in the ventral striatum correlated positively with a bias toward more model-based control (Deserno et al., 2015) and that experimentally induced increases in dopamine increase the degree of model-based control in the Daw two-step task (Sharp, Foerde, Daw, & Shohamy, 2015; Wunderlich, Smittenaar, & Dolan, 2012; see Chapter 11 by Sharpe and Schoenbaum). Together, these insights hint at the intriguing possibility that this effect of dopamine on model-based control may be viewed as the result of an alteration of the variables that enter the cost—benefit trade-off at the algorithmic level.

While the work we have reviewed in this chapter suggests a rich space of competition and cooperation between RL systems, we have in fact only skimmed the surface. New research suggests separate but interacting systems for Pavlovian (Dayan & Berridge, 2014) and episodic (Gershman & Daw, 2017) RL. One may reasonably worry that theorists are gleefully manufacturing theories to accommodate each new piece of data, without addressing how the systems act in concert as part of a larger cognitive architecture. What is needed is a theory of metacontrol that encompasses all of these systems. The development of such a theory will be a central project for the next generation of RL research.

REFERENCES

Akam, T., Costa, R., & Dayan, P. (2015). Simple plans or sophisticated habits? State, transition and learning interactions in the two-step task. *PLoS Computational Biology, 11*, e1004648.

Army Individual Test Battery. (1944). *Manual of directions and scoring.* Washington, DC: War Department, Adjutant General's Office.

Balleine, B. W., & O'Doherty, J. (2009). Human and rodent homologies in action control: Corticostriatal determinants of goal-directed and habitual action. *Neuropsychopharmacology: Official Publication of the American College of Neuropsychopharmacology, 35*, 48–69.

Botvinick, M. M., & Braver, T. (2015). Motivation and cognitive control: From behavior to neural mechanism. *Annual Review of Psychology, 66*, 83–113.

Boureau, Y.-L., Sokol-Hessner, P., & Daw, N. D. (2015). Deciding how to decide: Self-control and meta-decision making. *Trends in Cognitive Sciences, 19*, 700–710.

Carr, M. F., Jadhav, S. P., & Frank, L. M. (2011). Hippocampal replay in the awake state: A potential physiological substrate of memory consolidation and retrieval. *Nature Neuroscience, 14*, 147–153.

Cousins, M. S., Atherton, A., Turner, L., & Salamone, J. D. (1996). Nucleus accumbens dopamine depletions alter relative response allocation in a T-maze cost/benefit task. *Behavioural Brain Research, 74*, 189–197.

Cushman, F., & Morris, A. (2015). Habitual control of goal selection in humans. *Proceedings of the National Academy of Science of the United States of America, 112*, 13817–13822.

Daw, N. D., & Dayan, P. (2014). The algorithmic anatomy of model-based evaluation. *Philosophical Transactions of the Royal Society B: Biological Sciences, 369*, 20130478.

Daw, N. D., Gershman, S. J., Seymour, B., Dayan, P., & Dolan, R. J. (2011). Model-based influences on humans' choices and striatal prediction errors. *Neuron, 69*, 1204–1215.

Daw, N. D., Niv, Y., & Dayan, P. (2005). Uncertainty-based competition between prefrontal and dorsolateral striatal systems for behavioral control. *Nature Neuroscience, 8*, 1704–1711.

Dayan, P., & Berridge, K. C. (2014). Model-based and model-free Pavlovian reward learning: Revaluation, revision, and revelation. *Cognitive, Affective, & Behavioral Neuroscience, 14*, 473–492.

Decker, J. H., Otto, A. R., Daw, N. D., & Hartley, C. A. (2016). From creatures of habit to goal-directed learners: Tracking the developmental emergence of model-based reinforcement learning. *Psychological Science, 27*, 848–858.

Deserno, L., Huys, Q. J. M., Boehme, R., Buchert, R., Heinze, H.-J., Grace, A. A., ... Schlagenhauf, F. (2015). Ventral striatal dopamine reflects behavioral and neural signatures of model-based control during sequential decision making. *Proceedings of the National Academy of Sciences of the United States of America, 112*, 1595–1600.

Dezfouli, A., & Balleine, B. W. (2013). Actions, action sequences and habits: Evidence that goal-directed and habitual action control are hierarchically organized. *PLoS Computational Biology, 9*, e1003364.

Dickinson, A. (1985). Actions and habits: The development of behavioural autonomy. *Philosophical Transactions of the Royal Society B: Biological Sciences, 308*, 67–78.

Dolan, R. J., & Dayan, P. (2013). Goals and habits in the brain. *Neuron, 80*, 312–325.

Doll, B. B., & Daw, N. D. (2016). The expanding role of dopamine. *eLife, 5*, e15963.

Doll, B. B., Duncan, K. D., Simon, D. A., Shohamy, D., & Daw, N. D. (2015). Model-based choices involve prospective neural activity. *Nature Neuroscience, 18*, 767–772.

Economides, M., Kurth-Nelson, Z., Lübbert, A., Guitart-Masip, M., & Dolan, R. J. (2015). Model-based reasoning in humans becomes automatic with training. *PLoS Computational Biology, 11*, e1004463.

Fudenberg, D., & Levine, D. K. (2006). A dual self model of impulse control. *American Economic Review, 96*, 1449–1476.

Gershman, S. J. (2017). Reinforcement learning and causal models. In M. Waldmann (Ed.), *Oxford handbook of causal reasoning.* Oxford University Press.

Gershman, S. J., & Daw, N. (2012). Perception, action and utility: The tangled skein. In M. Rabinovich, K. Friston, & P. Varona (Eds.), *Principles of brain dynamics: Global state interactions.* Cambridge, MA: MIT Press.

Gershman, S. J., & Daw, N. D. (2017). Reinforcement learning and episodic memory in humans and animals: An integrative framework. *Annual Review of Psychology, 68*, 101–128.

Gershman, S. J., Horvitz, E. J., & Tenenbaum, J. B. (2015). Computational rationality: A converging paradigm for intelligence in brains, minds, and machines. *Science, 349*, 273–278.

Gershman, S. J., Markman, A. B., & Otto, A. R. (2014). Retrospective revaluation in sequential decision making: A tale of two systems. *Journal of Experimental Psychology: General, 143*, 182–194.

Gigerenzer, G., & Goldstein, D. G. (1996). Reasoning the fast and frugal way: Models of bounded rationality. *Psychological Review, 103*, 650–669.

Gillan, C. M., Kosinski, M., Whelan, R., Phelps, E. A., & Daw, N. D. (2016). Characterizing a psychiatric symptom dimension related to deficits in goal-directed control. *eLife, 5*, e11305.

Gillan, C. M., Otto, A. R., Phelps, E. A., & Daw, N. D. (2015). Model-based learning protects against forming habits. *Cognitive, Affective, & Behavioral Neuroscience, 15*, 523–536.

Gläscher, J., Daw, N., Dayan, P., & O'Doherty, J. (2010). States versus rewards: Dissociable neural prediction error signals underlying model-based and model-free reinforcement learning. *Neuron, 66*, 585–595.

Griffiths, T. L., Lieder, F., & Goodman, N. D. (2015). Rational use of cognitive resources: Levels of analysis between the computational and the algorithmic. *Topics in Cognitive Science, 7*, 217–229.

Hamid, A. A., Pettibone, J. R., Mabrouk, O. S., Hetrick, V. L., Schmidt, R., Vander Weele, C. M., … Berke, J. D. (2016). Mesolimbic dopamine signals the value of work. *Nature Publishing Group, 19*, 117–126.

Howard, R. (1966). Information value theory. *IEEE Transactions on Systems Science and Cybernetics, 2*.

Johnson, A., & Redish, A. D. (2005). Hippocampal replay contributes to within session learning in a temporal difference reinforcement learning model. *Neural Networks, 18*, 1163–1171.

Kahneman, D. (1973). *Attention and effort*. Englewood Cliffs, NJ: Prentice-Hall.

Keramati, M., Dezfouli, A., & Piray, P. (2011). Speed/accuracy trade-off between the habitual and the goal-directed processes. *PLoS Computational Biology, 7*, e1002055.

Keramati, M., Smittenaar, P., Dolan, R. J., & Dayan, P. (2016). Adaptive integration of habits into depth-limited planning defines a habitual-goal-directed spectrum. *Proceedings of the National Academy of Sciences of the United States of America, 113*, 12868–12873.

Kool, W., Cushman, F. A., & Gershman, S. J. (2016). When does model-based control pay off? *PLoS Computational Biology, 12*, e1005090.

Kool, W., Gershman, S. J., & Cushman, F. A. (2017). Cost-benefit arbitration between multiple reinforcement-learning systems. *Psychological Science, 28*, 1321–1333.

Kool, W., McGuire, J. T., Rosen, Z. B., & Botvinick, M. M. (2010). Decision making and the avoidance of cognitive demand. *Journal of Experimental Psychology: General, 139*, 665–682.

Korman, M., Raz, N., Flash, T., & Karni, A. (2003). Multiple shifts in the representation of a motor sequence during the acquisition of skilled performance. *Proceedings of the National Academy of Sciences of the United States of America, 100*, 12492–12497.

Kurzban, R., Duckworth, A. L., Kable, J. W., & Myers, J. (2013). An opportunity cost model of subjective effort and task performance. *Behavioral and Brain Sciences, 36*, 661–726.

Lee, S. W., Shimojo, S., & O'Doherty, J. P. (2014). Neural computations underlying arbitration between model-based and model-free learning. *Neuron, 81*, 687–699.

Ludvig, E. A., Mirian, M. S., Kehoe, E. J., & Sutton, R. S. (2017). Associative learning from replayed experience. *bioRxiv*. https://doi.org/10.1101/100800.

Miller, E. K., & Cohen, J. D. (2001). An integrative theory of prefrontal cortex function. *Annual Reviews in Neuroscience, 24*, 167–202.

Niv, Y., Daw, N., & Dayan, P. (2006). How fast to work: Response vigor, motivation and tonic dopamine. *Advances in Neural Information Processing Systems, 18*, 1019.

Otto, A. R., Gershman, S. J., Markman, A. B., & Daw, N. D. (2013). The curse of planning: Dissecting multiple reinforcement-learning systems by taxing the central executive. *Psychological Science, 24*, 751–761.

Otto, A. R., Raio, C. M., Chiang, A., Phelps, E., & Daw, N. (2013). Working-memory capacity protects model-based learning from stress. *Proceedings of the National Academy of Sciences of the United States of America, 110*, 20941–20946.

Otto, A. R., Skatova, A., Madlon-Kay, S., & Daw, N. D. (2015). Cognitive control predicts use of model-based reinforcement learning. *Journal of Cognitive Neuroscience, 27*, 319–333.

Oudiette, D., & Paller, K. A. (2013). Upgrading the sleeping brain with targeted memory reactivation. *Trends in Cognitive Sciences, 17*, 142–149.

Payne, J. W., Bettman, J. R., & Johnson, E. J. (1988). Adaptive strategy selection in decision making. *Journal of Experimental Psychology: Learning, Memory, and Cognition, 14*, 534–552.

Payne, J. W., Bettman, J. R., & Johnson, E. J. (1993). *The adaptive decision maker.* Cambridge, England: Cambridge University Press.

Pezzulo, G., Rigoli, F., & Chersi, F. (2013). The mixed instrumental controller: Using value of information to combine habitual choice and mental simulation. *Frontiers in Psychology, 4*, 92.

Rieskamp, J., & Otto, P. E. (2006). SSL: A theory of how people learn to select strategies. *Journal of Experimental Psychology: General, 135*, 207–236.

Sadacca, B. F., Jones, J. L., & Schoenbaum, G. (2016). Midbrain dopamine neurons compute inferred and cached value prediction errors in a common framework. *eLife, 5*, e13665.

Salamone, J. D., & Correa, M. (2012). The mysterious motivational functions of mesolimbic dopamine. *Neuron, 76*, 470–485.

Salamone, J. D., Correa, M., Farrar, A. M., Nunes, E. J., & Pardo, M. (2009). Dopamine, behavioral economics, and effort. *Frontiers in Behavioral Neuroscience, 3*, 2–12.

Schad, D. J., Jünger, E., Sebold, M., Garbusow, M., Bernhardt, N., Javadi, A.-H., … Huys, Q. J. M. (2014). Processing speed enhances model-based over model-free reinforcement learning in the presence of high working memory functioning. *Frontiers in Psychology, 5*, 1450.

Schouppe, N., Ridderinkhof, K. R., Verguts, T., & Notebaert, W. (2014). Context-specific control and context selection in conflict tasks. *Acta Psychologica, 146*, 63–66.

Schultz, W., Dayan, P., & Montague, P. R. (1997). A neural substrate of prediction and reward. *Science, 275*, 1593–1599.

Sebold, M., Deserno, L., Nebe, S., Nebe, S., Schad, D. J., Garbusow, M., … Huys, Q. J. M. (2014). Model-based and model-free decisions in alcohol dependence. *Neuropsychobiology, 70*, 122–131.

Sharp, M. E., Foerde, K., Daw, N. D., & Shohamy, D. (2015). Dopamine selectively remediates 'model-based' reward learning: A computational approach. *Brain: a Journal of Neurology, 139*, 355–364.

Shenhav, A., Botvinick, M. M., & Cohen, J. D. (2013). The expected value of control: An integrative theory of anterior cingulate cortex function. *Neuron, 79*, 217–240.

Skatova, A., Chan, P. A., & Daw, N. D. (2015). Extraversion differentiates between model-based and model-free strategies in a reinforcement learning task. *Frontiers in Human Neuroscience, 7*, 525.

Sloman, S. A. (1996). The empirical case for two systems of reasoning. *Psychological Bulletin, 119*, 3–22.

Smittenaar, P., FitzGerald, T. H. B., Romei, V., Wright, N. D., & Dolan, R. J. (2013). Disruption of dorsolateral prefrontal cortex decreases model-based in favor of model-free control in humans. *Neuron, 80*, 914–919.

Solway, A., & Botvinick, M. M. (2012). Goal-directed decision making as probabilistic inference: A computational framework and potential neural correlates. *Psychological Review, 119*, 120–154.

Stroop, J. R. (1935). Studies of interference in serial verbal reactions. *Journal of Experimental Psychology: General, 18*, 643–662.

Sutton, R. S. (1990). First results with Dyna, an interesting architecture for learning, planning, and reacting. In T. Miller, R. S. Sutton, & P. Werbos (Eds.), *Neural networks for control* (pp. 179–189). Cambridge, MA: MIT Press.

Sutton, R. S., & Barto, A. G. (1998). *Reinforcement learning: An introduction.* Cambridge, MA: MIT Press.

Thorndike, E. L. (1911). *Animal intelligence: Experimental studies.* New York: The Macmillan Company.

Walker, M. P., Brakefield, T., Morgan, A., Hobson, J. A., & Stickgold, R. (2002). Practice with sleep makes perfect: Sleep-dependent motor skill learning. *Neuron, 35*, 205–211.

Westbrook, A., & Braver, T. (2015). Cognitive effort: A neuroeconomic approach. *Cognitive, Affective, & Behavioral Neuroscience, 15*, 395–415.

Westbrook, A., & Braver, T. S. (2016). Dopamine does double duty in motivating cognitive effort. *Neuron, 89*, 695–710.

Westbrook, A., Kester, D., & Braver, T. S. (2013). What is the subjective cost of cognitive effort? Load, trait, and aging effects revealed by economic preference. *PLoS One, 22*, e68210.

Wunderlich, K., Smittenaar, P., & Dolan, R. (2012). Dopamine enhances model-based over model-free choice behavior. *Neuron, 75*, 418–424.

CHAPTER 8

Cortical Determinants of Goal-Directed Behavior

Etienne Coutureau[1,2], Shauna L. Parkes[1,2]

[1]CNRS, INCIA, UMR 5287, Bordeaux, France; [2]Université de Bordeaux, INCIA, UMR 5287, Bordeaux, France

INTRODUCTION

Appropriate decision-making is critical for adapting to a changing environment. Every day, we must make decisions based on internal goals, and the expectation that a given action will lead to goal achievement. Such decisions are experimentally defined as "goal-directed." Several regions of the mammalian cortex are involved in the integration of sensory, affective, and cognitive information to guide flexible choice between competing actions. Current evidence indicates that, in the rat, these regions principally involve the medial prefrontal cortex (mPFC), orbitofrontal cortex (OFC), and insular cortex (IC). Importantly, the emerging view is that each of these areas provides a distinct contribution to goal-directed behavior. Cortical coordination may therefore prove essential to flexible action control. This chapter outlines what we know about the involvement of these cortices in goal-directed behavior and proposes avenues for future research in the cortical control of choice.

We have focused our review on studies using free operant tasks (rather than stimulus-guided tasks) and causal interventions. However, it should be noted that much research exists on the responses of cortical neurons during goal-directed tasks (e.g., Furuyashiki, Holland, & Gallagher, 2008; Schoenbaum, Chiba, & Gallagher, 1998; Simon, Wood, & Moghaddam, 2015; Whitaker et al., 2017) and, while these studies are not reviewed here, correlational data from electrophysiological recordings and ex vivo imaging largely support our conclusions. The majority of the studies discussed in this chapter use the instrumental outcome devaluation paradigm (e.g., Adams & Dickinson, 1981; Colwill & Rescorla, 1985). This paradigm represents a powerful tool in the study of goal-directed behavior and is widely used across species. The reason for this is twofold. First, the paradigm offers a detailed behavioral analysis. It allows an experimental decomposition of action selection into its associative learning processes, including contingency learning (between actions and their consequences), learning the value of those consequences (in accord with one's motivational state, i.e., incentive learning) and, finally, integrating this information to guide choice. Second, our strong understanding of the associative learning processes involved in this task makes it amenable to stringent

Goal-Directed Decision Making
ISBN 978-0-12-812098-9, https://doi.org/10.1016/B978-0-12-812098-9.00008-5

neurobiological analyses. That is, a thorough understanding of the behavior can drive investigations into the brain regions and circuits underlying action control.

MEDIAL PREFRONTAL CORTEX

Anatomical considerations

The mPFC corresponds to the major portion of the medial wall of the anterior hemisphere and is dorsal to the genu of the corpus callosum. In rodents, this region has been traditionally divided into four distinct areas: the medial precentral (PrCm) or area Fr2, the anterior cingulate, the prelimbic (PL), and the infralimbic (IL) area. These latter two regions of the prefrontal cortex have a well-established role in instrumental behavior, and a number of elegant studies have described their distinct contribution to goal-directed action and choice.

Medial prefrontal cortex and goal-directed behavior

The anatomical heterogeneity of mPFC is reflected in the coordination of goal-directed and habitual responding. Rats with pretraining lesions or pharmacological inactivation affecting IL cortex maintain goal-directed responding under overtraining conditions that normally promote habit (stimulus-response) responding (Coutureau & Killcross, 2003; Killcross & Coutureau, 2003). By contrast, lesions of PL cortex result in performance that is insensitive to outcome devaluation under conditions that should support goal-directed (action–outcome) responding (Balleine & Dickinson, 1998; Corbit & Balleine, 2003; Coutureau, Marchand, & Di Scala, 2009; Killcross & Coutureau, 2003). Chronic stress also renders instrumental responding insensitive to outcome devaluation in mice, which might result from profound structural changes in the PL (Dias-Ferreira et al., 2009). The role of the PL in goal-directed behavior is also reflected at the cellular level since inhibition of Rho kinase in PL following action–outcome training maintains goal-directed control under training conditions known to produce habitual control (Swanson, DePoy, & Gourley, 2017).

Damage to mPFC also alters adaptation to changes in instrumental contingencies (Balleine & Dickinson, 1998; Corbit & Balleine, 2003; Swanson et al., 2017) but only under some circumstances. Indeed, whereas mPFC-lesioned (damage to dorsal and ventral regions) rats remain able to learn a shift to a negative contingency, they appear unable to correctly detect changes when the contingency is shifted to a null contingency (Coutureau, Esclassan, Di Scala, & Marchand, 2012). Importantly, this deficit does not reflect a different perception of the temporal relationship between the response and the outcome since mPFC-lesioned rats demonstrate normal sensitivity to changes in contiguity under delayed reward conditions (Coutureau et al., 2012). That is, both sham- and mPFC-lesioned rats maintain instrumental responding when rewards are immediate, but response rates in both groups decrease with increasing delays. Rats with lesions of the mPFC are therefore capable of some adaptation to shifts in instrumental contingency or contiguity, yet these rats have difficulty evaluating the balance between contingent and noncontingent reinforcement, as is the case in contingency degradation.

Follow-up studies have demonstrated that the role of PL in goal-directed action control is transient. It is required for the early stages of acquisition but not for the expression of goal-directed behavior (Ostlund & Balleine, 2005; Tran-Tu-Yen, Marchand, Pape, Di Scala, & Coutureau, 2009). Therefore, this brain region is selectively involved in the formation of action–outcome associations, but expression or storage of those associations occurs elsewhere. A key target of PL is the posterior dorsomedial striatum (pDMS). Disruption of pDMS function either pre- or posttraining impairs instrumental outcome devaluation and contingency degradation (Yin, Knowlton, & Balleine, 2005; Yin, Ostlund, Knowlton, & Balleine, 2005), thus suggesting that this region is critical for both the acquisition and performance of goal-directed actions and might be specifically involved in the storage of action–outcome learning (Shiflett & Balleine, 2011; Shiflett, Brown, & Balleine, 2010). The hypothesis that goal-directed behavior arises from functional interaction between PL and pDMS has received recent support. Increased MAPK/ERK phosphorylation (pERK), a marker of learning and memory, was revealed in distinct layers of posterior PL shortly after an instrumental training session, and this increase in neuronal activity was specific to prefrontal neurons projecting to pDMS (Hart & Balleine, 2016). In addition, bilateral disconnection of PL and pDMS impairs the acquisition of goal-directed actions (Hart & Balleine, 2018).

Medial prefrontal cortex summary

Current evidence indicates that mPFC plays a crucial role in the coordination of goal-directed and habitual responding. The anatomical characteristics of this brain region are critical for understanding its function; the dorsal portion of mPFC, including PL, appears specifically involved in learning goal-directed behavior while the ventral portion, including IL, mediates habit learning. Inputs from PL to pDMS are likely crucial for the striatal plasticity underlying goal-directed learning to occur (Hart & Balleine, 2016; Xiong, Znamenskiy, & Zador, 2015). The learning of habitual responding by ventral mPFC is far less understood but possibly results from complex interactions either within mPFC or with dorsolateral striatum, given that this region has repeatedly been reported to mediate habit learning (Quinn, Pittenger, Lee, Pierson, & Taylor, 2013; Shan, Christie, & Balleine, 2015; Tricomi, Balleine, & O'Doherty, 2009; Yin, Knowlton, & Balleine, 2004).

ORBITOFRONTAL CORTEX

Anatomical considerations

The OFC is situated rostral to the IC in the dorsal bank of the rhinal sulcus. Composed exclusively of agranular cortical areas, OFC receives sensory inputs from olfactory, gustatory, somatosensory, and visual areas and participates in high-level cognitive processes

(Ongur & Price, 2000). Several cytological divisions have been delineated, including ventral (VO), lateral (LO), and dorsolateral subregions as well as a medial area (MO) that is located in mPFC, below cingulate area 32 or IL (Paxinos & Watson, 2014). A specific ventrolateral region has also been defined (Van De Werd & Uylings, 2008) that receives dense innervation from the submedius nucleus in the medial thalamus (Reep, Corwin, & King, 1996; Tang, Qu, & Huo, 2009; Yoshida, Dostrovsky, & Chiang, 1992). Here, we will review evidence on the involvement of medial versus ventral and lateral areas in goal-directed behavior.

Orbitofrontal cortex and goal-directed behavior

The involvement of OFC in goal-directed behavior is a matter of debate. Much of this debate centers on the distinction between stimulus-guided versus outcome-guided behavior. One view is that VO and LO are involved in the former (Gallagher, McMahan, & Schoenbaum, 1999; Izquierdo & Murray, 2004, 2010; Izquierdo, Suda, & Murray, 2004; Machado & Bachevalier, 2007; Ostlund & Balleine, 2007b; Pickens, Saddoris, Gallagher, & Holland, 2005; Pickens et al., 2003; West, DesJardin, Gale, & Malkova, 2011) but not the latter (Balleine, Leung, & Ostlund, 2011; Fellows, 2011; Luk & Wallis, 2013; Ostlund & Balleine, 2007a, 2007b; Roberts, 2006; Rudebeck et al., 2008). For instance, in rats, lesions of VO and LO cause impairments in Pavlovian reinforcer devaluation (Gallagher et al., 1999) and specific Pavlovian-to-instrumental transfer (PIT) but leave sensitivity to instrumental outcome devaluation intact (Ostlund & Balleine, 2007b; Parkes et al., 2017).

However, emerging reports show that, in some instances, inhibition of VO and LO impairs instrumental outcome devaluation in both rodents and primates (Fiuzat, Rhodes, & Murray, 2017; Gremel & Costa, 2013; Gremel et al., 2016; Rhodes & Murray, 2013; Zimmermann, Yamin, Rainnie, Ressler, & Gourley, 2017), and we recently proposed that VO and LO subregions are recruited to resolve ambiguity in action—outcome associations (Parkes et al., 2017). Consistent with this suggestion, inhibition of VO and LO impairs goal-directed behavior when the subject is required to update previously established instrumental associations (Parkes et al., 2017) or when the performance of the same instrumental response differs depending on the context in which it is tested (Gremel & Costa, 2013; Gremel et al., 2016). These results suggest that VO and LO play a critical role in the online tracking of the current relationships between actions and their specific consequences and may help resolve the apparent inconsistencies in the literature regarding the role of VO and LO in goal-directed behavior.

By contrast, others have argued that it is MO, not VO and LO, which regulates goal-directed action (Bradfield, Dezfouli, van Holstein, Chieng, & Balleine, 2015; Gourley, Zimmermann, Allen, & Taylor, 2016). Lesions restricted to MO have been reported to produce deficits in instrumental tasks that rely on retrieving a representation of the outcome, including specific PIT and outcome devaluation (under extinction conditions), but lesions have no effect on outcome-selective reinstatement, contingency degradation, or outcome devaluation under rewarded conditions (Bradfield et al., 2015). However, using similar procedures, others have reported no effect of MO lesions on outcome devaluation (Gourley, Lee, Howell, Pittenger, & Taylor, 2010; Munster & Hauber, 2017) or specific PIT (Munster & Hauber, 2017). Nevertheless, a segregated view of OFC function has been proposed by some, which supports dissociable contributions of MO and VO/LO to reward-guided behavior; MO for action-dependent outcome retrieval and VO/LO for stimulus-dependent outcome retrieval (Bradfield et al., 2015; Rudebeck & Murray, 2011).

Our recent results provide clear evidence against this simple distinction and suggest a specific role for VO/LO in action-guided behavior but only when expectations are violated (Parkes et al., 2017). This is consistent with the view that behaviors relying on the explicit use of learned associations may be independent of VO and LO function (Schoenbaum & Roesch, 2005). Indeed, while VO and LO are not required for goal-directed behavior when instrumental contingencies remain stable (Ostlund & Balleine, 2007b; Parkes et al., 2017), this region is required for behavioral tasks that generate ambiguity, including Pavlovian reversal learning (Chudasama & Robbins, 2003; Dias, Robbins, & Roberts, 1997; Jones & Mishkin, 1972; Rudebeck & Murray, 2008; Schoenbaum, Nugent, Saddoris, & Setlow, 2002), contingency degradation (Alcaraz et al., 2015; Ostlund & Balleine, 2007b) and choice guided by learned taste aversions (Ramirez-Lugo, Penas-Rincon, Angeles-Duran, & Sotres-Bayon, 2016). It has also been recently proposed that more medial areas of OFC are involved in value-based processing, whereas the more lateral regions are critical for sensory processing of both conditioned and unconditioned stimuli (Izquierdo, 2017). This is largely consistent with the current literature including electrophysiological reports of MO neurons signaling value and the numerous studies illustrating impaired reversal learning in rats with VO and LO lesions (see Izquierdo, 2017 for a comprehensive review).

Finally, MO and VO/LO share distinct connections with other cortical areas and limbic regions, including the striatum (Hoover & Vertes, 2011). Neurons in both MO and VO/LO send projections to dorsal and ventral striatum; however, VO neurons do not project to nucleus accumbens (Hoover & Vertes, 2011). Indeed, chemogenetic-induced inhibition of the VO/LO to dorsal striatal pathway in mice abolishes instrumental outcome devaluation (Gremel et al., 2016), suggesting that activation of dorsal striatum-projecting vlOFC neurons is necessary for goal-directed control. The involvement of MO projections to striatum in action selection remains to be investigated.

Orbitofrontal cortex summary

In our view, there is clear evidence that both MO and VO/LO coordinate aspects of goal-directed action control. The distinction drawn between MO and VO/LO in the regulation of stimulus-dependent versus action-dependent outcome retrieval appears overly simplified. Indeed, a number of studies now implicate VO and LO in action-dependent tasks (Fiuzat et al., 2017; Gremel & Costa, 2013; Gremel et al., 2016; Zimmermann et al., 2017), particularly, in tracking the current relationship between actions and their consequences (Parkes et al., 2017). It should be noted that the aforementioned studies indiscriminately targeted both VO and LO subregions of OFC, but these regions might form a part of different anatomical and functional networks (Ongur & Price, 2000; Price, 2007). As such, manipulations that selectively manipulate VO versus LO will prove worthwhile to our understanding of OFC's involvement in goal-directed behavior (Izquierdo, 2017).

INSULAR CORTEX

Anatomical considerations

The rat IC is a longitudinal strip, occupying the dorsal bank of the rhinal sulcus. It extends ventrally to the piriform cortex and dorsally to the somatosensory cortex. In the rostro-caudal direction, IC spans from lateral OFC to perirhinal cortex (Saper, 1982). Its most rostral regions (termed here "anterior IC") are generally considered to represent the lateral prefrontal cortex. The insular region located caudal to the lateral prefrontal cortex contains gustatory, visceral, and somatosensory representations. Extensive cross talk exists in the rostrocaudal axis as well as between the distinct layers of IC (granular, dysgranular, and agranular) along the dorsoventral axis and, as such, the layers are typically not considered to be independent (Fujita, Adachi, Koshikawa, & Kobayashi, 2010; Fujita, Koshikawa, & Kobayashi, 2011; Mizoguchi, Fujita, Koshikawa, & Kobayashi, 2011). A myriad of approaches including electrophysiology (Kosar, Grill, & Norgren, 1986), anatomy (Krettek & Price, 1977; Saper, 1982; Shi & Cassell, 1998), and behavior (Accolla, Bathellier, Petersen, & Carleton, 2007; Nerad, Ramirez-Amaya, Ormsby, & Bermudez-Rattoni, 1996; Saddoris, Holland, & Gallagher, 2009) have localized a region of IC that is critical for encoding taste-related information, the so-called gustatory cortex (GC), which occupies approximately 15% of the total area of IC (Kosar et al., 1986; Yamamoto, Matsuo, & Kawamura, 1980). Neuroanatomical tracing studies show that GC is reciprocally connected to limbic structures involved in affective processing, including the prefrontal cortex, amygdala, and ventral striatum (Allen, Saper, Hurley, & Cechetto, 1991; Gabbott, Warner, Jays, & Bacon, 2003) making this region an ideal candidate to modulate goal-directed behavior.

Insular cortex and goal-directed behavior

Taste processing is arguably GC's most recognized function. The GC receives taste-related and visceral information via the thalamus and the parabrachial nucleus

(Allen et al., 1991; Cechetto & Saper, 1987), tactile information about the mouth and tongue from the somatosensory cortex and sensory information related to other modalities including olfaction and audition (for a review see Maffei, Haley, & Fontanini, 2012). However, beyond these sensory inputs, GC receives affective, anticipatory, and reward-related information from limbic areas including the mediodorsal thalamus, prefrontal cortex, and amygdala (Allen et al., 1991; Hoover & Vertes, 2011; Shi & Cassell, 1998).

In 2000, Balleine and Dickinson provided the first evidence that GC modulates goal-directed behavior for a food reward. Rats with pretraining excitotoxic lesions of GC showed impaired satiety-induced instrumental outcome devaluation when tested under unrewarded conditions. However, in a subsequent rewarded test, where performance on the actions delivered their associated outcomes, both sham- and GC-lesioned rats showed selective outcome devaluation and successfully biased their choice toward the lever associated with the valued outcome. GC lesions therefore induce a deficit in the ability to recall the representation of the devalued food and not a deficit in the detection of the primary taste itself or the assignment of incentive value to the taste (Balleine & Dickinson, 2000). More recent studies using temporary perturbation of GC function have confirmed its role in the retrieval of outcome value to guide choice. Outcome devaluation is impaired by pharmacological (Parkes & Balleine, 2013) or chemogenetic (Parkes et al., 2017) disruption of GC during a choice extinction test. By contrast, similar manipulations during the acquisition phase have no effect on instrumental learning per se (Parkes, Ferreira, & Coutureau, 2016; Parkes et al., 2017), and pretraining lesions leave contingency degradation intact (Balleine & Dickinson, 2000). Taken together, these studies provide clear evidence that GC is necessary to recall the current incentive value of food rewards, but it is not required to learn the association between actions and their corresponding outcomes.

GC likely achieves this behavioral modulation via its connections with the basolateral amygdala (BLA) (Parkes & Balleine, 2013) and the ventral striatum, particularly the core region of the nucleus accumbens (NAc) (Allen et al., 1991; Parkes, Bradfield, & Balleine, 2015), both of which have been implicated in goal-directed behavior. Using a sequential disconnection procedure, it was revealed that communication between the BLA and GC mediates the encoding and retrieval of outcome value (Parkes & Balleine, 2013). By contrast, interactions between GC and NAc core appear to mediate the effect of outcome value on the choice between actions. Indeed, outcome devaluation is attenuated following pretraining disconnection of GC and NAc core via asymmetrical excitotoxic lesions or by temporary, pharmacological disconnection during the choice test (Parkes et al., 2015). Therefore, the BLA-GC pathway mediates the encoding and retrieval of outcome value, whereas the GC-core pathway is involved in the retrieval of current outcome value and subsequent performance based on that value.

Insular cortex summary

The gustatory portion of IC (GC) plays a general role in guiding behavior based on the motivational value of expected food outcomes. GC modulates goal-directed behavior not by encoding the relationship between an action and its outcome but rather by recalling the current value of a specific sensory outcome to guide adaptive choice. It must be noted that the involvement of GC in goal-directed behavior may be limited to situations where the "goal" is a taste- or food-related reward. It is perhaps unlikely that the GC would recall the value of sexual or thermoregulatory rewards, for example, but another region of IC could be involved (Karama et al., 2002). At present, it is not clear if other regions of IC (particularly, anterior IC) are also involved in action selection but, given the existence of intrainsular connections, this requires attention.

FUTURE DIRECTIONS

Cortico–cortical interactions

Over the past few decades, investigations into the neural substrates of goal-directed behavior have revealed a crucial role for cortico–subcortical interactions, including corticostriatal and thalamocortical pathways (Balleine & O'Doherty, 2010; Hart, Leung, & Balleine, 2014; Jin & Costa, 2015). However, we do not yet understand how cortical regions might communicate to support complex cognitive functions, including action selection. While little causal research has examined the role of cortico–cortical pathways in choice, imaging studies in humans have revealed parallel phasic activation of mPFC and OFC during Pavlovian modulation of choice (Homayoun & Moghaddam, 2009) and cortico–cortical interactions appear to underlie adverse decision-making in human patients with obsessive compulsive disorder (Schlosser et al., 2010).

Rodent mPFC, OFC, and IC share a high degree of interconnectivity. Both dorsal and ventral mPFC send projections throughout IC, and regions of IC project back to mPFC, although direct connections from GC to mPFC have not been well studied (Allen et al., 1991; Gabbott et al., 2003). mPFC (dorsal and ventral subregions) and OFC (medial, ventral, and lateral subregions) are also reciprocally connected (Hoover & Vertes, 2011; Vertes, 2004) as are IC and OFC (Fujita et al., 2010; Hoover & Vertes, 2011; Shi & Cassell, 1998). These intrinsic cortico–cortical connections may play a critical role in choice behavior, but this remains to be investigated. For instance, given the strong intrainsular connections (Fujita et al., 2010) and the reciprocal projections between OFC and IC (Fujita et al., 2010; Hoover & Vertes, 2011; Shi & Cassell, 1998), communication between OFC and IC may be required to integrate knowledge about current action–outcome contingencies with goal value to guide action selection.

Direct cortico–cortical projections are often paralleled by transthalamic routes (Sherman, 2016; Theyel, Llano, & Sherman, 2010). Information can therefore be relayed directly between cortices or via connections with the various thalamic nuclei. Transthalamic pathways may allow information to be modulated by thalamic circuits and to be shared with subcortical sites; neither of which is possible via direct cortico—cortical communication (Sherman, 2016). Recently, the mediodorsal thalamic nucleus (MD) has been investigated as a critical site for goal-directed action control (Balleine, Morris, & Leung, 2015; Bradfield, Bertran-Gonzalez, Chieng, & Balleine, 2013; Parnaudeau et al., 2015). MD projects throughout prefrontal cortex, yet recent evidence indicates that highly segregated populations of MD neurons project to either mPFC or OFC (Alcaraz, Marchand, Courtand, Coutureau, & Wolff, 2016). Indeed, disconnection of MD and the dorsal region of mPFC by lesion or chemogenetics impairs the acquisition of goal-directed actions (Bradfield, Hart, & Balleine, 2013; Alcaraz et al., 2018). The role of these distinct projections in goal-directed behavior warrants further investigation.

Neuromodulatory systems

Thus far, we have focused our analysis on circuits of goal-directed behavior, without mentioning that these circuits are regulated by the major neuromodulatory systems. Dopaminergic modulation of the prefrontal cortex has long been shown to influence decision-making processes (Boureau & Dayan, 2011; Costa, 2007; Daw, Kakade, & Dayan, 2002; Dayan & Balleine, 2002; Floresco, 2007; Niv & Schoenbaum, 2008; Palmiter, 2008; Rutledge, Skandali, Dayan, & Dolan, 2015; Schultz, 2010; Wickens, Horvitz, Costa, & Killcross, 2007). In rodents, strong dopaminergic input from ventral tegmental area is observed throughout mPFC and in some regions of OFC, including medial OFC (Berger, Thierry, Tassin, & Moyne, 1976; Uylings, Groenewegen, & Kolb, 2003; Van De Werd & Uylings, 2008). The involvement of dopaminergic innervation of mPFC in goal-directed control has been clearly demonstrated. Using 6-OHDA lesions, it was shown that the loss of dopaminergic signaling in PL, but not IL, abolished the sensitivity of an instrumental response to contingency degradation, an effect replicated by the blockade of dopamine D1/D2 receptor at the time of contingency degradation (Naneix, Marchand, Di Scala, Pape, & Coutureau, 2009; but see; Lex & Hauber, 2010).

Follow-up studies have shown a time-dependent parallel maturation of the mesocortical dopaminergic system and goal-directed control throughout adolescence (Naneix, Marchand, Di Scala, Pape, & Coutureau, 2012) as well as a causal relationship between the two processes (Naneix, Marchand, Pichon, Pape, & Coutureau, 2013). Notably, the modulatory role of dopamine is not restricted to PL since microinfusion of dopamine directly into IL promotes the expression of goal-directed responding in conditions under which performance is normally habitual (Hitchcott, Quinn, & Taylor, 2007). The IC also receives dopaminergic input (Van De Werd & Uylings, 2008), and, given its role in goal-directed performance, the involvement of these inputs may also be functionally important.

The role of the other major neuromodulatory systems in the control of goal-directed behavior has received far less attention. However, noradrenergic inputs arising from locus coeruleus have long been thought to modulate the function of the prefrontal cortex, typically through attention or sensory gating (Nutt, Lalies, Lione, & Hudson, 1997), and this modulatory role has been integrated in recent theories of prefrontal function. More precisely, current models of OFC and the noradrenergic system (Sadacca, Wikenheiser, & Schoenbaum, 2017; Sara, 2016) suggest that both play a role in responding to environmental change, when flexibility and updating previous knowledge without unlearning is required in the face of a changing environment. Future research is therefore needed to understand the role of noradrenergic inputs to the prefrontal cortex in goal-directed behavior.

Homologous regions in rat and primate

We have limited our discussion to the involvement of rodent mPFC, OFC, and IC in goal-directed behavior. While an extensive comparative analysis is beyond the scope of this chapter, it should be noted that a wealth of research is aimed at understanding the degree of homology between rodent and primate cortices; a topic that remains the focus of intense discussion. Based on anatomical connectivity, density of connections, neurotransmitter types, embryological development, and cytoarchitectonics, it appears that a functional similarity exists between rodent mPFC and human ventromedial prefrontal—medial orbital cortex (Brown & Bowman, 2002; Uylings et al., 2003). Most notably, in the context of the present chapter, functional imaging studies using structurally similar tasks (see de Wit & Dickinson, 2009) indicate a considerable overlap in mPFC determinants of goal-directed actions in humans and rodents (Balleine & O'Doherty, 2010; O'Doherty, Cockburn, & Pauli, 2017). Importantly, current in-depth analyses of the cytoarchitectonic features of rodent mPFC propose a delineation based on cingulate areas rather than a distinction between PL and IL (Vogt, 2016). Consistent with this, the current edition of the widely used *Rat Brain Atlas in Stereotaxic Coordinates* (Paxinos & Watson, 2014) outlines a main subdivision of mPFC into a dorsal component, termed cingulate area 32 dorsal (A32D), which encompasses the FR2, dorsal anterior cingulate areas, and the dorsal part of PL, and a ventral component, termed cingulate area 32 ventral (A32V) that includes the ventral PL, IL, and some medial orbital areas (Paxinos & Watson, 2014; Vogt & Paxinos, 2014; Vogt et al., 2013). It is proposed that this dissociation (based on cingulate terminology) may prove more useful in tracking functional homologies between rodents and primates (Vogt, 2016).

Primate OFC has also been implicated in value-based behaviors. While there is some concern as to whether the rodent OFC (composed exclusively of agranular cortical areas) is homologous to the larger, granular OFC in primates (Preuss, 1995; Wise, 2008), equivalent areas have been identified on the basis of structural similarities as well as corticothalamic and corticostriatal connections (Price, 2007; Uylings et al., 2003). As in the rodent, functional differences within the primate OFC have been identified in both the mediolateral (Bouret & Richmond, 2010; Noonan et al., 2010; Rudebeck & Murray, 2011) and rostrocaudal axes (Murray, Moylan, Saleem, Basile, & Turchi, 2015). For example, in a particularly elegant outcome devaluation study in macaques, Murray and colleagues illustrated a dissociable role for anterior (area 11) versus posterior (area 13) OFC in action selection, with the former required for the choice between actions and the latter for encoding changes in outcome value (Murray et al., 2015).

Beyond the prefrontal cortex, similarities also exist between rodent and primate IC. The taste-processing cortical region in primates is also located in IC and receives afferents from the thalamus and several sensory cortices, is reciprocally connected with the amygdala, and projects to the ventral striatum (Gallay, Gallay, Jeanmonod, Rouiller, & Morel, 2012; Small, 2010). Functionally, neural activity in human insular GC reflects both the basic features of taste and valuation of taste-related rewards (Small, 2010). For example, human IC is more active during water intake when a subject is thirsty than when they are sated (de Araujo, Kringelbach, Rolls, & McGlone, 2003).

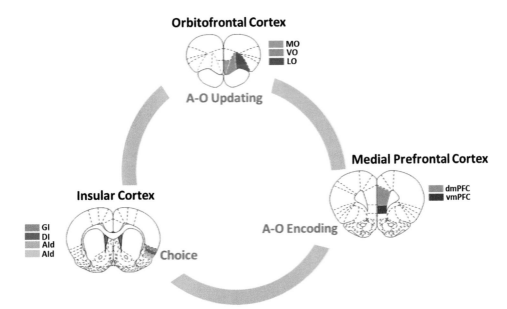

Figure 8.1 This schematic provides a simplified overview of the anatomical considerations and functional contribution of the three cortices to goal-directed behavior. The medial prefrontal cortex (mPFC), including dorsal (dmPFC) and ventral (vmPFC) subregions, is critical for action–outcome (A-O) encoding and the balance of action and habits. The orbitofrontal cortex (OFC) comprises a number of functionally distinct areas; namely, medial (MO), ventral (VO), and lateral (LO) areas, each of which plays a role in goal-directed behavior. Finally, IC, specifically the gustatory region (GC), is required for the retrieval of outcome value to guide choice. IC is composed of four layers that represented the gradual disappearance of the granular layer: granular (GI), dysgranular (DI), agranular dorsal (AId), and agranular ventral (AIv).

CONCLUSIONS

Distinct cortical regions are required for goal-directed behavior (see Fig. 8.1). While mPFC has traditionally been the focus of investigation, more recent studies have implicated both OFC and IC in the regulation of choice. Critically, different functional contributions within each cortex have also been identified. As previously mentioned, we have limited our discussion to studies using rodents. An important consideration is the extent to which these studies can be applied to other species. Evidence from other species, including humans and nonhuman primates, is generally supportive of the ideas discussed earlier, particularly insofar as acknowledging the distinct cortical contributions to goal-directed behavior (Balleine & O'Doherty, 2010; Rhodes & Murray, 2013; Rudebeck et al., 2008). We must therefore continue to investigate how the cortex modulates goal-directed control and the specific involvement of cortico—cortical interactions in this behavior.

REFERENCES

Accolla, R., Bathellier, B., Petersen, C. C., & Carleton, A. (2007). Differential spatial representation of taste modalities in the rat gustatory cortex. *The Journal of Neuroscience: The Official Journal of the Society for Neuroscience, 27*(6), 1396–1404. https://doi.org/10.1523/jneurosci.5188-06.2007.

Adams, C. D., & Dickinson, A. (1981). Instrumental responding following reinforcer devaluation. *The Quarterly Journal of Experimental Psychology Section B, 1981*(33), 109–122.

Alcaraz, F., Fresno, V., Marchand, A. R., Kremer, E. J., Coutureau, E., & Wolff, M. (2018). Thalamocortical and corticothalamic pathways differentially contribute to goal-directed behaviors in the rat. *Elife, 7*, e32517. https://doi.org/10.7554/eLife.32517.

Alcaraz, F., Marchand, A. R., Courtand, G., Coutureau, E., & Wolff, M. (2016). Parallel inputs from the mediodorsal thalamus to the prefrontal cortex in the rat. *The European Journal of Neuroscience, 44*(3), 1972–1986. https://doi.org/10.1111/ejn.13316.

Alcaraz, F., Marchand, A. R., Vidal, E., Guillou, A., Faugere, A., Coutureau, E., & Wolff, M. (2015). Flexible use of predictive cues beyond the orbitofrontal cortex: Role of the submedius thalamic nucleus. *The Journal of Neuroscience: The Official Journal of the Society for Neuroscience, 35*(38), 13183–13193. https://doi.org/10.1523/jneurosci.1237-15.2015.

Allen, G. V., Saper, C. B., Hurley, K. M., & Cechetto, D. F. (1991). Organization of visceral and limbic connections in the insular cortex of the rat. *The Journal of Comparative Neurology, 311*(1), 1–16. https://doi.org/10.1002/cne.903110102.

Balleine, B. W., & Dickinson, A. (1998). Goal-directed instrumental action: Contingency and incentive learning and their cortical substrates. *Neuropharmacology, 37*(4–5), 407–419.

Balleine, B. W., & Dickinson, A. (2000). The effect of lesions of the insular cortex on instrumental conditioning: Evidence for a role in incentive memory. *The Journal of Neuroscience: The Official Journal of the Society for Neuroscience, 20*(23), 8954–8964.

Balleine, B. W., Leung, B. K., & Ostlund, S. B. (2011). The orbitofrontal cortex, predicted value, and choice. *Annals of the New York Academy of Sciences, 1239*, 43–50. https://doi.org/10.1111/j.1749-6632.2011.06270.x.

Balleine, B. W., Morris, R. W., & Leung, B. K. (2015). Thalamocortical integration of instrumental learning and performance and their disintegration in addiction. *Brain Research, 1628*(Pt A), 104–116. https://doi.org/10.1016/j.brainres.2014.12.023.

Balleine, B. W., & O'Doherty, J. P. (2010). Human and rodent homologies in action control: Corticostriatal determinants of goal-directed and habitual action. *Neuropsychopharmacology: Official Publication of the American College of Neuropsychopharmacology, 35*(1), 48–69. https://doi.org/10.1038/npp.2009.131.

Berger, B., Thierry, A. M., Tassin, J. P., & Moyne, M. A. (1976). Dopaminergic innervation of the rat prefrontal cortex: A fluorescence histochemical study. *Brain Research, 106*(1), 133–145.

Boureau, Y. L., & Dayan, P. (2011). Opponency revisited: Competition and cooperation between dopamine and serotonin. *Neuropsychopharmacology: Official Publication of the American College of Neuropsychopharmacology, 36*(1), 74–97. https://doi.org/10.1038/npp.2010.151.

Bouret, S., & Richmond, B. J. (2010). Ventromedial and orbital prefrontal neurons differentially encode internally and externally driven motivational values in monkeys. *The Journal of Neuroscience: The Official Journal of the Society for Neuroscience, 30*(25), 8591–8601. https://doi.org/10.1523/jneurosci.0049-10.2010.

Bradfield, L. A., Bertran-Gonzalez, J., Chieng, B., & Balleine, B. W. (2013). The thalamostriatal pathway and cholinergic control of goal-directed action: Interlacing new with existing learning in the striatum. *Neuron, 79*(1), 153–166. https://doi.org/10.1016/j.neuron.2013.04.039.

Bradfield, L. A., Dezfouli, A., van Holstein, M., Chieng, B., & Balleine, B. W. (2015). Medial orbitofrontal cortex mediates outcome retrieval in partially observable task situations. *Neuron, 88*(6), 1268–1280. https://doi.org/10.1016/j.neuron.2015.10.044.

Bradfield, L. A., Hart, G., & Balleine, B. W. (2013). The role of the anterior, mediodorsal, and parafascicular thalamus in instrumental conditioning. *Frontiers in Systems Neuroscience, 7*, 51. https://doi.org/10.3389/fnsys.2013.00051.

Brown, V. J., & Bowman, E. M. (2002). Rodent models of prefrontal cortical function. *Trends in Neurosciences, 25*(7), 340–343.

Cechetto, D. F., & Saper, C. B. (1987). Evidence for a viscerotopic sensory representation in the cortex and thalamus in the rat. *The Journal of Comparative Neurology, 262*(1), 27–45. https://doi.org/10.1002/cne.902620104.

Chudasama, Y., & Robbins, T. W. (2003). Dissociable contributions of the orbitofrontal and infralimbic cortex to pavlovian autoshaping and discrimination reversal learning: Further evidence for the functional heterogeneity of the rodent frontal cortex. *The Journal of Neuroscience: The Official Journal of the Society for Neuroscience, 23*(25), 8771–8780.

Colwill, R. M., & Rescorla, R. A. (1985). Instrumental responding remains sensitive to reinforcer devaluation after extensive training. *Journal of Experimental Psychology: Animal Behavior Processes, 11*(4), 520–536. https://doi.org/10.1037/0097-7403.11.4.520.

Corbit, L. H., & Balleine, B. W. (2003). The role of prelimbic cortex in instrumental conditioning. *Behavioural Brain Research, 146*(1–2), 145–157.

Costa, R. M. (2007). Plastic corticostriatal circuits for action learning: what's dopamine got to do with it? *Annals of the New York Academy of Sciences, 1104*, 172–191. https://doi.org/10.1196/annals.1390.015.

Coutureau, E., Esclassan, F., Di Scala, G., & Marchand, A. R. (2012). The role of the rat medial prefrontal cortex in adapting to changes in instrumental contingency. *PLoS One, 7*(4), e33302. https://doi.org/10.1371/journal.pone.0033302.

Coutureau, E., & Killcross, S. (2003). Inactivation of the infralimbic prefrontal cortex reinstates goal-directed responding in overtrained rats. *Behavioural Brain Research, 146*(1–2), 167–174.

Coutureau, E., Marchand, A. R., & Di Scala, G. (2009). Goal-directed responding is sensitive to lesions to the prelimbic cortex or basolateral nucleus of the amygdala but not to their disconnection. *Behavioral Neuroscience, 123*(2), 443–448. https://doi.org/10.1037/a0014818.

Daw, N. D., Kakade, S., & Dayan, P. (2002). Opponent interactions between serotonin and dopamine. *Neural Networks: The Official Journal of the International Neural Network Society, 15*(4–6), 603–616.

Dayan, P., & Balleine, B. W. (2002). Reward, motivation, and reinforcement learning. *Neuron, 36*(2), 285–298.

de Araujo, I. E., Kringelbach, M. L., Rolls, E. T., & McGlone, F. (2003). Human cortical responses to water in the mouth, and the effects of thirst. *Journal of Neurophysiology, 90*(3), 1865–1876. https://doi.org/10.1152/jn.00297.2003.

de Wit, S., & Dickinson, A. (2009). Associative theories of goal-directed behaviour: A case for animal-human translational models. *Psychological Research, 73*(4), 463–476. https://doi.org/10.1007/s00426-009-0230-6.

Dias-Ferreira, E., Sousa, J. C., Melo, I., Morgado, P., Mesquita, A. R., Cerqueira, J. J., … Sousa, N. (2009). Chronic stress causes frontostriatal reorganization and affects decision-making. *Science, 325*(5940), 621–625. https://doi.org/10.1126/science.1171203.

Dias, R., Robbins, T. W., & Roberts, A. C. (1997). Dissociable forms of inhibitory control within prefrontal cortex with an analog of the Wisconsin card sort test: Restriction to novel situations and independence from "on-line" processing. *The Journal of Neuroscience: The Official Journal of the Society for Neuroscience, 17*(23), 9285—9297.

Fellows, L. K. (2011). Orbitofrontal contributions to value-based decision making: Evidence from humans with frontal lobe damage. *Annals of the New York Academy of Sciences, 1239,* 51—58. https://doi.org/10.1111/j.1749-6632.2011.06229.x.

Fiuzat, E. C., Rhodes, S. E., & Murray, E. A. (2017). The role of orbitofrontal-amygdala interactions in updating action-outcome valuations in macaques. *The Journal of Neuroscience: The Official Journal of the Society for Neuroscience.* https://doi.org/10.1523/jneurosci.1839-16.2017.

Floresco, S. B. (2007). Dopaminergic regulation of limbic-striatal interplay. *Journal of Psychiatry & Neuroscience, 32*(6), 400—411.

Fujita, S., Adachi, K., Koshikawa, N., & Kobayashi, M. (2010). Spatiotemporal dynamics of excitation in rat insular cortex: Intrinsic corticocortical circuit regulates caudal-rostro excitatory propagation from the insular to frontal cortex. *Neuroscience, 165*(1), 278—292. https://doi.org/10.1016/j.neuroscience.2009.09.073.

Fujita, S., Koshikawa, N., & Kobayashi, M. (2011). GABA(B) receptors accentuate neural excitation contrast in rat insular cortex. *Neuroscience, 199,* 259—271. https://doi.org/10.1016/j.neuroscience.2011.09.043.

Furuyashiki, T., Holland, P. C., & Gallagher, M. (2008). Rat orbitofrontal cortex separately encodes response and outcome information during performance of goal-directed behavior. *The Journal of Neuroscience: The Official Journal of the Society for Neuroscience, 28*(19), 5127—5138. https://doi.org/10.1523/jneurosci.0319-08.2008.

Gabbott, P. L., Warner, T. A., Jays, P. R., & Bacon, S. J. (2003). Areal and synaptic interconnectivity of prelimbic (area 32), infralimbic (area 25) and insular cortices in the rat. *Brain Research, 993*(1—2), 59—71.

Gallagher, M., McMahan, R. W., & Schoenbaum, G. (1999). Orbitofrontal cortex and representation of incentive value in associative learning. *The Journal of Neuroscience: The Official Journal of the Society for Neuroscience, 19*(15), 6610—6614.

Gallay, D. S., Gallay, M. N., Jeanmonod, D., Rouiller, E. M., & Morel, A. (2012). The insula of Reil revisited: Multiarchitectonic organization in macaque monkeys. *Cerebral Cortex, 22*(1), 175—190. https://doi.org/10.1093/cercor/bhr104.

Gourley, S. L., Lee, A. S., Howell, J. L., Pittenger, C., & Taylor, J. R. (2010). Dissociable regulation of instrumental action within mouse prefrontal cortex. *The European Journal of Neuroscience, 32*(10), 1726—1734. https://doi.org/10.1111/j.1460-9568.2010.07438.x.

Gourley, S. L., Zimmermann, K. S., Allen, A. G., & Taylor, J. R. (2016). The medial orbitofrontal cortex regulates sensitivity to outcome value. *The Journal of Neuroscience: The Official Journal of the Society for Neuroscience, 36*(16), 4600—4613. https://doi.org/10.1523/jneurosci.4253-15.2016.

Gremel, C. M., Chancey, J. H., Atwood, B. K., Luo, G., Neve, R., Ramakrishnan, C., ... Costa, R. M. (2016). Endocannabinoid modulation of orbitostriatal circuits gates habit formation. *Neuron, 90*(6), 1312—1324. https://doi.org/10.1016/j.neuron.2016.04.043.

Gremel, C. M., & Costa, R. M. (2013). Orbitofrontal and striatal circuits dynamically encode the shift between goal-directed and habitual actions. *Nature Communications, 4,* 2264. https://doi.org/10.1038/ncomms3264.

Hart, G., & Balleine, B. W. (2016). Consolidation of goal-directed action depends on MAPK/ERK signaling in rodent prelimbic cortex. *The Journal of Neuroscience: The Official Journal of the Society for Neuroscience, 36*(47), 11974—11986. https://doi.org/10.1523/JNEUROSCI.1772-16.2016.

Hart, G., & Balleine, B. W. (2018). Prefrontal corticostriatal disconnection blocks the acquisition of goal-directed action. *The Journal of Neuroscience: The Official Journal of the Society for Neuroscience, 38*(5), 1311—1322. https://doi.org/10.1523/jneurosci.2850-17.2017.

Hart, G., Leung, B. K., & Balleine, B. W. (2014). Dorsal and ventral streams: The distinct role of striatal subregions in the acquisition and performance of goal-directed actions. *Neurobiology of Learning and Memory, 108,* 104—118. https://doi.org/10.1016/j.nlm.2013.11.003.

Hitchcott, P. K., Quinn, J. J., & Taylor, J. R. (2007). Bidirectional modulation of goal-directed actions by prefrontal cortical dopamine. *Cerebral Cortex, 17*(12), 2820—2827. https://doi.org/10.1093/cercor/bhm010.

Homayoun, H., & Moghaddam, B. (2009). Differential representation of Pavlovian-instrumental transfer by prefrontal cortex subregions and striatum. *The European Journal of Neuroscience, 29*(7), 1461—1476. https://doi.org/10.1111/j.1460-9568.2009.06679.x.

Hoover, W. B., & Vertes, R. P. (2011). Projections of the medial orbital and ventral orbital cortex in the rat. *The Journal of Comparative Neurology, 519*(18), 3766—3801. https://doi.org/10.1002/cne.22733.

Izquierdo, A. (2017). Functional heterogeneity within rat orbitofrontal cortex in reward learning and decision making. *The Journal of Neuroscience: The Official Journal of the Society for Neuroscience, 37*(44), 10529—10540. https://doi.org/10.1523/jneurosci.1678-17.2017.

Izquierdo, A., & Murray, E. A. (2004). Combined unilateral lesions of the amygdala and orbital prefrontal cortex impair affective processing in rhesus monkeys. *Journal of Neurophysiology, 91*(5), 2023—2039. https://doi.org/10.1152/jn.00968.2003.

Izquierdo, A., & Murray, E. A. (2010). Functional interaction of medial mediodorsal thalamic nucleus but not nucleus accumbens with amygdala and orbital prefrontal cortex is essential for adaptive response selection after reinforcer devaluation. *The Journal of Neuroscience: The Official Journal of the Society for Neuroscience, 30*(2), 661—669. https://doi.org/10.1523/jneurosci.3795-09.2010.

Izquierdo, A., Suda, R. K., & Murray, E. A. (2004). Bilateral orbital prefrontal cortex lesions in rhesus monkeys disrupt choices guided by both reward value and reward contingency. *The Journal of Neuroscience: The Official Journal of the Society for Neuroscience, 24*(34), 7540—7548. https://doi.org/10.1523/jneurosci.1921-04.2004.

Jin, X., & Costa, R. M. (2015). Shaping action sequences in basal ganglia circuits. *Current Opinion in Neurobiology, 33*, 188—196. https://doi.org/10.1016/j.conb.2015.06.011.

Jones, B., & Mishkin, M. (1972). Limbic lesions and the problem of stimulus—reinforcement associations. *Experimental Neurology, 36*(2), 362—377.

Karama, S., Lecours, A. R., Leroux, J. M., Bourgouin, P., Beaudoin, G., Joubert, S., & Beauregard, M. (2002). Areas of brain activation in males and females during viewing of erotic film excerpts. *Human Brain Mapping, 16*(1), 1—13.

Killcross, S., & Coutureau, E. (2003). Coordination of actions and habits in the medial prefrontal cortex of rats. *Cerebral Cortex, 13*(4), 400—408.

Kosar, E., Grill, H. J., & Norgren, R. (1986). Gustatory cortex in the rat. I. Physiological properties and cytoarchitecture. *Brain Research, 379*(2), 329—341.

Krettek, J. E., & Price, J. L. (1977). Projections from the amygdaloid complex to the cerebral cortex and thalamus in the rat and cat. *The Journal of Comparative Neurology, 172*(4), 687—722. https://doi.org/10.1002/cne.901720408.

Lex, B., & Hauber, W. (2010). The role of dopamine in the prelimbic cortex and the dorsomedial striatum in instrumental conditioning. *Cerebral Cortex, 20*(4), 873—883. https://doi.org/10.1093/cercor/bhp151.

Luk, C. H., & Wallis, J. D. (2013). Choice coding in frontal cortex during stimulus-guided or action-guided decision-making. *The Journal of Neuroscience: The Official Journal of the Society for Neuroscience, 33*(5), 1864—1871. https://doi.org/10.1523/jneurosci.4920-12.2013.

Machado, C. J., & Bachevalier, J. (2007). The effects of selective amygdala, orbital frontal cortex or hippocampal formation lesions on reward assessment in nonhuman primates. *The European Journal of Neuroscience, 25*(9), 2885—2904. https://doi.org/10.1111/j.1460-9568.2007.05525.x.

Maffei, A., Haley, M., & Fontanini, A. (2012). Neural processing of gustatory information in insular circuits. *Current Opinion in Neurobiology, 22*(4), 709—716. https://doi.org/10.1016/j.conb.2012.04.001.

Mizoguchi, N., Fujita, S., Koshikawa, N., & Kobayashi, M. (2011). Spatiotemporal dynamics of long-term potentiation in rat insular cortex revealed by optical imaging. *Neurobiology of Learning and Memory, 96*(3), 468—478. https://doi.org/10.1016/j.nlm.2011.07.003.

Munster, A., & Hauber, W. (2017). Medial orbitofrontal cortex mediates effort-related responding in rats. *Cerebral Cortex,* 1—11. https://doi.org/10.1093/cercor/bhx293.

Murray, E. A., Moylan, E. J., Saleem, K. S., Basile, B. M., & Turchi, J. (2015). Specialized areas for value updating and goal selection in the primate orbitofrontal cortex. *eLife, 4.* https://doi.org/10.7554/eLife.11695.

Naneix, F., Marchand, A. R., Di Scala, G., Pape, J. R., & Coutureau, E. (2009). A role for medial prefrontal dopaminergic innervation in instrumental conditioning. *The Journal of Neuroscience: The Official Journal of the Society for Neuroscience, 29*(20), 6599—6606. https://doi.org/10.1523/JNEUROSCI.1234-09.2009.

Naneix, F., Marchand, A. R., Di Scala, G., Pape, J. R., & Coutureau, E. (2012). Parallel maturation of goal-directed behavior and dopaminergic systems during adolescence. *The Journal of Neuroscience: The Official Journal of the Society for Neuroscience, 32*(46), 16223–16232. https://doi.org/10.1523/JNEUROSCI.3080-12.2012.

Naneix, F., Marchand, A. R., Pichon, A., Pape, J. R., & Coutureau, E. (2013). Adolescent stimulation of D2 receptors alters the maturation of dopamine-dependent goal-directed behavior. *Neuropsychopharmacology: Official Publication of the American College of Neuropsychopharmacology, 38*(8), 1566–1574. https://doi.org/10.1038/npp.2013.55.

Nerad, L., Ramirez-Amaya, V., Ormsby, C. E., & Bermudez-Rattoni, F. (1996). Differential effects of anterior and posterior insular cortex lesions on the acquisition of conditioned taste aversion and spatial learning. *Neurobiology of Learning and Memory, 66*(1), 44–50.

Niv, Y., & Schoenbaum, G. (2008). Dialogues on prediction errors. *Trends in Cognitive Sciences, 12*(7), 265–272. https://doi.org/10.1016/j.tics.2008.03.006.

Noonan, M. P., Walton, M. E., Behrens, T. E., Sallet, J., Buckley, M. J., & Rushworth, M. F. (2010). Separate value comparison and learning mechanisms in macaque medial and lateral orbitofrontal cortex. *Proceedings of the National Academy of Sciences of the United States of America, 107*(47), 20547–20552. https://doi.org/10.1073/pnas.1012246107.

Nutt, D. J., Lalies, M. D., Lione, L. A., & Hudson, A. L. (1997). Noradrenergic mechanisms in the prefrontal cortex. *Journal of Psychopharmacology, 11*(2), 163–168.

O'Doherty, J. P., Cockburn, J., & Pauli, W. M. (2017). Learning, reward, and decision making. *Annual Review of Psychology, 68*, 73–100. https://doi.org/10.1146/annurev-psych-010416-044216.

Ongur, D., & Price, J. L. (2000). The organization of networks within the orbital and medial prefrontal cortex of rats, monkeys and humans. *Cerebral Cortex, 10*(3), 206–219.

Ostlund, S. B., & Balleine, B. W. (2005). Lesions of medial prefrontal cortex disrupt the acquisition but not the expression of goal-directed learning. *The Journal of Neuroscience: The Official Journal of the Society for Neuroscience, 25*(34), 7763–7770. https://doi.org/10.1523/JNEUROSCI.1921-05.2005.

Ostlund, S. B., & Balleine, B. W. (2007a). The contribution of orbitofrontal cortex to action selection. *Annals of the New York Academy of Sciences, 1121*, 174–192. https://doi.org/10.1196/annals.1401.033.

Ostlund, S. B., & Balleine, B. W. (2007b). Orbitofrontal cortex mediates outcome encoding in Pavlovian but not instrumental conditioning. *The Journal of Neuroscience: The Official Journal of the Society for Neuroscience, 27*(18), 4819–4825. https://doi.org/10.1523/jneurosci.5443-06.2007.

Palmiter, R. D. (2008). Dopamine signaling in the dorsal striatum is essential for motivated behaviors: Lessons from dopamine-deficient mice. *Annals of the New York Academy of Sciences, 1129*, 35–46. https://doi.org/10.1196/annals.1417.003.

Parkes, S. L., & Balleine, B. W. (2013). Incentive memory: Evidence the basolateral amygdala encodes and the insular cortex retrieves outcome values to guide choice between goal-directed actions. *The Journal of Neuroscience: The Official Journal of the Society for Neuroscience, 33*(20), 8753–8763. https://doi.org/10.1523/jneurosci.5071-12.2013.

Parkes, S. L., Bradfield, L. A., & Balleine, B. W. (2015). Interaction of insular cortex and ventral striatum mediates the effect of incentive memory on choice between goal-directed actions. *The Journal of Neuroscience: The Official Journal of the Society for Neuroscience, 35*(16), 6464–6471. https://doi.org/10.1523/jneurosci.4153-14.2015.

Parkes, S. L., Ferreira, G., & Coutureau, E. (2016). Acquisition of specific response-outcome associations requires NMDA receptor activation in the basolateral amygdala but not in the insular cortex. *Neurobiology of Learning and Memory, 128*, 40–45. https://doi.org/10.1016/j.nlm.2015.12.005.

Parkes, S. L., Ravassard, P. M., Cerpa, J. C., Wolff, M., Ferreira, G., & Coutureau, E. (2017). Insular and ventrolateral orbitofrontal cortices differentially contribute to goal-directed behavior in rodents. *Cerebral Cortex*, 1–13. https://doi.org/10.1093/cercor/bhx132.

Parnaudeau, S., Taylor, K., Bolkan, S. S., Ward, R. D., Balsam, P. D., & Kellendonk, C. (2015). Mediodorsal thalamus hypofunction impairs flexible goal-directed behavior. *Biological Psychiatry, 77*(5), 445–453. https://doi.org/10.1016/j.biopsych.2014.03.020.

Paxinos, G., & Watson, C. (2014). *The rat brain in stereotaxic coordinates*. Academic Press.

Pickens, C. L., Saddoris, M. P., Gallagher, M., & Holland, P. C. (2005). Orbitofrontal lesions impair use of cue-outcome associations in a devaluation task. *Behavioral Neuroscience, 119*(1), 317—322. https://doi.org/10.1037/0735-7044.119.1.317.

Pickens, C. L., Saddoris, M. P., Setlow, B., Gallagher, M., Holland, P. C., & Schoenbaum, G. (2003). Different roles for orbitofrontal cortex and basolateral amygdala in a reinforcer devaluation task. *The Journal of Neuroscience: The Official Journal of the Society for Neuroscience, 23*(35), 11078—11084.

Preuss, T. M. (1995). Do rats have prefrontal cortex? The rose-woolsey-akert program reconsidered. *Journal of Cognitive Neuroscience, 7*(1), 1—24. https://doi.org/10.1162/jocn.1995.7.1.1.

Price, J. L. (2007). Definition of the orbital cortex in relation to specific connections with limbic and visceral structures and other cortical regions. *Annals of the New York Academy of Sciences, 1121*, 54—71. https://doi.org/10.1196/annals.1401.008.

Quinn, J. J., Pittenger, C., Lee, A. S., Pierson, J. L., & Taylor, J. R. (2013). Striatum-dependent habits are insensitive to both increases and decreases in reinforcer value in mice. *The European Journal of Neuroscience, 37*(6), 1012—1021. https://doi.org/10.1111/ejn.12106.

Ramirez-Lugo, L., Penas-Rincon, A., Angeles-Duran, S., & Sotres-Bayon, F. (2016). Choice behavior guided by learned, but not innate, taste aversion recruits the orbitofrontal cortex. *The Journal of Neuroscience: The Official Journal of the Society for Neuroscience, 36*(41), 10574—10583. https://doi.org/10.1523/jneurosci.0796-16.2016.

Reep, R. L., Corwin, J. V., & King, V. (1996). Neuronal connections of orbital cortex in rats: Topography of cortical and thalamic afferents. *Experimental Brain Research, 111*(2), 215—232.

Rhodes, S. E., & Murray, E. A. (2013). Differential effects of amygdala, orbital prefrontal cortex, and prelimbic cortex lesions on goal-directed behavior in rhesus macaques. *The Journal of Neuroscience: The Official Journal of the Society for Neuroscience, 33*(8), 3380—3389. https://doi.org/10.1523/jneurosci.4374-12.2013.

Roberts, A. C. (2006). Primate orbitofrontal cortex and adaptive behaviour. *Trends in Cognitive Sciences, 10*(2), 83—90. https://doi.org/10.1016/j.tics.2005.12.002.

Rudebeck, P. H., Behrens, T. E., Kennerley, S. W., Baxter, M. G., Buckley, M. J., Walton, M. E., & Rushworth, M. F. (2008). Frontal cortex subregions play distinct roles in choices between actions and stimuli. *The Journal of Neuroscience: The Official Journal of the Society for Neuroscience, 28*(51), 13775—13785. https://doi.org/10.1523/jneurosci.3541-08.2008.

Rudebeck, P. H., & Murray, E. A. (2008). Amygdala and orbitofrontal cortex lesions differentially influence choices during object reversal learning. *The Journal of Neuroscience: The Official Journal of the Society for Neuroscience, 28*(33), 8338—8343. https://doi.org/10.1523/jneurosci.2272-08.2008.

Rudebeck, P. H., & Murray, E. A. (2011). Balkanizing the primate orbitofrontal cortex: Distinct subregions for comparing and contrasting values. *Annals of the New York Academy of Sciences, 1239*, 1—13. https://doi.org/10.1111/j.1749-6632.2011.06267.x.

Rutledge, R. B., Skandali, N., Dayan, P., & Dolan, R. J. (2015). Dopaminergic modulation of decision making and subjective well-being. *The Journal of Neuroscience: The Official Journal of the Society for Neuroscience, 35*(27), 9811—9822. https://doi.org/10.1523/JNEUROSCI.0702-15.2015.

Sadacca, B. F., Wikenheiser, A. M., & Schoenbaum, G. (2017). Toward a theoretical role for tonic norepinephrine in the orbitofrontal cortex in facilitating flexible learning. *Neuroscience, 345*, 124—129. https://doi.org/10.1016/j.neuroscience.2016.04.017.

Saddoris, M. P., Holland, P. C., & Gallagher, M. (2009). Associatively learned representations of taste outcomes activate taste-encoding neural ensembles in gustatory cortex. *The Journal of Neuroscience: The Official Journal of the Society for Neuroscience, 29*(49), 15386—15396. https://doi.org/10.1523/jneurosci.3233-09.2009.

Saper, C. B. (1982). Convergence of autonomic and limbic connections in the insular cortex of the rat. *The Journal of Comparative Neurology, 210*(2), 163—173. https://doi.org/10.1002/cne.902100207.

Sara, S. J. (2016). Locus coeruleus reports changes in environmental contingencies. *Behavioral and Brain Sciences, 39*, e223. https://doi.org/10.1017/s0140525x15001946.

Schlosser, R. G., Wagner, G., Schachtzabel, C., Peikert, G., Koch, K., Reichenbach, J. R., & Sauer, H. (2010). Fronto-cingulate effective connectivity in obsessive compulsive disorder: A study with fMRI and dynamic causal modeling. *Human Brain Mapping, 31*(12), 1834—1850. https://doi.org/10.1002/hbm.20980.

Schoenbaum, G., Chiba, A. A., & Gallagher, M. (1998). Orbitofrontal cortex and basolateral amygdala encode expected outcomes during learning. *Nature Neuroscience, 1*(2), 155–159. https://doi.org/10.1038/407.

Schoenbaum, G., Nugent, S. L., Saddoris, M. P., & Setlow, B. (2002). Orbitofrontal lesions in rats impair reversal but not acquisition of go, no-go odor discriminations. *Neuroreport, 13*(6), 885–890.

Schoenbaum, G., & Roesch, M. (2005). Orbitofrontal cortex, associative learning, and expectancies. *Neuron, 47*(5), 633–636. https://doi.org/10.1016/j.neuron.2005.07.018.

Schultz, W. (2010). Dopamine signals for reward value and risk: Basic and recent data. *Behavioral and Brain Functions, 6*, 24. https://doi.org/10.1186/1744-9081-6-24.

Shan, Q., Christie, M. J., & Balleine, B. W. (2015). Plasticity in striatopallidal projection neurons mediates the acquisition of habitual actions. *The European Journal of Neuroscience, 42*(4), 2097–2104. https://doi.org/10.1111/ejn.12971.

Sherman, S. M. (2016). Thalamus plays a central role in ongoing cortical functioning. *Nature Neuroscience, 19*(4), 533–541. https://doi.org/10.1038/nn.4269.

Shi, C. J., & Cassell, M. D. (1998). Cortical, thalamic, and amygdaloid connections of the anterior and posterior insular cortices. *The Journal of Comparative Neurology, 399*(4), 440–468.

Shiflett, M. W., & Balleine, B. W. (2011). Contributions of ERK signaling in the striatum to instrumental learning and performance. *Behavioural Brain Research, 218*(1), 240–247. https://doi.org/10.1016/j.bbr.2010.12.010.

Shiflett, M. W., Brown, R. A., & Balleine, B. W. (2010). Acquisition and performance of goal-directed instrumental actions depends on ERK signaling in distinct regions of dorsal striatum in rats. *The Journal of Neuroscience: The Official Journal of the Society for Neuroscience, 30*(8), 2951–2959. https://doi.org/10.1523/jneurosci.1778-09.2010.

Simon, N. W., Wood, J., & Moghaddam, B. (2015). Action-outcome relationships are represented differently by medial prefrontal and orbitofrontal cortex neurons during action execution. *Journal of Neurophysiology, 114*(6), 3374–3385. https://doi.org/10.1152/jn.00884.2015.

Small, D. M. (2010). Taste representation in the human insula. *Brain Structure & Function, 214*(5–6), 551–561. https://doi.org/10.1007/s00429-010-0266-9.

Swanson, A. M., DePoy, L. M., & Gourley, S. L. (2017). Inhibiting Rho kinase promotes goal-directed decision making and blocks habitual responding for cocaine. *Nature Communications, 8*(1), 1861. https://doi.org/10.1038/s41467-017-01915-4.

Tang, J. S., Qu, C. L., & Huo, F. Q. (2009). The thalamic nucleus submedius and ventrolateral orbital cortex are involved in nociceptive modulation: A novel pain modulation pathway. *Progress in Neurobiology, 89*(4), 383–389. https://doi.org/10.1016/j.pneurobio.2009.10.002.

Theyel, B. B., Llano, D. A., & Sherman, S. M. (2010). The corticothalamocortical circuit drives higher-order cortex in the mouse. *Nature Neuroscience, 13*(1), 84–88. https://doi.org/10.1038/nn.2449.

Tran-Tu-Yen, D. A. S., Marchand, A. R., Pape, J. R., Di Scala, G., & Coutureau, E. (2009). Transient role of the rat prelimbic cortex in goal-directed behaviour. *The European Journal of Neuroscience, 30*, 464–471.

Tricomi, E., Balleine, B. W., & O'Doherty, J. P. (2009). A specific role for posterior dorsolateral striatum in human habit learning. *The European Journal of Neuroscience, 29*(11), 2225–2232. https://doi.org/10.1111/j.1460-9568.2009.06796.x.

Uylings, H. B., Groenewegen, H. J., & Kolb, B. (2003). Do rats have a prefrontal cortex? *Behavioural Brain Research, 146*(1–2), 3–17.

Van De Werd, H. J., & Uylings, H. B. (2008). The rat orbital and agranular insular prefrontal cortical areas: A cytoarchitectonic and chemoarchitectonic study. *Brain Structure & Function, 212*(5), 387–401. https://doi.org/10.1007/s00429-007-0164-y.

Vertes, R. P. (2004). Differential projections of the infralimbic and prelimbic cortex in the rat. *Synapse, 51*(1), 32–58. https://doi.org/10.1002/syn.10279.

Vogt, B. A. (2016). Midcingulate cortex: Structure, connections, homologies, functions and diseases. *Journal of Chemical Neuroanatomy, 74*, 28–46. https://doi.org/10.1016/j.jchemneu.2016.01.010.

Vogt, B. A., Hof, P. R., Zilles, K., Vogt, L. J., Herold, C., & Palomero-Gallagher, N. (2013). Cingulate area 32 homologies in mouse, rat, macaque and human: Cytoarchitecture and receptor architecture. *The Journal of Comparative Neurology, 521*(18), 4189–4204. https://doi.org/10.1002/cne.23409.

Vogt, B. A., & Paxinos, G. (2014). Cytoarchitecture of mouse and rat cingulate cortex with human homologies. *Brain Structure & Function, 219*(1), 185—192. https://doi.org/10.1007/s00429-012-0493-3.

West, E. A., DesJardin, J. T., Gale, K., & Malkova, L. (2011). Transient inactivation of orbitofrontal cortex blocks reinforcer devaluation in macaques. *The Journal of Neuroscience: The Official Journal of the Society for Neuroscience, 31*(42), 15128—15135. https://doi.org/10.1523/jneurosci.3295-11.2011.

Whitaker, L. R., Warren, B. L., Venniro, M., Harte, T. C., McPherson, K. B., Beidel, J., ... Hope, B. T. (2017). Bidirectional modulation of intrinsic excitability in rat prelimbic cortex neuronal ensembles and non-ensembles after operant learning. *The Journal of Neuroscience: The Official Journal of the Society for Neuroscience, 37*(36), 8845—8856. https://doi.org/10.1523/jneurosci.3761-16.2017.

Wickens, J. R., Horvitz, J. C., Costa, R. M., & Killcross, S. (2007). Dopaminergic mechanisms in actions and habits. *The Journal of Neuroscience: The Official Journal of the Society for Neuroscience, 27*(31), 8181—8183. https://doi.org/10.1523/JNEUROSCI.1671-07.2007.

Wise, S. P. (2008). Forward frontal fields: Phylogeny and fundamental function. *Trends in Neurosciences, 31*(12), 599—608. https://doi.org/10.1016/j.tins.2008.08.008.

Xiong, Q., Znamenskiy, P., & Zador, A. M. (2015). Selective corticostriatal plasticity during acquisition of an auditory discrimination task. *Nature, 521*(7552), 348—351. https://doi.org/10.1038/nature14225.

Yamamoto, T., Matsuo, R., & Kawamura, Y. (1980). Localization of cortical gustatory area in rats and its role in taste discrimination. *Journal of Neurophysiology, 44*(3), 440—455.

Yin, H. H., Knowlton, B. J., & Balleine, B. W. (2004). Lesions of dorsolateral striatum preserve outcome expectancy but disrupt habit formation in instrumental learning. *The European Journal of Neuroscience, 19*(1), 181—189.

Yin, H. H., Knowlton, B. J., & Balleine, B. W. (2005). Blockade of NMDA receptors in the dorsomedial striatum prevents action-outcome learning in instrumental conditioning. *The European Journal of Neuroscience, 22*(2), 505—512. https://doi.org/10.1111/j.1460-9568.2005.04219.x.

Yin, H. H., Ostlund, S. B., Knowlton, B. J., & Balleine, B. W. (2005). The role of the dorsomedial striatum in instrumental conditioning. *The European Journal of Neuroscience, 22*(2), 513—523. https://doi.org/10.1111/j.1460-9568.2005.04218.x.

Yoshida, A., Dostrovsky, J. O., & Chiang, C. Y. (1992). The afferent and efferent connections of the nucleus submedius in the rat. *The Journal of Comparative Neurology, 324*(1), 115—133. https://doi.org/10.1002/cne.903240109.

Zimmermann, K. S., Yamin, J. A., Rainnie, D. G., Ressler, K. J., & Gourley, S. L. (2017). Connections of the mouse orbitofrontal cortex and regulation of goal-directed action selection by brain-derived neurotrophic factor. *Biological Psychiatry, 81*(4), 366—377. https://doi.org/10.1016/j.biopsych.2015.10.026.

CHAPTER 9

Distinct Functional Microcircuits in the Nucleus Accumbens Underlying Goal-Directed Decision-Making

Elizabeth A. West, Travis M. Moschak, Regina M. Carelli
Department of Psychology and Neuroscience, University of North Carolina, Chapel Hill, NC, United States

INTRODUCTION

Animals depend on the ability to learn about their environment in order to seek out the outcomes necessary for survival (e.g., food, drink, sex). This ability requires learning about predictive stimuli and the actions needed to achieve a desired goal, in addition to judging outcome and cost to direct future actions toward that goal. Furthermore, the ability to modify actions in response to environmental changes is critical to ensure access to necessary resources. Importantly, in humans, certain psychiatric disorders (e.g., drug addiction) are characterized by a disruption of goal-directed decision-making (Lubman, Yucel, & Pantelis, 2004; see also Morris, Chapter 17), which results in a decreased ability to change behaviors associated with long-term negative consequences. Therefore, understanding and characterizing the underlying neural circuits that control goal-directed decision-making is critical to understanding and treating psychiatric disorders.

Although numerous brain regions are involved in goal-directed decision-making, the nucleus accumbens (NAc) is well situated to play a role in this process. The NAc is postulated to function as a limbic—motor interface (Mogenson, Jones, & Yim, 1980), receiving afferent information from limbic (and prefrontal) structures involved in memory, drive, and motivation and projecting to motor regions to guide behavior. Therefore, the NAc likely plays a major role in integrating information about goals and predictive cues and guiding goal-directed behaviors. Further, its primary subregions, the core and shell, appear to play unique, but complimentary roles in guiding goal-directed behavior. Specifically, the core appears involved in learning and action during goal-directed behavior (Carelli, 2004; Saddoris, Sugam, Cacciapaglia, & Carelli, 2013), while the shell processes hedonic or motivational value (Castro, Cole, & Berridge, 2015; Kelley, 2004; Saddoris, Cacciapaglia, Wightman, & Carelli, 2015). Both computational processes are critical for appropriate goal-directed decision-making.

Goal-Directed Decision Making
ISBN 978-0-12-812098-9, https://doi.org/10.1016/B978-0-12-812098-9.00009-7

199

ANATOMICAL ORGANIZATION OF THE NUCLEUS ACCUMBENS CORE AND SHELL

Both the NAc core and shell consist predominantly (>95%) of medium spiny neurons. However, there are distinct morphological and anatomical connectivity differences between the two subregions that may contribute to their unique functional properties (discussed elsewhere, see Meredith, Baldo, Andrezjewski, & Kelley, 2008). Morphologically, the NAc core has a denser distribution of spines than the NAc shell. In addition, dopaminergic inputs, which are necessary for glutamatergic activity-dependent plasticity (Wolf, Mangiavacchi, & Sun, 2003), are predominately found on spines in the core (Zahm, 1992). Conversely, these inputs are found in dendrites instead of spines in the shell (Hara & Pickel, 2005). Thus, core neurons seem more amenable to activity-dependent synaptic plasticity, the basis for learning and memory, than shell neurons.

Despite similarities in afferent and efferent projections of the core and shell, there are some notable differences that may account, in part, for their distinct behavioral functions. For example, the ventral subiculum of the hippocampus projects predominately to the shell, while the dorsal subiculum projects mainly to the core (reviewed in Kelley, 2004). The ventral subiculum plays a stronger role in drug seeking and anxiety while the dorsal hippocampus is more involved in memory and spatial navigation (Fanselow & Dong, 2010; Sun & Rebec, 2003). In the rat, the medial prefrontal cortex (mPFC) is comprised of dorsal (prelimbic; PrL) and ventral (infralimbic; IL) areas (Heidbreder & Groenewegen, 2003; Ongur & Price, 2000) that selectively project to the NAc core and shell, respectively. Importantly, these two mPFC subregions are involved in goal-directed behavior (more PrL) and extinction and habitlike behavior (more IL; Gourley & Taylor, 2016). The basolateral amygdala (BLA), known for its involvement in associative learning, sends projections to both the core and shell, but those projections diverge from its anterior (core) and posterior (shell) parts, respectively (Brog, Salyapongse, Deutch, & Zahm, 1993; Wright, Beijer, & Groenewegen, 1996). In addition, the shell, but not the core, receives direct reciprocal projections from the lateral hypothalamus (LH) (Heimer, Zahm, Churchill, Kalivas, & Wohltmann, 1991), thus receiving important information about homeostatic satiety levels.

In addition to distinct afferent inputs to the NAc core and shell, their efferent projections also differ. For example, the NAc core sends projections predominantly to the motor regions important in executing action, such as the ventral pallidum and subthalamic nucleus (Heimer et al., 1991; Mogenson et al., 1980; Zahm, 1999). The shell, however, predominantly sends projections to limbic areas such as the LH, bed nucleus stria terminalis, central amygdala, and ventromedial ventral pallidum (Heimer et al., 1991; Kelley, 2004). As such, the core has been proposed to have greater involvement in goal-directed actions (e.g., necessary for learning about the environment) while the shell has been proposed to have greater involvement with emotionally relevant information necessary for encoding the value of stimuli in the environment. Critically, it is these differences in the morphology and anatomy that may account for the distinct functional properties of core and shell in goal-directed decision-making.

DISTINCT FUNCTIONAL ROLES OF THE NUCLEUS ACCUMBENS CORE VERSUS SHELL IN GOAL-DIRECTED BEHAVIOR

In support, early studies demonstrated differing functional properties for the core and shell. For example, NMDA receptor blockade in the core reduced locomotor activity and novel object exploration (Maldonado-Irizarry & Kelley, 1994), while similar injections in the shell had no effect. In contrast, blockade of non-NMDA glutamate receptors in the shell (but not core) produced increased feeding behavior (Maldonado-Irizarry, Swanson, & Kelley, 1995). These early findings support the view that the shell is more involved in hedonic processing while the core is more involved in goal-directed exploratory activity. In Table 9.1, we have outlined studies that determine whether the core or shell is necessary for particular behaviors. Critically, a lack of necessity of a subregion in a behavioral task does not preclude the neurons or dopamine (DA) release in the subregion from tracking this information during behavior.

Table 9.1 The involvement in core or shell across behavioral tasks.

	Core	Shell
Reinforcement learning	● Parkinson et al. (2000) and Hernandez et al. (2005)	X Parkinson et al. (2000) and Hernandez et al. (2005)
Feeding behavior	X Kelley and Swanson (1997) and Stratford and Kelley (1997)	● Kelley and Swanson (1997) and Stratford and Kelley (1997)
Reinforcer devaluation		
Pavlovian	● Singh et al. (2010)	● Singh et al. (2010)
Instrumental	≃ Corbit et al. (2001)	X Corbit et al. (2001)
Reversal learning	X Schoenbaum and Setlow (2003) and Dalton et al. (2014)	X Dalton et al. (2014)
Set-shifting	● Floresco et al. (2006)	X Floresco et al. (2006)
Delay discounting	● Cardinal et al. (2001) and Pothuizen et al. (2005)	≃ Pothuizen et al. (2005) and Feja et al. (2014)
Effort discounting	● Bezzina et al. (2008) and Ghods-Sharifi and Floresco (2010)	≃ Day et al. (2011)
Risk (probabilistic) discounting	≃ Cardinal and Howes (2005) and Yang and Liao (2015)	● Dalton et al. (2014) and Stopper and Floresco (2011)

X, not required; ●, required/involved; ≃, potentially required/involved (see text for more information).

Nucleus accumbens involvement in hedonic processing

Both NAc core and NAc shell neurons respond to tastes with specific hedonic value. For example, neurons in both subregions show differential firing patterns to appetitive and aversive taste stimuli (Roitman, Wheeler, & Carelli, 2005). Furthermore, both core and shell neurons encode the devaluation of natural reinforcers following conditioned taste aversion (Roitman, Wheeler, Tiesinga, Roitman, & Carelli, 2010) and delayed cocaine access (Wheeler et al., 2008) that reflect the change in hedonic value from positive to negative.

However, while both regions play a role in hedonic processing, evidence suggests a stronger role for the NAc shell. Critically, shell neurons, but not core, specifically track the motivational value of a reward depending on physiological state (Loriaux, Roitman, & Roitman, 2011). In addition, NAc shell DA, but not core DA, tracks the devaluation of a natural reward by delayed, but impending cocaine (Roitman, Wheeler, Wightman, & Carelli, 2008; Wheeler et al., 2011). Furthermore, DA in the shell, but not core, encodes the relative value of appetitive rewards such as one versus two sucrose pellets (Sackett, Saddoris, & Carelli, 2017; Saddoris, Sugam, & Carelli, 2017). In addition, NAc shell DA tracks cues signaling different reward magnitudes (Beyene, Carelli, & Wightman, 2010; Sackett et al., 2017), and NAc shell inactivation disrupts the ability of rats to judge reward magnitude options (Stopper & Floresco, 2011). These findings are consistent with earlier work demonstrating that pharmacological inactivation (either AMPA glutamate blockade or stimulation of GABA) of the shell (but not the core) induces feeding behavior (Kelley & Swanson, 1997; Stratford & Kelley, 1997), suggesting an involvement of the shell in hedonic value and homeostatic regulation.

Interestingly, the NAc shell is a heterogeneous structure that differs in function along its rostral—caudal axis with respect to hedonic value processing. For example, pharmacological inactivation in the rostral shell promotes appetitive responses, but in the caudal shell yields aversive taste behaviors (Berridge & Kringelbach, 2015; Castro et al., 2015). These findings led to the idea that an "affective keyboard" exists along the rostral—caudal NAc shell; the rostral shell promotes appetitive, positive value states while the caudal shell elicits aversive, negative value states (Berridge & Kringelbach, 2015). In addition, optical stimulation of DA terminals in the rostral shell prevented cocaine-induced devaluation of reward, while activation of terminals in the caudal shell elicited a pronounced enhancement of the devaluation (Hurley, West, & Carelli, 2017). Thus, the shell seems to be involved in processing hedonic value with the rostral—caudal axis playing distinct roles in the particular valence (appetitive vs. aversive).

Nucleus accumbens involvement in reward learning

Both the NAc core and shell play a role in reward learning. Similar to reward processing, core and shell neurons both encode associative information about outcome-predictive cues (Carelli, 2000, 2004; Day, Roitman, Wightman, & Carelli, 2007;

Day, Wheeler, Roitman, & Carelli, 2006) and actions (Carelli & Ijames, 2000; Hollander, Ijames, Roop, & Carelli, 2002). Specifically, NAc (core and shell) neurons show distinct patterned discharges (either an increase or decrease in cell firing) at the onset of a salient event (e.g., a reward-predictive cue) termed "phasic" activity. Indeed, neurons in both subregions show phasic responsiveness to cues paired with food rewards such as sucrose (Jones, Day, Wheeler, & Carelli, 2010; Saddoris & Carelli, 2014; Saddoris, Stamatakis, & Carelli, 2011; Setlow, Schoenbaum, & Gallagher, 2003) or drug rewards such as cocaine (Hollander & Carelli, 2007), ethanol (Robinson & Carelli, 2008), and heroin (Chang, Janak, & Woodward, 1998), as well as responses to the food and drugs themselves (Hollander & Carelli, 2007; Saddoris et al., 2011; Stuber, Roitman, Phillips, Carelli, & Wightman, 2005). In addition, elevated rapid DA signaling is also observed in the core and shell to predictive cues and their respective rewards (Day et al., 2007; Owesson-White, Cheer, Beyene, Carelli, & Wightman, 2008; Roitman, Stuber, Phillips, Wightman, & Carelli, 2004; Stuber et al., 2005).

Even though both the core and shell encode information about reward-predictive cues, evidence suggests a stronger role for the core in this process. Indeed, early work showed that the NAc core, not shell, is necessary for the development and expression of a conditioned response to reward-predictive cues (Parkinson et al., 2000). Glutamate receptor blockade, disruption of protein synthesis, or D1 antagonism in the core (but not shell) impairs operant conditioning during acquisition, but not after the learning has already occurred (Hernandez, Andrzejewski, Sadeghian, Panksepp, & Kelley, 2005; Hernandez, Sadeghian, & Kelley, 2002). Additionally, core neurons are more likely to selectively encode either the cue or the response in a cue-guided operant task, while shell neurons are more likely to respond to both the cue and the response (Cacciapaglia, Saddoris, Wightman, & Carelli, 2012). Similarly, we have observed that core DA elevates specifically to the most proximal reward-predictive action after learning a chain schedule (Saddoris, Sugam, et al., 2015), similar to prediction error signals in putative DA neurons in the ventral tegmental area (VTA) (Mirenowicz & Schultz, 1996). In contrast, the shell shows elevated DA to all predictive actions and the outcomes (Saddoris, Cacciapaglia, et al., 2015). Thus, although rapid DA signaling in both the core and shell respond to rewards and their predictors, recent evidence suggests that these subregions contribute distinct but complementary roles in motivated behaviors. One view we postulated is that DA signaling in the NAc core shows more prediction errorlike encoding, while the shell reflects more incentive salience (i.e., motivational aspects of the task; Saddoris, Cacciapaglia, et al., 2015). These circuits work in parallel, thus helping animals optimize their goal-directed behaviors.

Nucleus accumbens in flexible goal-directed behavior

While survival depends on the ability to engage in optimal goal-directed behaviors, animals also depend on the ability to flexibly alter their behavior in the face of changing consequences. A behavior that at one time produced a positive outcome can at a later time produce a neutral or even negative outcome, making it important to adjust behavior in a changing environment. Many different laboratory tasks can be used to measure the different roles of the NAc core and shell in behavioral flexibility including reversal learning, set-shifting, and reinforcer devaluation.

One measure of behavioral flexibility is reversal learning, the ability to learn that a previously unrewarded stimulus or action now predicts reward (or, conversely, that a previously rewarded stimulus no longer predicts reward). Neither subregion of the NAc is necessary for simple reversal learning (Dalton, Phillips, & Floresco, 2014; Schoenbaum & Setlow, 2003). Nonetheless, NAc core activity (shell unknown) tracks cue–outcome associations as they change during reversal learning (Setlow et al., 2003). Specifically, a neuron that responded to a cue that predicted sucrose no longer responds once the contingencies change (e.g., when the cue predicts quinine instead of sucrose). Furthermore, DA D1 or D2 antagonism in the NAc does not impair reversal learning (Calaminus & Hauber, 2007), even though NAc core DA tracks the switching contingencies of cue–outcome associations during reversal learning (Klanker et al., 2015) similar to NAc core neurons. These findings suggest that the subpopulation of NAc core neurons that track reversal learning is not critical to the ability of the rat to select the appropriate response, and other brain systems most likely compensate for the loss of the NAc in reversal learning (e.g., BLA). Nonetheless, these cue–outcome associations formed in the core can greatly impact behavioral flexibility, discussed below.

Another measure of behavioral flexibility, attentional set-shifting, measures an animal's ability to update a stimulus–action–outcome association following a shift in the modality of the stimulus. In contrast to reversal learning, the core (but not shell) *is* necessary for set-shifting (Floresco, Ghods-Sharifi, Vexelman, & Magyar, 2006). Furthermore, disruption of this task by core lesions is not due to perseverative errors, but rather to a disruption of the acquisition and maintenance of a new behavior strategy (i.e., relearning). One possible explanation for the different role of the core in reversal learning and set-shifting is that set-shifting may be a more challenging task than reversal learning. Thus, other brain regions may take over control in the absence of the NAc to perform reversal learning that cannot do so for set-shifting. Unlike the core, the NAc shell does not appear to be involved in either of these tasks and only becomes necessary under conditions of ambiguity (i.e., probability/risk, discussed below).

Finally, the canonical test for determining whether behavior is goal-directed (vs. a habitual stimulus–response strategy) is the reinforcer devaluation task (Adams & Dickinson, 1981; Dickinson, 2012; see also Dickinson, Chapter 1). In this task, animals (1) form an association between a cue/action and an outcome, (2) register the decrease in value of the reinforcer after its devaluation, and (3) integrate the learned cue/action–outcome association with the decreased outcome value to direct behavior. Testing is performed under extinction; thus, rats must use an internal representation of the previously learned association and alter behavior based on the newly computed expected outcome value (Lucantonio, Caprioli, & Schoenbaum, 2014). Critically, core or shell lesions, or core DA depletion (DA shell depletion is unknown) disrupts performance in reinforcer devaluation tasks when rats learn to make associations between Pavlovian cues and rewards (Lex & Hauber, 2010; Singh, McDannald, Haney, Cerri, & Schoenbaum, 2010). A similarly clear role is not well established for *instrumental* devaluation tasks. NAc core lesions or DA depletion induce a general reduction in responding in these tasks (Corbit, Muir, & Balleine, 2001; Lex & Hauber, 2010). As these tasks use variable or random ratio schedules of reinforcement, which necessitate a fair degree of effort, this reduction in response may be due to the critical role the NAc core plays in effortful responding (discussed below), thus making it difficult to determine the core's role in instrumental reinforcer devaluation. However, lesion of the NAc shell does not have such profound effects on effortful responding, and studies have been able to clearly demonstrate that it is not necessary for instrumental reinforcer devaluation tasks (Corbit et al., 2001). Thus, the role of the NAc in reinforcer devaluation may depend on the method by which it is assessed (Pavlovian or instrumental).

A number of studies have shed light on how the NAc may be acting to influence reinforcer devaluation with cues. As mentioned earlier, both core and shell neurons encode information regarding relative outcome values and behavioral responses (Day, Jones, & Carelli, 2011; Roesch, Singh, Brown, Mullins, & Schoenbaum, 2009; Sugam, Saddoris, & Carelli, 2014). Furthermore, when the outcome value of a predictive cue changes from appetitive (e.g., sucrose) to aversive (e.g., quinine) following reversal learning, NAc neurons dynamically switch responding to reflect the newly learned cue—outcome associations (Setlow et al., 2003). In addition, core neurons encode information across different aspects of choice behavior (proximity to lever, reward magnitude, and effort) but rarely encode integration of expected outcome value (Morrison & Nicola, 2014). Our lab recently examined the differing roles of the core and shell during both initial learning of the cue—outcome association and during behavior following reinforcer devaluation. We found that the degree to which core cells encoded reward-associated cues during training reliably predicted the ability of rats to later suppress responding for devalued outcomes (West & Carelli, 2016). This finding may suggest that the animals that formed the strongest cue—outcome association during training used those associations to improve performance. Although we found that core encoding during training predicted performance after devaluation, the core did not encode the shift in outcome value itself, consistent with the idea that core does not encode expected outcome.

In contrast to the NAc core, we found that the shell neurons exhibited no relationship between neural encoding during training and subsequent devaluation. However, there were significantly fewer cue-encoding cells in the shell following reinforcer devaluation (West & Carelli, 2016). This suggests that, unlike the core, the NAc shell *does* encode devaluation and therefore likely tracks reward-predictive cues relative to the expected outcome. Indeed, the shell (but not core) is necessary for outcome-specific enhancement of goal-directed behavior by Pavlovian cues, suggesting that the shell may be processing information about specific outcomes (Corbit & Balleine, 2011). Furthermore, the shell is more involved than the core in motivationally potentiated behavior in a Pavlovian-to-instrumental-transfer task (Saddoris et al., 2011). In addition, c-fos activity in the shell (but not core) was increased following exposure to the reward-predictive cue after outcome devaluation (Kerfoot, Agarwal, Lee, & Holland, 2007). These findings suggest that the NAc shell (but not core) dynamically encodes outcome-selective information about predictive cues based on the current value of that reward relative to the animal's motivational state (Saddoris, Cacciapaglia, et al., 2015).

Nucleus accumbens in decision-making

In order to navigate the real world, animals often need to make decisions that not only require value encoding, learning, and flexible behaviors but also require integrating additional aspects of their environment such as cost, risk, or changing outcome values. Animals prefer reward options that require lower cost and risk and that are acquired immediately. However, ideal options do not always exist, and organisms often need to weigh all of these different components to pick the best outcome. In the laboratory, we can measure some of these different components by manipulating delay, effort, willingness to take risks, and probability of reward.

To investigate the impact of delay on reward processing, animals are often tested for their willingness to wait for a reward. This test is presented as a choice: The animal can choose to wait for a large reward or choose a small reward that is delivered immediately. These procedures are known as delay discounting tasks because the animal "discounts", or lowers, the subjective value of the larger reward as a result of the delay. Animals for which the subjective value of reward is strongly decreased by delay are said to have high levels of delay discounting and are considered to be more impulsive. NAc core lesions or pharmacological inactivation increase preference for the immediate lower magnitude option more than control animals (i.e., lesioned animals are not willing to wait for the more valuable option), suggesting an increase in impulsivity (Bezzina et al., 2008; Cardinal, Pennicott, Sugathapala, Robbins, & Everitt, 2001; Feja, Hayn, & Koch, 2014; Galtress & Kirkpatrick, 2010; Pothuizen, Jongen-Relo, Feldon, & Yee, 2005; but see; Acheson et al., 2006; Moschak & Mitchell, 2014). Few studies have examined the NAc shell's role in delay discounting, and the findings are mixed. One study found that shell inactivation led to increased impulsivity (i.e., animals did not want to wait for the delayed more-valued option); however, this effect was not as strong as core inactivation (Feja et al., 2014). Another study found no effect of shell lesions at all (Pothuizen et al., 2005). Thus, the evidence suggests that the NAc core plays a greater role in delay discounting than the shell.

Although the role of the NAc core in delay discounting seems to be larger than the shell, both the shell and the core encode neural activity to delay and magnitude cues (Day et al., 2011; Roesch et al., 2009). Similarly, in vivo phasic DA release in the NAc core also tracks delay and magnitude cues (Day, Jones, Wightman, & Carelli, 2010; Saddoris, Sugam, et al., 2015). Indeed, animals that are classified as high discounters (i.e., animals that select the immediate reward more rapidly) show blunted NAc DA release induced by stimulation (shell and core), drugs (shell and core), or cues (only core tested) (Diergaarde et al., 2008; Moschak & Carelli, 2017; Zeeb, Soko, Ji, & Fletcher, 2016), even though there are no differences in DA-related proteins in either subregion in animals that are classified as either high (more impulsive) or low discounters (less impulsive) (Loos et al., 2010; Simon et al., 2013). In spite of these findings, DA in the NAc as a whole is not necessary for delay discounting (Winstanley, Theobald, Dalley, & Robbins, 2005), suggesting it has a modulatory role in this process. However, although NAc DA is not necessary for delay discounting, enhancing DA release in the NAc as a whole using optogenetic manipulations alters delay sensitivity (Saddoris, Sugam, et al., 2015), specifically shifting rats' preferences to select the delay option more than controls. It remains to be seen if manipulating DA specifically in either the core or shell alone has a unique effect on delay discounting. In total, the core and shell show similar neural and DA responses during delay discounting, but the core appears more critically involved based on lesion/inactivation studies.

In addition to delay sensitivity, the NAc also plays a role in the amount of effort animals are willing to exert to receive a reward. Indeed, the role of the core and shell in effortful decision-making is similar to that seen in delay discounting. Core, but not shell, lesions or inactivation renders animals less willing to expend effort to receive a reward (Bezzina et al., 2008; Ghods-Sharifi & Floresco, 2010; Hauber & Sommer, 2009). Importantly, NAc DA is necessary for effortful decision-making, as loss of this DA greatly decreases choice of a highly palatable, high-effort reward in favor of a low-palatable, low-effort reward (Aberman & Salamone, 1999; Salamone, Wisniecki, Carlson, & Correa, 2001). Critically, this effect seems to be subregion-specific, since DA depletion in the core (Mai, Sommer, & Hauber, 2012), but not the shell (Sokolowski & Salamone, 1998), is necessary for effortful decision-making. Additionally, D1 or D2 antagonism in the core decreases effortful responding; this effect is weakened or absent in the shell (Bari & Pierce, 2005; Nowend, Arizzi, Carlson, & Salamone, 2001). Furthermore, phasic NAc core DA tracks the value of the cue based on the animals preferred reward option (i.e., less effort), but again this effect is weaker (only a trend) in the shell (Day et al., 2010). Similarly, changes in tonic DA in the core correlate with required effort; a similar but weaker and nonsignificant correlation was seen in the shell (Ostlund, Wassum, Murphy, Balleine, & Maidment, 2011). Together, these findings suggest that the DA in the core (both tonic and phasic) tracks effort decision-making. Interestingly, high-effort responders exhibit more DA-related signal transduction in the core but not in the shell (Randall et al., 2012), suggesting that DA in the core may also modulate individual differences in willingness to work toward a reward. Unlike DA, neural activity tracks effortful responding/cues equally in both the core and the shell (Day et al., 2011). Nonetheless, the data suggest that the NAc core, and especially core DA, is critical for effortful responding, while the shell has a lesser role.

The aforementioned literature suggests that the NAc core is more involved than the shell in decision-making processes that involve a cost, such as effort or delay. However, the role of the NAc in risky decision-making is more complicated. In these types of tasks, animals are often asked to decide between a safe option (e.g., one pellet delivered 100% of the time) and a "risky" option (e.g., two pellets delivered 50% of the time). This type of decision-making does not entail a cost per se, but rather an individual's subjective preference to incur a risk. Lesion of the core makes rats more risk-averse, selecting the safe option more than controls (Cardinal & Howes, 2005; Yang & Liao, 2015); although lesion of the entire NAc (i.e., core and shell) has no effect (Acheson et al., 2006) and the effect of lesioning the shell alone is unknown. However, pharmacological inactivation of the shell also makes rats more risk-averse and disrupts learning in risky decision-making tasks based on probability; surprisingly, inactivation of the core has no effect (Dalton et al., 2014; Stopper & Floresco, 2011). The differences seen in core and shell with regard to the lesion and inactivation studies could be due to the different methods used. With lesions, the brain is allowed time to compensate for the loss of the neural structure; as such, a different region could serve the particular functions of the lesioned region.

The shell and core also exhibit differences in neural firing during risky decision-making, as the shell, but not the core, encodes the preferred option during forced-choice trials, while shell and core both encode the preferred option during free-choice trials (Sugam et al., 2014). Furthermore, the core and shell have opposite patterns of encoding omissions of reward as a function of risk preference. Specifically, in more risk-preferring rats, the core is predominantly excited by omissions, while the shell is predominantly inhibited by omissions (Sugam et al., 2014). Phasic DA tracks value in the core but not the shell during risky decision-making (Sugam, Day, Wightman, & Carelli, 2012), although tonic shell DA does track value over the course of a session (St Onge, Ahn, Phillips, & Floresco, 2012). Similar to that observed with delay discounting, DA in the core and shell is not necessary for risky decision-making (Mai, Sommer, & Hauber, 2015; Mai et al., 2012). However, DA has been shown to modulate this behavior. Specifically, antagonism of D1 receptors in the NAc (core and shell together) made rats risk-averse, D1 agonists optimized risky behavior to maximize reward, and D2 manipulation was without effect (Stopper, Khayambashi, & Floresco, 2013). Furthermore, a combined D1/D2 antagonist in the NAc shell made rats risk-averse, but the same compound had no effect in the NAc core (Mai et al., 2015). A few studies have found relationships with risky preference and DA receptor expression in the shell, although the findings are complex. Specifically, one study found that expression of D1 mRNA in the shell predicted risky choice (no relationships with D2; Simon et al., 2011), but a separate study found a relationship with D2 mRNA and risky preference in the shell, but not D1 mRNA (Mitchell et al., 2014); notably, neither study found any relationships with core mRNA. However, the authors noted that the first study was done in adults while the second study was performed in adolescents. Thus, the different findings could reflect a developmental shift in DA from adolescence to adulthood. Generally, these studies support a slightly stronger role for the shell than the core in risky decision-making, although the literature is mixed and likely suggests a complicated relationship.

PSYCHOSTIMULANT EFFECTS ON NUCLEUS ACCUMBENS PROCESSING AND BEHAVIOR

Substance abuse disorders are characterized by an inability to stop drug seeking despite maladaptive negative consequences (see also Corbit, Chapter 16). As such, impairments in goal-directed behavior often accompany this disorder, with an overreliance on drug-directed behavior at the expense of behavior toward other goals. Indeed, patients suffering from addiction have altered reward processing and altered brain circuitry, including the NAc (Volkow, Fowler, Wang, & Goldstein, 2002).

The NAc is critically involved in psychostimulant-related learning and behavior. Both core and shell neurons and DA signaling respond to lever pressing for intravenous cocaine and to drug-associated cues (Carelli & Ijames, 2000; Phillips, Stuber, Heien, Wightman, & Carelli, 2003). Interestingly, tonic core DA (but not shell) responds to cues that predict cocaine (Ito, Dalley, Howes, Robbins, & Everitt, 2000). In addition, the core, but not the shell, is necessary for autoshaping and conditioned reinforcement for drug reinforcers (Everitt et al., 1999). In contrast, the shell, but not core, is necessary for amphetamine-induced enhancement of conditioned responding (Ito, Robbins, & Everitt, 2004). In addition, the shell is necessary for the rewarding effects of psychostimulant (measured by conditioned place preference), whereas the NAc core contributes to enhanced locomotor activity following psychostimulant treatment (Ito et al., 2004; Sellings & Clarke, 2003). These findings suggest that the rewarding and locomotor-stimulating effects of the drug are anatomically dissociated in the NAc subregions; the shell is more involved in rewarding effects, whereas the core is more involved in behavioral activation. Thus, it seems that the differences in core and shell function that were noted previously in this chapter also exist for drug rewards, i.e., drug-related learning depends on the core, while the shell seems to be involved in the rewarding (and enhancing) properties of psychostimulants.

While the NAc seems to be involved in reward processing and learning about drugs of abuse, the drugs themselves also alter NAc core and shell processing (Dreyer, Vander Weele, Lovic, & Aragona, 2016; Saddoris, 2016) and behavior

(Saddoris, Wang, Sugam, & Carelli, 2016; Saddoris et al., 2017). A history of cocaine followed by drug withdrawal and abstinence alters rats' decision-making strategies. Indeed, individuals with a history of psychostimulant use have steeper delay discounting (more impulsivity) than controls (Coffey, Gudleski, Saladin, & Brady, 2003; Hoffman et al., 2006). In addition, rats with steeper discounting subsequently consume more drug (Poulos, Le, & Parker, 1995), acquire drug self-administration faster (Anker, Zlebnik, Gliddon, & Carroll, 2009), and show increased reinstatement for the drug after a period of abstinence (Broos, Diergaarde, Schoffelmeer, Pattij, & De Vries, 2012; Diergaarde et al., 2008). Finally, both experimenter-administered and self-administered cocaine have been shown to increase delay discounting following withdrawal and abstinence (Hernandez et al., 2014; Mendez et al., 2010; but see; Broos et al., 2012; Moschak & Carelli, 2017). In addition to delay discounting, it has been shown that a history of cocaine during withdrawal and abstinence increases risky decision-making such that animals are more likely to select the risky option over the safer option (Ferland & Winstanley, 2016; Mitchell et al., 2014).

A history of cocaine also impairs the ability of rats to use previously learned information to drive new learning and flexible goal-directed behavior. For example, cocaine-experienced rats have no difficulty learning that a cue or action predicts a particular reward (Saddoris & Carelli, 2014; Schoenbaum & Setlow, 2005). However, if these rats are then required to use those learned relationships in more complex behavior, they show profound deficits. For example, rats with a history of cocaine are impaired in reversal learning (Schoenbaum, Saddoris, Ramus, Shaham, & Setlow, 2004) and higher-order learning such as second-order conditioning (Saddoris & Carelli, 2014). In addition, a history of psychostimulant (cocaine or amphetamine) impairs the ability of rats to direct behavior postdevaluation (Nelson & Killcross, 2006; Schoenbaum & Setlow, 2005). Specifically, animals continue to respond to a cue that is associated with an outcome that has been devalued. Importantly, this deficit only occurs when rats experience psychostimulants prior to learning action–outcome associations; flexible goal-directed behavior is intact if psychostimulant treatment is given after action–outcome associations are already learned (Nelson & Killcross, 2006). This suggests that a history of psychostimulants interferes with how cue/action–outcome associations are processed. Furthermore, following cocaine exposure, rats show a loss of NAc cue-encoding to cues predicting natural (nondrug) rewards (Saddoris & Carelli, 2014). Interestingly, a history of cocaine alters DA signaling in the NAc core in response to reward-predictive cues (Saddoris et al., 2016), while it impairs DA signaling in the NAc shell in response to differently valued rewards (i.e., one vs. two pellets) (Saddoris et al., 2017). In addition, prior cocaine experience also impairs the ability of rats to learn that different levers correspond to differently valued rewards (Saddoris et al., 2017). These findings suggest that a history of cocaine interferes with both learning (perhaps via interference with NAc core DA signaling) and the ability to process value (perhaps via interference with NAc shell DA signaling).

BEYOND THE NUCLEUS ACCUMBENS

The NAc does not function in isolation but is part of a larger neural circuit involved in processing hedonics, learning, flexible behavior, and decision-making, which collectively direct goal-directed behaviors (see also Chapter 8 by Parkes and Coutureau). Indeed, distinct NAc efferent and afferent projections highlighted above may account for the different, but complimentary roles the core and shell play in specific aspects of goal-directed behavior. For example, one key neural region linked to the shell (but not core) that may explain, in part, its specific involvement in hedonic value is the LH. For example, pharmacological inactivation of the shell induces feeding (Kelley & Swanson, 1997) as observed during LH stimulation (Kelley, Baldo, & Pratt, 2005). Further, LH input to the shell is necessary for the enhancement of appetitive behavior following glutamate blockade in the shell (Kelley, 2004; Maldonado-Irizarry et al., 1995). As such, it is possible that the shell receives outcome-specific satiety signals from the LH that it can use to alter encoding information about a reward-associated cue to reflect the updated motivational value. Thus, our prior study showing that the shell processes devaluation to reward-predictive cues (West & Carelli, 2016) may be dependent upon LH modulation, especially since LH neurons encode sensory-specific satiety (Rolls, Murzi, Yaxley, Thorpe, & Simpson, 1986).

In contrast, the mPFC sends projections to both the core and shell, but within distinct parallel circuits (i.e., the PrL projects to the core while the IL sends projections to the shell) (Heidbreder & Groenewegen, 2003). Inactivation of the PrL cortex before learning disrupts performance in reinforcer devaluation tasks, but the PrL cortex is not necessary for the expression of the devaluation effect after learning has occurred (Tran-Tu-Yen, Marchand, Pape, Di Scala, & Coutureau, 2009). This is similar to our finding that the core encodes information during learning that is necessary for performance but not information related to the performance itself (West & Carelli, 2016). Thus, the PrL and NAc core may interact to form the cue—outcome associations required for goal-directed behavior. In addition, inactivation of the mPFC (aimed at the PrL) led rats to become more impulsive by affecting preference for smaller immediate over larger delayed rewards, similar to what is observed with core manipulations (Churchwell, Morris, Heurtelou, & Kesner, 2009). However, many decision-making studies have not distinguished the PrL from the IL, which projects to the shell. Thus, future studies should incorporate these regional distinctions into their studies, especially considering the divergent roles of the core and shell in this behavior.

CONCLUSIONS

Both the NAc core and shell play critical roles in guiding goal-directed decision-making. Overall, it seems that shell plays a larger role in hedonic value processing and the core plays a stronger role in learning, and both these processes are necessary for guiding complex decisions. As with most observations within neuroscience, the core versus shell dichotomy in learning versus value encoding is likely an oversimplified view of the functional roles of these subregions. The shell is certainly involved in complex decision-making (e.g., risky decision-making and reinforcer devaluation), and the NAc core plays a critical role in processing beyond simple learning of cue—outcome contingencies (e.g., effortful responding). Importantly, studies examining the core and shell separately have shown that these two regions play distinct and complimentary roles in goal-directed decision-making. As such, they highlight the importance of examining these two distinct NAc subregions independently, but ultimately as part of a larger neural network, to understand how the brain processes reward value, learning, and goal-directed behavior.

REFERENCES

Aberman, J. E., & Salamone, J. D. (1999). Nucleus accumbens dopamine depletions make rats more sensitive to high ratio requirements but do not impair primary food reinforcement. *Neuroscience, 92*(2), 545—552.

Acheson, A., Farrar, A. M., Patak, M., Hausknecht, K. A., Kieres, A. K., Choi, S., & Richards, J. B. (2006). Nucleus accumbens lesions decrease sensitivity to rapid changes in the delay to reinforcement. *Behavioural Brain Research, 173*(2), 217—228. https://doi.org/10.1016/j.bbr.2006.06.024.

Adams, C. D., & Dickinson, A. (1981). Instrumental responding following reinforcer devaluation. *Quarterly Journal of Experimental Psychology Section B: Comparative and Physiological Psychology, 33*(May), 109—121.

Anker, J. J., Zlebnik, N. E., Gliddon, L. A., & Carroll, M. E. (2009). Performance under a Go/No-go task in rats selected for high and low impulsivity with a delay-discounting procedure. *Behavioural Pharmacology, 20*(5—6), 406—414. https://doi.org/10.1097/FBP.0b013e3283305ea2.

Bari, A. A., & Pierce, R. C. (2005). D1-like and D2 dopamine receptor antagonists administered into the shell subregion of the rat nucleus accumbens decrease cocaine, but not food, reinforcement. *Neuroscience, 135*(3), 959—968. https://doi.org/10.1016/j.neuroscience.2005.06.048.

Berridge, K. C., & Kringelbach, M. L. (2015). Pleasure systems in the brain. *Neuron, 86*(3), 646—664. https://doi.org/10.1016/j.neuron.2015.02.018.

Beyene, M., Carelli, R. M., & Wightman, R. M. (2010). Cue-evoked dopamine release in the nucleus accumbens shell tracks reinforcer magnitude during intracranial self-stimulation. *Neuroscience, 169*(4), 1682—1688. https://doi.org/10.1016/j.neuroscience.2010.06.047.

Bezzina, G., Body, S., Cheung, T. H., Hampson, C. L., Deakin, J. F., Anderson, I. M., … Bradshaw, C. M. (2008). Effect of quinolinic acid-induced lesions of the nucleus accumbens core on performance on a progressive ratio schedule of reinforcement: Implications for inter-temporal choice. *Psychopharmacology, 197*(2), 339—350. https://doi.org/10.1007/s00213-007-1036-0.

Brog, J. S., Salyapongse, A., Deutch, A. Y., & Zahm, D. S. (1993). The patterns of afferent innervation of the core and shell in the "accumbens" part of the rat ventral striatum: Immunohistochemical detection of retrogradely transported fluoro-gold. *The Journal of Comparative Neurology, 338*(2), 255—278. https://doi.org/10.1002/cne.903380209.

Broos, N., Diergaarde, L., Schoffelmeer, A. N., Pattij, T., & De Vries, T. J. (2012). Trait impulsive choice predicts resistance to extinction and propensity to relapse to cocaine seeking: A bidirectional investigation. *Neuropsychopharmacology: Official Publication of the American College of Neuropsychopharmacology, 37*(6), 1377—1386. https://doi.org/10.1038/npp.2011.323.

Cacciapaglia, F., Saddoris, M. P., Wightman, R. M., & Carelli, R. M. (2012). Differential dopamine release dynamics in the nucleus accumbens core and shell track distinct aspects of goal-directed behavior for sucrose. *Neuropharmacology, 62*(5—6), 2050—2056. https://doi.org/10.1016/j.neuropharm.2011.12.027.

Calaminus, C., & Hauber, W. (2007). Intact discrimination reversal learning but slowed responding to reward-predictive cues after dopamine D1 and D2 receptor blockade in the nucleus accumbens of rats. *Psychopharmacology, 191*(3), 551—566. https://doi.org/10.1007/s00213-006-0532-y.

Cardinal, R. N., & Howes, N. J. (2005). Effects of lesions of the nucleus accumbens core on choice between small certain rewards and large uncertain rewards in rats. *BMC Neuroscience, 6*, 37. https://doi.org/10.1186/1471-2202-6-37.

Cardinal, R. N., Pennicott, D. R., Sugathapala, C. L., Robbins, T. W., & Everitt, B. J. (2001). Impulsive choice induced in rats by lesions of the nucleus accumbens core. *Science, 292*(5526), 2499–2501. https://doi.org/10.1126/science.1060818.

Carelli, R. M. (2000). Activation of accumbens cell firing by stimuli associated with cocaine delivery during self-administration. *Synapse, 35*(3), 238–242. https://doi.org/10.1002/(SICI)1098-2396(20000301)35:3<238::AID-SYN10>3.0.CO;2-Y.

Carelli, R. M. (2004). Nucleus accumbens cell firing and rapid dopamine signaling during goal-directed behaviors in rats. *Neuropharmacology, 47*(Suppl. 1), 180–189. https://doi.org/10.1016/j.neuropharm.2004.07.017. pii:S0028390804002114.

Carelli, R. M., & Ijames, S. G. (2000). Nucleus accumbens cell firing during maintenance, extinction, and reinstatement of cocaine self-administration behavior in rats. *Brain Research, 866*(1–2), 44–54.

Castro, D. C., Cole, S. L., & Berridge, K. C. (2015). Lateral hypothalamus, nucleus accumbens, and ventral pallidum roles in eating and hunger: Interactions between homeostatic and reward circuitry. *Frontiers in Systems Neuroscience, 9*, 90. https://doi.org/10.3389/fnsys.2015.00090.

Chang, J. Y., Janak, P. H., & Woodward, D. J. (1998). Comparison of mesocorticolimbic neuronal responses during cocaine and heroin self-administration in freely moving rats. *The Journal of Neuroscience: The Official Journal of the Society for Neuroscience, 18*(8), 3098–3115.

Churchwell, J. C., Morris, A. M., Heurtelou, N. M., & Kesner, R. P. (2009). Interactions between the prefrontal cortex and amygdala during delay discounting and reversal. *Behavioral Neuroscience, 123*(6), 1185–1196. https://doi.org/10.1037/a0017734.

Coffey, S. F., Gudleski, G. D., Saladin, M. E., & Brady, K. T. (2003). Impulsivity and rapid discounting of delayed hypothetical rewards in cocaine-dependent individuals. *Experimental and Clinical Psychopharmacology, 11*(1), 18–25.

Corbit, L. H., & Balleine, B. W. (2011). The general and outcome-specific forms of Pavlovian-instrumental transfer are differentially mediated by the nucleus accumbens core and shell. *The Journal of Neuroscience: The Official Journal of the Society for Neuroscience, 31*(33), 11786–11794. https://doi.org/10.1523/JNEUROSCI.2711-11.2011.

Corbit, L. H., Muir, J. L., & Balleine, B. W. (2001). The role of the nucleus accumbens in instrumental conditioning: Evidence of a functional dissociation between accumbens core and shell. *The Journal of Neuroscience: The Official Journal of the Society for Neuroscience, 21*(9), 3251–3260.

Dalton, G. L., Phillips, A. G., & Floresco, S. B. (2014). Preferential involvement by nucleus accumbens shell in mediating probabilistic learning and reversal shifts. *The Journal of Neuroscience: The Official Journal of the Society for Neuroscience, 34*(13), 4618–4626. https://doi.org/10.1523/JNEUROSCI.5058-13.2014.

Day, J. J., Jones, J. L., & Carelli, R. M. (2011). Nucleus accumbens neurons encode predicted and ongoing reward costs in rats. *The European Journal of Neuroscience, 33*(2), 308–321. https://doi.org/10.1111/j.1460-9568.2010.07531.x.

Day, J. J., Jones, J. L., Wightman, R. M., & Carelli, R. M. (2010). Phasic nucleus accumbens dopamine release encodes effort- and delay-related costs. *Biological Psychiatry, 68*(3), 306–309. https://doi.org/10.1016/j.biopsych.2010.03.026.

Day, J. J., Roitman, M. F., Wightman, R. M., & Carelli, R. M. (2007). Associative learning mediates dynamic shifts in dopamine signaling in the nucleus accumbens. *Nature Neuroscience, 10*(8), 1020–1028. https://doi.org/10.1038/nn1923.

Day, J. J., Wheeler, R. A., Roitman, M. F., & Carelli, R. M. (2006). Nucleus accumbens neurons encode Pavlovian approach behaviors: Evidence from an autoshaping paradigm. *The European Journal of Neuroscience, 23*(5), 1341–1351. https://doi.org/10.1111/j.1460-9568.2006.04654.x.

Dickinson, A. (2012). Associative learning and animal cognition. *Philosophical Transactions of the Royal Society B: Biological Sciences, 367*(1603), 2733–2742. https://doi.org/10.1098/rstb.2012.0220.

Diergaarde, L., Pattij, T., Poortvliet, I., Hogenboom, F., de Vries, W., Schoffelmeer, A. N., & De Vries, T. J. (2008). Impulsive choice and impulsive action predict vulnerability to distinct stages of nicotine seeking in rats. *Biological Psychiatry, 63*(3), 301–308. https://doi.org/10.1016/j.biopsych.2007.07.011.

Dreyer, J. K., Vander Weele, C. M., Lovic, V., & Aragona, B. J. (2016). Functionally distinct dopamine signals in nucleus accumbens core and shell in the freely moving rat. *The Journal of Neuroscience: The Official Journal of the Society for Neuroscience, 36*(1), 98−112. https://doi.org/10.1523/JNEUROSCI. 2326-15.2016.

Everitt, B. J., Parkinson, J. A., Olmstead, M. C., Arroyo, M., Robledo, P., & Robbins, T. W. (1999). Associative processes in addiction and reward. The role of amygdala-ventral striatal subsystems. *Annals of the New York Academy of Sciences, 877,* 412−438.

Fanselow, M. S., & Dong, H. W. (2010). Are the dorsal and ventral hippocampus functionally distinct structures? *Neuron, 65*(1), 7−19. https://doi.org/10.1016/j.neuron.2009.11.031.

Feja, M., Hayn, L., & Koch, M. (2014). Nucleus accumbens core and shell inactivation differentially affects impulsive behaviours in rats. *Progress in Neuro-Psychopharmacology & Biological Psychiatry, 54,* 31−42. https://doi.org/10.1016/j.pnpbp.2014.04.012.

Ferland, J. N., & Winstanley, C. A. (2016). Risk-preferring rats make worse decisions and show increased incubation of craving after cocaine self-administration. *Addiction Biology.* https://doi.org/10.1111/adb.12388.

Floresco, S. B., Ghods-Sharifi, S., Vexelman, C., & Magyar, O. (2006). Dissociable roles for the nucleus accumbens core and shell in regulating set shifting. *The Journal of Neuroscience: The Official Journal of the Society for Neuroscience, 26*(9), 2449−2457. https://doi.org/10.1523/JNEUROSCI.4431-05.2006.

Galtress, T., & Kirkpatrick, K. (2010). The role of the nucleus accumbens core in impulsive choice, timing, and reward processing. *Behavioral Neuroscience, 124*(1), 26−43. https://doi.org/10.1037/a0018464.

Ghods-Sharifi, S., & Floresco, S. B. (2010). Differential effects on effort discounting induced by inactivations of the nucleus accumbens core or shell. *Behavioral Neuroscience, 124*(2), 179−191. https://doi.org/10.1037/a0018932.

Gourley, S. L., & Taylor, J. R. (2016). Going and stopping: Dichotomies in behavioral control by the prefrontal cortex. *Nature Neuroscience, 19*(6), 656−664. https://doi.org/10.1038/nn.4275.

Hara, Y., & Pickel, V. M. (2005). Overlapping intracellular and differential synaptic distributions of dopamine D1 and glutamate N-methyl-D-aspartate receptors in rat nucleus accumbens. *The Journal of Comparative Neurology, 492*(4), 442−455. https://doi.org/10.1002/cne.20740.

Hauber, W., & Sommer, S. (2009). Prefrontostriatal circuitry regulates effort-related decision making. *Cerebral Cortex, 19*(10), 2240−2247. https://doi.org/10.1093/cercor/bhn241.

Heidbreder, C. A., & Groenewegen, H. J. (2003). The medial prefrontal cortex in the rat: Evidence for a dorso-ventral distinction based upon functional and anatomical characteristics. *Neuroscience and Biobehavioral Reviews, 27*(6), 555−579.

Heimer, L., Zahm, D. S., Churchill, L., Kalivas, P. W., & Wohltmann, C. (1991). Specificity in the projection patterns of accumbal core and shell in the rat. *Neuroscience, 41*(1), 89−125.

Hernandez, P. J., Andrzejewski, M. E., Sadeghian, K., Panksepp, J. B., & Kelley, A. E. (2005). AMPA/kainate, NMDA, and dopamine D1 receptor function in the nucleus accumbens core: A context-limited role in the encoding and consolidation of instrumental memory. *Learning & Memory, 12*(3), 285−295. https://doi.org/10.1101/lm.93105.

Hernandez, G., Oleson, E. B., Gentry, R. N., Abbas, Z., Bernstein, D. L., Arvanitogiannis, A., & Cheer, J. F. (2014). Endocannabinoids promote cocaine-induced impulsivity and its rapid dopaminergic correlates. *Biological Psychiatry, 75*(6), 487−498. https://doi.org/10.1016/j.biopsych.2013.09.005.

Hernandez, P. J., Sadeghian, K., & Kelley, A. E. (2002). Early consolidation of instrumental learning requires protein synthesis in the nucleus accumbens. *Nature Neuroscience, 5*(12), 1327−1331. https://doi.org/10.1038/nn973.

Hoffman, W. F., Moore, M., Templin, R., McFarland, B., Hitzemann, R. J., & Mitchell, S. H. (2006). Neuropsychological function and delay discounting in methamphetamine-dependent individuals. *Psychopharmacology, 188*(2), 162−170. https://doi.org/10.1007/s00213-006-0494-0.

Hollander, J. A., & Carelli, R. M. (2007). Cocaine-associated stimuli increase cocaine seeking and activate accumbens core neurons after abstinence. *The Journal of Neuroscience: The Official Journal of the Society for Neuroscience, 27*(13), 3535−3539. https://doi.org/10.1523/JNEUROSCI.3667-06.2007.

Hollander, J. A., Ijames, S. G., Roop, R. G., & Carelli, R. M. (2002). An examination of nucleus accumbens cell firing during extinction and reinstatement of water reinforcement behavior in rats. *Brain Research, 929*(2), 226−235.

Hurley, S. W., West, E. A., & Carelli, R. M. (2017). Opposing roles of rapid dopamine signaling across the rostral-caudal axis of the nucleus accumbens shell in drug-induced negative affect. *Biological Psychiatry, 82*(11), 839−846. https://doi.org/10.1016/j.biopsych.2017.05.009.

Ito, R., Dalley, J. W., Howes, S. R., Robbins, T. W., & Everitt, B. J. (2000). Dissociation in conditioned dopamine release in the nucleus accumbens core and shell in response to cocaine cues and during cocaine-seeking behavior in rats. *The Journal of Neuroscience: The Official Journal of the Society for Neuroscience, 20*(19), 7489−7495.

Ito, R., Robbins, T. W., & Everitt, B. J. (2004). Differential control over cocaine-seeking behavior by nucleus accumbens core and shell. *Nature Neuroscience, 7*(4), 389−397. https://doi.org/10.1038/nn1217.

Jones, J. L., Day, J. J., Wheeler, R. A., & Carelli, R. M. (2010). The basolateral amygdala differentially regulates conditioned neural responses within the nucleus accumbens core and shell. *Neuroscience, 169*(3), 1186−1198. https://doi.org/10.1016/j.neuroscience.2010.05.073.

Kelley, A. E. (2004). Ventral striatal control of appetitive motivation: Role in ingestive behavior and reward-related learning. *Neuroscience and Biobehavioral Reviews, 27*(8), 765−776. https://doi.org/10.1016/j.neubiorev.2003.11.015.

Kelley, A. E., Baldo, B. A., & Pratt, W. E. (2005). A proposed hypothalamic-thalamic-striatal axis for the integration of energy balance, arousal, and food reward. *The Journal of Comparative Neurology, 493*(1), 72−85. https://doi.org/10.1002/cne.20769.

Kelley, A. E., & Swanson, C. J. (1997). Feeding induced by blockade of AMPA and kainate receptors within the ventral striatum: A microinfusion mapping study. *Behavioural Brain Research, 89*(1−2), 107−113.

Kerfoot, E. C., Agarwal, I., Lee, H. J., & Holland, P. C. (2007). Control of appetitive and aversive taste-reactivity responses by an auditory conditioned stimulus in a devaluation task: A FOS and behavioral analysis. *Learning & Memory, 14*(9), 581−589. https://doi.org/10.1101/lm.627007.

Klanker, M., Sandberg, T., Joosten, R., Willuhn, I., Feenstra, M., & Denys, D. (2015). Phasic dopamine release induced by positive feedback predicts individual differences in reversal learning. *Neurobiology of Learning and Memory, 125*, 135−145. https://doi.org/10.1016/j.nlm.2015.08.011.

Lex, B., & Hauber, W. (2010). The role of nucleus accumbens dopamine in outcome encoding in instrumental and Pavlovian conditioning. *Neurobiology of Learning and Memory, 93*(2), 283−290. https://doi.org/10.1016/j.nlm.2009.11.002. pii:S1074-7427(09)00212-3.

Loos, M., Pattij, T., Janssen, M. C., Counotte, D. S., Schoffelmeer, A. N., Smit, A. B., … van Gaalen, M. M. (2010). Dopamine receptor D1/D5 gene expression in the medial prefrontal cortex predicts impulsive choice in rats. *Cerebral Cortex, 20*(5), 1064−1070. https://doi.org/10.1093/cercor/bhp167.

Loriaux, A. L., Roitman, J. D., & Roitman, M. F. (2011). Nucleus accumbens shell, but not core, tracks motivational value of salt. *Journal of Neurophysiology, 106*(3), 1537−1544. https://doi.org/10.1152/jn.00153.2011.

Lubman, D. I., Yucel, M., & Pantelis, C. (2004). Addiction, a condition of compulsive behaviour? Neuroimaging and neuropsychological evidence of inhibitory dysregulation. *Addiction, 99*(12), 1491−1502. https://doi.org/10.1111/j.1360-0443.2004.00808.x.

Lucantonio, F., Caprioli, D., & Schoenbaum, G. (2014). Transition from 'model-based' to 'model-free' behavioral control in addiction: Involvement of the orbitofrontal cortex and dorsolateral striatum. *Neuropharmacology, 76*(Pt B), 407−415. https://doi.org/10.1016/j.neuropharm.2013.05.033.

Mai, B., Sommer, S., & Hauber, W. (2012). Motivational states influence effort-based decision making in rats: The role of dopamine in the nucleus accumbens. *Cognitive, Affective & Behavioral Neuroscience, 12*(1), 74−84. https://doi.org/10.3758/s13415-011-0068-4.

Mai, B., Sommer, S., & Hauber, W. (2015). Dopamine D1/D2 receptor activity in the nucleus accumbens core but not in the nucleus accumbens shell and orbitofrontal cortex modulates risk-based decision making. *The International Journal of Neuropsychopharmacology, 18*(10), pyv043. https://doi.org/10.1093/ijnp/pyv043.

Maldonado-Irizarry, C. S., & Kelley, A. E. (1994). Differential behavioral effects following microinjection of an NMDA antagonist into nucleus accumbens subregions. *Psychopharmacology, 116*(1), 65−72.

Maldonado-Irizarry, C. S., Swanson, C. J., & Kelley, A. E. (1995). Glutamate receptors in the nucleus accumbens shell control feeding behavior via the lateral hypothalamus. *The Journal of Neuroscience: The Official Journal of the Society for Neuroscience, 15*(10), 6779−6788.

Mendez, I. A., Simon, N. W., Hart, N., Mitchell, M. R., Nation, J. R., Wellman, P. J., & Setlow, B. (2010). Self-administered cocaine causes long-lasting increases in impulsive choice in a delay discounting task. *Behavioral Neuroscience, 124*(4), 470–477. https://doi.org/10.1037/a0020458.

Meredith, G. E., Baldo, B. A., Andrzejewski, M. E., & Kelley, A. E. (2008). The structural basis for mapping behavior onto the ventral striatum and its subdivisions. *Brain Structure & Function, 213*(1–2), 17–27. https://doi.org/10.1007/s00429-008-0175-3.

Mirenowicz, J., & Schultz, W. (1996). Preferential activation of midbrain dopamine neurons by appetitive rather than aversive stimuli. *Nature, 379*(6564), 449–451. https://doi.org/10.1038/379449a0.

Mitchell, M. R., Weiss, V. G., Beas, B. S., Morgan, D., Bizon, J. L., & Setlow, B. (2014). Adolescent risk taking, cocaine self-administration, and striatal dopamine signaling. *Neuropsychopharmacology: Official Publication of the American College of Neuropsychopharmacology, 39*(4), 955–962. https://doi.org/10.1038/npp.2013.295.

Mogenson, G. J., Jones, D. L., & Yim, C. Y. (1980). From motivation to action: Functional interface between the limbic system and the motor system. *Progress in Neurobiology, 14*(2–3), 69–97. https://doi.org/10.1016/0301-0082(80)90018-0.

Morrison, S. E., & Nicola, S. M. (2014). Neurons in the nucleus accumbens promote selection bias for nearer objects. *The Journal of Neuroscience: The Official Journal of the Society for Neuroscience, 34*(42), 14147–14162. https://doi.org/10.1523/JNEUROSCI.2197-14.2014.

Moschak, T. M., & Carelli, R. M. (2017). Impulsive rats exhibit blunted dopamine release dynamics during a delay discounting task independent of cocaine history. *eNeuro*. https://doi.org/10.1523/ENEURO.0119-17.2017.

Moschak, T. M., & Mitchell, S. H. (2014). Partial inactivation of nucleus accumbens core decreases delay discounting in rats without affecting sensitivity to delay or magnitude. *Behavioural Brain Research, 268*, 159–168. https://doi.org/10.1016/j.bbr.2014.03.044.

Nelson, A., & Killcross, S. (2006). Amphetamine exposure enhances habit formation. *The Journal of Neuroscience: The Official Journal of the Society for Neuroscience, 26*(14), 3805–3812. https://doi.org/10.1523/JNEUROSCI.4305-05.2006.

Nowend, K. L., Arizzi, M., Carlson, B. B., & Salamone, J. D. (2001). D1 or D2 antagonism in nucleus accumbens core or dorsomedial shell suppresses lever pressing for food but leads to compensatory increases in chow consumption. *Pharmacology, Biochemistry, and Behavior, 69*(3–4), 373–382.

Ongur, D., & Price, J. L. (2000). The organization of networks within the orbital and medial prefrontal cortex of rats, monkeys and humans. *Cerebral Cortex, 10*(3), 206–219.

Ostlund, S. B., Wassum, K. M., Murphy, N. P., Balleine, B. W., & Maidment, N. T. (2011). Extracellular dopamine levels in striatal subregions track shifts in motivation and response cost during instrumental conditioning. *The Journal of Neuroscience: The Official Journal of the Society for Neuroscience, 31*(1), 200–207. https://doi.org/10.1523/JNEUROSCI.4759-10.2011.

Owesson-White, C. A., Cheer, J. F., Beyene, M., Carelli, R. M., & Wightman, R. M. (2008). Dynamic changes in accumbens dopamine correlate with learning during intracranial self-stimulation. *Proceedings of the National Academy of Sciences of the United States of America, 105*(33), 11957–11962. https://doi.org/10.1073/pnas.0803896105.

Parkinson, J. A., Cardinal, R. N., & Everitt, B. J. (2000). Limbic cortical-ventral striatal systems underlying appetitive conditioning. *Progress in Brain Research, 126*, 263–285. https://doi.org/10.1016/S0079-6123(00)26019-6.

Phillips, P. E., Stuber, G. D., Heien, M. L., Wightman, R. M., & Carelli, R. M. (2003). Subsecond dopamine release promotes cocaine seeking. *Nature, 422*(6932), 614–618. https://doi.org/10.1038/nature01476.

Pothuizen, H. H., Jongen-Relo, A. L., Feldon, J., & Yee, B. K. (2005). Double dissociation of the effects of selective nucleus accumbens core and shell lesions on impulsive-choice behaviour and salience learning in rats. *The European Journal of Neuroscience, 22*(10), 2605–2616. https://doi.org/10.1111/j.1460-9568.2005.04388.x.

Poulos, C. X., Le, A. D., & Parker, J. L. (1995). Impulsivity predicts individual susceptibility to high levels of alcohol self-administration. *Behavioural Pharmacology, 6*(8), 810–814.

Randall, P. A., Pardo, M., Nunes, E. J., Lopez Cruz, L., Vemuri, V. K., Makriyannis, A., … Salamone, J. D. (2012). Dopaminergic modulation of effort-related choice behavior as assessed by a progressive ratio chow feeding choice task: Pharmacological studies and the role of individual differences. *PLoS One, 7*(10), e47934. https://doi.org/10.1371/journal.pone.0047934.

Robinson, D. L., & Carelli, R. M. (2008). Distinct subsets of nucleus accumbens neurons encode operant responding for ethanol versus water. *The European Journal of Neuroscience, 28*(9), 1887−1894. https://doi.org/10.1111/j.1460-9568.2008.06464.x.

Roesch, M. R., Singh, T., Brown, P. L., Mullins, S. E., & Schoenbaum, G. (2009). Ventral striatal neurons encode the value of the chosen action in rats deciding between differently delayed or sized rewards. *The Journal of Neuroscience: The Official Journal of the Society for Neuroscience, 29*(42), 13365−13376. https://doi.org/10.1523/JNEUROSCI.2572-09.2009.

Roitman, M. F., Stuber, G. D., Phillips, P. E., Wightman, R. M., & Carelli, R. M. (2004). Dopamine operates as a subsecond modulator of food seeking. *The Journal of Neuroscience: The Official Journal of the Society for Neuroscience, 24*(6), 1265−1271. https://doi.org/10.1523/JNEUROSCI.3823-03.2004.

Roitman, M. F., Wheeler, R. A., & Carelli, R. M. (2005). Nucleus accumbens neurons are innately tuned for rewarding and aversive taste stimuli, encode their predictors, and are linked to motor output. *Neuron, 45*(4), 587−597. https://doi.org/10.1016/j.neuron.2004.12.055.

Roitman, M. F., Wheeler, R. A., Tiesinga, P. H., Roitman, J. D., & Carelli, R. M. (2010). Hedonic and nucleus accumbens neural responses to a natural reward are regulated by aversive conditioning. *Learning & Memory, 17*(11), 539−546. https://doi.org/10.1101/lm.1869710. pii:17/11/539.

Roitman, M. F., Wheeler, R. A., Wightman, R. M., & Carelli, R. M. (2008). Real-time chemical responses in the nucleus accumbens differentiate rewarding and aversive stimuli. *Nature Neuroscience, 11*(12), 1376−1377. https://doi.org/10.1038/nn.2219.

Rolls, E. T., Murzi, E., Yaxley, S., Thorpe, S. J., & Simpson, S. J. (1986). Sensory-specific satiety: Food-specific reduction in responsiveness of ventral forebrain neurons after feeding in the monkey. *Brain Research, 368*(1), 79−86.

Sackett, D. S., Saddoris, M. P., & Carelli, R. M. (2017). Nucleus accumbens shell dopamine preferentially tracks information related to outcome value of reward. *eNeuro, 4*(3). https://doi.org/10.1523/ENEURO.0058-17. pii:ENEURO.0058−17.2017.

Saddoris, M. P. (2016). Terminal dopamine release kinetics in the accumbens core and shell are distinctly altered after withdrawal from cocaine self-administration. *eNeuro, 3*(5). https://doi.org/10.1523/ENEURO.0274-16.2016.

Saddoris, M. P., Cacciapaglia, F., Wightman, R. M., & Carelli, R. M. (2015). Differential dopamine release dynamics in the nucleus accumbens core and shell reveal complementary signals for error prediction and incentive motivation. *The Journal of Neuroscience: The Official Journal of the Society for Neuroscience, 35*(33), 11572−11582. https://doi.org/10.1523/JNEUROSCI.2344-15.2015.

Saddoris, M. P., & Carelli, R. M. (2014). Cocaine self-administration abolishes associative neural encoding in the nucleus accumbens necessary for higher-order learning. *Biological Psychiatry, 75*(2), 156−164. https://doi.org/10.1016/j.biopsych.2013.07.037.

Saddoris, M. P., Stamatakis, A., & Carelli, R. M. (2011). Neural correlates of Pavlovian-to-instrumental transfer in the nucleus accumbens shell are selectively potentiated following cocaine self-administration. *The European Journal of Neuroscience, 33*(12), 2274−2287. https://doi.org/10.1111/j.1460-9568.2011.07683.x.

Saddoris, M. P., Sugam, J. A., Cacciapaglia, F., & Carelli, R. M. (2013). Rapid dopamine dynamics in the accumbens core and shell: Learning and action. *Frontiers in Bioscience (Elite Edition), 5*, 273−288.

Saddoris, M. P., Sugam, J. A., & Carelli, R. M. (2017). Prior cocaine experience impairs normal phasic dopamine signals of reward value in accumbens shell. *Neuropsychopharmacology: Official Publication of the American College of Neuropsychopharmacology, 42*(3), 766−773. https://doi.org/10.1038/npp.2016.189.

Saddoris, M. P., Sugam, J. A., Stuber, G. D., Witten, I. B., Deisseroth, K., & Carelli, R. M. (2015). Mesolimbic dopamine dynamically tracks, and is causally linked to, discrete aspects of value-based decision making. *Biological Psychiatry, 77*(10), 903−911. https://doi.org/10.1016/j.biopsych.2014.10.024.

Saddoris, M. P., Wang, X., Sugam, J. A., & Carelli, R. M. (2016). Cocaine self-administration experience induces pathological phasic accumbens dopamine signals and abnormal incentive behaviors in drug-abstinent rats. *The Journal of Neuroscience: The Official Journal of the Society for Neuroscience, 36*(1), 235–250. https://doi.org/10.1523/JNEUROSCI.3468-15.2016.

Salamone, J. D., Wisniecki, A., Carlson, B. B., & Correa, M. (2001). Nucleus accumbens dopamine depletions make animals highly sensitive to high fixed ratio requirements but do not impair primary food reinforcement. *Neuroscience, 105*(4), 863–870.

Schoenbaum, G., Saddoris, M. P., Ramus, S. J., Shaham, Y., & Setlow, B. (2004). Cocaine-experienced rats exhibit learning deficits in a task sensitive to orbitofrontal cortex lesions. *The European Journal of Neuroscience, 19*(7), 1997–2002. https://doi.org/10.1111/j.1460-9568.2004.03274.x.

Schoenbaum, G., & Setlow, B. (2003). Lesions of nucleus accumbens disrupt learning about aversive outcomes. *The Journal of Neuroscience: The Official Journal of the Society for Neuroscience, 23*(30), 9833–9841.

Schoenbaum, G., & Setlow, B. (2005). Cocaine makes actions insensitive to outcomes but not extinction: Implications for altered orbitofrontal-amygdalar function. *Cerebral Cortex, 15*(8), 1162–1169. https://doi.org/10.1093/cercor/bhh216.

Sellings, L. H., & Clarke, P. B. (2003). Segregation of amphetamine reward and locomotor stimulation between nucleus accumbens medial shell and core. *The Journal of Neuroscience: The Official Journal of the Society for Neuroscience, 23*(15), 6295–6303.

Setlow, B., Schoenbaum, G., & Gallagher, M. (2003). Neural encoding in ventral striatum during olfactory discrimination learning. *Neuron, 38*(4), 625–636.

Simon, N. W., Beas, B. S., Montgomery, K. S., Haberman, R. P., Bizon, J. L., & Setlow, B. (2013). Prefrontal cortical-striatal dopamine receptor mRNA expression predicts distinct forms of impulsivity. *The European Journal of Neuroscience, 37*(11), 1779–1788. https://doi.org/10.1111/ejn.12191.

Simon, N. W., Montgomery, K. S., Beas, B. S., Mitchell, M. R., LaSarge, C. L., Mendez, I. A., Bañuelos, C., Vokes, C. M., Taylor, A. B., Haberman, R. P., Bizon, J. L., & Setlow, B. (2011). Dopaminergic modulation of risky decision-making. *J Neurosci., 31*(48), 17460–17470. https://doi.org/10.1523/JNEUROSCI.3772-11.2011.

Singh, T., McDannald, M. A., Haney, R. Z., Cerri, D. H., & Schoenbaum, G. (2010). Nucleus accumbens core and shell are necessary for reinforcer devaluation effects on Pavlovian conditioned responding. *Frontiers in Integrative Neuroscience, 4*, 126. https://doi.org/10.3389/fnint.2010.00126.

Sokolowski, J. D., & Salamone, J. D. (1998). The role of accumbens dopamine in lever pressing and response allocation: Effects of 6-OHDA injected into core and dorsomedial shell. *Pharmacology, Biochemistry, and Behavior, 59*(3), 557–566.

St Onge, J. R., Ahn, S., Phillips, A. G., & Floresco, S. B. (2012). Dynamic fluctuations in dopamine efflux in the prefrontal cortex and nucleus accumbens during risk-based decision making. *The Journal of Neuroscience: The Official Journal of the Society for Neuroscience, 32*(47), 16880–16891. https://doi.org/10.1523/JNEUROSCI.3807-12.2012.

Stopper, C. M., & Floresco, S. B. (2011). Contributions of the nucleus accumbens and its subregions to different aspects of risk-based decision making. *Cognitive, Affective & Behavioral Neuroscience, 11*(1), 97–112. https://doi.org/10.3758/s13415-010-0015-9.

Stopper, C. M., Khayambashi, S., & Floresco, S. B. (2013). Receptor-specific modulation of risk-based decision making by nucleus accumbens dopamine. *Neuropsychopharmacology: Official Publication of the American College of Neuropsychopharmacology, 38*(5), 715–728. https://doi.org/10.1038/npp.2012.240.

Stratford, T. R., & Kelley, A. E. (1997). GABA in the nucleus accumbens shell participates in the central regulation of feeding behavior. *The Journal of Neuroscience: The Official Journal of the Society for Neuroscience, 17*(11), 4434–4440.

Stuber, G. D., Roitman, M. F., Phillips, P. E., Carelli, R. M., & Wightman, R. M. (2005). Rapid dopamine signaling in the nucleus accumbens during contingent and noncontingent cocaine administration. *Neuropsychopharmacology: Official Publication of the American College of Neuropsychopharmacology, 30*(5), 853–863. https://doi.org/10.1038/sj.npp.1300619.

Sugam, J. A., Day, J. J., Wightman, R. M., & Carelli, R. M. (2012). Phasic nucleus accumbens dopamine encodes risk-based decision-making behavior. *Biological Psychiatry, 71*(3), 199–205. https://doi.org/10.1016/j.biopsych.2011.09.029.

Sugam, J. A., Saddoris, M. P., & Carelli, R. M. (2014). Nucleus accumbens neurons track behavioral preferences and reward outcomes during risky decision making. *Biological Psychiatry, 75*(10), 807–816. https://doi.org/10.1016/j.biopsych.2013.09.010.

Sun, W., & Rebec, G. V. (2003). Lidocaine inactivation of ventral subiculum attenuates cocaine-seeking behavior in rats. *The Journal of Neuroscience: The Official Journal of the Society for Neuroscience, 23*(32), 10258–10264.

Tran-Tu-Yen, D. A., Marchand, A. R., Pape, J. R., Di Scala, G., & Coutureau, E. (2009). Transient role of the rat prelimbic cortex in goal-directed behaviour. *The European Journal of Neuroscience, 30*(3), 464–471. https://doi.org/10.1111/j.1460-9568.2009.06834.x.

Volkow, N. D., Fowler, J. S., Wang, G. J., & Goldstein, R. Z. (2002). Role of dopamine, the frontal cortex and memory circuits in drug addiction: Insight from imaging studies. *Neurobiology of Learning and Memory, 78*(3), 610–624.

West, E. A., & Carelli, R. M. (2016). Nucleus accumbens core and shell differentially encode reward-associated cues after reinforcer devaluation. *The Journal of Neuroscience: The Official Journal of the Society for Neuroscience, 36*(4), 1128–1139. https://doi.org/10.1523/JNEUROSCI.2976-15.2016.

Wheeler, R. A., Aragona, B. J., Fuhrmann, K. A., Jones, J. L., Day, J. J., Cacciapaglia, F., ... Carelli, R. M. (2011). Cocaine cues drive opposing context-dependent shifts in reward processing and emotional state. *Biological Psychiatry, 69*(11), 1067–1074. https://doi.org/10.1016/j.biopsych.2011.02.014.

Wheeler, R. A., Twining, R. C., Jones, J. L., Slater, J. M., Grigson, P. S., & Carelli, R. M. (2008). Behavioral and electrophysiological indices of negative affect predict cocaine self-administration. *Neuron, 57*(5), 774–785. https://doi.org/10.1016/j.neuron.2008.01.024.

Winstanley, C. A., Theobald, D. E., Dalley, J. W., & Robbins, T. W. (2005). Interactions between serotonin and dopamine in the control of impulsive choice in rats: Therapeutic implications for impulse control disorders. *Neuropsychopharmacology: Official Publication of the American College of Neuropsychopharmacology, 30*(4), 669–682. https://doi.org/10.1038/sj.npp.1300610.

Wolf, M. E., Mangiavacchi, S., & Sun, X. (2003). Mechanisms by which dopamine receptors may influence synaptic plasticity. *Annals of the New York Academy of Sciences, 1003*, 241–249.

Wright, C. I., Beijer, A. V., & Groenewegen, H. J. (1996). Basal amygdaloid complex afferents to the rat nucleus accumbens are compartmentally organized. *The Journal of Neuroscience: The Official Journal of the Society for Neuroscience, 16*(5), 1877–1893.

Yang, J. H., & Liao, R. M. (2015). Dissociable contribution of nucleus accumbens and dorsolateral striatum to the acquisition of risk choice behavior in the rat. *Neurobiology of Learning and Memory, 126*, 67–77. https://doi.org/10.1016/j.nlm.2015.11.002.

Zahm, D. S. (1992). An electron microscopic morphometric comparison of tyrosine hydroxylase immunoreactive innervation in the neostriatum and the nucleus accumbens core and shell. *Brain Research, 575*(2), 341–346.

Zahm, D. S. (1999). Functional-anatomical implications of the nucleus accumbens core and shell subterritories. *Annals of the New York Academy of Sciences, 877*, 113–128.

Zeeb, F. D., Soko, A. D., Ji, X., & Fletcher, P. J. (2016). Low impulsive action, but not impulsive choice, predicts greater conditioned reinforcer salience and augmented nucleus accumbens dopamine release. *Neuropsychopharmacology: Official Publication of the American College of Neuropsychopharmacology, 41*(8), 2091–2100. https://doi.org/10.1038/npp.2016.9.

CHAPTER 10

Studying Integrative Processing and Prospected Plasticity in Cholinergic Interneurons: The Importance of Pavlovian—Instrumental Transfer

Jesus Bertran-Gonzalez, Vincent Laurent
Decision Neuroscience Laboratory, School of Psychology, University of New South Wales, Sydney, NSW, Australia

INTRODUCTION

Every day, we make countless decisions, and we choose to engage in certain courses of action while ignoring others. In many circumstances, these choices are controlled by the value that we attribute to the consequences of all available actions (Dickinson & Balleine, 1994). On a hot summer day, we buy and drink a fresh lemonade, and we avoid a warm coffee beverage. In other circumstances, our choices are influenced by the presence of environmental cues that we have learned predict particular outcomes (Colwill & Rescorla, 1988). For example, the smell of lemons will often drive us toward procuring and consuming a fresh lemonade. Although the latter choice may well be perceived as being somewhat forced upon us, it confers clear adaptive advantages. The predictive cue (i.e., the smell of lemon) informs us about the likelihood of the outcome being available (i.e., the lemonade) and accordingly, we favor a course of actions that allows us to secure this highly probable outcome. However, cue-based choices can result in detrimental and maladaptive behaviors. For instance, predictive cues influence choice between actions regardless of the value associated with the consequences of these actions: the smell of lemons could lead us to drink a lemonade even though we are not thirsty (Corbit, Janak, & Balleine, 2007). Similar suboptimal decision-making processes are believed to play a central role in many pathologies such as drug addiction or obesity (Belin, Jonkman, Dickinson, Robbins, & Everitt, 2009; Colagiuri & Lovibond, 2015; Lamb, Schindler, & Pinkston, 2016), highlighting the need to gain a better understanding of the psychological and neurobiological mechanisms underlying cue-based choices. The best approach to model these choices in animals and humans is known as Pavlovian—instrumental transfer (PIT), and as we discuss here, PIT is a key paradigm to understanding the neuronal mechanisms underlying cue-driven decision-making.

The PIT paradigm can be implemented in both the aversive and appetitive domains (Campese, McCue, Lázaro-Muñoz, LeDoux, & Cain, 2013; Cartoni, Balleine, & Baldassarre, 2016; Holmes, Marchand, & Coutureau, 2010), but the latter remains more prominent and as a consequence, the present chapter will focus on appetitive forms of PIT. The appetitive paradigm typically entails two separate training stages (see Fig. 10.1A). One involves Pavlovian conditioning during which various cues, or stimuli, are trained to predict the delivery of distinct food outcomes (i.e., stimulus—outcome or Pavlovian contingencies). The other training stage consists of instrumental conditioning where the same food outcomes can be earned by engaging in specific actions (i.e., action—outcome or instrumental contingencies). Then, a Pavlovian—instrumental transfer test is used to assess the influence of the predictive stimuli learned in the Pavlovian phase on performing the actions trained in the instrumental phase. Importantly, such test reveals the existence of two distinct forms of PIT. The "general" form demonstrates that a stimulus predicting a food outcome increases performance on an action delivering food. This increase is evidenced by greater responding on the action in the presence of the predictive stimulus than in its absence and/or in the presence of a nonpredictive stimulus

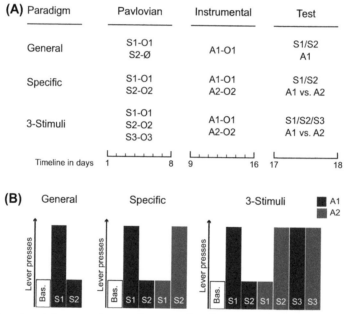

Figure 10.1 *Experimental approaches used to examine the various forms of the Pavlovian instrumental transfer (PIT) paradigm.* (A). Three designs are presented that study general and/or specific PIT. All involved three stages. The first stage is Pavlovian conditioning during which several stimuli (S1, S2, S3) predict distinct food outcomes (O1, O2, O3). Note that the general PIT paradigm includes a control stimulus (S2) that is trained as neutral (i.e., it did not predict any outcome). Instrumental conditioning is then administered and involved training one or two actions (A1, A2) to deliver the previously used food outcomes (O1, O2). Finally, a PIT test assesses responding on the instrumental actions in the presence or absence of the various stimuli. (B). Schematic graphs that illustrate the typical pattern of instrumental performance observed using the three distinct paradigms. *Bas.,* baseline responding in the absence of the stimuli.

(Fig. 10.1B). Critically, general PIT is observed regardless of the specific type of appetitive outcomes predicted by the stimulus and earned by the action. In other words, a stimulus that signals the delivery of a fresh lemonade energizes performance on an action that delivers a warm coffee beverage. In striking contrast, the "specific" form of PIT shows that the influence of predictive stimuli on instrumental actions can be highly selective. Indeed, specific PIT demonstrates that a stimulus predicting a particular outcome biases choice toward actions earning the same outcome (Fig. 10.1B). That is, a stimulus that signals the delivery of a fresh lemonade increases performance on an action earning lemonade but leaves unaffected performance on a copresent action earning a warm coffee beverage. Thus, general PIT usually involves a single instrumental action, whereas at least two actions must be trained to observe specific PIT.

The two forms of PIT have been described in humans, monkeys, rats, mice, and equines (Cartoni et al., 2016; Colagiuri & Lovibond, 2015; Holmes et al., 2010), demonstrating their robustness, validity, and significance to understanding how environmental cues guide actions in our daily life. Although general PIT could be perceived as a simpler approach to study cue-based choice, specific PIT has received far more attention in the literature. At the behavioral level, specific PIT has been shown to present two remarkable characteristics. The first is its resistance to manipulations that hinder the stimulus—outcome contingencies established across Pavlovian conditioning and the action—outcome contingencies produced by instrumental conditioning. Thus, extinction of the former (Delamater, 1996; Laurent, Chieng, & Balleine, 2016) and degradation of the latter (Rescorla, 1992) fail to remove the specific influence of predictive stimulus on choice between actions. A second characteristic is that specific PIT occurs even though the outcomes predicted by the stimuli and earned by the actions are not desirable. That is, a stimulus predicting an outcome that is being paired with sickness, and is therefore somewhat aversive, remains able to guide choice toward an action delivering that devalued outcome (Holland, 2004; Rescorla, 1994). In a similar fashion, animals will persevere in showing appetitive-specific PIT even though they are not hungry (Corbit et al., 2007). If anything, the influence of predictive stimuli on choice between actions therefore appears to override that produced by outcome value. The two characteristics just mentioned may well explain the increased interest in specific PIT over the past few decades. Indeed, they provide some insights as to why drug addictive behaviors are so resistant to cognitive treatments and are prone to relapse. Further, these characteristics also mimic the observation that obese individuals may engage in food consumption even though they are not hungry.

In this chapter, we intend to provide some insights into the importance of the neurobiology underlying appetitive PIT. Several reviews are available that explain how the study of cue-based choice may help our understanding of human pathologies, such as obesity and drug addiction (Belin et al., 2009; Hogarth, Balleine, Corbit, & Killcross, 2013; Johnson, 2013; Lamb et al., 2016). Here, we will take another approach and will argue that PIT deserves attention for two crucial reasons. The first is that it is an extremely powerful tool to establish the specific roles played by various brain regions in the integration of two main forms of conditioning: Pavlovian and instrumental. This will be demonstrated in the first section of this chapter that summarizes current knowledge about the neural circuitry supporting PIT. The second reason justifying the importance of PIT is that this paradigm is extremely well suited to confirm a central tenet on neuronal plasticity: the way the brain anticipates and adjusts to future biologically significant events. Specifically, the second part of this chapter will present evidence that Pavlovian conditioning triggers highly specific neuronal changes that are not necessary for that present conditioning but rather are essential in the future, when it exerts an influence over choice between actions.

THE NEUROANATOMY OF CUE-BASED CHOICE

PIT demonstrates that Pavlovian and instrumental learning interact with each other to control adaptive behavior. Accordingly, it is not surprising that most of the brain regions implicated in PIT are also known to process some aspects of the stimulus–outcome (i.e., S-O) contingencies established by Pavlovian conditioning and/or the action–outcome (i.e., A-O) associations produced by instrumental conditioning. This section will therefore describe the neuroanatomy of PIT with respect to the role played by brain regions in either Pavlovian or instrumental conditioning. However, we will end this section by arguing that the ventral striatum plays little role in either form of conditioning. Rather, we will propose that it integrates information from brain regions supporting the two forms of conditioning in order to promote the PIT effect.

Neuronal circuitry involved in stimulus–outcome learning

Compelling evidence indicates that the orbitofrontal cortex (OFC) is involved in specific PIT. However, the nature of this involvement critically depends on which subregion of the OFC is being considered. Lesion encompassing both the ventral and lateral OFC has been shown to abolish specific PIT whether it is completed before Pavlovian and instrumental training or after these two training stages (Ostlund & Balleine, 2005, 2007). Further, the same lesion was found to impair the updating of the S-O associations that are established across Pavlovian conditioning. Specifically, animals with such lesion fail to cease responding to a stimulus that is no longer a reliable predictor of its previously associated outcome (i.e., contingency degradation). Contrasting with this impairment, damage to the ventral or lateral portion of the OFC does not disrupt encoding or retrieval of the A-O associations produced by instrumental conditioning. For instance, animals carrying such damage are perfectly able to select an action according to the value associated with its outcome. These data therefore suggest that activity in the ventral and lateral

OFC is required during specific PIT because it processes some aspects of the S-O contingencies that develop across Pavlovian conditioning.

The amygdala was one of the first brain regions to be identified as being critical for cue-based choice. Yet, early lesion studies resulted in some discrepancies with respect to which particular subnucleus of the amygdala contributes to the PIT effect. Some studies revealed a crucial role for the central nucleus of the amygdala (CeA) while others highlighted the importance of its basolateral complex (BLA) (Blundell, Hall, & Killcross, 2001; J. Hall, Parkinson, Connor, Dickinson, & Everitt, 2001; P.C. Holland & Gallagher, 2003). A potential explanation for these early inconsistencies comes from the observation that these studies were assessing distinct forms of PIT, with some investigating general PIT and others examining specific PIT. This explanation was cleverly confirmed with the introduction of the so-called three-stimuli design (see Fig. 10.1A) by Corbit and Balleine (2005). This design consists of training three stimuli (i.e., S1, S2, and S3) to predict the delivery of three distinct outcomes (i.e., O1, O2, and O3) across Pavlovian conditioning (i.e., S1-O1, S2-O2, and S3-O3). Then, instrumental conditioning is administered during which O1 and O2 can be earned by performing actions A1 and A2, respectively (i.e., A1-O1 and A2-O2). A final PIT test examines the effects of the three stimuli on the two instrumental actions. This design is extremely powerful because it allows to observe the two forms of PIT during the same test in a within-subject fashion. That is, control subjects typically exhibit specific PIT as S1 increases responding on A1 but not A2, whereas S2 elevates performance on A2 but not A1. At the same time, these subjects display general PIT as S3 energizes responding on both A1 and A2. Using this design, it was found that BLA lesion removes the effect of S1 and S2 on A1 and A2, but it leaves unaffected the energizing impact of S3 on the two actions. Conversely, CeA lesion spares the bias in responding produced by S1 and S2 on A1 and A2, but it abolishes the effect of S3 on these two actions. These findings convincingly demonstrate that general PIT requires activity in the CeA but not the BLA. By contrast, specific PIT relies on BLA activity but not CeA activity. Although these data were obtained in rodents, it must be emphasized that a similar dissociation at the level of the amygdala has been revealed using functional imaging in human subjects that were submitted to a general or specific PIT task (Prévost, Liljeholm, Tyszka, & O'Doherty, 2012). The ability to translate the rodent findings to humans clearly demonstrates the significance of studying the PIT paradigm to understand how environmental cues guide actions in our daily life.

It must be acknowledged that the involvement of the amygdala in the two forms of PIT is somewhat unsurprising. The BLA and the CeA have repeatedly been implicated in the processing of the S-O associations that develop across Pavlovian conditioning (Holland & Gallagher, 1999; Petrovich, 2011). It is therefore likely that disrupting BLA activity across a PIT test would prevent retrieval of the specific S-O contingencies (i.e., S1-O1 and S2-O2) established across Pavlovian conditioning. Given that specific

PIT requires the integration of these contingencies with the A–O relationships produced by instrumental conditioning, it is clear that BLA disruption will abolish the PIT effect. However, the dissociation between the roles played by the BLA and the CeA in specific and general PIT remains particularly valuable to our understanding of how these two nuclei contribute to Pavlovian conditioning. This understanding has largely been dominated by findings employing the aversive form of Pavlovian conditioning (i.e., the outcome is aversive). For decades, the BLA has been viewed as the site of encoding, storage, and retrieval of the S–O associations. Meanwhile, the CeA was defined as a simple relay that, when triggered by the BLA, coordinates the various conditioned responses reflecting the nature of the outcome predicted by the stimulus (LeDoux, 2000; Maren, 2001). Clearly, the PIT findings are at odds with such serial model of amygdala functioning in Pavlovian conditioning. For instance, the PIT findings previously described are commonly interpreted as showing that the CeA is necessary for processing and retrieving the general motivational property (e.g., an appetitive or aversive) of a predicted outcome, while the BLA would compute and retrieve information about the sensory-specific property (i.e., its shape, odor, or texture) of that same outcome (Corbit & Balleine, 2005). Indeed, the former property is sufficient to trigger general PIT, whereas the latter is required for specific PIT. Accordingly, it has been proposed that, rather than functioning in a serial manner, the CeA and BLA work in parallel to incorporate the various aspects (i.e., motivational and sensory-specific) of the S–O associations that develop across Pavlovian conditioning (Balleine & Killcross, 2006). Importantly, this proposal has now received substantial support, even in the aversive form of Pavlovian conditioning. For example, it has been shown that the acquisition and consolidation of Pavlovian fear conditioning is disrupted by pharmacological inactivation of either the BLA or the CeA (Wilensky, Schafe, Kristensen, & LeDoux, 2006). There is also evidence for experience-dependent plasticity changes occurring in the CeA that are necessary for the formation of a fear memory (Li et al., 2013). Perhaps more strikingly, a recent study has revealed that the CeA controls some aspects of the syntactic strengthening in BLA neurons that has long been argued to underlie the predictive relationships between a stimulus and an aversive stimulation (Yu et al., 2017). Although the findings on PIT may not have been the sole driver for reevaluating how the CeA and BLA interact to support Pavlovian conditioning, they do highlight the strength of the paradigm in getting a better knowledge as to how various brain regions contribute to this form of conditioning.

Neuronal circuitry involved in action–outcome learning

As mentioned previously, the involvement of the OFC in specific PIT critically depends on which of its subregion is being studied. Thus, unlike its lateral and ventral portion, the medial portion of the OFC appears to be required for specific PIT because it plays an important role in computing some aspects of the A-O relationships (Bradfield, Dezfouli, van Holstein, Chieng, & Balleine, 2015). More specifically, it has been proposed that the medial OFC is critical for retrieving sensory-specific information about nonobservable outcomes and directing choice between actions according to that information. It is noteworthy that the use of the specific PIT task was central in pinpointing the precise role played by the medial OFC in instrumental outcome retrieval.

Beyond the OFC, two other prefrontal regions have been suspected to play an important role in cue-based choice. These two regions are the prelimbic (PL) and infralimbic (IL) cortices. A potential involvement of the PL and the IL is particularly relevant, as they have been shown to contribute to instrumental conditioning (Corbit & Balleine, 2003; Coutureau & Killcross, 2003). Specifically, the PL is known to be necessary for the acquisition of A-O associations. However, this role appears to be relatively transient and thereby, it is not surprising that manipulations of PL activity have failed to interfere with the expression of either general or specific PIT (Corbit & Balleine, 2003). In contrast to the PL, the IL cortex is believed to contribute to the establishment of habits that develop with extensive instrumental training (Coutureau & Killcross, 2003). Habits are behaviors that are characterized by their inflexibility, as evidenced by a loss of sensitivity to manipulations of A-O contingencies or instrumental outcome value (Balleine, Liljeholm, & Ostlund, 2009). As noted previously, this loss of sensitivity is also evident in the specific PIT paradigm (Holland, 2004; Rescorla, 1994). Accordingly, one study has revealed that disrupting IL activity abolishes specific PIT (Keistler, Barker, & Taylor, 2015). However, another study failed to reproduce such abolishment (Laurent et al., 2016). The reason for this failure remains unclear, but it must be noted that the latter study involved manipulations of S-O contingencies prior to testing for specific PIT. These manipulations may have impacted how the IL contributes to cue-based choice. Regardless, it is clear that a putative role of the IL in specific PIT could stem from its involvement in the development of habits across extensive instrumental conditioning.

The dorsal striatum is another brain region that has been implicated in cue-based choice. As with the amygdala, the use of the PIT paradigm has allowed to dissociate the role played by distinct subregions of the dorsal striatum. Thus, inactivation of the dorsolateral striatum (DLS) during a specific PIT test has been shown to generally reduce instrumental performance (Corbit & Janak, 2007). However, the inactivation left the specific PIT effect intact. That is, a stimulus predicting a particular outcome was still able to bias choice toward an action earning that same outcome. This bias was, however, absent when the dorsomedial part of the striatum (DMS) was inactivated at test, even though overall instrumental performance remained similar to controls. Similar to the findings on the amygdala, the involvement of the DMS in specific PIT should not necessarily been interpreted as evidenced that this brain region is critical for cue-based choice per se. As noted, specific PIT requires the integration of the S-O and A-O associations, and there is ample evidence that the DMS is crucial for establishing the latter type of associations (Balleine et al., 2009). Thus, the specific PIT impairment produced by DMS inactivation is likely to reveal an inability to retrieve the specific A-O contingencies established across instrumental conditioning. However, these results provide valuable and new information on how the DLS and DMS functions in parallel to influence behavior. The former appears to have a general energizing effect on instrumental performance, whereas the latter may allow outcome-specific expression of the performance. Although this may only be the case when predictive stimuli are present, it could well be a general feature of the role played by these two brain regions during instrumental conditioning.

Neuronal integration of Pavlovian–instrumental transfer

Similar to the dorsal striatum, the role of the ventral striatum in choice between actions depends on the subregion that is considered. This dissociation was convincingly demonstrated using the three-stimuli design that was employed to establish the specific contribution of the CeA and BLA during the PIT paradigm. Thus, using this design, it was shown that the nucleus accumbens core promotes general PIT, whereas the nucleus accumbens shell (NAc-S) is necessary for specific PIT (Corbit & Balleine, 2011; Corbit, Muir & Balleine, 2001). Interestingly, it has also been shown that IL (Keistler et al., 2015) and BLA (Shiflett & Balleine, 2010) inputs onto the NAc-S are required for the specific PIT effect.

The role of NAc-S in specific PIT is particularly interesting as, unlike other brain regions, it appears to be restricted to the expression of specific PIT per se. Several observations are consistent with such statement. First, rats with NAc-S lesion do not exhibit any deficit across Pavlovian or instrumental conditioning (Corbit, Muir, Balleine, & Balleine, 2001). Second, these rats are able to stop responding to a stimulus when the contingent relationship with its outcome is degraded (unpublished data from our laboratory), implying that this brain region plays little role in establishing specific S-O associations. Further, the NAc-S is not involved in instrumental conditioning: local lesion spares the sensitivity of A-O associations to change in contingencies or in outcome value (Corbit et al., 2001). Thus, we propose that the NAc-S plays very little role, if any, in the establishment of S-O and A-O associations. Rather, we argue that the NAc-S is essential later for integrating the specific S-O and A-O relationships that mediate specific PIT.

In summary, several brain regions have been found to be essential in promoting general and specific PIT. In addition to enhancing our knowledge about the neurobiology underlying cue-based choice, we have provided evidence that the use of the PIT paradigm allows a better understanding of the specific role played by various brain regions in Pavlovian and instrumental conditioning. For example, we have seen that PIT provided evidence for a parallel functioning of two main subnuclei of the amygdala. It revealed the critical role played by the medial OFC in retrieving information about unobservable outcomes. It also delivered insights about the distinct roles played by the DLS and DMS in action initiation and performance. The tremendous advantage of employing PIT is not limited to neural advances, as it has improved our understanding of several behavioral phenomena such as extinction (Delamater, 1996; Laurent et al., 2016) or inhibitory learning (Laurent & Balleine, 2015; Quail, Laurent, & Balleine, 2017), by providing an accurate approach to distinguish between motivational and outcome-specific information. Regardless, one striking observation from the literature described above concerns the highly precise role played by the NAc-S in specific PIT. Unlike other brain regions, the NAc-S appears to be specifically involved in integrating S-O and A-O contingencies to promote cue-based choice in specific PIT. This unique anticipatory feature of the NAc-S has motivated the completion of a large number studies in our laboratory that aimed at describing how this brain region computes the multidimensional integration necessary for cue-based choice. This process, which involves a complex interaction between different converging neuromodulatory signals in distinct local circuits in the NAc-S, will be described in the second part of this chapter. A central aim of this part is to illustrate how the specific PIT paradigm sets the optimal environment to study a central feature of neuronal plasticity: how the brain circuitry arranges throughout learning in order to prepare for future significant events.

THE NEUROBIOLOGY OF CUE-BASED CHOICE

Neuromodulation signals in the nucleus accumbens shell

The integration of neuronal signals is a ubiquitous property of the brain. Virtually, all structural levels of the brain are designed to integrate signals, from intracellular molecular programs detecting fluctuations in the extracellular environment to promote changes in transcription, to multidomain neuronal clusters funnlling down sparse information states to compute coherent behavioral responses. It is therefore difficult to pinpoint a "hotspot" for integration across learning, unless two or more parallel sources of information processing (and integration on the way) clash in a very defined point in time and space. This optimal niche for large-scale integration is what specific PIT promotes at the moment of test, and it appears to take place in the NAc-S. If one looks into the specifics of the NAc-S neuronal organization, it becomes immediately evident how this structure—as the rest of the striatum—is specifically designed to integrate large amounts of signals from converging, functionally related circuits (Hunnicutt et al., 2016; Voorn, Vanderschuren, Groenewegen, Robbins, & Pennartz, 2004). The first revealing feature for such a function is its main population of neurons, the so-called spiny projection neurons (SPNs, also known as medium-sized spiny neurons or MSNs). These GABAergic neurons comprise over 95% of the total neuronal population in the NAc-S and are characterized by the very high spine densities in their dendrites and their reasonably long-range projections, both rare features among GABAergic neuronal types. These cells are therefore adapted to receiving large loads of upstream glutamatergic information (e.g., the cortex, thalamus, amygdala), which is then funneled down to a much lower number of cells. SPNs are thus carrying out one of the most important step-down processes in the brain, and they do it while transitioning from a fundamentally glutamatergic to an essentially GABAergic environment. These and other properties of SPNs that escape from the scope of this chapter provide perspective on the scale of integration that these cells mediate.

One factor strongly influencing the integrative capacity of SPNs in the NAc-S is their sharp subdivision into two subpopulations that, while completely intermingled in the tissue, are governed by opposing dopamine-dependent signaling (Bertran-Gonzalez, Hervé, Girault, & Valjent, 2010). The D1-SPN subpopulation initiates extensive molecular signaling programs in response to dopamine (through intracellular cyclic AMP signal [cAMP]—promoting D1 dopamine receptors), while the D2-SPN population derepresses similar molecular signaling loads in response to dopamine clearing from the synapse (through cAMP-ceasing D2 dopamine receptors) (Bertran-Gonzalez et al., 2008). These opposing patterns of signaling activity in response to dopamine oscillations are key to understanding the way integration of first-hand glutamatergic information occurs in the NAc-S, but they are not the full story. An important character modulating the information flow onto SPNs—which has drawn substantial attention in recent years—is the striatal cholinergic system, which depends on a very particular type of neuron known as cholinergic interneuron (CIN). CINs are among the largest neurons in the mammalian CNS, and although they only account for less than 2% of the neurons in the striatum, they provide the largest acetylcholine load in the brain. They do so through extensive local axonal fields and their explicitly sparse spatial distribution, which covers every corner of striatal tissue (Bolam, Wainer, & Smith, 1984). Importantly, modern quantitative studies of the CIN population revealed particularly high densities of neuronal bodies and dendritic processes in the NAc-S (Matamales, Götz, & Bertran-Gonzalez, 2016a), which highlights the importance of the cholinergic system in this region of the ventral striatum.

Although less well studied in the context of associative learning, a third important modulation signal in the striatum is the opioid system, which counts with various types of 7-transmembrane G protein-coupled receptors that are activated mostly by endogenous opioid peptides (Kieffer & Evans, 2009). One particularly important receptor in the ventral striatum, based on its enrichment in the NAc-S, is the delta-opioid receptor (DOR), the ligand of which, enkephalin (ENK), is produced in enormous quantities in the striatum (Carlyle et al., 2017), and its release depends exclusively on D2-SPNs (Steiner & Gerfen, 1998). The first hint of the importance of the ENK-DOR system for predictive learning was obtained in our laboratory, when we found that both the genetic deletion of DOR, as well as the pharmacological blockade of DOR in the NAc-S, abolished specific PIT (Laurent, Leung, Maidment, & Balleine, 2012). This study thus revealed the necessity of this receptor for the integration processing undertaken in the NAc-S at the moment of PIT; however, given that predictive learning and the choices it informs take place at different times, it remained unknown how the integration of Pavlovian and instrumental learning emerged from the NAc-S circuitry.

Overall, the NAc-S nests a series of neuromodulation signals (i.e., dopamine, acetylcholine, and opioids at least) that, across learning, interact with one another to modulate incoming glutamatergic information and mediate the integration of Pavlovian and instrumental learning. The mechanism by which this occurs at the level of molecules and circuits is complex and is the focus of the following sections of this chapter.

The cholinergic system as an integration machinery in the striatum

Mounting evidence in the last few years positions CINs as central to all integration processing occurring in SPNs of the striatum, from stimulus detection to reinforcement learning and associative processing (Apicella, 2007; Goldberg & Reynolds, 2011; Yamanaka et al., 2017). These interneurons are also known as tonically active neurons (TANs), as they are characterized by their intrinsic capacity to fire tonically (Ben D Bennett & Wilson, 1999) and they have the intriguing ability of pausing and rebounding their firing during Pavlovian stimulus—outcome conditioning (Aosaki et al., 1994). Decades of intense research on these "conditioned pause" or "burst/pause" responses have revealed their close relationship to the phasic activity of midbrain dopamine neurons during conditioning (e.g., Morris, Arkadir, Nevet, Vaadia, & Bergman, 2004), but the mechanism by which these timely firing patterns are initiated and maintained are still debated (Brown et al., 2012; Ding, Guzman, Peterson, Goldberg, & Surmeier, 2010; Goldberg & Reynolds, 2011; Zhang & Cragg, 2017).

Given the antiphasic relationship between midbrain dopamine neurons and CINs during predictive learning, dopamine itself has been proposed to critically participate in driving firing pauses (Ding et al., 2010; Reynolds, Hyland, & Wickens, 2004). However, this modulation could be indirect, since pause—rebound behavior is observable in the absence of phasic dopamine, and dopamine alone cannot explain the multiple phases of this response (Zhang & Cragg, 2017). Another force proposed to critically control these burst/pause and rebound responses in CINs is the glutamatergic drive from corticostriatal and/or thalamostriatal projections, both abundantly synapsing with CINs (Bradfield, Bertran-Gonzalez, Chieng, & Balleine, 2013; Doig, Magill, Apicella, Bolam, & Sharott, 2014; Lapper & Bolam, 1992; Matamales et al., 2016b). However, it is still unclear how purely excitatory inputs from glutamatergic systems can solely explain each phase of the burst/pause—rebound response, so highly synchronized interactions with other modulation systems (such as dopamine) are the most likely scenario (Cachope et al., 2012; Ding et al., 2010; Threlfell et al., 2012). Finally, Gamma-Aminobutyric acid (GABA) has also been suggested as candidate orchestrator of the pause response (Sullivan, Chen, & Morikawa, 2008), also during Pavlovian conditioning (Brown et al., 2012). Yet again, precipitation of the burst/pause and rebound response, with all its phases and time variabilities, is difficult to explain based solely on GABAergic inputs.

Most likely, the emergence and fine-tuning of these responses during conditioning relies on the combination of multiple signals that build up synchrony as learning progresses. Given the fast nature of the pause—rebound response, the striatum offers a propitious environment for generating large but focal oscillations of acetylcholine that could define domains of targeted corticostriatal plasticity and information integration. In the case of the integrational role played by the NAc-S on cue-based choice, it is reasonable to think that important neurophysiological adaptations occur during the initial stages of predictive learning, and the burst/pause—rebound activity in CINs of the NAc-S is very likely to emerge during the initial Pavlovian phase. Incidentally, new research in monkeys has revealed that TANs in the ventral striatum display larger and longer pauses than dorsal striatal counterparts in response to predictive stimuli during Pavlovian conditioning (Marche, Martel, & Apicella, 2017), suggesting that pause behavior in ventral regions is important and can be subjected to additional sources of neuromodulation.

Modulating the modulator: how CINs in the NAc-S get ready for cue-driven choice

One particularly interesting feature of CINs during predictive learning is, as was initially described (Aosaki et al., 1994), their ability to "become" sensitive to the conditioned pause response as Pavlovian training progresses. In their first study in monkeys, Aosaki and colleagues found that the proportion of TANs displaying such modes of firing increased substantially over training, ranging from around 17% at the onset of training to 50%—70% by the end of training. Importantly, competent CINs were able to fire in this mode after long periods of training intermission and were able to rapidly switch their sensory properties after a change in conditioning rules. It is therefore clear that some sort of long-lasting adaptation emerges in these neurons during learning, but to date the way these neurons build their capacity to pause their behavior remains unknown. As discussed above, one possibility is that, with learning, synchrony between different neurotransmitter systems builds around CINs and orchestrates their firing and pausing. An alternative possibility is that these neurons become competent for burst/pause firing through intrinsic adaptations promoted by learning. In other words, CINs could be undergoing plastic changes that provide them with the physiological properties required to behave in burst/pause modes.

In our laboratory, we described a striking molecular adaptation occurring in CINs of the NAc-S in mice trained to appetitive Pavlovian conditioning (Bertran-Gonzalez, Laurent, Chieng, Christie, & Balleine, 2013). We found that S-O contingencies established during the initial Pavlovian training specifically promoted the accumulation of DOR in the membrane of CINs, something that remained for at least 10 days until the mice were submitted to the PIT test (Fig. 10.2A,B). Importantly, the extent of DOR accumulation correlated with performance during conditioned responding (Pavlovian phase) and also stimulus–based choice (PIT test), and animals exposed to the same number of stimuli and reward but in a noncontingent manner did not develop membrane DOR in their CINs (Fig. 10.2A,B). We also found that CINs in NAc-S slices from animals with Pavlovian contingent training were more susceptible to irregular firing and burst firing than those in noncontingently trained mice, a mode of firing that was further precipitated with DOR agonist treatment (Fig. 10.2C). These data showed for the first time a molecular adaptation in an interneuronal system induced purely by learning: CINs strategically accumulated inhibitory DORs in their somatic membrane to develop sensitivity to ENK modulation, an adaptive feature that is well suited to favor the expression of timely burst/pause and rebound responses emerged during conditioning. This adaptation was enduring, as it stayed for a long time after its formation (up to several weeks in our unpublished experiments) and, once formed, did not decrease after initially formed contingencies were behaviorally degraded (unpublished data). Importantly, one critical property of this adaptation is its anticipatory nature: neither DOR receptor activity nor the entire NAc-S function is required for conditioned responding during Pavlovian learning (Corbit et al., 2001; Laurent et al., 2012). This suggests that whatever neuronal plasticity the specific learning of stimulus—outcome contingencies is promoting in the NAc-S during initial conditioning, it is to be "used" prospectively in future

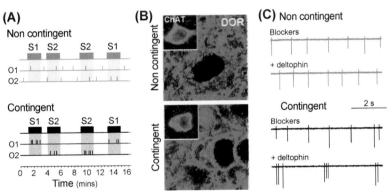

Figure 10.2 *Predictive learning induces opioid plasticity in cholinergic interneurons (CINs) of the nucleus accumbens shell (NAc-S).* (A) In the Pavlovian phase, learning of two parallel stimulus— outcome contingencies were trained in mice and rats by associating two stimuli (S1 and S2) with two different outcomes (O1 and O2) (see Fig. 10.1A). In the contingent group, O1 and O2 were delivered coinciding with the presentation of the stimuli S1 and S2, respectively. In the noncontingent group, delivery of O1 and O2 occurred irrespective of the presentation of S1 and S2. (B) Only contingently trained mice showed a sustained accumulation of delta-opioid receptors (DORs) in the somatic membrane of CINs of the NAc-S (marked with ChAT staining). This adaptation lasted for several weeks after training. (C) CINs of contingently trained rats showing enhanced sensitivity to the DOR agonist deltorphin in brain slices. Deltorphin promoted burst/pause firing in a larger proportion of CINs in animals undergoing contingent Pavlovian training. *(Data are from Bertran-Gonzalez et al. (2013).)*

circumstances, where, for example, these contingencies need to drive choice between actions. In other words, modulatory systems in the NAc-S appear to be getting ready for future decision-making, and they do so by introducing molecular changes in modulatory systems themselves.

The next question our lab aimed at addressing was which mechanisms could be promoting membrane DOR accumulation in CINs during conditioning. We first hypothesized that phasic dopamine release expected to occur during Pavlovian learning could be causing sustained reductions in ENK release from D2-SPNs, something that could be sensitizing DORs and therefore drive their expression in the membrane. However, reducing the tone of D2-SPNs with specific D2 agonists during conditioning did not affect DOR accumulation (unpublished research). We next reasoned that substance P (SP), a neuropeptide that is specifically released by D1-SPNs, could be mediating DOR accumulation in response to learning-related phasic dopamine. Using a series of pharmacological manipulations in vivo, it was found that SP was key to drive the accumulation of DOR in the membrane of CINs, and it did that through its natural receptors (NK1Rs), which are highly enriched in CINs (Gerfen, 1991; Heath, Chieng, Christie, & Balleine, 2017). Although direct experimental evidence is still lacking, these results suggested that SP released as a consequence of sustained D1-SPN activity during conditioning could be an important contributor to the enrichment of DORs in the membrane of CINs (Fig. 10.3A). The way the SP signal reconciles with sensory-specific and contingency-specific signals broadcasted in the brain, as learning is established remains unknown but sets the grounds for intensive research in the future.

Confluence of signals in the nucleus accumbens shell during cue-driven choice

As we have seen in this chapter, the learning necessary for specific PIT gathers complex processing, where extensive circuitry and neurotransmitter systems involved in generating stimulus—outcome and action—outcome learning mature over training and converge in the NAc-S to promote cue-driven action selection. Our next question was, therefore, how the different neurotransmitter systems that had undergone plastic adaptations during learning behaved at the moment of test. Strikingly, we found that the NAc-S of mice undergoing PIT showed a very high proportion of neurons with intracellular signaling activation, and these neurons were D1-SPNs (Laurent, Bertran-Gonzalez, Chieng, & Balleine, 2014). Consistently, blocking dopamine D1 receptors (D1Rs) but not D2R in the NAc-S prevented specific PIT. Further, NAc-S blockade of D1Rs in one side and DORs in the other also impaired specific PIT, suggesting that dopamine (through D1Rs) and ENK (through DORs) cooperate in the NAc-S to mediate stimulus-based choice. Since the modulatory adaptations observed during Pavlovian training occurred in CINs (Bertran-Gonzalez et al., 2013), we hypothesized that the link between CINs and D1-SPNs had to be acetylcholinergic. We speculated

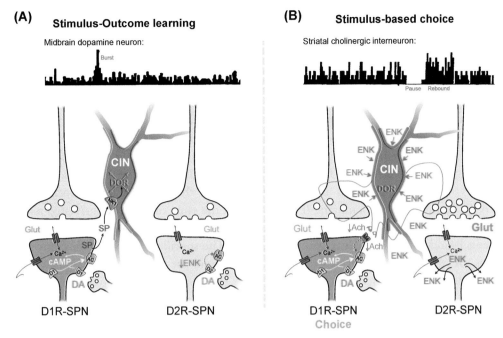

Figure 10.3 *Model of the cooperative interactions between opioidergic, acetylcholinergic, and dopaminergic systems in the NAc-S in stimulus—outcome learning and stimulus-based choice.* Each diagram (A—C) exemplifies the molecular interactions occurring in a cortico-D1-SPN synapse (left), a CIN (center), and a cortico-D2-SPN (right). (A) Neurotransmitter interactions during stimulus—outcome learning (Pavlovian conditioning). Top trace is a representation of the firing of a dopamine neuron in the midbrain. The firing bursts due to stimuli presentation and reward prediction errors during Pavlovian training are translated into phasic dopamine in the NAc-S. (B) Neurotransmitter interactions during stimulus-based choice (Pavlovian-instrumental transfer). Top trace is a representation of the firing of a CIN in the NAc-S. The burst/pause and rebound mode of firing in response to predictive stimuli is the result of a high sensitivity to ENK. The green contour represents the accumulated DORs in the somatic region of the CIN due to prior Pavlovian conditioning (see Bertran-Gonzalez et al., 2013). Blue color represents overactivity. Red color represents hypoactivity. See text for explanatory details. Top traces are representations of the original recordings in a dopamine neuron (A) (Schultz, Dayan, & Montague, 1997) and a CIN (B) (Aosaki et al., 1994). *AC*, adenylate cyclase; *Ach*, acetylcholine; *Ca²⁺*; intracellular calcium signal; *cAMP*, intracellular cyclic AMP signal; *CIN*, cholinergic interneuron; *DA*, dopamine; *DOR*, δ-opioid receptor; *ENK*, enkephalin; *Glut*, glutamate; *NAc-S*, nucleus accumbens shell; *SP*, substance P; *SPN*, spiny projection neuron.

that focal drops of acetylcholine during burst/pause—rebound responses in CINs at the moment of PIT could be disengaging the acetylcholine muscarinic receptors type 4 (M4R), which are coexpressed with and specifically counteract D1R function. Therefore, the intense intracellular signaling observed in D1-SPNs of the NAc-S during stimulus-driven choice could be a consequence of a permissive acetylcholinergic signal provided by timely modulations of burst/pause behavior in CINs, a firing mode that appeared to be promoted by DORs accumulated in their somatic membrane (Fig. 10.3B). In direct support of this model, we found that specific blockade of M4R in the NAc-S rescued specific PIT in rats bearing pharmacological inactivation of DORs (Laurent et al., 2014), suggesting that local pauses of acetylcholine release can remove the cholinergic "clamp" usually exerted on corticostriatal plasticity.

Considering all these studies, we propose a functional model by which the different neurotransmitter systems cooperate in the NAc-S to integrate stimulus–outcome and action–outcome processing to encode stimulus-guided choice (Fig. 10.3). During the initial Pavlovian training, frequent phasic dopamine around stimulus–outcome conditioning promotes SP release from D1R-SPNs through stimulation of D1Rs. SP then binds to NK1Rs expressed in CINs and initiates the progressive translocation of DORs to the somatic membrane (Fig. 10.3A). At the same time, learning-related phasic dopamine can also reduce the ENK tone during conditioning, which in addition prevents ligand-mediated internalization of DORs. The specific circuitry feeding this process in the NAc-S remains unknown but could be related to afferents from stimulus–outcome encoding areas such as the BLA or the OFC (see above). By the end of Pavlovian training, the proportion of CINs bearing DOR in the membrane increases in the NAc-S, which can increase their susceptibility to burst/pause firing when the neurochemical environment is favorable.

Such an environment is found at the moment of PIT, when S-O learning drives vigorous instrumental responding toward the outcome that is predicted by each stimulus. This "intense" behavioral state promotes major corticostriatal activity from S-O-related and A-O-related circuitry in the form of glutamate release onto SPNs as well as CINs in the NAc-S (Fig. 10.3B). Abundant glutamate stimulation in D2-SPNs triggers the release of large amounts of ENK, which is found in large quantities in the striatum (Carlyle et al., 2017). ENK then stimulates the preenriched DORs in CINs, which together with externally sourced glutamate and GABA signals, promotes tightly modulated burst/pause firing. These physiological responses of CINs generate high–low–high patterns of acetylcholine availability, which is fast degraded by extracellular acetylcholinesterase (Quinn, 1987). In synapses formed by CINs and D1-SPNs, local drops of acetylcholine disengage the M4Rs, thus derepressing dopamine and glutamate-induced intracellular signaling in D1-SPNs. The afferent circuitry participating in CIN modulation in the NAc-S can be complex (see discussion above) but is likely to cooperate with the local neurochemical environment to timely modulate acetylcholine release and clearing in the NAc-S.

CONCLUSION

Here, we have argued that PIT is a powerful paradigm because it provides the means to establish the precise role played by various brain regions in the two main forms of conditioning: Pavlovian and instrumental. For example, we have shown that the use of PIT has motivated the development of a more refined model of amygdala functioning, in which separate subnuclei mediate distinct aspects of Pavlovian information. We have also shown how the use of PIT was central in gaining a better understanding of the various contributions of dorsostriatal and orbitofrontal regions in instrumental conditioning. Beyond this, it is obvious that Pavlovian and instrumental information processed by these brain regions are ultimately integrated together to mediate the PIT effect. We have argued that this integration occurs in the NAc-S and that the role of this brain region remains limited to that integration in the context of specific PIT. We then described how CINs of the NAc-S develop sensitivity to opioid modulations during Pavlovian conditioning, and how this sensitivity is later used to guide choice between actions. We have argued that this finding is consistent with a central tenet on neuronal plasticity by revealing how the brain circuitry arranges throughout learning in order to prepare for future significant events. Given that PIT has been observed in many species including humans, we believe that it provides a fantastic opportunity to understand how predictive stimuli influence decision-making processes.

REFERENCES

Aosaki, T., Tsubokawa, H., Ishida, A., Watanabe, K., Graybiel, A. M., & Kimura, M. (1994). Responses of tonically active neurons in the primate's striatum undergo systematic changes during behavioral sensorimotor conditioning. *The Journal of Neuroscience: The Official Journal of the Society for Neuroscience, 14*(6), 3969–3984.

Apicella, P. (2007). Leading tonically active neurons of the striatum from reward detection to context recognition. *Trends in Neurosciences, 30*(6), 299–306.

Balleine, B. W., & Killcross, S. (2006). Parallel incentive processing: An integrated view of amygdala function. *Trends in Neurosciences, 29*(5), 272–279.

Balleine, B. W., Liljeholm, M., & Ostlund, S. B. (2009). The integrative function of the basal ganglia in instrumental conditioning. *Behavioural Brain Research, 199*(1), 43–52.

Belin, D., Jonkman, S., Dickinson, A., Robbins, T. W., & Everitt, B. J. (2009). Parallel and interactive learning processes within the basal ganglia: Relevance for the understanding of addiction. *Behavioural Brain Research, 199*(1), 89–102.

Bennett, B. D., & Wilson, C. J. (1999). Spontaneous activity of neostriatal cholinergic interneurons in vitro. *The Journal of Neuroscience: The Official Journal of the Society for Neuroscience, 19*(13), 5586–5596.

Bertran-Gonzalez, J., Bosch, C., Maroteaux, M., Matamales, M., Hervé, D., Valjent, E., & Girault, J.-A. (2008). Opposing patterns of signaling activation in dopamine D1 and D2 receptor-expressing striatal neurons in response to cocaine and haloperidol. *The Journal of Neuroscience: The Official Journal of the Society for Neuroscience, 28*(22), 5671–5685. https://doi.org/10.1523/JNEUROSCI.1039-08.2008.

Bertran-Gonzalez, J., Hervé, D., Girault, J.-A., & Valjent, E. (2010). What is the degree of segregation between striatonigral and striatopallidal projections? *Frontiers in Neuroanatomy, 4*.

Bertran-Gonzalez, J., Laurent, V., Chieng, B. C., Christie, M. J., & Balleine, B. W. (2013). Learning-related translocation of δ-opioid receptors on ventral striatal cholinergic interneurons mediates choice between goal-directed actions. *The Journal of Neuroscience: The Official Journal of the Society for Neuroscience, 33*(41), 16060–16071.

Blundell, P., Hall, G., & Killcross, S. (2001). Lesions of the basolateral amygdala disrupt selective aspects of reinforcer representation in rats. *The Journal of Neuroscience: The Official Journal of the Society for Neuroscience, 21*(22), 9018–9026.

Bolam, J. P., Wainer, B. H., & Smith, A. D. (1984). Characterization of cholinergic neurons in the rat neo-striatum. A combination of choline acetyltransferase immunocytochemistry, Golgi-impregnation and electron microscopy. *Neuroscience, 12*(3), 711−718.

Bradfield, L. A., Bertran-Gonzalez, J., Chieng, B., & Balleine, B. W. (2013). The thalamostriatal pathway and cholinergic control of goal-directed action: Interlacing new with existing learning in the striatum. *Neuron, 79*(1), 153−166.

Bradfield, L. A., Dezfouli, A., van Holstein, M., Chieng, B., & Balleine, B. W. (2015). Medial orbitofrontal cortex mediates outcome retrieval in partially observable task situations. *Neuron, 88*(6), 1268−1280.

Brown, M. T. C., Tan, K. R., O'Connor, E. C., Nikonenko, I., Muller, D., & Lüscher, C. (2012). Ventral tegmental area GABA projections pause accumbal cholinergic interneurons to enhance associative learning. *Nature, 492*(7429), 452−456.

Cachope, R., Mateo, Y., Mathur, B. N., Irving, J., Wang, H.-L., Morales, M., et al. (2012). Selective activation of cholinergic interneurons enhances accumbal phasic dopamine Release: Setting the tone for reward processing. *Cell Reports, 2*(1), 33−41.

Campese, V., McCue, M., Lázaro-Muñoz, G., LeDoux, J. E., & Cain, C. K. (2013). Development of an aversive Pavlovian-to-instrumental transfer task in rat. *Frontiers in Behavioral Neuroscience, 7.*

Carlyle, B. C., Kitchen, R. R., Kanyo, J. E., Voss, E. Z., Pletikos, M., Sousa, A. M. M., et al. (2017). A multiregional proteomic survey of the postnatal human brain. *Nature Neuroscience, 20*(12), 1787−1795.

Cartoni, E., Balleine, B., & Baldassarre, G. (2016). Appetitive Pavlovian-instrumental transfer: A review. *Neuroscience and Biobehavioral Reviews, 71.*

Colagiuri, B., & Lovibond, P. F. (2015). How food cues can enhance and inhibit motivation to obtain and consume food. *Appetite, 84*, 79−87.

Colwill, R. M., & Rescorla, R. A. (1988). Associations between the discriminative stimulus and the reinforcer in instrumental learning. *Journal of Experimental Psychology: Animal Behavior Processes, 14*(2), 155−164.

Corbit, L. H., & Balleine, B. W. (2003). The role of prelimbic cortex in instrumental conditioning. *Behavioural Brain Research, 146*(1−2), 145−157.

Corbit, L. H., & Balleine, B. W. (2005). Double dissociation of basolateral and central amygdala lesions on the general and outcome-specific forms of Pavlovian-instrumental transfer. *The Journal of Neuroscience: The Official Journal of the Society for Neuroscience, 25*(4), 962−970.

Corbit, L. H., & Balleine, B. W. (2011). The general and outcome-specific forms of Pavlovian-instrumental transfer are differentially mediated by the nucleus accumbens core and shell. *The Journal of Neuroscience: The Official Journal of the Society for Neuroscience, 31*(33), 11786−11794.

Corbit, L. H., & Janak, P. H. (2007). Inactivation of the lateral but not medial dorsal striatum eliminates the excitatory impact of Pavlovian stimuli on instrumental responding. *The Journal of Neuroscience: The Official Journal of the Society for Neuroscience, 27*(51), 13977−13981.

Corbit, L. H., Janak, P. H., & Balleine, B. W. (2007). General and outcome-specific forms of Pavlovian-instrumental transfer: The effect of shifts in motivational state and inactivation of the ventral tegmental area. *The European Journal of Neuroscience, 26*(11), 3141−3149.

Corbit, L. H., Muir, J. L., Balleine, B. W., & Balleine, B. W. (2001). The role of the nucleus accumbens in instrumental conditioning: Evidence of a functional dissociation between accumbens core and shell. *The Journal of Neuroscience: The Official Journal of the Society for Neuroscience, 21*(9), 3251−3260.

Coutureau, E., & Killcross, S. (2003). Inactivation of the infralimbic prefrontal cortex reinstates goal-directed responding in overtrained rats. *Behavioural Brain Research, 146*(1−2), 167−174.

Delamater, A. R. (1996). Effects of several extinction treatments upon the integrity of Pavlovian stimulus-outcome associations. *Animal Learning & Behavior, 24*(4), 437−449.

Dickinson, A., & Balleine, B. (1994). Motivational control of goal-directed action. *Animal Learning & Behavior, 22*(1), 1−18.

Ding, J. B., Guzman, J. N., Peterson, J. D., Goldberg, J. A., & Surmeier, D. J. (2010). Thalamic gating of corticostriatal signaling by cholinergic interneurons. *Neuron, 67*(2), 294−307.

Doig, N. M., Magill, P. J., Apicella, P., Bolam, J. P., & Sharott, A. (2014). Cortical and thalamic excitation mediate the multiphasic responses of striatal cholinergic interneurons to motivationally salient stimuli. *The Journal of Neuroscience: The Official Journal of the Society for Neuroscience, 34*(8), 3101—3117.

Gerfen, C. R. (1991). Substance P (neurokinin-1) receptor mRNA is selectively expressed in cholinergic neurons in the striatum and basal forebrain. *Brain Research, 556*(1), 165—170.

Goldberg, J. A., & Reynolds, J. N. J. (2011). Spontaneous firing and evoked pauses in the tonically active cholinergic interneurons of the striatum. *Neuroscience, 198*, 27—43.

Hall, J., Parkinson, J. A., Connor, T. M., Dickinson, A., & Everitt, B. J. (2001). Involvement of the central nucleus of the amygdala and nucleus accumbens core in mediating Pavlovian influences on instrumental behaviour. *The European Journal of Neuroscience, 13*(10), 1984—1992.

Heath, E., Chieng, B., Christie, M. J., & Balleine, B. W. (2017). Substance P and dopamine interact to modulate the distribution of delta-opioid receptors on cholinergic interneurons in the striatum. *The European Journal of Neuroscience.* https://doi.org/10.1111/ejn.13750.

Hogarth, L., Balleine, B. W., Corbit, L. H., & Killcross, S. (2013). Associative learning mechanisms underpinning the transition from recreational drug use to addiction. *Annals of the New York Academy of Sciences, 1282*(1), 12—24.

Holland, P. C. (2004). Relations between Pavlovian-instrumental transfer and reinforcer devaluation. *Journal of Experimental Psychology: Animal Behavior Processes, 30*(2), 104—117.

Holland, P. C., & Gallagher, M. (1999). Amygdala circuitry in attentional and representational processes. *Trends in Cognitive Sciences, 3*(2), 65—73.

Holland, P. C., & Gallagher, M. (2003). Double dissociation of the effects of lesions of basolateral and central amygdala on conditioned stimulus-potentiated feeding and Pavlovian-instrumental transfer. *The European Journal of Neuroscience, 17*(8), 1680—1694.

Holmes, N. M., Marchand, A. R., & Coutureau, E. (2010). Pavlovian to instrumental transfer: A neurobehavioural perspective. *Neuroscience and Biobehavioral Reviews, 34*(8), 1277—1295.

Hunnicutt, B. J., Jongbloets, B. C., Birdsong, W. T., Gertz, K. J., Zhong, H., & Mao, T. (2016). A comprehensive excitatory input map of the striatum reveals novel functional organization. *eLife, 5*, e19103.

Johnson, A. W. (2013). Eating beyond metabolic need: How environmental cues influence feeding behavior. *Trends in Neurosciences, 36*(2), 101—109.

Keistler, C., Barker, J. M., & Taylor, J. R. (2015). Infralimbic prefrontal cortex interacts with nucleus accumbens shell to unmask expression of outcome-selective Pavlovian-to-instrumental transfer. *Learning & Memory, 22*(10), 509—513.

Kieffer, B. L., & Evans, C. J. (2009). Opioid receptors: From binding sites to visible molecules in vivo. *Neuropharmacology, 56*, 205—212.

Lamb, R. J., Schindler, C. W., & Pinkston, J. W. (2016). Conditioned stimuli's role in relapse: Preclinical research on Pavlovian-instrumental-transfer. *Psychopharmacology, 233*(10), 1933—1944.

Lapper, S. R., & Bolam, J. P. (1992). Input from the frontal cortex and the parafascicular nucleus to cholinergic interneurons in the dorsal striatum of the rat. *Neuroscience, 51*(3), 533—545.

Laurent, V., & Balleine, B. W. (2015). Factual and counterfactual action-outcome mappings control choice between goal-directed actions in rats. *Current Biology, 25*(8), 1074—1079.

Laurent, V., Bertran-Gonzalez, J., Chieng, B. C., & Balleine, B. W. (2014). δ-opioid and dopaminergic processes in accumbens shell modulate the cholinergic control of predictive learning and choice. *The Journal of Neuroscience: The Official Journal of the Society for Neuroscience, 34*(4), 1358—1369.

Laurent, V., Chieng, B., & Balleine, B. W. (2016). Extinction generates outcome-specific conditioned inhibition. *Current Biology, 26*(23), 3169—3175.

Laurent, V., Leung, B., Maidment, N., & Balleine, B. W. (2012). μ- and δ-opioid-related processes in the accumbens core and shell differentially mediate the influence of reward-guided and stimulus-guided decisions on choice. *The Journal of Neuroscience: The Official Journal of the Society for Neuroscience, 32*(5), 1875—1883.

LeDoux, J. E. (2000). Emotion circuits in the brain. *Annual Review of Neuroscience, 23*, 155—184. Annual Reviews 4139 El Camino Way, P.O. Box 10139, Palo Alto, CA 94303-0139, USA.

Li, H., Penzo, M. A., Taniguchi, H., Kopec, C. D., Huang, Z. J., & Li, B. (2013). Experience-dependent modification of a central amygdala fear circuit. *Nature Neuroscience, 16*(3), 332–339.

Marche, K., Martel, A.-C., & Apicella, P. (2017). Differences between dorsal and ventral striatum in the sensitivity of tonically active neurons to rewarding events. *Frontiers in Systems Neuroscience, 11*, 11–52.

Maren, S. (2001). Neurobiology of Pavlovian fear conditioning. *Annual Review of Neuroscience, 24*(1), 897–931.

Matamales, M., Götz, J., & Bertran-Gonzalez, J. (2016a). Quantitative imaging of cholinergic interneurons reveals a distinctive spatial organization and a functional gradient across the mouse striatum. *PLoS One, 11*(6), e0157682.

Matamales, M., Skrbis, Z., Hatch, R. J., Balleine, B. W., Götz, J., & Bertran-Gonzalez, J. (2016b). Aging-related dysfunction of striatal cholinergic interneurons produces conflict in action selection. *Neuron, 90*(2), 362–373.

Morris, G., Arkadir, D., Nevet, A., Vaadia, E., & Bergman, H. (2004). Coincident but distinct messages of midbrain dopamine and striatal tonically active neurons. *Neuron, 43*(1), 133–143.

Ostlund, S. B., & Balleine, B. W. (2005). Lesions of medial prefrontal cortex disrupt the acquisition but not the expression of goal-directed learning. *The Journal of Neuroscience: The Official Journal of the Society for Neuroscience, 25*(34), 7763–7770.

Ostlund, S. B., & Balleine, B. W. (2007). Orbitofrontal cortex mediates outcome encoding in Pavlovian but not instrumental conditioning. *The Journal of Neuroscience: The Official Journal of the Society for Neuroscience, 27*(18), 4819–4825.

Petrovich, G. D. (2011). Forebrain circuits and control of feeding by learned cues. *Neurobiology of Learning and Memory, 95*(2), 152–158.

Prévost, C., Liljeholm, M., Tyszka, J. M., & O'Doherty, J. P. (2012). Neural correlates of specific and general Pavlovian-to-instrumental transfer within human amygdalar subregions: A high-resolution fMRI study. *The Journal of Neuroscience: The Official Journal of the Society for Neuroscience, 32*(24), 8383–8390.

Quail, S. L., Laurent, V., & Balleine, B. W. (2017). Inhibitory Pavlovian–instrumental transfer in humans. *Journal of Experimental Psychology: Animal Learning and Cognition*, 1–11.

Quinn, D. M. (1987). Acetylcholinesterase: Enzyme structure, reaction dynamics, and virtual transition states. *Chemical Reviews, 87*(5), 955–979.

Rescorla, R. A. (1992). Response-independent outcome presentation can leave instrumental RO associations intact. *Animal Learning & Behavior, 20*(2), 104–111.

Rescorla, R. A. (1994). Transfer of instrumental control mediated by a devalued outcome. *Animal Learning & Behavior, 22*(1), 27–33.

Reynolds, J. N. J., Hyland, B. I., & Wickens, J. R. (2004). Modulation of an afterhyperpolarization by the substantia nigra induces pauses in the tonic firing of striatal cholinergic interneurons. *The Journal of Neuroscience: The Official Journal of the Society for Neuroscience, 24*(44), 9870–9877.

Schultz, W., Dayan, P., & Montague, P. R. (1997). A neural substrate of prediction and reward. *Science, 275*(5306), 1593–1599.

Shiflett, M. W., & Balleine, B. W. (2010). At the limbic-motor interface: Disconnection of basolateral amygdala from nucleus accumbens core and shell reveals dissociable components of incentive motivation. *The European Journal of Neuroscience, 32*(10), 1735–1743.

Steiner, H., & Gerfen, C. R. (1998). Role of dynorphin and enkephalin in the regulation of striatal output pathways and behavior. *Experimental Brain Research, 123*(1–2), 60–76.

Sullivan, M. A., Chen, H., & Morikawa, H. (2008). Recurrent inhibitory network among striatal cholinergic interneurons. *The Journal of Neuroscience: The Official Journal of the Society for Neuroscience, 28*(35), 8682–8690.

Threlfell, S., Lalic, T., Platt, N. J., Jennings, K. A., Deisseroth, K., & Cragg, S. J. (2012). Striatal dopamine release is triggered by synchronized activity in cholinergic interneurons. *Neuron, 75*(1), 58–64.

Voorn, P., Vanderschuren, L. J. M. J., Groenewegen, H. J., Robbins, T. W., & Pennartz, C. M. A. (2004). Putting a spin on the dorsal–ventral divide of the striatum. *Trends in Neurosciences, 27*(8), 468–474.

Wilensky, A. E., Schafe, G. E., Kristensen, M. P., & LeDoux, J. E. (2006). Rethinking the fear circuit: The central nucleus of the amygdala is required for the acquisition, consolidation, and expression of Pavlovian fear conditioning. *The Journal of Neuroscience: The Official Journal of the Society for Neuroscience, 26*(48), 12387–12396.

Yamanaka, K., Hori, Y., Minamimoto, T., Yamada, H., Matsumoto, N., Enomoto, K., et al. (2017). Roles of centromedian parafascicular nuclei of thalamus and cholinergic interneurons in the dorsal striatum in associative learning of environmental events. *Journal of Neural Transmission, 54*(Pt 2).

Yu, K., Ahrens, S., Zhang, X., Schiff, H., Ramakrishnan, C., Fenno, L., et al. (2017). The central amygdala controls learning in the lateral amygdala. *Nature Neuroscience, 17*(12), 1680–1685.

Zhang, Y.-F., & Cragg, S. J. (2017). Pauses in striatal cholinergic Interneurons: What is revealed by their common themes and variations? *Frontiers in Systems Neuroscience, 11*, 9424.

CHAPTER 11

Does the Dopaminergic Error Signal Act Like a Cached-Value Prediction Error?

Melissa J. Sharpe[1,2,3], Geoffrey Schoenbaum[1,4,5]

[1]National Institute on Drug Abuse, National Institute of Health, Baltimore, MD, United States; [2]Princeton Neuroscience Institute, Princeton University, Princeton, NJ, United States; [3]School of Psychology, UNSW Australia, Sydney, NSW, Australia; [4]Department of Anatomy and Neurobiology, University of Maryland School of Medicine, Baltimore, MD, United States; [5]Solomon H. Snyder Department of Neuroscience, The John Hopkins University, Baltimore, MD, United States

INTRODUCTION

The finding that dopaminergic neurons in the midbrain signal errors in prediction when an unexpected reward is delivered has transformed the study of behavioral neuroscience. This is because the concept of prediction errors had been the lynch pin of models of reinforcement learning for 50 years before this signal was discovered in the brain (Bush & Mosteller, 1951; Estes, 1950; Rescorla & Wagner, 1972; Sutton & Barto, 1981). Errors in outcome prediction—colloquially referred to as a "surprise" signal—are argued to drive learning about the antecedent stimuli that predict their occurrence. Essentially, the prediction error in these models acts as the teaching signal, which underlies the development of complex relationships between events in our environment (Holland & Rescorla, 1975; Miller & Matzel, 1988; Rescorla, 1973; Rescorla & Wagner, 1972; Wagner & Rescorla, 1972). The discovery that this signal actually exists in the brain gave some street credibility to associative models of reinforcement learning.

Yet when this prediction error signal was discovered in the midbrain, it was interpreted in a manner that diverged from the concept of driving real-world associations. Instead, the neural instantiation of this signal was taken to be synonymous with the error contained in the model-free reinforcement algorithm described by Sutton and Barto (1981, 1987, 1998). Here, reinforcement learning consists of the transference of what is termed "cached" value back from the reward to the stimulus, which reliably predicts reward occurrence. Cached value is the quantitative representation of the value presumed to be inherent in the reward. Essentially, an idea of how good a reward is to the subject, divorced from any specific knowledge of the identity or sensory properties of the reward itself. This allows the cue to become endowed with the scalar value, which drives motivated behavior in response to the reward-predictive cue in the future. However, this mechanism does not envision the development of an explicit association between the cue and the reward it predicts.

Goal-Directed Decision Making
ISBN 978-0-12-812098-9, https://doi.org/10.1016/B978-0-12-812098-9.00011-5

There is now a host of studies in which phasic activity of dopamine neurons appears to reflect or correlate with errors in cached value. However, in each study that has shown this correlation, it is also possible that these signals reflect errors in the prediction of the specific features of the reward, which are related to value, but exist independent of it. Indeed, dopaminergic error signals have recently been shown to cues that could only acquire an expectation for reward through the deployment of rich associative models of the world (Daw, Gershman, Seymour, Dayan, & Dolan, 2011; Sadacca, Jones, & Schoenbaum, 2016). Further, dopamine prediction errors appear to be both sufficient and necessary for acquisition of these more complex model-based associations (Sharpe et al., 2017). Such research suggests that the dopamine prediction error is not synonymous with the model-free reinforcement learning algorithm described by Sutton and Barto (1987). Instead, research implicating the dopamine prediction error in these more complex learning phenomena encourages a step back to the spirit of traditional models of reinforcement learning, which perceive the prediction error as the catalyst for learning about relationships between events in the world.

DOPAMINE NEURONS IN THE MIDBRAIN RESPOND TO REWARD AND REWARD-PAIRED CUES

Perhaps, the first suggestion that dopamine plays a role in reward processing was the finding that rodents will perform an action to receive intracranial stimulation of dopamine neurons in the midbrain (Olds & Milner, 1954; Wise & Rompre, 1989). The propensity to respond was proportional to the frequency of stimulation; rodents responded at a higher frequency for a delivery of a higher frequency of stimulation. Such findings led to the wide-held belief that dopamine functions in the brain to allow natural rewards to possess powerful control over behavior, a concept supported by findings showing that selective dopaminergic lesions of the ventral tegmental area (VTA) reduced subjects desire to pursue natural rewards (Robbins & Everitt, 1982; Spyraki, Fibiger, & Phillips, 1982; Stricker & Zigmond, 1974; Ungerstedt, 1971). The midbrain dopamine system was conceptualized as a system, which registers the receipt of something valuable to a subject, like food or water in a deprived state, which increases the likelihood that a subject will seek out these items in the future and promote survival (Wise, 1987; Wise & Rompre, 1989; Yokel & Wise, 1975).

Early recording studies in nonhuman primates corroborated a role for dopamine in registering the reinforcing effects of natural rewards (Romo & Schultz, 1990; Schultz, 1986). Such studies showed that dopamine neurons in the VTA respond when subjects receive food in a food-restricted state (Romo & Schultz, 1990; Schultz, 1986). Further, the response of dopamine neurons to receipt of reward correlated with the magnitude of the reward received (Romo & Schultz, 1990; Schultz, 1986; Schultz, Apicella, & Ljungberg, 1993). That is, if a subject received a greater amount of reward, dopamine neurons increased their response accordingly. And since these early studies, it has been demonstrated that dopamine neurons encode not only the magnitude of reward but also the subjective preference of reward (Fiorillo, Tobler, & Schultz, 2003; Lak, Stauffer, & Schultz, 2014; Stauffer, Lak, & Schultz, 2014). Dopamine neurons showed a preference for a particular reward over another in a manner that reflects the subject's observed choice preference for reward, regardless of caloric content. Such studies demonstrated that dopamine neurons register something about reward in a manner that reflects the subject's desire for that reward.

However, recording activity in dopamine neurons across the course of learning painted a more complex picture of the role of dopamine in reward. Specifically, recordings made across the course of conditioning showed that while dopamine neurons exhibit a phasic increase in activity during reward when nonhuman primates first experience reward (Mirenowicz & Schultz, 1994; Schultz et al., 1993), across time this phasic signal-to-reward receipt waned, appearing to transfer to the predictive cue with learning (Hollerman & Schultz, 1998; Ljungberg, Apicella, & Schultz, 1992; Mirenowicz & Schultz, 1994). Further, this transfer of phasic activity was progressive and occurred over successive pairings of the cue with reward, proportionate to increases in the appetitive response to the reward-paired cue (Mirenowicz & Schultz, 1994). Finally, once the subject learned the relationship between the cue and the reward, the omission of the expected reward elicited a depression in firing of these neurons at the time the reward was expected to occur (Mirenowicz & Schultz, 1994). These recording studies suggested that dopamine does not encode reward per se. Rather, it appeared that dopamine encodes some aspects of the predictive relationship between cues in the environment and rewards.

PREDICTION ERRORS IN MODELS OF REINFORCEMENT LEARNING

At the same time that researchers began to posit that dopamine was important for processing rewards, a parallel literature was being formed, which attempted to understand how natural rewards can function to reinforce responses result in procurement of those rewards. Some of the earliest mathematical models of this nature were developed to explain instrumental conditioning, where subjects learn to perform a particular action, which results in delivery of a particular reward (Bush & Mosteller, 1951; Estes, 1950; Estes & Burke, 1953). These models emerged out of the tradition of Thorndike (1898), in which it was argued that the pleasure derived from reward receipt increased the probability of a response being made in the presence of a particular stimulus. That is, procurement of reward was thought to make an agent more likely to perform a response again due to the reward's reinforcement of the association between the stimulus and the response.

Early mathematical models were simply an attempt to reconcile this idea with the incremental nature of the learning that takes place in instrumental tasks (Bush & Mosteller, 1951;

Estes, 1950; Estes & Burke, 1953; Thorndike, 1898). To do this, they used the standard linear operator, in which learning is governed by the discrepancy between the current probability of making a response and the maximal response probability (Bush & Mosteller, 1951). Here, the maximal response probability is determined by the reinforcer resulting from performance of the response and the effort required to procure that reinforcer. The learning rate parameter acted on the linear operator to govern the proportion of learning occurring on any one trial, allowing for a gradual increase in the probability of a response across the course of learning. Significantly, such models assumed that an increment in one particular response is completely independent of learning another response—even if both lead to the same eventual goal (Bush & Mosteller, 1951). Thus, a reliable correlation between a response and reward in the presence of a particular stimulus was enough to stamp in an association between that particular stimulus and the response, regardless of how many other responses could be made to procure the same reward or if that particular response was causally related to producing a reward. Rather, if a response was produced and was followed by reward, the subject would be more likely to make that particular response again.

Not long thereafter, however, the Kamin (1969) blocking experiments demonstrated that a correlation between events was not sufficient to produce learning. In the prototypical example, Kamin (1969) showed that pairing one stimulus (e.g., cue A) with reward would subsequently block learning about a second stimulus (e.g., cue X) if they were presented in compound (cue AX) with reward. In this example, cue A would have a high probability of producing a response when tested later separately, whereas cue X had a low probability. With these experiments, the notion of "surprise" began to enter the associative conversation. That is, in order for an association between events to form, the subject had to be surprised by the consequence. Learning required an error in the prediction of the reward. And so was born the notion that learning is driven by errors in prediction.

This notion initially took the form of the Rescorla and Wagner (1972) theory. This theory of Pavlovian conditioning deviated from its instrumental predecessors in two important ways. Firstly, the language used to describe the learning changed. Rather than talking in terms of increments in the probability of a response, this theory, instead, discussed learning in terms of associative strength—the strength of associations between events that allow a subject to make predictions about future rewards. Secondly, while the Rescorla-Wagner (1972) model used the same linear operator as that employed by Bush and Mosteller (1951), the update on any one trial according to the Rescorla-Wagner (1972) model was equal to the discrepancy between the maximal associative strength supported by the reward and the associative strength that has already accrued toward all present stimuli. Thus, all present stimuli essentially compete for associative strength with the reward they predict. This allowed the model to explain reports, described by Kamin (1969) and others, in which stimuli conditioned in compound shared the associative strength derived from reward. All stimuli that are currently present can contribute to the upcoming prediction for reward. Therefore, an error in prediction will only occur if the upcoming reward is not predicted by the summed expectation for reward provided by all present stimuli.

While this model did not elaborate on the specific nature of representations of the events being associated, focusing instead on the conditions for learning (e.g., *when* learning will take place rather than *what* is learned), it had important consequences for how associative theorists have since conceptualized learning. Specifically, the emphasis on changing the strength of associations changed the conversation in associative learning theory from one about strength of the response to one about the underlying associative framework underlying the likelihood of a response. This tradition has continued with the elaboration of empirically derived accounts, designed to explain more complex behaviors, which share in spirit the idea that errors in prediction drive changes in our understanding of the causal structure of our environment (Balleine & Dickinson, 1991; Colwill & Rescorla, 1985; Holland & Rescorla, 1975; Miller & Matzel, 1988; Rescorla, 1973; Wagner, Spear, & Miller, 1981).

This brings us to the model currently applied to interpret the dopaminergic prediction error, developed by Sutton and Barto (1981, 1987). While this reinforcement learning model arose out of the field of machine learning, it was influenced by the work described above. Specifically, this model attempted to bridge the concepts derived from the associative learning literature described above with that observed in the field of neurobiology. Essentially, Sutton and Barto (1981) wanted to apply the concepts of associative learning to Hebbian synaptic plasticity—the biological principal that if neuron A repeatedly evokes firing of another neuron B, then the firing of neuron A will be more efficacious in inducing activity in neuron B in the future (Hebb, 1949, 2005). Synaptic plasticity was essentially taken as a neurobiological instantiation of associative learning. In the earliest version of this model, Sutton and Barto (1981) argued that a reward acts to strengthen the ability of the stimulus to elicit a response, going back to the Thorndike (1898) tradition. According to this model, the response is the same response usually elicited by the reward, which comes to be controlled by the predictive stimulus as the associative strength between the stimulus and response increases. The rules for how the reward acts to strengthen the weights between the stimulus and response were adopted from the Rescorla—Wagner model (1972), where learning was driven by the discrepancy between the expected reward predicted by all stimuli present and the actual reward received. In essence, the prediction error acted to bias the response to obtain the maximum amount of reward. Critically, while the Sutton and Barto (1981) model borrowed some aspects of the Rescorla-Wagner (1972) model, in the form of the use of an error incorporating a summed prediction, the content of what could be learned narrowed sharply from the budding notions of a cognitive framework of relationships between events. In the place of this idea, the Sutton and Barto (1981) model substituted a somewhat Thorndikian conception of stimulus—response associations governing learning and behavior.

Later versions of the Sutton and Barto (1981) model moved even further away from the concept of associative strength between stimuli and responses and instead argued that a reward-predictive cue actually acquires the value inherent in the reward, through a process whereby the value inherent in the reward backpropagated to the antecedent cue (Sutton & Barto, 1987, 1998). Referred to as temporal difference reinforcement learning (TDRL), this iteration of the model conceptualized a stimulus (or, more generally, a "state") as segregated into multiple consecutive time steps. Each time step of a stimulus was associated with its own scalar value estimate that allows a subject to track the expectation of future sum of rewards through time. This development allowed the model to make accurate value predictions from stimulus onset despite the delay until reward delivery, as reward value backpropagates to the initial time step. Further, it estimated when the reward will occur as time steps of the stimulus that are closer to reward will acquire greater value. These value estimates can also be used to choose appropriate actions, where an action can be elected on the basis of the associated state value. Thus, TDRL was a time-derivative model of the original Sutton and Barto (1981), where it was argued that the value inherent in the reward propagates back to the state, which preceded reward delivery to subsequently influence a choice between future actions (Sutton & Barto, 1987, 1998). Importantly, the prediction error was divorced from the system that arbitrates between action choices. Rather, it just acted to endow the reward-predictive cue with the value inherent in the reward, which drove selection of the response associated with that stimulus.

THEORETICAL INTERPRETATIONS OF THE DOPAMINE PREDICTION ERROR

As noted above, Schultz, Dayan, and Montague (1997) hypothesized that dopamine signals acted as the prediction error postulated by the model-free reinforcement learning algorithm described by Sutton and Barto (1981, 1987, 1998). Their proposal was prompted by the close correspondence between several specific features of the model and the pattern of firing in these neurons. Most importantly, the observation that the dopamine response to reward transferred back to onset of the reward-predictive cue seemed to fulfill critical predictions of the Sutton and Barto model (1987), in which the value inherent in reward transferred back to the cue, which predicts its occurrence. Thus, phasic dopamine activity was taken to be the value signal hypothesized by the Sutton and Barto model (1981, 1987), which drives backpropagation of reward value to the stimulus (Schultz et al., 1997). Effectively phasic dopaminergic activity was argued to provide the learning signal that allows the state preceding reward delivery to acquire the value approaching the sum of future rewards, subsequently serving to bias an organism toward a choice of an action, which would result in the procurement of maximal reward.

This interpretation of phasic dopamine activity as a cached-value error signal has been remarkably influential across the decades since its proposal. The general idea has permeated how the field of neuroscience views not only dopamine function but also the functions of a host of brain regions that interact with the midbrain and are known to be involved in associative learning. This pervasiveness is despite our growing understanding of the complexity of learning. We now know unequivocally that organisms respond to stimuli in the environment as a consequence of relationships between events either in addition or in spite of any sort of static or cached value that may have accrued toward them in prior training (Balleine & Dickinson, 1991, 1998; Colwill & Rescorla, 1985; Delamater, 1996; Dickinson & Balleine, 1994; Holland & Rescorla, 1975; Killcross & Coutureau, 2003; Rescorla, 1973; Rescorla & Wagner, 1972). Indeed, humans and other animals are capable of developing rich models of the associative relationships between events in the world that are used in the absence of direct experience to influence ongoing and future behavior (Brogden, 1939; Colwill & Rescorla, 1985; Tolman, 1948). Yet the interpretation that the dopamine prediction error functions only to endow reward-predictive cues with a scalar value precludes the involvement of dopamine in the development of these more complex models of the environment. So, could dopamine also be involved in more complex forms of associative learning that transcend cached value or is it restricted to the model-free reinforcement learning algorithm described by Sutton and Barto (1981, 1987, 1998)?

The hypothesis that phasic dopamine acts only as a cached-value prediction error (Sutton & Barto, 1981, 1998) makes three notable predictions about when changes in phasic dopaminergic activity should be seen and what sorts of learning this phasic activity can support. Firstly, this theory predicts that stimulation or inhibition of dopamine neurons should act as a value signal to produce increments and decrements in responding toward reward-paired cues. Secondly, such manipulations of dopamine activity should not produce learning about the relationships between events of the world outside of a scalar expectation of value. Finally, phasic activity of dopamine neurons should not be evident in response to valueless changes in reward or to cues which have come to predict reward indirectly. We will now discuss these predictions in light of several recent studies that we believe provide particularly strong tests of their validity.

PREDICTION ONE: PHASIC STIMULATION OR INHIBITION OF DOPAMINE NEURONS SHOULD SUBSTITUTE AS A CACHED-VALUE PREDICTION ERROR TO DRIVE LEARNING

According to Schultz et al. (1997), an increase in phasic activity of dopamine neurons should serve to increase the value attributed toward a reward-paired cue, whereas a phasic decrease should reduce the value attributed to a cue. The advent of optogenetics has afforded us the temporal specificity to manipulate putative dopamine neurons in manner that allows us to causally test this hypothesis (Deisseroth, 2011; Deisseroth et al., 2006). Ideally, such experiments should arrange the learning materials so that all that is lacking is contingency—or an error in reward prediction—and then attempt to restore that error by manipulating the dopamine neurons precisely when it would be expected to occur. For example, Steinberg et al. (2013) used an optogenetic approach to mimic a positive reward prediction error during the blocking task first described by Kamin (1969). Specifically, rats were first presented with an auditory stimulus (e.g., A), which predicted reward. Following this training, a novel visual cue (e.g., X) was presented in compound with A and followed by delivery of the same reward. Rats in the control group failed to learn about cue X, presumably because cue X was blocked by prior training with cue A and reward. However, stimulation of dopaminergic neurons in the VTA during reward delivery after presentation of the AX compound unblocked learning about cue X. This was evident in an increase in rats' responses to the food port when cue X was presented alone in extinction. These data are consistent with the value hypothesis. Specifically, the artificial prediction error could be construed as attaching excess value to cue X despite the predictability of the reward, allowing cue X to become associated with the particular response being made at the time to enter the food port during presentation of the conditioned cue AX.

However, this study does not rule out a simple alternative, which is that the dopamine signal is increasing the salience of the preceding cue, which would also be expected to cause learning. If the dopamine signal acts in this manner, then inhibiting it would result in less learning. If, on the other hand, it acts as a cached-value error signal then phasic inhibition of dopamine should cause extinction learning, essentially decreasing the value attributed to a cue (Schultz et al., 1997). To test this question, Chang et al. (2016) optogenetically produced a brief negative error in VTA dopamine neurons during an over-expectation task. Overexpectation usually involves first pairing two cues (e.g., A and X) individually with reward (e.g., three food pellets). Then, these two cues are paired in compound with the same magnitude of reward. Usually, learning to cue X will decrease as the reward is now "overexpected" by the sum of the expectations elicited by cue A and X (e.g., six food pellets). However, in a modified version of this task Chang et al. (2016) delivered the expected reward (e.g., six pellets) during the compound phase, in order to eliminate the normal negative prediction error and prevent extinction. In half of the rats, dopamine activity in the VTA was inhibited during delivery of the final three pellets in the compound stage. Chang et al. (2016) found that brief inhibition of dopamine neurons during pellet delivery in the compound phase of this modified overexpectation task restored the normal extinction learning to cue X. That is, responding to X decreased with introduction of a brief inhibition of dopamine neurons during reward receipt. Again, these results are consistent with the hypothesis that VTA DA acts as a bidirectional value signal described in the model-free reinforcement learning algorithm postulated by Sutton and Barto (1981, 1987; Schultz et al., 1997). Specifically, that phasic inhibition of dopamine can act to decrease the value attributed to a cue and therefore reduce the response associated with that cue state.

PREDICTION TWO: WHAT IS LEARNED OR STAMPED IN BY THE PHASIC DOPAMINE SIGNAL SHOULD BE RELATED TO GENERIC OR CACHED VALUE

Experiments showing that optogenetic inhibition or stimulation can drive increases or decreases in responding to reward-predictive cues are consistent with the idea that this signal constitutes a scalar value, which increases or decreases the value attributed to a reward-paired cue. However, in both the studies described above (Chang et al., 2016; Steinberg et al., 2013), the learning induced by manipulating the firing of the dopamine neurons could consist of either general value or the formation of a more detailed associations between the cue and reward in the case of unblocking, and the cue and reward omission in the case of extinction. The former would constitute a learning mechanism consistent with that described in the model-free reinforcement algorithm postulated by Sutton and Barto (1981, 1998), where the latter is a more complex association between events that transcends the backpropagation of value to the reward-predictive cue.

To test whether dopamine transients are sufficient for associative learning beyond value, we used sensory preconditioning. Sensory preconditioning normally entails first pairing two neutral cues together in close succession such that an association forms between them (e.g., C→X). The development of this association can be revealed if one of those cues is later paired with reward. Specifically, if cue X is paired with reward, both cue C and X will elicit an expectation for reward when presented individually under extinction conditions. As cue C has never been directly paired with reward, it can only acquire an association with reward via its association with cue X, which allows it to enter into an association with the reward. This is supported by studies that have shown that cue C will not support conditioned reinforcement—rats will press a lever for cue X but not for cue C (Sharpe, Batchelor, & Schoenbaum, 2017). This suggests that cue C does not have any value independent of the food that it predicts, and so rats will not exert effort to obtain presentations of that stimulus alone. Thus, the sensory preconditioning procedure is well suited to an investigation of whether phasic dopamine may also support the development of rich associations between events in a manner that transcends cached value.

Using a modified version of the sensory preconditioning procedure, we aimed to assess whether optogenetic stimulation of dopamine neurons in the VTA could support associative learning beyond the transfer of cached value (Sharpe et al., 2017).

To do this, we first reduced the likelihood that rats would form an association between the two neutral cues C and X by pairing cue A with X (A→X). Subsequently, cues A and C were presented in compound prior to presentation of cue X (AC→X). Cue X was later paired with reward. In controls, we found that learning about cue C was blocked in this design, similarly to the original blocking studies shown with cues and rewards (Kamin, 1969). However, in our experimental group, stimulation of phasic dopamine at the beginning of X following the AC compound (i.e., AC→X trials) unblocked learning about cue C. These rats entered the magazine when cue C was presented in the final probe test as though they expected delivery of the reward. This suggests that triggering the dopamine neurons to fire at the start of X served to facilitate the formation of an association of a relationship between C and X, which allowed cue C to enter into a direct relationship with reward paired with X. This was confirmed by subsequent tests, which revealed that responding to C was sensitive to devaluation of the reward, showing rats responded to C because they desired the particular food reward. We also do not believe our results can be accounted for by salience, since there was no change in the rate of learning about X during conditioning, after the dopamine stimulation, nor was there any increase in learning about A (which would be evident in our design as stronger blocking of cue D; Sharpe et al., 2017). Thus, overall, these data suggest that the dopamine prediction error is capable of supporting the development of more complex associations than that envisioned by model-free reinforcement learning algorithms (Schultz et al., 1997; Sutton & Barto, 1981, 1998).

PREDICTION THREE: PHASIC CHANGES IN DOPAMINE SHOULD NOT REFLECT INFORMATION ABOUT CUE–REWARD RELATIONSHIPS THAT DOES NOT REFLECT DIRECT EXPERIENCE

If dopamine transients signal the prediction error is that contained in the model-free reinforcement algorithm described by Sutton and Barto (1981, 1987), then phasic dopaminergic activity should not reflect associations that have been inferred from prior associative relationships or a change in current state of the environment. This is because the error contained in the model-free reinforcement learning algorithm only receives predictions based on the value that backpropagates from the reward to the cue after the cue and reward have been paired in close succession (Sutton & Barto, 1981, 1998). This cannot happen if no direct association has been experienced. While the findings from Sharpe et al. (2017) suggest that dopamine can support the acquisition of complex associations between events (rewarding or otherwise), this does not require that the content of the information encoded in the prediction error signal itself go beyond errors in cached value. That is, stimulation or inhibition of dopamine could be allowing other neural structures to form more complex associations about the relationship between events, yet phasic activity in dopaminergic neurons may be ignorant of these associations under normal circumstances, operating only in response to cached-value errors.

Assessing whether the dopamine prediction error has access to information about the relationship between events requires examining how dopamine neurons or dopamine release changes in response to errors that reflect such associative information. There are now a growing number of studies that do this (Aitken, Greenfield, & Wassum, 2016; Bromberg-Martin & Hikosaka, 2009; Nakahara, Itoh, Kawagoe, Takikawa, & Hikosaka, 2004; Papageorgiou, Baudonnat, Cucca, & Walton, 2016; Sadacca et al., 2016; Takahashi et al., 2011). For example, dopamine activity to reward-paired cues changes depending on the physiological state of the subject (Aitken et al., 2016; Papageorgiou et al., 2016). In one study, Papageorgiou et al. (2016) monitored dopamine release using fast scan voltammetry in the nucleus accumbens (NaCC), as rats were performing an instrumental learning task. Here, rats had a choice of pressing one of two levers for one of two rewards (R1→O1 or R2→O2). On some of the trials, rats were presented with one lever option (forced trials; R1 or R2) while on others they could make a choice between pressing either one of the two levers (choice trials; R1 and R2). Prior to test sessions, rats were given free access to one of the rewards (e.g., devaluing O1). Subsequently, rats exhibited a preference for the lever associated with the nondevalued reward they had not had access to prior to the session (R2→O2). Papageorgiou et al. (2016) found that dopamine release to the reward-paired cues (i.e., the insertion of the lever into the behavioral chamber) was modulated by outcome devaluation prior to the rats experiencing the lever producing the now devalued outcome. That is, the dopamine response to lever presentation on forced trials reflected the new value of the devalued reward before it had been experienced with the lever-press response. Further, the dopaminergic response to presentation of the other lever was increased, showing an increased preference for the nondevalued option. This demonstrates that dopamine responses to reward-paired cues can update in response to the current physiological state of the subject without the subject directly experiencing the association between the cue and now devalued reward. These data are at odds with an interpretation of the dopamine signal as the model-free reinforcement learning algorithm described by Sutton and Barto (1981, 1998), since the cue and the devalued reward have never been paired, and so the new value of the reward cannot be attributed to the cue.

The data from Papageorgiou et al. (2016) beg the question of whether the phasic dopamine signal might also reflect information about an entirely new association developed in the absence of experience. In line with this possibility, Sadacca et al. (2016) showed that phasic activity of dopamine neurons can reflect associations between cues and rewards that have been inferred from prior knowledge of associative relationships in the experimental context. Specifically, Sadacca et al. (2016) recorded the activity of putative dopamine neurons in the VTA during sensory preconditioning. In this study, rats were first presented with two neutral cues in close temporal succession (A→B). Following this training, one of these cues was paired with reward (B→US). During conditioning, putative dopamine neurons exhibited the expected reward prediction error correlates, firing to reward early in conditioning, and transferring this response back to the cue later in learning. After conditioning, in the probe test in which both cues A and B were presented in the absence of reward, putative dopamine neurons continued to exhibit increased firing to B — the cue paired with reward — while also now firing to A, the cue paired with B in the preconditioning phase. Further, dopamine neuron firing to A and B was correlated, suggesting that the information signaled in response to A was the same as what was signaled in response to B. The simplest interpretation of these data is that dopamine neurons in the VTA signal reward prediction errors similarly whether they are based on directly experienced associations or whether they require inference. Again, this is not accommodated by a theory which argues that the dopamine signal reflects value that has backpropagated from the reward to a cue from their pairing (a notion reinforced by data showing a preconditioned cue does not acquire general value during the preconditioning procedure; This is the Sharpe, Batchelor, and Schoenbaum, 2017, eLife paper again. Rather, these data suggest that dopamine neurons may make more general predictions about the nature of upcoming rewards, garnered from associative model of the world and based on past experience.

CONCLUSIONS

In this chapter, we have discussed data that are problematic for the hypothesis that phasic dopamine signals encode a scalar cached-value signal, which allows a state preceding reward to acquire the value inherent in the reward and motivate behavior. Specifically, optogenetic stimulation or inhibition of has been found not only to increase or decrease responding to a cue preceding reward but also to facilitate the acquisition of associations between two neutral stimuli. Further, changes in phasic dopaminergic activity have been observed in response to a change in the physiological state of the subject despite the subject never experiencing pairing of the cue with the outcome that has been devalued in that state. Finally, the phasic dopaminergic response has also been seen in response to cues that have come to predict reward via prior knowledge of associative relationships in the experimental context, without being directly paired with reward. Such data challenge the conception that transient changes in dopamine carry the cached-value prediction error contained in model-free reinforcement learning algorithms (Sutton & Barto, 1981, 1998). Specifically, such evidence is outside the realm of a theory which argues that a cue only acquires a dopaminergic response via the backpropagation of value. Value cannot transfer back to a cue, which has not been paired with something valuable, and a value signal cannot facilitate the acquisition of associations between neutral stimuli.

So where to now? One theory that warrants consideration is that put forward by Nakahara (2014). Nakahara (2014) argues that dopamine prediction errors can be influenced by more than the expectation elicited from the current state. That is, a prediction error does not need to be calculated on the basis of current sensory information. Rather, a prediction error can be calculated on the basis of hidden states derived from prior experience. Further, the dopamine prediction error in this model can also be used to update these internal models of the environment. However, while the calculation of prediction errors can utilize information garnered from internal models of the world to generate a prediction about upcoming rewards, the signal itself still reflects the discrepancy between the value of the reward expected and that received. Thus, while Nakahara (2014) extends what the dopamine prediction error can use to make predictions, the prediction itself is still one of value which similarly serves to update the expected value of future rewards.

An alternative proposal is that dopamine transients reflect errors in event prediction more generally and that they are also involved in supporting learning about future events whether those events are the delivery of a particular reward, presentation of a neutral stimulus, or even absence of some stimuli or some other events. This would constitute a return to thinking about the prediction error in associative theory as driving real-world associations between events, as described in earlier theories of associative learning (Colwill & Rescorla, 1985; Holland & Rescorla, 1975; Miller & Matzel, 1988; Rescorla, 1973; Rescorla & Wagner, 1972; Wagner & Rescorla, 1972; Wagner et al., 1981) but somewhat abandoned by the world of neuroscience with the advent of TDRL (Sutton & Barto, 1981, 1987, 1998). Data from our lab already show that dopamine transients are both sufficient and also necessary for learning associations between neutral cues that inherently have no value (Sharpe et al., 2017). This is one prediction of the account described above. Below, we consider several more that might be examined in the future to support this hypothesis.

FUTURE DIRECTIONS

Conceptualizing the dopamine prediction error as a signal that detects a discrepancy between expected and actual events makes some testable predictions about when phasic activity in dopamine neurons should be observed. While Sadacca et al. (2016) and Papageorgiou et al. (2016) showed that dopamine activity to reward-paired cues can change as a result of knowledge not acquired through direct experience, in each case dopaminergic activity still signaled an upcoming prediction that could be construed as being about reward value. However, the alternative proposal made here suggests that changes in phasic dopaminergic activity would also be seen as a result of other changes in the predicted event that do not constitute a shift in value. For example, an increase in dopaminergic signaling should occur in response to a change in the identity of a reward. That is, if a cue previously paired with a particular reward was unexpectedly presented with a different reward that was equally valuable, we would expect to see a prediction error in dopaminergic neurons. While a few studies have looked at errors in response to change value of different rewards and have claimed not to see evidence of such a signal, these studies do not examine how dopamine signals change when identity is altered independent of value or reward preference (Lak et al., 2014; Stauffer et al., 2014). Thus, while their results show that the value error is similar across different identity rewards, they do not address whether changes in identity evoke error signals. It is also worth noting that it may be necessary to move beyond single unit correlates to appreciate more subtle error signaling functions of the dopamine system. As in other brain areas, meaningful signals may be carried in the pattern of finding across an ensemble of dopamine neurons, which may not be evident in individual "grandmother" neurons. In this regard, value errors may be a particularly amazing example of a more general population function. In any event, whether in individual units or ensemble responses, such a finding would represent strong evidence that the dopamine prediction error accesses information about the content of what is expected, independent of its meaning with regard to value.

Further, future research may also search for the presence of a dopaminergic error signal when a more general associative relationship between neutral stimuli is violated even in the absence of rewards. It is well-established that dopamine neurons in the midbrain fire when a novel stimulus is first presented unexpectedly, (Schultz, 1998). While this has been interpreted in the literature as a "novelty bonus" (Kakade & Dayan, 2002), it is also possible that this is an error signal in response to the appearance of an unexpected stimulus. It would be valuable to assess in an appropriately controlled environment whether these dopamine signals are seen when the contingency between neutral stimuli is manipulated such that expectation about upcoming stimuli is violated. Such research would support the hypothesis that the dopamine prediction error may reflect a more general signal for detecting the discrepancy between actual and expected events. Experiments like these would be useful since positive findings would open up new possibilities for how this biological signal may support associative learning in these and other contexts.

REFERENCES

Aitken, T. J., Greenfield, V. Y., & Wassum, K. M. (2016). Nucleus accumbens core dopamine signaling tracks the need-based motivational value of food-paired cues. *Journal of Neurochemistry, 136*(5), 1026–1036.

Balleine, B., & Dickinson, A. (1991). Instrumental performance following reinforcer devaluation depends upon incentive learning. *The Quarterly Journal of Experimental Psychology, 43*(3), 279–296.

Balleine, B. W., & Dickinson, A. (1998). Goal-directed instrumental action: Contingency and incentive learning and their cortical substrates. *Neuropharmacology, 37*(4), 407–419.

Blundell, P., Hall, G., & Killcross, S. (2003). Preserved sensitivity to outcome value after lesions of the basolateral amygdala. *Journal of Neuroscience, 23*(20), 7702–7709.

Brogden, W. (1939). Sensory pre-conditioning. *Journal of Experimental Psychology, 25*(4), 323.

Bromberg-Martin, E. S., & Hikosaka, O. (2009). Midbrain dopamine neurons signal preference for advance information about upcoming rewards. *Neuron, 63*(1), 119–126.

Bush, R. R., & Mosteller, F. (1951). A mathematical model for simple learning. *Psychological Review, 58*(5), 313.

Chang, C. Y., Esber, G. R., Marrero-Garcia, Y., Yau, H.-J., Bonci, A., & Schoenbaum, G. (2016). Brief optogenetic inhibition of dopamine neurons mimics endogenous negative reward prediction errors. *Nature Neuroscience, 19*(1), 111–116.

Colwill, R. M., & Rescorla, R. A. (1985). Postconditioning devaluation of a reinforcer affects instrumental responding. *Journal of Experimental Psychology: Animal Behavior Processes, 11*(1), 120.

Daw, N. D., Gershman, S. J., Seymour, B., Dayan, P., & Dolan, R. J. (2011). Model-based influences on humans' choices and striatal prediction errors. *Neuron, 69*(6), 1204–1215.

Deisseroth, K. (2011). Optogenetics. *Nature Methods, 8*(1), 26–29.

Deisseroth, K., Feng, G., Majewska, A. K., Miesenböck, G., Ting, A., & Schnitzer, M. J. (2006). Next-generation optical technologies for illuminating genetically targeted brain circuits. *Journal of Neuroscience, 26*.

Delamater, A. R. (1996). Effects of several extinction treatments upon the integrity of Pavlovian stimulus-outcome associations. *Learning & Behavior, 24*(4), 437–449.

Dickinson, A., & Balleine, B. (1994). Motivational control of goal-directed action. *Animal Learning & Behavior, 22*(1), 1–18.

Estes, W. K. (1950). Effects of competing reactions on the conditioning curve for bar pressing. *Journal of Experimental Psychology, 40*(2), 200.

Estes, W. K., & Burke, C. J. (1953). A theory of stimulus variability in learning. *Psychological Review, 60*(4), 276.

Fiorillo, C. D., Tobler, P. N., & Schultz, W. (2003). Discrete coding of reward probability and uncertainty by dopamine neurons. *Science, 299*(5614), 1898–1902.

Hebb, D. O. (1949). *The organization of behavior: A neuropsychological approach.* John Wiley & Sons.

Hebb, D. O. (2005). *The organization of behavior: A neuropsychological theory.* Psychology Press.

Holland, P. C., & Rescorla, R. A. (1975). The effect of two ways of devaluing the unconditioned stimulus after first-and second-order appetitive conditioning. *Journal of Experimental Psychology: Animal Behavior Processes, 1*(4), 355.

Hollerman, J. R., & Schultz, W. (1998). Dopamine neurons report an error in the temporal prediction of reward during learning. *Nature Neuroscience, 1*(4), 304–309.

Kakade, S., & Dayan, P. (2002). Dopamine: Generalization and bonuses. *Neural Networks, 15*(4), 549–559.

Kamin, L. J. (1969). Predictability, surprise, attention, and conditioning. *Punishment and Aversive Behavior,* 279–296.

Killcross, S., & Coutureau, E. (2003). Coordination of actions and habits in the medial prefrontal cortex of rats. *Cerebral Cortex, 13*(4), 400–408.

Lak, A., Stauffer, W. R., & Schultz, W. (2014). Dopamine prediction error responses integrate subjective value from different reward dimensions. *Proceedings of the National Academy of Sciences of the United States of America, 111*(6), 2343–2348.

Ljungberg, T., Apicella, P., & Schultz, W. (1992). Responses of monkey dopamine neurons during learning of behavioral reactions. *Journal of Neurophysiology, 67*(1), 145–163.

Miller, R. R., & Matzel, L. D. (1988). The comparator hypothesis: A response rule for the expression of associations. *Psychology of Learning and Motivation, 22*, 51–92.

Mirenowicz, J., & Schultz, W. (1994). Importance of unpredictability for reward responses in primate dopamine neurons. *Journal of Neurophysiology, 72*(2), 1024–1027.

Nakahara, H. (2014). Multiplexing signals in reinforcement learning with internal models and dopamine. *Current Opinion in Neurobiology, 25*, 123–129.

Nakahara, H., Itoh, H., Kawagoe, R., Takikawa, Y., & Hikosaka, O. (2004). Dopamine neurons can represent context-dependent prediction error. *Neuron, 41*(2), 269–280.

Olds, J., & Milner, P. (1954). Positive reinforcement produced by electrical stimulation of septal area and other regions of rat brain. *Journal of Comparative and Physiological Psychology, 47*(6), 419.

Papageorgiou, G. K., Baudonnat, M., Cucca, F., & Walton, M. E. (2016). Mesolimbic dopamine encodes prediction errors in a state-dependent manner. *Cell Reports, 15*(2), 221–228.

Rescorla, R. A. (1973). Effects of US habituation following conditioning. *Journal of Comparative and Physiological Psychology, 82*(1), 137.

Rescorla, R. A., & Wagner, A. R. (1972). A theory of Pavlovian conditioning: Variations in the effectiveness of reinforcement and nonreinforcement. *Classical Conditioning II: Current Research and Theory, 2*, 64–99.

Robbins, T., & Everitt, B. (1982). Functional studies of the central catecholamines. *International Review of Neurobiology, 23*, 303–365.

Romo, R., & Schultz, W. (1990). Dopamine neurons of the monkey midbrain: Contingencies of responses to active touch during self-initiated arm movements. *Journal of Neurophysiology, 63*(3), 592–606.

Sadacca, B. F., Jones, J. L., & Schoenbaum, G. (2016). Midbrain dopamine neurons compute inferred and cached value prediction errors in a common framework. *eLife, 5*, e13665.

Schultz, W. (1986). Responses of midbrain dopamine neurons to behavioral trigger stimuli in the monkey. *Journal of Neurophysiology, 56*(5), 1439–1461.

Schultz, W. (1998). Predictive reward signal of dopamine neurons. *Journal of Neurophysiology, 80*(1), 1–27.

Schultz, W., Apicella, P., & Ljungberg, T. (1993). Responses of monkey dopamine neurons to reward and conditioned stimuli during successive steps of learning a delayed response task. *Journal of Neuroscience, 13*(3), 900–913.

Schultz, W., Dayan, P., & Montague, P. R. (1997). A neural substrate of prediction and reward. *Science, 275*(5306), 1593–1599.

Sharpe, M. J., Chang, C. Y., Liu, M. A., Batchelor, H. M., Mueller, L. E., Jones, J. L., … Schoenbaum, G. (2017). Dopamine transients are sufficient and necessary for acquisition of model-based associations. *Nature Neuroscience, 20*.

Sharpe, M. J., Batchelor, H. M., & Schoenbaum, G. (2017). Preconditioned cues have no value. *Elife, 6*.

Spyraki, C., Fibiger, H. C., & Phillips, A. G. (1982). Dopaminergic substrates of amphetamine-induced place preference conditioning. *Brain Research, 253*(1), 185–193.

Stauffer, W. R., Lak, A., & Schultz, W. (2014). Dopamine reward prediction error responses reflect marginal utility. *Current Biology, 24*(21), 2491–2500.

Steinberg, E. E., Keiflin, R., Boivin, J. R., Witten, I. B., Deisseroth, K., & Janak, P. H. (2013). A causal link between prediction errors, dopamine neurons and learning. *Nature Neuroscience, 16*(7), 966–973.

Stricker, E. M., & Zigmond, M. J. (1974). Effects on homeostasis of intraventricular injections of 6-hydroxydopamine in rats. *Journal of Comparative and Physiological Psychology, 86*(6), 973.

Sutton, R. S., & Barto, A. G. (1981). Toward a modern theory of adaptive networks: Expectation and prediction. *Psychological Review, 88*(2), 135–170.

Sutton, R. S., & Barto, A. G. (1987). A temporal-difference model of classical conditioning. In *Paper presented at the proceedings of the ninth annual conference of the cognitive science society.*

Sutton, R. S., & Barto, A. G. (1998). *Reinforcement learning: An introduction* (Vol. 1). MIT Press Cambridge.

Takahashi, Y. K., Roesch, M. R., Wilson, R. C., Toreson, K., O'donnell, P., Niv, Y., & Schoenbaum, G. (2011). Expectancy-related changes in firing of dopamine neurons depend on orbitofrontal cortex. *Nature Neuroscience, 14*(12), 1590–1597.

Thorndike, E. L. (1898). Review of animal intelligence: An experimental study of the associative processes in animals. *The Psychological Review: Monograph Supplements, 2*.

Tolman, E. C. (1948). *Cognitive maps in rats and men*. American Psychological Association.

Ungerstedt, U. (1971). Adipsia and aphagia after 6-hydroxydopamine induced degeneration of the nigro-striatal dopamine system. *Acta Physiologica Scandinavica, 82*(S367), 95–122.

Wagner, A., & Rescorla, R. (1972). Inhibition in Pavlovian conditioning: Application of a theory. *Inhibition and Learning*, 301–336.

Wagner, A. R., Spear, N., & Miller, R. (1981). SOP: A model of automatic memory processing in animal behavior. *Information Processing in Animals: Memory Mechanisms, 85*, 5–47.

Wise, R. A. (1987). The role of reward pathways in the development of drug dependence. *Pharmacology & Therapeutics, 35*(1–2), 227–263.

Wise, R. A., & Rompre, P.-P. (1989). Brain dopamine and reward. *Annual Review of Psychology, 40*(1), 191–225.

Yokel, R. A., & Wise, R. A. (1975). Increased lever pressing for amphetamine after pimozide in rats: Implications for a dopamine theory of reward. *Science, 187*(4176), 547–549.

CHAPTER 12

A State Representation for Reinforcement Learning and Decision-Making in the Orbitofrontal Cortex

Nicolas W. Schuck[1,2], Robert Wilson[3], Yael Niv[2]

[1]Max Planck Research Group NeuroCode, Max Planck Institute for Human Development, Berlin, Germany; [2]Princeton Neuroscience Institute & Department of Psychology, Princeton University, Princeton, NJ, United States; [3]Department of Psychology, University of Arizona, Tucson, AZ, United States

INTRODUCTION

The orbitofrontal cortex (OFC) is an intensely studied brain area. Pubmed currently lists over 1000 publications with the word "orbitofrontal" in the title, reflecting more than six decades of research that has mainly sought to answer one question: What mental functions are subserved by the OFC? Still, an integration of existing knowledge about this brain area has proven difficult (Cavada & Schultz, 2000; Stalnaker, Cooch, & Schoenbaum, 2015). Lesion studies have pointed to a plethora of often subtle and complex impairments, but what mental operation is common to all these impairments has remained unclear. Neural recordings have shown that a large variety of different aspects of the current environment are encoded in orbitofrontal activity but have not yet explained why these variables are jointly represented in OFC or how they are integrated. Finally, the study of OFC's anatomy has uncovered a complex internal organization of subregions, which has made the identification of homologies in other mammals difficult, and raised questions about the functional division of labor associated with these subregions.

Despite this diversity of findings and the lack of consensus on their interpretation, research has converged on the idea that many of OFC's functions must lie within the domains of decision-making and reinforcement learning (RL). In this chapter, we will provide a selective overview of the studies supporting this broader idea and describe a novel theoretical framework for understanding the role of the OFC in RL and decision-making. This framework, the State-Space Theory of OFC, proposes that the OFC represents, at any given time, the specific information needed in order to maximize reward on the current task.

We will begin the chapter with a brief overview of OFC anatomy, focusing mostly on its most salient aspects in primates. We will then discuss the State-Space Theory of OFC in more detail and evaluate how well it can accommodate current knowledge. We will follow with a discussion of other major theoretical accounts of the OFC and how they might be integrated into the State-Space Theory. Finally, we will discuss which findings

Goal-Directed Decision Making
ISBN 978-0-12-812098-9, https://doi.org/10.1016/B978-0-12-812098-9.00012-7

about OFC lie outside the scope of our framework and highlight some areas for future research. Given the large number of investigations, our overview remains necessarily incomplete, and we refer the reader to several recent excellent reviews on this topic for more information (Kringelbach, 2005; Murray, O'Doherty, & Schoenbaum, 2007; Rudebeck & Murray, 2011; Rushworth, Noonan, Boorman, Walton, & Behrens, 2011; Schoenbaum, Takahashi, Liu, & McDannald, 2011; Stalnaker et al., 2015).

WHAT IS THE ORBITOFRONTAL CORTEX?

The primate OFC is a large cortical area located at the most ventral surface of the prefrontal cortex (PFC), directly above the orbit of the eyes (hence the name), and including parts of the medial wall between the hemispheres (see Fig. 12.1A). It is defined as the part of PFC that receives input from the medial magnocellular nucleus of the mediodorsal thalamus (Fuster, 1997) and consists of Brodmann areas 10, 11, and 47. Brodmann's initial classification, however, was unfinished and showed inconsistencies between humans and primates, possibly reflecting the heterogeneity of sulcal folding patterns in OFC (Kringelbach, 2005). Later cytoarchitectonic work has refined this classification, and today's widely accepted parcellations are based on five subdivisions known as Walker's areas 10, 11, 47/12, 13, and 14 (e.g., Öngür, Ferry, & Price, 2003; Glasser et al., 2016, see Fig. 12.1B,C). One unusual aspect of the primate OFC is its mixed cytoarchitecture that is partly five-layered (agranular) and partly six-layered (granular). Because granular cortex emerged later in evolution than agranular cortex, this suggests that some parts of OFC are phylogenetically older than other parts of frontal cortex and complicates comparisons between primates and nonprimates, whose OFC is entirely agranular (Passingham & Wise, 2012; Preuss, 1995). Based on these differences, Wise and

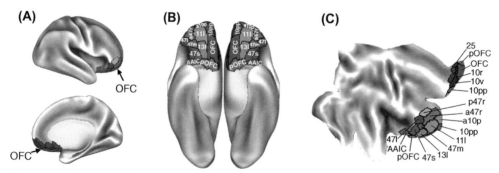

Figure 12.1 *Anatomy of the human orbitofrontal cortex (OFC).* (A): The location of the OFC on an inflated brain is highlighted by the brown shaded areas in a lateral (upper) and medial (lower) view. (B): Subdivisions of the OFC, shown on the ventral surface, according to most recent parcellation proposed by Glasser et al. (2016). (C): OFC subareas shown on a flat map of the left hemisphere, with the same color coding as in (A) and (B). *AAIC*, anterior agranular insular complex. *(Figure made using data from Glasser et al. (2016) and the Connectome Workbench.)*

colleagues have proposed that nonprimate mammals have no homologue of what is the granular OFC in primates (Wise, 2008). Many controversies about this topic are still ongoing, for instance, about whether the medial OFC network does or does not have a homologue in rodents (Heilbronner, Rodriguez-Romaguera, Quirk, Groenewegen, & Haber, 2016; Rudebeck & Murray, 2011; Schoenbaum, Roesch, & Stalnaker, 2006).

Another noteworthy feature of OFC's anatomy is its connectivity. OFC has remarkably close connections to all sensory areas (often only bi- or trisynaptic) in addition to widespread connections to other parts of the frontal cortex, striatum, amygdala, and hippocampus, among others. Connectivity patterns to these areas highlight a distinction between a medial and a lateral subnetwork in the OFC (Cavada, Company, Tejedor, Cruz-Rizzolo, & Reinoso-Suárez, 2000; Kahnt, Chang, Park, Heinzle, & Haynes, 2012), a difference that is often assumed to have functional implications (Elliott, Dolan, & Frith, 2000; Noonan, Walton, et al., 2010; Rudebeck & Murray, 2011; Walton, Behrens, Noonan, & Rushworth, 2011). Specifically, the lateral network has been shown to have many connections to lateral orbital areas as well as the amygdala and receives connections from sensory areas related to olfactory, gustatory, visual, somatic/sensory, and visceral processing. The medial network, in contrast, connects to areas along the medial wall (Brodmann areas 25, 24, and 32) and receives input from the amygdala, the mediodorsal thalamus, various regions in the medial temporal lobe (hippocampus, parahippocampus, rhinal cortex), ventral striatum, hypothalamus, and periaqueductal gray (Cavada et al., 2000). For the remainder of this chapter, we will consider both networks as OFC, following definitions of the orbital medial PFC (Öngür et al., 2003). Note that the area commonly referred to as ventromedial PFC is therefore partly included in our definition of OFC.

In summary, the OFC represents a remarkably densely connected brain area with links to all sensory domains, learning and memory structures like the striatum, amygdala, and hippocampus as well as several frontal subregions. The OFC is also a highly heterogenous brain area, containing two broadly distinct subnetworks (medial vs. lateral), cytoarchitectonic diverse subregions (granular vs. agranular cortex), and large interindividual differences in sulcal folding patterns.

THE STATE-SPACE THEORY OF ORBITOFRONTAL CORTEX

At the heart of studying decision-making is the quest to understand how the brain answers the following question: Given the state of the world, which actions promise to yield the best outcomes? Much research has focused on how expected outcomes are learned and represented or how actions are selected based on these expectations. In contrast, the other major aspect of the question, namely how decisions depend on the current environment, and what the animal considers "the state of the world," has received much less attention. Because natural environments of animals are often rich in sensory information and complex temporal dependencies, how the state of the environment is represented in the brain is crucial for successful decision-making. Below, we will define more precisely what we mean by "state of the world" and provide more detail on how the representation of the state of the world is shaped by the requirements of the decision-making process. Then we will propose a specific role for the OFC in representing this information during decision-making.

The computational theory of RL (Sutton & Barto, 1998) relies on a representation of all the information that is relevant for the current decision, referred to as *the state*. The state is not just a one-to-one reflection of the physical state of the environment, but rather a reflection of what information about the world the decision-making agent represents at the moment the decision is made. How precisely should an agent represent its environment to optimally support decision-making and learning? As we will see below, this is not an easy question, and answering it requires a good understanding of the problem at hand. Consider, for instance, an RL agent trying to learn to balance a pole hinged to a cart that can be moved either left or right (a classic benchmark task in RL, Michie & Chambers, 1968). From an RL perspective, an optimal policy for this problem can only be computed if the current state contains all information that is sufficient to fully predict the immediate future state of the cart, once a certain action is taken. This characteristic is known as the *Markovian* property and effectively means that the conditional probabilities of future states depend only on the current state and action, but not the past states. In the case of the pole problem, this means that representing the cart's position and the angle of the pole as the state are not enough because these variables alone are insufficient to predict which way the pole is moving and to infer how to move the cart. Instead, the cart's velocity and the rate at which the angle between the pole and the cart is changing are needed. Representing these variables as part of the current state will allow one to learn a much better behavioral control policy than if they were omitted from the state representation.

This requirement for the state representation raises another problem: Some variables are not easily extracted from the information the agent gets from its sensors but require memory of past sensory information and further computation. For example, the velocity-related variables must be inferred by comparing past and current sensory inputs and thus require memory. If states need to reflect information beyond what is accessible through current sensory input and there is uncertainty regarding their true underlying value, the states are called *partially observable*.

Finally, not all aspects of the current sensory input are relevant. Lighting conditions, for instance, do not need to be included into the state, as they are irrelevant for the policy even if they change the sensory signals. Including unnecessary aspects in the state representation will lead to slower learning due to the need to separately learn a policy for states that seem different but are effectively equivalent, a phenomenon known as *the curse of dimensionality*. A good state representation is therefore one that solves two problems: It deals with partial observability and non-Markovian environments by supplementing sensory information with the necessary unobservables, and it filters the sensory input to only include relevant aspects in order to avoid the curse of dimensionality.

While pole balancing itself is a rare activity for humans, the curse of dimensionality and the problem of partial observability of states are ubiquitous. A brain area well suited to solve this problem would need to be able to access sensory cortices as well as brain areas relevant for episodic memory and selective attention processes (Niv & Langdon, 2016). On purely anatomical grounds, the OFC is a good candidate for this representation: It is unique among areas in the PFC in its close connectivity to all five sensory modalities, and it has bidirectional connections to brain areas relevant for memory and decision-making such as the hippocampus and the striatum. In addition to these general considerations, a review of decades of studies on the function of the OFC has recently led us to propose that the role of the OFC in decision-making is to represent partially observable states of the environment when they are needed to perform the task at hand. Specifically, in Wilson, Takahashi, Schoenbaum, and Niv (2014), we investigated how changes in the way states are represented would affect behavior in tasks that are known to be impacted by OFC lesions. The central idea was that in many cases the state space of a task must include partially observable information, but OFC-lesioned animals might be incapable of integrating the necessary observable and unobservable information. In order to test this idea theoretically, we used an RL modeling framework and manipulated how states were represented. To simulate OFC-lesioned animals, all states of the task that are associated with identical sensory input were therefore modeled as the same state, whereas healthy animals were modeled as having the ability to disambiguate states that involve identical sensory input based, for instance, on past events. Strikingly, this manipulation caused subtle but pervasive impairments of the model's ability to perform exactly those tasks that are known to be impacted by OFC lesions.

One example is the delayed alternation task, which is known to be impaired by OFC lesions in both animals as well as humans (e.g., Freedman, Black, Ebert, & Binns, 1998; Mishkin, Vest, Waxler, & Rosvold, 1969). In this task, two simple actions (say, pressing a left or right lever) can lead to reward. Specifically, the delivery of reward is coupled to the previous choice such that on each trial, only the action that was *not* chosen on the previous trial is rewarded. To solve this task, the states corresponding to the two options need to be supplemented by the previous choice: Although all trials look similar in terms of the externally available stimuli, when the previous choice was A, the best action is B, and vice versa (see Fig. 12.2A,B). If OFC lesions impair the ability to distinguish between two identically looking states based on unobservable context, one would expect that OFC-lesioned animals would represent the task as having only one state (Fig. 12.2B). As a consequence, the animal would be severely impaired in its ability to correctly perform the task, which indeed has been shown (specifically, performance went down to chance due to the lesion, but only if trials were separated by a delay that rendered

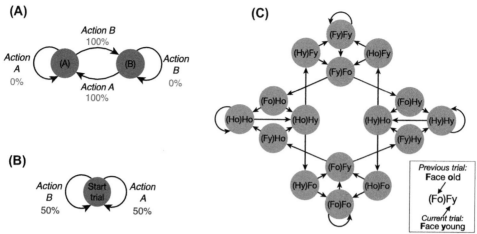

Figure 12.2 *State spaces of the delayed alternation task and the Schuck et al. task.* (A): In a delayed alternation task, performance is rewarded if the previously unchosen action is selected. A suitable state representation therefore must distinguish between trials in which action A was previously performed versus trials in which action B was previously taken. We denote the two possible actions as "A" and "B" and label the states accordingly "(A)" and "(B)" to denote the action on the previous trial. In the diagram, state transitions depend on the action taken. The probability of reward on each trial, which depends on the transition/action chosen, is denoted in gray for each transition. With such a state representation, different values can be assigned to an action depending on whether is was preceded by the same or a different action, thus allowing the agent to learn the optimal policy that leads to 100% reward (alternating choices). (B): If, on the other hand, an agent is unable to differentiate between states based on the unobservable choice history, then the environment is perceived as having only one state, in which each action yields reward on only 50% of trials. According to the theory, complete OFC lesions would result in this reduced state representation, and consequently, performance would be at chance accuracy, as is indeed seen empirically (Mishkin et al., 1969). (C):The state space used in the Schuck et al. task described in the text. "Hy" indicates a trial in which the relevant category was **H**ouse, and the correct response was **y**oung. For simplicity, only transitions for correct actions are shown (the wrong action leads to repetition of the trial). State-relevant information from the previous trial is denoted in brackets, such that "(Fo)Fy" indicates a young Face trial (the "Fy" part of the state) that was preceded by an old Face trial (the "(Fo)" part of the state, see legend).

the previous choice unobservable; Mishkin et al., 1969). In Wilson et al. (2014), we showed that a variety of behavioral consequences of OFC lesions can indeed be explained by an impairment in the state space underlying performance on the task. We also showed that changes in dopaminergic firing following OFC lesions can be explained as a consequence of impaired state differentiation (Takahashi et al., 2011).

In a follow-up study in humans, we used a task specifically designed to test our hypothesis, to investigate orbitofrontal representations during decision-making (Schuck, Cai, Wilson, & Niv, 2016). On each trial of the task, participants had to judge whether either a face or a house (presented overlaid as a compound stimulus) was old or young. Crucially, to determine whether they should be judging the house or the face, participants had to continuously monitor both the current and previous trial: Whenever the age response on the previous trial was different from that on the current trial, the category to be judged on the following trial was switched. Otherwise, the next trial's category was the same as the current trial. Given these rules, the task required a complex state space with 16 different partially observable states (Fig. 12.2C). Using multivariate pattern analysis techniques, we investigated what aspects of the state information were encoded within OFC. This analysis showed that on each trial, the OFC contained information about all partially observable aspects of the state: the previous age, the previous category, and the current category. A whole-brain analysis suggested that medial OFC was the only region in which all necessary unobservable information could be decoded. Still, information about events two trials in the past, which was not relevant to correct performance, could not be decoded in OFC (in contrast to other brain areas where we could decode not only some of the relevant unobservable state components but also some irrelevant information such as the category from two trials back). Finally, in an error-locked analysis of single-trial information, we found that errors during the task were preceded by a deterioration of the state representation in OFC. These results provide strong support for the above-outlined hypothesis of the representational role of OFC in decision-making.

Several other studies have come to similar conclusions. Recording from neurons in lateral OFC in rodents, Nogueira et al. (2017) reported that task-relevant but unobservable information from the previous trial was integrated with the current sensory input. A study from our lab used a task in which participants had to infer the current state based on a series of past observations and found that activity patterns in OFC reflect the posterior probability distribution over unobserved states, given the observed sequence of events (Chan, Niv, & Norman, 2016). Bradfield, Dezfouli, van Holstein, Chieng, and Balleine (2015) reported that bilateral excitotoxic lesions and designer-drug-induced inactivations of the rat medial OFC led to an inability to retrieve or anticipate unobservable outcomes across a range of tasks. Finally, Stalnaker, Berg, Aujla, and Schoenbaum (2016) studied rats in a task in which outcome magnitudes and identities were occasionally reversed. They found that the unobservable state of the task (that is, the "block identity") could be decoded from activity of cholinergic interneurons in the dorsomedial striatum, and importantly, that this information vanished when the OFC was lesioned. Taken together, these studies support the idea that within OFC, task-relevant information is combined into a state representation that facilitates efficient decision-making in the face of partial observability and non-Markovian environments.

A ROLE FOR THE ORBITOFRONTAL CORTEX IN STATE INFERENCE AND BELIEF STATES?

If the OFC is involved in representing partially observable information in the service of decision-making, one important question is whether OFC is also involved in *inferring* the state from observations. As we highlighted above, a useful state representation is not simply a reflection of the current sensory input, but rather can be viewed as the (often hidden) collection of attributes that causally determine future rewards and state transitions. For example, in the young/old task described above, the current stimulus is not sufficient for determining which action will be rewarded. Similarly, when deciding whether the cab you requested is still on its way, or they have forgotten your request and you should call the company again, the current observation of "no cab here" is not sufficient and you must make use of information such as how long you have already been waiting, what is the time, and what time did you request the cab for. This implies that the current state must often be *inferred* from more than current observations, and in many cases, there is considerable uncertainty about the current state. RL theory has shown that an optimal way to learn under such uncertainty is to use Bayesian inference to estimate the probability distribution over possible unobservable states given the observations, and use this quantity (referred to as the *belief state*) as the current state of the task (Daw, Courville, & Touretzky, 2006; Dayan & Daw, 2008; Kaelbling, Littman, & Moore, 1996; Rao, 2010; Rodriguez, Parr, & Koller, 1999; Samejima & Doya, 2007).

Although it is still unclear whether the brain indeed performs a similar inference process and if the OFC is representing a belief state distribution rather than a single (for instance, most likely) state, recent evidence has pointed in that direction. For example, in a recent study, we studied the process of inferring the true state in a task in which observations were only probabilistically related to states (Chan et al., 2016). Using a representational similarity approach, we found that the similarity of neural patterns in medial OFC was related to the similarity of probabilistic state distributions predicted by a Bayesian inference model. In line with this idea, other studies have indicated that OFC also represents the confidence with which animals make a choice (Kepecs, Uchida, Zariwala, & Mainen, 2008; Lak et al., 2014), a quantity that also affects dopaminergic midbrain activity and may reflect belief states (Lak, Nomoto, Keramati, Sakagami, & Kepecs, 2017). Other studies investigating dopaminergic prediction error signals have shown that reward predictions are based on a state inference process rather than purely sensory states (Langdon, Sharpe, Schoenbaum, & Niv, 2018; Starkweather, Babayan, Uchida, & Gershman, 2017). In particular, the passage of time provides a ubiquitous cue for inferring transitions between states that may be externally similar (as in the cab example above), and recent work suggests that prediction error signals rely on input from the ventral striatum to reflect such time-based state inference (Takahashi, Langdon, Niv, & Schoenbaum, 2016). Our previous work has shown that dopaminergic prediction errors indeed depend on state representations in the OFC (Takahashi et al., 2011; Wilson et al., 2014). Together with this recent work that indirectly suggests the existence of belief states elsewhere in the brain, these findings support the idea that a state inference process results in a representation of an entire belief-state distribution within the OFC. More evidence is needed, however, and previous work has suggested that belief states may be encoded either in lateral PFC (Samejima & Doya, 2007) or sensory cortices (Daw et al., 2006; van Bergen, Ji Ma, Pratte, & Jehee, 2015).

In addition, many questions about the inference process itself remain unanswered. Of particular importance is the question about the role of state transitions in the state inference process. The abovementioned theoretical approaches to belief states indicate that inferring the current belief state relies on two quantities: current and past observations on the one hand and the previous belief state on the other hand (see Fig. 12.3). Previous belief states influence the current belief state through the state transition function: Similar to how knowledge about someone's previous location will constrain where that person could possibly be one time step later, knowledge about possible state transitions and the previous state will influence the estimate of the current state. Several of the tasks in which

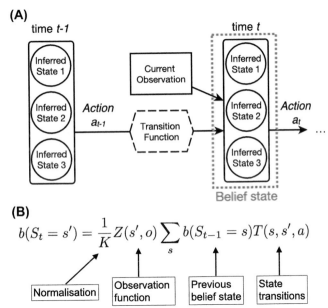

(B)

$$b(S_t = s') = \frac{1}{K} Z(s', o) \sum_s b(S_{t-1} = s) T(s, s', a)$$

Normalisation

Observation function

Previous belief state

State transitions

Figure 12.3 *Belief state representations and state inference.* (A): The diagram illustrates how uncertainty at different stages is incorporated into a belief state representation (*blue box*). On each trial, because of partial observability, we assume that the sensory input ("current observation") cannot be deterministically mapped onto states, but rather leads to a probabilistic estimate of how likely different states are (captured by the *observation function* $Z(s', o)$). In addition, the distribution over states on the previous trial is combined with the (presumably known) state transition function to further constrain the likely current state. In analogy to a spatial setting, this process corresponds to estimating your current location based on what is currently observed, in combination with your previous belief regarding where you have just been, which determines the locations that were adjacent to you and reachable in one step. Together, the current sensory input and your previous beliefs regarding your likely location constrain your current location estimate. (B): The key belief state—updating equation relates the above-described quantities to yield the probability of being in a specific state s' at a specific time t, denoted as $b(S_t = s')$.

OFC lesions lead to behavioral impairments, such as devaluation (Gallagher, McMahan, & Schoenbaum, 1999; Pickens et al., 2003) and Pavlovian—instrumental transfer (Bradfield et al., 2015) tasks, require not only state representations but also inference about expected outcomes based on knowledge of state transitions. Furthermore, neuro-imaging studies have found that the OFC may be involved in updating knowledge about transitions between cues and outcome identities (Boorman, Rajendran, O'Reilly, & Behrens, 2016), which could reflect a role in representing state transitions more generally. Other studies have pointed to a prominent role of the hippocampus in storing state transitions (Schapiro, Kustner, & Turk-Browne, 2012).

Finally, an important aspect of state inference is that the environmental features that are predictive of the outcome might change over time, or previously unknown relations between sensory input and outcomes might only be discovered after a period of learning. In our previous work, we have begun to investigate both of these cases (Niv et al., 2015; Schuck et al., 2015). This work has so far pointed not toward the OFC but rather suggested a role for a frontoparietal attention network in controlling the adaptation of state representations and of medial PFC in covertly preparing the updated state representations.

In summary, current research bears only indirectly on the possible role of OFC in the representation of belief states, state transitions, and state updating processes, and it seems unlikely that all these functions would be performed by a single neural circuit. Future research therefore needs to address these questions more directly and also investigate how different brain areas such as the hippocampus and medial PFC might perform these complex computations in cooperation with the OFC (e.g., Kaplan, Schuck, & Doeller, 2017). Moreover, one important question is where and how the relations between states, as reflected in the possible transitions between them, are represented in the brain. Building on the existing work about hippocampal representations of transitions between spatial and observable states (Schapiro et al., 2012; Stachenfeld, Botvinick, & Gershman, 2017), future work needs to investigate the neural representation of transitions between partially observable states. Such research needs to carefully take into account the methodological hurdles of estimating representational similarity with fMRI, however (Cai, Schuck, Pillow, & Niv, 2016).

ORBITOFRONTAL VALUE SIGNALS AND THEIR ROLE IN DECISION-MAKING

While the work reviewed in the previous sections provides support for the State-Space Theory of OFC, these findings are only a small part of the large literature on the OFC, and other authors have proposed different accounts of OFC function. Perhaps the most influential alternative is that OFC represents economic value associated with a given choice, in particular, value that has to be calculated on the fly rather than learned from experience. This theory has its roots in recordings of neural activity in the OFC of monkeys making choices between different food options (Padoa-Schioppa & Assad, 2006; Tremblay & Schultz, 1999). In particular, in one influential study, Padoa-Schioppa and Assad (2006) recorded activity in the OFC (area 13) of monkeys while the monkeys chose between two different kinds of juice. On each trial, different amounts of each juice were offered through the display of visual cues, and the animals could freely decide which option they preferred. Following standard economic theory, the authors calculated the subjective value of each juice based on the monkeys' choices and showed that a proportion of recorded neurons showed firing activity that varied linearly (some increasing and some decreasing) with the subjective value of the chosen juice, regardless of which option was chosen (they additionally reported neurons that responded to the identity of the chosen option and the value of each of the two offers). This general finding, a value or reward representation that is independent of sensory or motor aspects of the option, is supported by a number of similar studies in humans, monkeys, and rodents (e.g., Gottfried, O'Doherty, & Dolan, 2003; Hare, O'Doherty, Camerer, Schultz, & Rangel, 2008; Howard, Gottfried, Tobler, & Kahnt, 2015; Plassmann, O'Doherty, & Rangel, 2007; Schoenbaum & Eichenbaum, 1995; Thorpe, Rolls, & Maddison, 1983; Tremblay & Schultz, 1999). Moreover, subsequent studies have suggested that these value representations do exhibit a number of properties in line with the value account of OFC function. For example, Padoa-Schioppa and Assad (2008), found that OFC value signals are invariant to the other available options (a property called transitivity; but see Tremblay & Schultz, 1999).

What these findings imply about OFC function is not as clear as it might seem, however. First, due to the long recording durations and many experimental sessions involved in monkey electrophysiology, it is difficult to claim with certainty that the monkeys are computing the values of the alternatives on the fly and that this is what OFC is necessary for. Moreover, other findings have cast doubt on the claim that OFC's primary function is to represent values of choices during decision-making in general (Schoenbaum et al., 2011). For example, studies of OFC–lesioned animals have shown no impairment in general learning abilities during initial value acquisition (e.g., Butter, 1969; Chudasama & Robbins, 2003; Chudasama, Kralik, & Murray, 2007; Izquierdo, Suda, & Murray, 2004; O'Doherty, Critchley, Deichmann, & Dolan, 2003; West, DesJardin, Gale, & Malkova, 2011) and inconsistent results after a reversal of cue—outcome contingencies (Kazama & Bachevalier, 2009; Rudebeck, Saunders, Prescott, Chau, & Murray, 2013; Stalnaker, Franz, Singh, & Schoenbaum, 2007). Other studies have shown that OFC lesions do not impact monkeys' ability to make value-dependent choices even when values are constantly changing (Walton, Behrens, Buckley, Rudebeck, & Rushworth, 2010). A recent proposal by Walton et al. (2011) has highlighted that the nature of the decision-making impairment depends on whether the lesion affected medial or lateral OFC. Specifically, Walton and colleagues suggested that lateral OFC is necessary for correctly assigning credit for outcomes to previous actions for the purpose of learning, whereas the medial OFC is important for basing decisions on the highest-valued option while ignoring the irrelevant other options. This idea might indeed explain why decision-making impairments are seen under some circumstances but not others.

Yet, values (and choices) are not the only task-related quantity that is encoded in OFC. Several reports have shown that OFC encodes many variables that are related to the current task but are independent of value, including for instance outcome identities (Howard et al., 2015; McDannald et al., 2014, 2011; Stalnaker et al., 2014), salience (Ogawa et al., 2013), confidence signals (Kepecs et al., 2008; Lak et al., 2014), and even social category of faces (Watson & Platt, 2012) or spatial context (Farovik et al., 2015). In fact, these value-free signals may represent the vast majority of information encoded in OFC—one study found only 8% of lateral OFC neurons coding value in a linear manner (Lopatina et al., 2015), a number that is not out of line with other reports that often analyze only small subsets of the recorded neurons. This raises the possibility that firing patterns of orbitofrontal neurons reflect value only in the context of, or as part of, the current state. This interpretation is also supported by the above-described study, in which we found state signals in the OFC in the absence of any overt rewards or values (Schuck et al., 2016).

This idea that OFC value representations are embedded in a more general state signal in the OFC is supported by a recent study from Rich and Wallis (2016). In this experiment, OFC neurons were recorded while monkeys deliberated between two choices leading to differently valued outcomes. While activity patterns during deliberation were predictive of the value of the later chosen option, the authors also report that the value encoded by single neurons was dependent on a network-represented state: The same neuron encoded the value of pictures on the right when the rest of the network (not including this neuron) was in the "right state" (i.e., signaling that something was shown on the right), and the value of pictures on the left when the network was in the "left state."

The notion that the unique function of OFC is to integrate information about partially observable task states (and perhaps their resulting values) is also supported by research on flexible, goal-directed behavior in the so-called devaluation paradigm. In this task, animals are first trained to perform actions (say, pressing on one of several bars) in order to obtain desired outcomes (e.g., different types of food) such that each action is associated with a particular outcome. Subsequently, in a separate setting, one of the possible outcomes is devalued by satiation or pairing the outcome with food poison. A test then assesses whether the animal can use the new (lower) value of the outcome to guide its behavior. In this, the animal is once again allowed to make outcome-earning actions (although no outcomes are actually delivered in this phase), with the main question being whether it will continue to perform the action that previously led to the now-devalued outcome. While healthy animals that have not been overtrained to the point that the actions have become habitual adapt their behavior appropriately, OFC-lesioned animals continue to chose the wrong option, a behavior that has been suggested to reflect their inability to update outcome expectations (Gallagher et al., 1999; Pickens et al., 2003), rather than mere failures to inhibit prepotent reposes (Chudasama et al., 2007; Walton et al., 2010). In another devaluation study, value representations in the OFC as well as the amygdala were found to be changed following devaluation (Gottfried et al., 2003), suggesting that the values encoded in OFC can be updated offline based on knowledge of state transitions and the newly experienced outcomes. Interestingly, consistent with the State-Space Theory, similar studies looking at the representations of value-*independent* outcome identities in the OFC found that these representations differed based on the current goal, consistent with the state-space representation being task dependent (Critchley & Rolls, 1996; Howard & Kahnt, 2017). Thus, the role that OFC seems to play in value-based decision-making seems to be at the junction of representing values and correctly inferring partially observable current or future states.

Other studies using unblocking paradigms support similar conclusions (Burke, Franz, Miller, & Schoenbaum, 2008; McDannald, Lucantonio, Burke, Niv, & Schoenbaum, 2011). In one variant of this task, animals learn to discriminate different odors that predict different quantities and flavors of milk. After learning, additional odors are added to the original odor and either the same outcome is presented, the size of the outcome is changed, or the flavor of the outcome is changed. Afterward, the degree to which an association between the novel odors and the outcomes was learned is assessed. Because value-based learning is driven by the difference between one's expectations and the outcomes (the so-called prediction error), learning theory suggests that no association should be learned for the novel odor when it led to the same outcome as before (hence the term "blocking" as the association between the old odor and the outcome blocks a new association from forming Kamin, 1969). The novel odors associated with a change in reward size, however, should be followed by a prediction error that would trigger learning ("unblocking"). This learning can be based purely on value signals, as found in areas such as the striatum. A very different and interesting prediction arises when considering trials in which the milk flavor was changed. Because the two milk flavors were matched for overall value, learning about novel cues predicting flavor changes cannot be driven by a value mismatch error. Rather, for changes in the outcome identity to trigger learning about the new odor-specific knowledge about the expected identity of the outcome (and its violation) is required, as found in the OFC (Critchley & Rolls, 1996; Howard & Kahnt, 2017; McDannald et al., 2014; Stalnaker et al., 2014). Indeed, lesion studies have shown that OFC is critical for this type of outcome identity—dependent unblocking but not for unblocking due to changes in the value of the outcome in this task (McDannald et al., 2011).

In summary, while these findings suggest that orbitofrontal neurons encode information about the values of different options, OFC's function is not readily captured by the proposal that the sole or primary role of this area is value comparisons. From the perspective of the State-Space Theory, these findings rather point to a more holistic integration of decision-relevant information in the OFC, ranging from partially observable context that is necessary for solving the task to the expected sensory aspects of the outcomes and the values associated with them. Interestingly, the above-described lesion and inactivation studies all suggest that this representation is only necessary when changes in the contingencies between states require to reassess the value of different choices.

BEYOND LEARNING AND DECISION-MAKING

Although the literature on the function of the OFC has mainly addressed its role in decision-making, some investigations have focused on other potentially important aspects of OFC function. In particular, observations from human patients with OFC damage have often led to reports about post-lesion changes in their "personality" (e.g., Cicerone & Tanenbaum, 1997; Galleguillos, Parrao, & Delgado, 2011). These clinical impressions are corroborated by studies that have established links between OFC damage and activity and aggressive behavior (e.g., Beyer, Muente, Goettlich, & Kraemer, 2015; Butter, Snyder, & McDonald, 1970; Raleigh, Steklis, Ervin, Kling, & McGuire, 1979), processing of social information (e.g., Azzi, Sirigu, & Duhamel, 2012; Ishai, 2007; O'Doherty, Winston, et al., 2003; Perry et al., 2016), emotional information (e.g., Bechara, 2004; Izquierdo, Suda, & Murray, 2005; Kumfor, Irish, Hodges, & Piguet, 2013; Schutter & van Honk, 2006), and risky or impulsive behavior (Bechara, Damasio, & Damasio, 2000). In addition, some studies have indicated that OFC may be important for long-term memory (Frey & Petrides, 2002; Meunier, Bachevalier, & Mishkin, 1997; Petrides, 2007) and working memory (Barbey, Koenigs, & Grafman, 2011).

Given the complex anatomy of the OFC, such diversity of findings is not surprising. Previous research has shown that lesion effects can be the result of damage to passing fibers rather than damage to OFC per se (Rudebeck et al., 2013). Moreover, some effects may result from codamage to other areas (Noonan, Sallet, Rudebeck, Buckley, & Rushworth, 2010), and incorporating the effects of connected areas is generally an important approach for understanding OFC's function (Rempel-Clower, 2007). In addition, it is certainly possible that some subregions of OFC have functions outside of the domain of decision-making.

While our framework does not aim to account for the entirety of the function of this large brain area, it is noteworthy that some of the effects mentioned above are not orthogonal to our proposal. Studies investigating the processing of emotional and facial information have used stimuli that could also be interpreted as either having positive (smiling, attractiveness) or negative value (angry facial expression) (Chien, Wiehler, Spezio, & Gläscher, 2016; Winston, O'Doherty, Kilner, Perrett, & Dolan, 2007).

In general, inferring others' emotions and intent are the epitome of inference of a partially observable state. Indeed, these tasks require participants to process "subtle social and emotional cues required for the appropriate interpretation of events" (Cicerone & Tanenbaum, 1997, abstract), which could be affected by lesioned patients' inability to integrate partially observable information and current sensory input into a suitable state representation. Moreover, the reported association with working memory seemed to be specific to n-back tasks (Barbey et al., 2011) that also require decision-making and are similar to the task we used to test our theory in humans (Schuck et al., 2016).

Apart from these issues of integrating all available evidence, one important avenue for future research is to specify predictions about the OFC representations that drive decision-making in different tasks and how these OFC representations are affected by lesions and disease. Existing computational models therefore need to be specified in order to predict which hidden and observable aspects of the environment need to be incorporated into the current state in order to solve the task sufficiently (see Chapter 5 by Collins). Ideally, models and theory would also yield predictions about how these state representations develop during task learning, depending on the specific history of choices and experiences of each subject (Gershman, Norman, & Niv, 2015; Niv et al., 2015).

SUMMARY

In this review, we have focused on work that links the OFC to many aspects of decision-making. Several prominent findings have reported that individual neurons in OFC linearly increase their firing with the value of the chosen option. Other studies, however, have highlighted that many different aspects of the ongoing task are encoded in OFC's neural activity, that value neurons dynamically change which option they are encoding depending on the network state, and that value neurons make up only a small proportion of the OFC population in certain circumstances. Moreover, studies of OFC lesions are not consistent with the claim that the OFC is the brain's sole site of performing basic value-based decision-making. Rather, behavioral impairments seem to occur only under certain circumstances, e.g., when a change of previously learned associations is prompted by drastic changes in outcome value, when task rules require rapid switching between values or when learning must focus on the sensory properties of the outcome rather than its value. In all these cases, performance depends critically on the ability to *infer* a partially observable state of the environment, and maladaptive behavior becomes visible when the state changes in the absence of any sensory cues, and previously learned values no longer apply.

We therefore suggest that the role of the OFC in decision-making is to represent the current state of the environment, in particular, if that state is partially observable. We have presented several lines of evidence, including lesion studies, electrophysiological recordings, computational modeling, and fMRI, that support our framework. We also outlined avenues for future research that should seek to directly investigate to what extent previously reported associations between OFC and cognitive functions outside of the domain of decision-making could also result from changes in partially observable state representations (e.g., working memory or personality changes). In addition, taking into account anatomical diversity within OFC, cross-species anatomical differences and, in the case of lesion studies, careful scrutiny of the nature of OFC insults (e.g., if passing fibers have been damaged) might clarify the origins of some of the diversity of functions associated with the OFC. Crucially, we argued that computational models that specify the representations underlying successful decision-making need to be advanced and tested against empirical data. Together with the existing evidence, these efforts promise to yield unprecedented insight into the functions of an elusive brain area.

ACKNOWLEDGMENTS

We thank Matthew Glasser for advice in making the OFC map shown in Fig. 12.1 and Sam Chien for feedback on this manuscript. This work was supported by the grant 1R01DA042065 from the National Institution on Drug Abuse at NIH awarded to YN.

REFERENCES

Azzi, J. C. B., Sirigu, A., & Duhamel, J.-R. (2012). Modulation of value representation by social context in the primate orbitofrontal cortex. *Proceedings of the National Academy of Sciences of the United States of America, 109*(6), 2126–2131.

Barbey, A. K., Koenigs, M., & Grafman, J. (2011). Orbitofrontal contributions to human working memory. *Cerebral Cortex, 21*(4), 789–795.

Bechara, A. (2004). The role of emotion in decision-making: Evidence from neurological patients with orbitofrontal damage. *Brain and Cognition, 55*(1), 30–40.

Bechara, A., Damasio, H., & Damasio, A. R. (2000). Emotion, decision making and the orbitofrontal cortex. *Cerebral Cortex, 10*(3), 295–307.

Beyer, F., Muente, T. F., Goettlich, M., & Kraemer, U. M. (2015). Orbitofrontal cortex reactivity to angry facial expression in a social interaction correlates with aggressive behavior. *Cerebral Cortex, 25*(9), 3057–3063.

Boorman, E. D., Rajendran, V. G., O'Reilly, J. X., & Behrens, T. E. (2016). Two anatomically and computationally distinct learning signals predict changes to stimulus-outcome associations in hippocampus. *Neuron, 89*(6), 1343–1354.

Bradfield, L. A., Dezfouli, A., van Holstein, M., Chieng, B., & Balleine, B. W. (2015). Medial orbitofrontal cortex mediates outcome retrieval in partially observable task situations. *Neuron*, 1–13.

Burke, K. A., Franz, T. M., Miller, D. N., & Schoenbaum, G. (2008). The role of the orbitofrontal cortex in the pursuit of happiness and more specific rewards. *Nature, 454*(7202), 340–344.

Butter, C. M. (1969). Perseveration in extinction and in discrimination reversal tasks following selective frontal ablations in *Macaca mulatta*. *Physiology & Behavior, 4*, 163–171.

Butter, C. M., Snyder, D. R., & McDonald, J. A. (1970). Effects of orbital frontal lesions on aversive and aggressive behaviors in rhesus monkeys. *Journal of Comparative and Physiological Psychology, 72*(1), 132–144.

Cai, M. B., Schuck, N. W., Pillow, J. W., & Niv, Y. (2016). A Bayesian method for reducing bias in neural representational similarity analysis. *Neural Information Processing Systems, 29*.

Cavada, C., Compañy, T., Tejedor, J., Cruz-Rizzolo, R. J., & Reinoso-Suárez, F. (2000). The anatomical connections of the macaque monkey orbitofrontal cortex. A review. *Cerebral Cortex, 10*(3), 220–242.

Cavada, C., & Schultz, W. (2000). The mysterious orbitofrontal cortex. Foreword. *Cerebral Cortex, 10*(3), 205.

Chan, S. C. Y., Niv, Y., & Norman, K. A. (2016). A probability distribution over latent causes, in the orbitofrontal cortex. *Journal of Neuroscience, 36*(30), 7817–7828.

Chien, S., Wiehler, A., Spezio, M., & Gläscher, J. (2016). Congruence of inherent and acquired values facilitates reward-based decision-making. *The Journal of Neuroscience, 36*(18), 5003–5012.

Chudasama, Y., Kralik, J. D., & Murray, E. A. (2007). Rhesus monkeys with orbital prefrontal cortex lesions can learn to inhibit prepotent responses in the reversed reward contingency task. *Cerebral Cortex, 17*(5), 1154–1159.

Chudasama, Y., & Robbins, T. W. (2003). Dissociable contributions of the orbitofrontal and infralimbic cortex to Pavlovian autoshaping and discrimination reversal Learning: Further evidence for the functional heterogeneity of the rodent frontal cortex. *Journal of Neuroscience, 23*(25), 8771–8780.

Cicerone, K. D., & Tanenbaum, L. N. (1997). Disturbance of social cognition after traumatic orbitofrontal brain injury. *Archives of Clinical Neuropsychology, 12*(2), 173–188.

Critchley, H. D., & Rolls, E. T. (1996). Hunger and satiety modify the responses of olfactory and visual neurons in the primate orbitofrontal cortex. *Journal of Neurophysiology, 75*(4), 1673–1686.

Daw, N. D., Courville, A. C., & Touretzky, D. S. (2006). Representation and timing in theories of the dopamine system. *Neural Computation, 18*, 1637–1677.

Dayan, P., & Daw, N. D. (2008). Decision theory, reinforcement learning, and the brain. *Cognitive, Affective, & Behavioral Neuroscience, 8*(4), 429–453.

Elliott, R., Dolan, R. J., & Frith, C. D. (2000). Dissociable functions in the medial and lateral orbitofrontal cortex: Evidence from human neuroimaging studies. *Cerebral Cortex, 10*, 308–317.

Farovik, A., Place, R. J., Mckenzie, S., Porter, B., Munro, C. E., & Eichenbaum, H. (2015). Orbitofrontal cortex encodes memories within value-based schemas and represents contexts that guide memory retrieval. *Journal of Neuroscience, 35*(21), 8333–8344.

Freedman, M., Black, S., Ebert, P., & Binns, M. (1998). Orbitofrontal function, object alternation and perseveration. *Cerebral Cortex, 8*(1), 18–27.

Frey, S., & Petrides, M. (2002). Orbitofrontal cortex and memory formation. *Neuron, 36*(1), 171–176.

Fuster, J. M. (1997). *The prefrontal cortex*. New York: Raven Press.

Gallagher, M., McMahan, R. W., & Schoenbaum, G. (1999). Orbitofrontal cortex and representation of incentive value in associative learning. *The Journal of Neuroscience, 19*(15), 6610–6614.

Galleguillos, L., Parrao, T., & Delgado, C. (2011). Personality disorder related to an acute orbitofrontal lesion in multiple sclerosis. *Journal of Neuropsychiatry and Clinical Neurosciences, 23*(4), 7.

Gershman, S. J., Norman, K. A., & Niv, Y. (2015). Discovering latent causes in reinforcement learning. *Current Opinion in Behavioral Sciences, 5*, 43–50.

Glasser, M. F., Coalson, T. S., Robinson, E. C., Hacker, C. D., Harwell, J., Yacoub, E., … Van Essen, D. C. (2016). A multi-modal parcellation of human cerebral cortex. *Nature, 536*(7615), 171–178.

Gottfried, J. A., O'Doherty, J., & Dolan, R. J. (2003). Encoding predictive reward value in human amygdala and orbitofrontal cortex. *Science, 301*(5636), 1104–1107.

Hare, T. A., O'Doherty, J., Camerer, C. F., Schultz, W., & Rangel, A. (2008). Dissociating the role of the orbitofrontal cortex and the striatum in the computation of goal values and prediction errors. *Journal of Neuroscience, 28*(22), 5623–5630.

Heilbronner, S. R., Rodriguez-Romaguera, J., Quirk, G. J., Groenewegen, H. J., & Haber, S. N. (2016). Circuit-based corticostriatal homologies between rat and primate. *Biological Psychiatry, 80*(7), 509–521.

Howard, J. D., Gottfried, J. A., Tobler, P. N., & Kahnt, T. (2015). Identity-specific coding of future rewards in the human orbitofrontal cortex. *Proceedings of the National Academy of Sciences of the United States of America, 112*(16), 5195–5200.

Howard, J. D., & Kahnt, T. (2017). Identity-specific reward representations in orbitofrontal cortex are modulated by selective devaluation. *Journal of Neuroscience, 37*(10), 3473–16.

Ishai, A. (2007). Sex, beauty and the orbitofrontal cortex. *International Journal of Psychophysiology, 63*(2), 181–185.

Izquierdo, A., Suda, R. K., & Murray, E. A. (2004). Bilateral orbital prefrontal cortex lesions in rhesus monkeys disrupt choices guided by both reward value and reward contingency. *Journal of Neuroscience, 24*(34), 7540–7548.

Izquierdo, A., Suda, R. K., & Murray, E. A. (2005). Comparison of the effects of bilateral orbital prefrontal cortex lesions and amygdala lesions on emotional responses in rhesus monkeys. *Journal of Neuroscience, 25*(37), 8534—8542.

Kaelbling, L. P., Littman, M. L., & Moore, A. W. (1996). Reinforcement learning: A survey. *Journal of Artificial Intelligence Research, 4*, 237—285.

Kahnt, T., Chang, L. J., Park, S. Q., Heinzle, J., & Haynes, J.-D. (2012). Connectivity-based parcellation of the human orbitofrontal cortex. *Journal of Neuroscience, 32*(18), 6240—6250.

Kamin, L. J. (1969). *Predictability, surprise, attention and conditioning.*

Kaplan, R., Schuck, N. W., & Doeller, C. F. (2017). The role of mental maps in decision-making. *Trends in Neurosciences, 40*(5), 256—259.

Kazama, A., & Bachevalier, J. (2009). Selective aspiration or neurotoxic lesions of orbital frontal areas 11 and 13 spared monkeys' performance on the object discrimination reversal task. *Journal of Neuroscience, 29*(9), 2794—2804.

Kepecs, A., Uchida, N., Zariwala, H. A., & Mainen, Z. F. (2008). Neural correlates, computation and behavioural impact of decision confidence. *Nature, 455*(7210), 227—231.

Kringelbach, M. L. (2005). The human orbitofrontal cortex: Linking reward to hedonic experience. *Nature Reviews Neuroscience, 6*(9), 691—702.

Kumfor, F., Irish, M., Hodges, J. R., & Piguet, O. (2013). The orbitofrontal cortex is involved in emotional enhancement of memory: Evidence from the dementias. *Brain: A Journal of Neurology, 136*(10), 2992—3003.

Lak, A., Costa, G. M., Romberg, E., Koulakov, A. A., Mainen, Z. F., & Kepecs, A. (2014). Orbitofrontal cortex is required for optimal waiting based on decision confidence. *Neuron, 84*(1), 190—201.

Lak, A., Nomoto, K., Keramati, M., Sakagami, M., & Kepecs, A. (2017). Midbrain dopamine neurons signal belief in choice accuracy during a perceptual decision. *Current Biology, 27*(6), 821—832.

Langdon, A. J., Sharpe, M., Schoenbaum, G., & Niv, Y. (2018). Model-based predictions for dopamine. *Current Opinion in Neurobiology, 49*, 1—7.

Lopatina, N., McDannald, M. A., Styer, C. V., Sadacca, B. F., Cheer, J. F., & Schoenbaum, G. (2015). Lateral orbitofrontal neurons acquire responses to up-shifted, downshifted, or blocked cues during unblocking. *eLife, 4*, 1—17.

McDannald, M. A., Esber, G. R., Wegener, M. A., Wied, H. M., Liu, T.-L., Stalnaker, T. A., ... Schoenbaum, G. (2014). Orbitofrontal neurons acquire responses to valueless' Pavlovian cues during unblocking. *eLife, 3*, e02653.

McDannald, M. A., Lucantonio, F., Burke, K. A., Niv, Y., & Schoenbaum, G. (2011). Ventral striatum and orbitofrontal cortex are both required for model-based, but not model-free, reinforcement learning. *Journal of Neuroscience, 31*(7), 2700—2705.

Meunier, M., Bachevalier, J., & Mishkin, M. (1997). Effects of orbital frontal and anterior cingulate lesions on object and spatial memory in rhesus monkeys. *Neuropsychologia, 35*(7), 999—1015.

Michie, D., & Chambers, R. A. (1968). BOXES: An experiment in adaptive control. In E. Dale, & D. Michie (Eds.), *Machine intelligence 2* (pp. 137—152). Edinburgh: Oliver and Boyd.

Mishkin, M., Vest, B., Waxler, M., & Rosvold, H. (1969). A re-examination of the effects of frontal lesions on object alternation. *Neuropsychologia, 7*, 357—363.

Murray, E. A., O'Doherty, J. P., & Schoenbaum, G. (2007). What we know and do not know about the functions of the orbitofrontal cortex after 20 years of cross-species studies. *Journal of Neuroscience, 27*(31), 8166—8169.

Niv, Y., Daniel, R., Geana, A., Gershman, S. J., Leong, Y. C., Radulescu, A., & Wilson, R. C. (2015). Reinforcement learning in multidimensional environments relies on attention mechanisms. *Journal of Neuroscience, 35*(21), 8145—8157.

Niv, Y., & Langdon, A. (2016). Reinforcement learning with Marr. *Current Opinion in Behavioral Sciences, 11*, 67—73.

Nogueira, R., Abolafia, J. M., Drugowitsch, J., Balaguer-Ballester, E., Sanchez-Vives, M. V., & Moreno-Bote, R. (2017). Lateral orbitofrontal cortex anticipates choices and integrates prior with current information. *Nature Communications, 8*, 14823.

Noonan, M. P., Sallet, J., Rudebeck, P. H., Buckley, M. J., & Rushworth, M. F. (2010). Does the medial orbitofrontal cortex have a role in social valuation? *European Journal of Neuroscience, 31*(12), 2341–2351.

Noonan, M. P., Walton, M. E., Behrens, T. E. J., Sallet, J., Buckley, M. J., & Rushworth, M. F. S. (2010). Separate value comparison and learning mechanisms in macaque medial and lateral orbitofrontal cortex. *Proceedings of the National Academy of Sciences of the United States of America, 107*(47), 20547–20552.

Ogawa, M., van der Meer, M. A. A., Esber, G. R., Cerri, D. H., Stalnaker, T. A., & Schoenbaum, G. (2013). Risk-responsive orbitofrontal neurons track acquired salience. *Neuron, 77*(2), 251–258.

Öngür, D., Ferry, A. T., & Price, J. L. (2003). Architectonic subdivision of the human orbital and medial prefrontal cortex. *Journal of Comparative Neurology, 460*(3), 425–449.

O'Doherty, J., Critchley, H., Deichmann, R., & Dolan, R. J. (2003). Dissociating valence of outcome from behavioral control in human orbital and ventral prefrontal cortices. *Journal of Neuroscience, 23*(21), 7931–7939.

O'Doherty, J., Winston, J., Critchley, H., Perrett, D., Burt, D. M., & Dolan, R. J. (2003). Beauty in a smile: The role of medial orbitofrontal cortex in facial attractiveness. *Neuropsychologia, 41*(2), 147–155.

Padoa-Schioppa, C., & Assad, J. A. (2006). Neurons in the orbitofrontal cortex encode economic value. *Nature, 441*, 223–226.

Padoa-Schioppa, C., & Assad, J. A. (2008). The representation of economic value in the orbitofrontal cortex is invariant for changes of menu. *Nature Neuroscience, 11*(1), 95–102.

Passingham, R. E., & Wise, S. E. (2012). *The Neurobiology of the Prefrontal Cortex*. New York: Oxford University Press.

Perry, A., Lwi, S. J., Verstaen, A., Dewar, C., Levenson, R. W., & Knight, R. T. (2016). The role of the orbitofrontal cortex in regulation of interpersonal space: Evidence from frontal lesion and frontotemporal dementia patients. *Social Cognitive and Affective Neuroscience, 11*(8), 1894–1901.

Petrides, M. (2007). The orbitofrontal cortex: Novelty, deviation from expectation, and memory. *Annals of the New York Academy of Sciences, 1121*, 33–53.

Pickens, C. L., Saddoris, M. P., Setlow, B., Gallagher, M., Holland, P. C., & Schoenbaum, G. (2003). Different roles for orbitofrontal cortex and basolateral amygdala in a reinforcer devaluation task. *Journal of Neuroscience, 23*(35), 11078–11084.

Plassmann, H., O'Doherty, J., & Rangel, A. (2007). Orbitofrontal cortex encodes willingness to pay in everyday economic transactions. *Journal of Neuroscience, 27*(37), 9984–9988.

Preuss, T. M. (1995). Do rats have prefrontal Cortex? The Rose-Woolsey-Akert program reconsidered. *Journal of Cognitive Neuroscience, 7*(1), 1–24.

Raleigh, M. J., Steklis, H. D., Ervin, F. R., Kling, A. S., & McGuire, M. T. (1979). The effects of orbitofrontal lesions on the aggressive behavior of vervet monkeys (*Cercopithecus aethiops sabaeus*). *Experimental Neurology, 66*(1), 158–168.

Rao, R. P. N. (2010). Decision making under uncertainty: A neural model based on partially observable Markov decision processes. *Frontiers in Computational Neuroscience, 4*(11), 146.

Rempel-Clower, N. L. (2007). Role of orbitofrontal cortex connections in emotion. *Annals of the New York Academy of Sciences, 1121*, 72–86.

Rich, E. L., & Wallis, J. D. (2016). Decoding subjective decisions from orbitofrontal cortex. *Nature Neuroscience, 19*(7), 973–980.

Rodriguez, A., Parr, R., & Koller, D. (1999). Reinforcement learning using approximate belief states. *Advances in Neural Information Processing Systems, 12*, 1036–1042.

Rudebeck, P. H., & Murray, E. A. (2011). Balkanizing the primate orbitofrontal cortex: Distinct subregions for comparing and contrasting values. *Annals of the New York Academy of Sciences, 1239*, 1–13.

Rudebeck, P. H., Saunders, R. C., Prescott, A. T., Chau, L. S., & Murray, E. A. (2013). Prefrontal mechanisms of behavioral flexibility, emotion regulation and value updating. *Nature Neuroscience, 16*(8), 1140–1145.

Rushworth, M. F. S., Noonan, M. P., Boorman, E. D., Walton, M. E., & Behrens, T. E. (2011). Frontal cortex and reward-guided learning and decision-making. *Neuron, 70*(6), 1054–1069.

Samejima, K., & Doya, K. (2007). Multiple representations of belief states and action values in corticobasal ganglia loops. *Annals of the New York Academy of Sciences, 1104*, 213–228.

Schapiro, A. C., Kustner, L. V., & Turk-Browne, N. B. (2012). Shaping of object representations in the human medial temporal lobe based on temporal regularities. *Current Biology, 22*(17), 1622–1627.

Schoenbaum, G., & Eichenbaum, H. (1995). Information coding in the rodent prefrontal cortex. II. Ensemble activity in orbitofrontal cortex. *Journal of Neurophysiology, 74*(2), 751–762.

Schoenbaum, G., Roesch, M. R., & Stalnaker, T. A. (2006). Orbitofrontal cortex, decision-making and drug addiction. *Trends in Neurosciences, 29*(2), 116–124.

Schoenbaum, G., Takahashi, Y., Liu, T.-L., & McDannald, M. A. (2011). Does the orbitofrontal cortex signal value? *Annals of the New York Academy of Sciences, 1239,* 87–99.

Schuck, N. W., Cai, M. B., Wilson, R. C., & Niv, Y. (2016). Human orbitofrontal cortex represents a cognitive map of state space. *Neuron, 91*(6), 1402–1412.

Schuck, N. W., Gaschler, R., Wenke, D., Heinzle, J., Frensch, P. A., Haynes, J.-D., & Reverberi, C. (2015). Medial prefrontal cortex predicts internally driven strategy shifts. *Neuron, 86*(1), 331–340.

Schutter, D. J. L. G., & van Honk, J. (2006). Increased positive emotional memory after repetitive transcranial magnetic stimulation over the orbitofrontal cortex. *Journal of Psychiatry & Neuroscience, 31*(2), 101–104.

Stachenfeld, K. L., Botvinick, M. M., & Gershman, S. J. (2017). The hippocampus as a predictive map. *Nature Neuroscience, 20*(11), 1643–1653.

Stalnaker, T. A., Berg, B., Aujla, N., & Schoenbaum, G. (2016). Cholinergic interneurons use orbitofrontal input to track beliefs about current state. *Journal of Neuroscience, 36*(23), 6242–6257.

Stalnaker, T. A., Cooch, N. K., McDannald, M. A., Liu, T.-L., Wied, H., & Schoenbaum, G. (2014). Orbitofrontal neurons infer the value and identity of predicted outcomes. *Nature Communications, 5,* 3926.

Stalnaker, T. A., Cooch, N. K., & Schoenbaum, G. (2015). What the orbitofrontal cortex does not do. *Nature Neuroscience, 18*(5), 620–627.

Stalnaker, T. A., Franz, T. M., Singh, T., & Schoenbaum, G. (2007). Basolateral amygdala lesions abolish orbitofrontal-dependent reversal impairments. *Neuron, 54*(1), 51–58.

Starkweather, C. K., Babayan, B. M., Uchida, N., & Gershman, S. J. (2017). Dopamine reward prediction errors reflect hidden-state inference across time. *Nature Neuroscience, 2*(July 2016), 1–11.

Sutton, R. S., & Barto, A. G. (1998). *Reinforcement learning: An introduction.* Cambridge: Cambridge University Press.

Takahashi, Y. K., Langdon, A. J., Niv, Y., & Schoenbaum, G. (2016). Temporal specificity of reward prediction errors signaled by putative dopamine neurons in rat VTA depends on ventral striatum. *Neuron, 91*(1), 182–193.

Takahashi, Y. K., Roesch, M. R., Wilson, R. C., Toreson, K., O'Donnell, P., Niv, Y., & Schoenbaum, G. (2011). Expectancy-related changes in firing of dopamine neurons depend on orbitofrontal cortex. *Nature Neuroscience, 14*(12), 1590–1597.

Thorpe, S. J., Rolls, E. T., & Maddison, S. (1983). The orbitofrontal cortex: Neuronal activity in the behaving monkey. *Experimental Brain Research, 49*(1), 93–115.

Tremblay, L., & Schultz, W. (1999). Relative reward preference in primate orbitofrontal cortex. *Nature, 398*(6729), 704–708.

van Bergen, R. S., Ji Ma, W., Pratte, M. S., & Jehee, J. F. (2015). Sensory uncertainty decoded from visual cortex predicts behavior. *Nature Neuroscience, 18*(12), 1728–1730.

Walton, M. E., Behrens, T. E. J., Buckley, M. J., Rudebeck, P. H., & Rushworth, M. F. S. (2010). Separable learning systems in the macaque brain and the role of orbitofrontal cortex in contingent learning. *Neuron, 65*(6), 927–939.

Walton, M. E., Behrens, T. E. J., Noonan, M. P., & Rushworth, M. F. S. (2011). Giving credit where credit is due: Orbitofrontal cortex and valuation in an uncertain world. *Annals of the New York Academy of Sciences, 1239*(1), 14–24.

Watson, K. K., & Platt, M. L. (2012). Social signals in primate orbitofrontal cortex. *Current Biology, 22*(23), 2268–2273.

West, E. A., DesJardin, J. T., Gale, K., & Malkova, L. (2011). Transient inactivation of orbitofrontal cortex blocks reinforcer devaluation in macaques. *Journal of Neuroscience, 31*(42), 15128–15135.

Wilson, R. C., Takahashi, Y. K., Schoenbaum, G., & Niv, Y. (2014). Orbitofrontal cortex as a cognitive map of task space. *Neuron, 81*(2), 267–279.

Winston, J. S., O'Doherty, J., Kilner, J. M., Perrett, D. I., & Dolan, R. J. (2007). Brain systems for assessing facial attractiveness. *Neuropsychologia, 45*(1), 195–206.

Wise, S. P. (2008). Forward frontal fields: Phylogeny and fundamental function. *Trends in Neurosciences, 31*(12), 599–608.

CHAPTER 13

The Development of Goal-Directed Decision-Making

Hillary A. Raab[1], Catherine A. Hartley[1,2]

[1]Department of Psychology, New York University, New York, NY, United States; [2]Center for Neural Science, New York University, New York, NY, United States

INTRODUCTION

In recent years, an extensive literature stemming from the study of learning in animal models has begun to elucidate the mechanisms that underpin goal-directed behavior (Dolan & Dayan, 2013). This conceptual framework distinguishes two types of "instrumental" learning, or ways in which actions can be facilitated based on their resulting rewards or punishments. A "goal-directed" action is a behavior driven by an expectation that it is likely to bring about a desired outcome (Dickinson, 1985). Goal-directed decisions leverage causal knowledge of the potential consequences of actions to flexibly pursue a current goal. Such deliberate action selection is distinguished from "habitual" behavior, in which an action is reflexively elicited by the cues or contexts associated with its prior successful performance. Whereas goal-directed actions are selected based on expectations of their consequent outcomes, habits are thought to stem from the formation of stimulus—response associations, reinforced by reward and automatically elicited by their antecedent stimuli.

This distinction between two types of instrumental action has its origins in a historic scientific debate. The dominant behaviorist account of instrumental behavior in the early 20th century proposed that instrumental action reflected an assembly of stimulus—response associations, stamped in through reinforcement, and reflexively elicited by environmental cues or contexts. In the mid-20th century Edward Tolman advocated for an alternative view, proposing instead that animals and humans form mental models of their environments (cognitive models or "maps") that can be flexibly consulted and recruited to pursue a current goal (Tolman, 1948). In the ensuing decades, a large interdisciplinary empirical literature has provided support for this conceptualization of goal-directed learning. Behavioral tasks have been designed to test for the key features of goal-directed behavior and examine the experimental factors that influence the balance between goal-directed versus habitual action (Dickinson, 1985). Convergent computational, psychological, and neuroscientific literatures have begun to characterize the diverse algorithmic, cognitive, and neurobiological processes that enable goal-directed action (Dolan & Dayan, 2013; Doll, Simon, & Daw, 2012).

Goal-Directed Decision Making
ISBN 978-0-12-812098-9, https://doi.org/10.1016/B978-0-12-812098-9.00013-9
279

To date, the vast majority of this empirical literature has examined goal-directed behavior in adult humans and animals. In contrast, the typical development of goal-directed decision-making and the neurocognitive processes that underlie its developmental trajectory have not been widely studied. In this chapter, we will review empirical findings that illustrate marked changes in goal-directed behavior over the course of development from infancy to young adulthood. We will discuss developmental changes in the component cognitive processes and underlying neural circuits that may contribute to these shifts in behavior, highlighting the current gaps in our understanding. We begin by describing tasks designed to dissociate goal-directed behavior from habitual action. We then present findings from developmental studies employing these tasks, which reveal age-related changes in goal-directed decision-making. While we focus primarily on data exploring human development, we also integrate relevant developmental studies in animal models. We briefly outline a provisional model of the neurocircuitry implicated in goal-directed decision-making from studies in adult humans and animals and give an overview of the dynamic changes that occur within these circuits from childhood to adulthood. We then discuss what is known about the development of the component cognitive processes involved in the construction and use of the cognitive models that underpin goal-directed action, as well as the neural correlates of these developmental changes.

ASSAYS OF GOAL-DIRECTED BEHAVIOR

The goal-directedness of instrumental behavior can be assessed in experimental paradigms gauging the degree to which a decision to take an action is governed by knowledge of its likely causal consequences (i.e., action—outcome contingencies) as well as the current desirability of these expected outcomes (i.e., outcome value). Behavior that stems from the formation of habit-like stimulus—response associations is insensitive to manipulations of either of these properties, allowing for the dissociation between goal-directed and habitual behavior.

One such experimental manipulation is outcome revaluation. In these tasks, an animal learns actions that can effectively bring about the delivery of rewards. For example, an animal might learn that pressing one lever will yield a food pellet, while pressing a second lever will deliver water from a spout. In the next stage of the study, the value of one outcome is altered. This change in value may be brought about through changes in the animal's motivational state (e.g., altering the animal's degree of hunger or thirst) or through manipulations that alter the intrinsic value of the outcome (e.g., pairing a food with pharmacologically induced illness or revealing a new rewarding use for an object). The effect of this manipulation on performance of the instrumental action is then tested. Importantly, this test is performed in extinction (i.e., when no further outcomes are delivered), which ensures that instrumental behavior can only be informed by outcome knowledge learned during initial training, as well as by any change in representation of the current value of the outcome. To the extent that an action is driven by consideration of the current desirability of its likely outcome, a decrease in outcome value should result in an attenuation of the instrumental response, whereas an increase in value should yield a corresponding increase in performance of the action. In contrast, habitual actions, which are thought to involve no consideration of the likely outcome, are insensitive to such changes in value. For example, an animal that continues to press a lever associated with food delivery after having eaten to satiety would reveal its actions to be habitual, whereas a cessation of lever pressing would reflect a goal-directed evaluation.

A second assay of the goal-directedness of action involves altering the learned contingency between an action and a desired outcome. A behavior is only an effective means of bringing about a desired outcome if the probability of obtaining a reinforcer when a specific action is performed is greater than the probability of reinforcement when that action is not taken. Such action—outcome contingencies can be degraded through the provision of noncontingent reward (i.e., delivering a food reinforcer in the absence of any instrumental action). If an action is habitual, the lack of consideration of action—outcome relationships will reduce sensitivity to the causal ineffectiveness of the action. Thus, a behavior that persists when its action—outcome contingency has been degraded is considered to be habitual.

Two classes of reinforcement learning algorithms have been proposed to approximate the neural computations underlying goal-directed and habitual behavior and to reproduce their key behavioral properties (Daw, Niv, & Dayan, 2005). "Model-based" algorithms select actions via a flexible but computationally and representationally intensive "tree-search" process of evaluating potential state transitions and outcomes to determine the action most likely to yield reward. In contrast, "model-free" algorithms recruit trial-and-error feedback to update a stored action value associated with a stimulus, allowing the most highly valued action to be readily elicited when the stimulus is encountered.

A sequential decision-making task (the "two-step task") was designed to dissociate these two learning processes (Daw, Gershman, Seymour, Dayan, & Dolan, 2011) and has recently been adapted to a child-friendly format (Decker, Otto, Daw, & Hartley, 2016). On each trial of the task (Fig. 13.1A), participants make a first-stage choice between two stimuli (spaceships), which is followed by a probabilistic transition to one of two second-stage states (a red or purple planet). In stage two, participants choose between two second-stage choice options (red or purple aliens), each of which is associated with a probability of yielding reward (space treasure). Reward probabilities change slowly and independently, encouraging participants to explore second-stage options throughout the task. Importantly, the probabilistic transition between first- and second-stage states (i.e., each spaceship commonly (70%) goes to one planet, and rarely (30%) to the other) creates a task structure that enables the distinction between goal-directed (model-based) and habitual (model-free) choices. Whereas a goal-directed chooser uses a cognitive

(A)

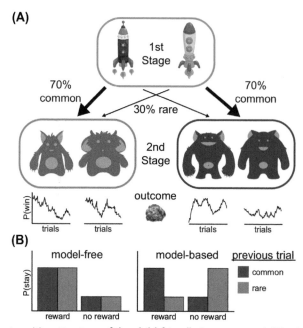

Figure 13.1 (A) The transition structure of the child-friendly "two-step task." Participants make a first-stage choice between two stimuli (spaceships), which is followed by a probabilistic transition to one of two second-stage states (a red or purple planet). They then choose between two second-stage choice options (red or purple aliens), each of which is associated with a slowly varying probability of receiving reward (space treasure). (B) The probability of repeating a first-stage choice for an idealized model-free (left) or model-based (right) chooser is shown as a function of previous transition type (common or rare) and outcome (rewarded or unrewarded). *(From Decker, J. H., Otto, A. R., Daw, N. D., & Hartley, C. A. (2016). From creatures of habit to goal-directed learners tracking the developmental emergence of model-based reinforcement learning.* Psychological Science, 27, 848–858.)

model of the transitions and rewards in the task to select actions, a habitual model-free chooser, who does not recruit such a cognitive model, simply repeats previously rewarded actions. Thus, the influence of the previous trial on the first-stage choice of the next trial depends on one's learning strategy (Fig. 13.1B). A habitual, model-free learner is likely to repeat a previously rewarded first-stage choice regardless of the transition type that led to the reward (i.e., a win-stay effect; Fig. 13.1B, left). In contrast, a goal-directed chooser takes into account the state transition structure, reflected by an interactive effect of transition type (common vs. rare) and reward on "stay" decisions (Fig. 13.1B, right). For example, a habitual chooser is more likely to repeat a first-stage choice following a rare transition that led to a reward, whereas a goal-directed chooser is more likely to switch, choosing instead the spaceship that is most likely to lead to the previously rewarded planet.

Whereas tests of outcome revaluation and contingency degradation rely on a small number of trials following the task manipulation to assess goal-directedness, the two-step task can derive estimates of the degree to which learning is model-based or model-free using all trials of the task. Moreover, this paradigm lends itself to trial-by-trial computational modeling of the distinct model-based and model-free algorithms for action evaluation. These features of the task make it particularly useful for human neuroimaging analyses investigating the brain mechanisms underlying each learning strategy (Daw et al., 2011), as well as studies examining the effects of cognitive or affective manipulations on learning strategies (e.g., Otto, Gershman, Markman, & Daw, 2013; Otto, Raio, Chiang, Phelps, & Daw, 2013).

Each of these experimental assays of goal-directed action has been used in recent studies in adult humans to better understand the principles governing goal-directed and habitual learning and action selection. Below, we present findings from the few studies that have leveraged these tasks to begin to characterize changes in goal-directed decision-making across development.

DEVELOPMENT OF GOAL-DIRECTED INSTRUMENTAL ACTION

Recent studies have used experimental assays including outcome revaluation, contingency degradation, and sequential reinforcement learning tasks in both humans and animals at differing ages to better characterize developmental changes in the learning and reliance upon goal-directed versus habitual action.

In one study examining the sensitivity of learned actions to outcome devaluation (Kenward, Folke, Holmberg, Johansson, & Gredebäck, 2009), young children (aged 14, 19, or 24 months) learned to press a button to release an object from a box. In a second stage of the experiment, the children learned how to use an object as part of an enjoyable game (either a ball could be placed into a series of chutes, or a wooden block could be used to activate a music box). For only one of the two groups of children, the object used in the play demonstration was the same object that the child previously learned to release from the box. Thus, learning that the object could be used for play constituted an increase in its value. In a test phase, the objects used in the play demonstration were removed so that children's responses could be assessed in extinction, and children's button presses on the box from the first stage were measured as an index of goal-directed action. The group of 24-month-old children, for whom the game involved the object that they had learned to obtain through a button press, exhibited a shorter latency to press the button and a greater number of button presses compared to 24-month-old children in the control group, who had not learned any valuable use for the object they had learned to release from the box. In contrast, 14- and 19-month-old children showed no effect of the outcome revaluation on their button press behavior, suggesting that their actions were not influenced by outcome expectancies.

Another outcome revaluation study tested children ranging in age from 18 to 48 months (Klossek, Russell, & Dickinson, 2008). The children learned that by making distinct responses on a touch screen, they could view an animated video clip from one of two cartoon series. After the responses were acquired, one video outcome was devalued by presenting children with four repeated viewings of each clip from that cartoon series. During the subsequent test phase, children were able to make responses on the screen, but no video clips were presented as a result of their actions (i.e., responses were tested in extinction). Children aged approximately 3 years and older reduced their performance of the action associated with the cartoon series that had been played repeatedly, relative to the nondevalued series, reflecting consideration of the decreased desirability of the extensively viewed cartoon series when making choices. In contrast, younger children showed no differentiation in their responses to the two cartoon series, suggesting that their actions were insensitive to this change in outcome value, and instead were driven by the previously learned reward associations.

Collectively, these studies suggest that the propensity to prospectively consider the likely outcome of an action and its current value when making choices increases with age. While in the first study we discussed (Kenward et al., 2009), sensitivity to changes in outcome value was only evident in 24-month-old children, in the second study (Klossek et al., 2008) this signature of goal-directedness only emerged in children at approximately 3 years of age. One factor that may contribute to these age differences in the emergence of goal-directed choice is task complexity. The cartoon task (Klossek et al., 2008), in which outcome value sensitivity became evident at a later age, required children to consider two potential actions that differed in value following devaluation, rather than a single action associated with a revalued outcome, as in the first study (Kenward et al., 2009). These results suggest that for tasks in which greater cognitive demand is required to bring to mind the outcomes of potential actions, goal-directed behavior may only be evident at later developmental time points.

A recent study examining sensitivity to outcome devaluation and contingency degradation in adolescent and adult rodents provides additional support for this proposal, suggesting that age-related increases in goal-directed behavior extend beyond adolescence (Naneix, Marchand, Scala, Pape, & Coutureau, 2012). Adolescent and adult rodents were trained to lever-press to obtain a food reward. Rodents in both age groups exhibited sensitivity to reward devaluation, decreasing their lever pressing after having unrestricted access to the food reward. However, when the action—outcome contingency was degraded through noncontingent delivery of the food reward, lever pressing in adult, but not adolescent, rodents decreased. When these adolescent rodents subsequently underwent the same procedure upon reaching adulthood, they exhibited sensitivity to contingency degradation at this later developmental time point. These findings suggest that there may be a more protracted developmental trajectory of goal-directed learning for behaviors that rely on greater cognitive demands. Another implication of this study is that sensitivity to outcome revaluation is dissociable from sensitivity to contingency degradation, and that outcome value sensitivity may be evident at earlier developmental time points than sensitivity to changes in the causal efficacy of an action.

A recent study in humans corroborates this evidence of continued increases in goal-directed evaluation into adulthood (Decker et al., 2016). Children, adolescents, and adults, aged 8 to 25 years, completed a child-friendly adaptation of the "two-step task," a sequential reinforcement learning task designed by Daw et al. (2011) to disentangle goal-directed and habitual learning strategies (see Fig. 13.1 above for task details). In this task, participants can pursue reward either by simply repeating previously rewarded actions (a strategy reflected by a main effect of previous reward on first-stage choices) or by using knowledge of the transition structure of the task to select an action most likely to lead to a goal (a strategy reflected by an interaction effect of previous reward and transition type on first-stage choices).

Whereas participants across ages were equally likely to demonstrate a main effect of reward from the previous trial on first-stage choices (the behavioral signature of model-free learning), the interaction effect, indicating use of the previous trial's transition structure and outcome to pursue reward (the model-based behavioral signature), increased with age (Fig. 13.2A and B). A computational analysis was conducted to determine whether participants' trial-by-trial choices in the task were better captured by a model-based evaluation algorithm, which selects actions via a forward search through a mental model of actions and outcomes, or a model-free algorithm that recruits trial-and-error feedback to efficiently update a cached value associated with an action, but maintains no representation of the outcomes themselves. Corroborating the behavioral finding that only considers the effect of the previous trial on subsequent choices, the computational analysis that incorporates learning from the full history of task trials showed that the tendency to engage in model-based, but not model-free, computations of value increased with age.

Figure 13.2 (A) Evidence of model-free learning (significant main effect of reward) is present from childhood onward, whereas model-based learning (significant reward × transition interaction effect) is only evident in adolescents and adults. (B) Model-based behavior increases with age from childhood into adulthood ($P < .001$). *(From Decker, J. H., Otto, A. R., Daw, N. D., & Hartley, C. A. (2016). From creatures of habit to goal-directed learners tracking the developmental emergence of model-based reinforcement learning. Psychological Science, 27, 848–858.)*

Collectively, these studies suggest that across a diverse set of tasks, the learning of motivated behaviors in individuals at younger ages is more likely to result in habit-like behaviors than goal-directed actions that are deliberately engaged to achieve a desired outcome. Based on a large literature in adult humans and animals, instrumental learning is proposed to initially begin as goal-directed but typically become habitual over time through extensive training of the instrumental response (Graybiel, 2008; Yin & Knowlton, 2006). The developmental studies reviewed above suggest that ontogenetically, this sequential progression from goal-directed to habitual behavior may be reversed. Behavioral responses may initially be acquired through the formation of stimulus—response associations that are reinforced by rewarding outcomes. Through repetition of these actions, causal understanding of action—outcome contingencies may be acquired and ultimately used to enable the selection of actions likely to achieve one's present goals. The developmental time point at which an individual shifts toward employing a goal-directed strategy may depend on both the intrinsic complexity of the task at hand and the development of the myriad cognitive processes involved in the formation and recruitment of a mental model of that task. In the following sections, we discuss developmental changes in the neural and cognitive processes that support goal-directed decision-making.

NEUROCIRCUITRY UNDERPINNING GOAL-DIRECTED BEHAVIOR AND ITS DEVELOPMENT

A provisional model of the neural substrates of goal-directed action has emerged from a convergent body of research in adult animals and humans. Below, we outline the neural circuitry implicated in goal-directed learning in adulthood and discuss how these circuits change over the course of development from childhood to adulthood.

Goal-directed behavior involves selecting and performing an action based on the current value of its outcome. The striatum, a subcortical region of the brain, is centrally implicated in the evaluation and selection of actions (Balleine & O'Doherty, 2010). Dopaminergic input to the striatum is proposed to encode a computational reward prediction error signal, reflecting the degree to which an experienced outcome is better or worse than expected. This signal can support a feedback-driven learning process through which action values, which differ according to one's learning strategy, are estimated (McClure, Berns, & Montague, 2003; O'Doherty, Dayan, Friston, Critchley, & Dolan, 2003; Schultz, Dayan, & Montague, 1997). The prediction error signals that support model-free and model-based learning depend on action values and thus vary by learning strategy. Correlates of both model-free and model-based prediction errors can be observed in the ventral striatum and are associated respectively with a greater tendency toward habitual or goal-directed choice behavior (Daw et al., 2011). Distinct subregions of the dorsal striatum underpin the selection of goal-directed and habitual actions (Balleine & O'Doherty, 2010; Balleine, Delgado, & Hikosaka, 2007; Yin & Knowlton, 2006).

The caudate nucleus (dorsomedial striatum) is implicated in the learning of contingent action—outcome associations that are central to goal-directed behavior (Haruno et al., 2004; Tricomi, Delgado, & Fiez, 2004; Yin, Ostlund, Knowlton, & Balleine, 2005). The putamen (dorsolateral striatum) plays a central role in the acquisition and expression of habitual actions (Tricomi, Balleine, & O'Doherty, 2009).

The action—outcome evaluations underlying goal-directed learning additionally integrate information about states and outcomes stemming from a more extensive network of regions, including the prefrontal cortex. The prefrontal cortex exhibits strong connectivity to the striatum, and lesions to corticostriatal connections disrupt the acquisition of goal-directed behavior (Hart, Bradfield, & Balleine, 2018). A number of functions critical for goal-directed action are thought to be supported by distinct areas of the prefrontal cortex. The ventromedial prefrontal cortex (vmPFC) is broadly implicated in the representation of value signals that inform motivated behavior (Dayan, Niv, Seymour, & Daw, 2006; Gläscher, Hampton, & O'Doherty, 2009; Hampton, Bossaerts, & O'Doherty, 2006; Tanaka et al., 2006; Valentin, Dickinson, & O'Doherty, 2007). The orbitofrontal cortex (OFC) has been proposed to represent associations between actions and their specific outcomes, forming a cognitive model of the task (McDannald et al., 2012; Schuck, Cai, Wilson, & Niv, 2016; Stalnaker, Cooch, & Schoenbaum, 2015; Valentin et al., 2007; Wilson, Takahashi, Schoenbaum, & Niv, 2014). Recruitment of this cognitive model enables expectations of a specific outcome to guide behavior. The OFC has been found to mediate goal-directed behavior in studies using assays such as contingency degradation and outcome revaluation (Gottfried, O'Doherty, & Dolan, 2003; Izquierdo, Suda, & Murray, 2004; McDannald et al., 2012; Valentin et al., 2007). The dorsolateral prefrontal cortex is also engaged during goal-directed learning (Smittenaar, FitzGerald, Romei, Wright, & Dolan, 2013) and may reflect the contribution of working memory and cognitive control processes, which are critical for the timely retrieval and maintenance of action—outcome associations recruited to obtain a current goal (Barch et al., 1997; Miller & Cohen, 2001). Thus, the prefrontal cortex supports diverse cognitive processes that underpin goal-directed behavior.

The hippocampus is also widely implicated in goal-directed behavior due to its central role in the learning of relationships between stimuli, events, and contexts (Shohamy & Turk-Browne, 2013; Zeithamova, Schlichting, & Preston, 2012). Learning and memory processes supported by the hippocampus are critical for the construction of the mental models of the environment that underpin goal-directed action (Pennartz, Ito, Verschure, Battaglia, & Robbins, 2011; Pfeiffer & Foster, 2013). Connectivity between the hippocampus, striatum, and vmPFC appears to support the integration of state information with knowledge about potential rewards or goal states during choice (Pennartz et al., 2011; Wimmer & Shohamy, 2012).

Over the course of development from childhood to adulthood, the neural circuitry underlying goal-directed behavior undergoes striking changes. The volume of cortical gray matter in the brain increases in early childhood, during which there is an overproduction of synapses (Huttenlocher, 1990). This proliferative period is followed by decreases in the thickness of cortex beginning in middle childhood (approximately 8 years of age) that continue into late adolescence or early adulthood, depending on the cortical region (Huttenlocher & Dabholkar, 1997; Mills et al., 2016). Cortical thinning, thought to reflect synaptic pruning as well as other cellular changes, occurs in a topographically organized and hierarchical manner. Sensory and motor regions undergo thinning first, followed by higher-order association cortices, with anterior and lateral regions of the prefrontal cortex exhibiting continued thinning into young adulthood (Gogtay et al., 2004; Shaw et al., 2008). In contrast to the marked decline in cortical gray matter from childhood to adulthood, the nonlinear volumetric changes in gray matter volume of subcortical structures—including the striatum and the hippocampus—are less pronounced in magnitude, and highly individually variable (Raznahan et al., 2014; Wierenga, Langen, Oranje, & Durston, 2014).

While subcortical structures exhibit greater volumetric stability, patterns of white matter connectivity between cortical and subcortical structures exhibit dynamic changes over the course of development. White matter volume in the brain increases into young adulthood (Giedd et al., 1999; Mills et al., 2016). These increases are thought to reflect myelination of white matter tracts, which increases their speed of information transmission. Studies examining the functional consequences of these developmental changes suggest that connectivity between the prefrontal cortex and striatum increases from childhood to young adulthood and contributes to age-related improvements in cognitive processes that underpin goal-directed behavior (van den Bos, Cohen, Kahnt, & Crone, 2012; van Duijvenvoorde, Achterberg, Braams, Peters, & Crone, 2016; Somerville & Casey, 2010; Somerville, Hare, & Casey, 2011). Connectivity between the prefrontal cortex and subcortical structures, including the amygdala and the hippocampus, exhibits similar age-related increases (Blankenship, Redcay, Dougherty, & Riggins, 2017; Gabard-Durnam et al., 2014). Network analyses of whole brain dynamics suggest that subcortical connectivity patterns become increasingly differentiated from childhood to adulthood (Gu et al., 2015).

Developmental changes are also evident in the dopaminergic system (Wahlstrom, White, & Luciana, 2010), which is thought to modulate behavioral flexibility (Grace, Floresco, Goto, & Lodge, 2007). Dopaminergic neurons innervate many regions of the brain that have been implicated in cognitive processes that support goal-directed behavior, including the hippocampus, striatum, and prefrontal cortex (Grace et al., 2007; Roshan Cools, 2008; Shohamy & Adcock, 2010), and this neuromodulatory system undergoes dynamic changes over the course of development (Wahlstrom et al., 2010). As modulation of dopaminergic signaling is implicated in the balance between model-based and model-free choice in adults (Deserno et al., 2015; Wunderlich, Smittenaar, & Dolan, 2012), changes in dopamine neurotransmission may also play a role in differences in behavioral control across developmental time points.

While studies to date have identified substantial structural and functional development in the brain circuits implicated in goal-directed learning, few studies have directly related these changes to behavioral indices of goal-directed versus habitual action selection. However, insights into the neurocognitive development of goal-directed decision-making can be gleaned from studies examining age-related changes in the cognitive processes that support goal-directed action.

In the following sections, we first discuss developmental trajectories in the learning processes that support the construction of cognitive models, including discussion of the neural mechanisms underpinning these changes wherever possible. We then turn our attention to the cognitive processes involved in the *use* of cognitive models, focusing on the central role of proactive cognitive control.

THE CONSTRUCTION OF COGNITIVE MODELS

Goal-directed actions are characterized by sensitivity to their contingent outcomes. Thus, integration of a mental representation of the structure of the environment with one's current goals is critical for the development of goal-directed decision-making. The capacity to infer causal relationships between one's actions and contingent outcomes and use this knowledge to explain or predict such outcomes is evident in children as young as 2 years (Gopnik, 2004). However, the formation of a cognitive model that can support planning in complex environments involves not only the capacity for causal inference but also a diverse array of learning and memory processes that enable the identification and representation of regularities in the environment. Events that tend to cooccur or follow in sequence can be learned through experience. Such relational associations discovered in distinct learning episodes can be assembled together, continually augmenting one's mental model with newly learned information. Knowledge of the rewarding properties of a given event, stimulus, or state can be used to prioritize that outcome as a goal. This reward information can be integrated with causal knowledge about which actions might lead to this desired outcome or to generalize reward value to states that are similar to one in which reward has been directly experienced. Below, we discuss studies that explore the development of these learning and memory processes, which provide a foundation for goal-directed action.

Through statistical learning, individuals can recognize events that tend to occur in sequence or covary with high probability. The ability to discover statistical regularities from a continuous stream of sensory experience is evident from infancy (Fiser & Aslin, 2002; Saffran, Aslin, & Newport, 1996) and has been proposed to play a central role in functions such as language acquisition. Learning of these regularities enables prediction of upcoming events, a necessary precursor to goal-directed actions that depend on those anticipated outcomes. While some studies have observed equivalent performance on statistical learning tasks from childhood to adulthood (Amso & Davidow, 2012), in other studies, performance has been found to improve with age (Potter, Bryce, & Hartley, 2017; Schlichting, Guarino, Schapiro, Turk-Browne, & Preston, 2017), suggesting that learning of more complex sequential structures may improve over development. Consistent with extensive evidence in adults that statistical learning of sequential information depends on the hippocampus and other medial temporal lobe structures (Covington, Brown-Schmidt, & Duff, 2018; Davachi & DuBrow, 2015; Preston, Shrager, Dudukovic, & Gabrieli, 2004; Schapiro, Gregory, Landau, McCloskey, & Turk-Browne, 2014), developmental improvements in statistical learning parallel the structural development of the hippocampus (Schlichting et al., 2017).

Beyond the ability to directly extract regularities from a single learning experience, associations can also be inferred across learning episodes. The ability to associate distinct learning experiences is critical for generalizing knowledge derived from past experience to new situations. Developmental studies using paradigms testing associative inference suggest that the ability to infer relations between overlapping experiences improves with age. In one such study (Schlichting et al., 2017), participants aged 6 to 30 years completed an associative inference task in which several sets of novel object triads (e.g., A, B, and C) were presented in pairs that shared an overlapping element (e.g., pair AB and pair BC). In a subsequent test phase, participants were shown one of the items previously paired with the overlapping element (e.g., item A) and asked to select the object that was indirectly related to it through their common association. The choice set included the target object (i.e., item C), as well as two unassociated objects (e.g., items D and E). Inference improved with age, even when accuracy of recall for the direct item pairings was taken into account. This suggests that the ability to flexibly integrate the shared aspects of distinct learned relationships—a key process underlying abstract knowledge representation (Preston, Molitor, Pudhiyidath, & Schlichting, 2017)—improves into adulthood. Moreover, associative inference and statistical learning, assessed in the same cohort, were positively correlated, and both were associated with developmental changes in hippocampal structure (Schlichting et al., 2017). Studies in adults suggest that associative inference involves functional integration of the hippocampus and vmPFC (Zeithamova, Dominick, & Preston, 2012). Thus, developmental changes in the ability to flexibly integrate learned associations likely stem in part from refinement of functional connectivity between the hippocampus and the prefrontal cortex, which increases into adulthood (Menon, Boyett-Anderson, & Reiss, 2005; Ofen, Chai, Schuil, Whitfield-Gabrieli, & Gabrieli, 2012).

Other studies have similarly found that the ability to transitively integrate relational premises (e.g., "A comes before B and B comes before C. Which comes first, A or C?") also improves with age (Halford, 1984) and that the age at which children show competence at such inferences depends on the degree of relational complexity (i.e., the number of premises to be integrated) involved in the judgment (Halford, Andrews, Dalton, Boag, & Zielinski, 2002). Medial temporal lobe regions and the rostrolateral prefrontal cortex are implicated in the learning and integration of premises during transitive inference in adults (Wendelken & Bunge, 2010). Thus, as with associative inference, developmental changes in prefrontal—hippocampal connectivity might also contribute to age-related improvements in integrating learned relations through transitive inference.

The recognition that stimuli or contexts are related can promote the generalization of associations or responses learned in one context to another. Studies of acquired equivalence (or "functional equivalence") directly assess such generalization. In these tasks, two stimuli or contexts are associated with the same outcomes or responses. Then, participants learn a novel association for one of the stimuli, and the transfer of this knowledge to the second stimulus is assessed in a subsequent generalization test phase. In one study examining acquired equivalence in 4- and 5-year-olds (Smeets, Barnes, & Roche, 1997), many children exhibited poor learning of the initial associations. However, if these associations were learned, children typically exhibited generalization in the test phase. A more recent study of participants, aged 3 to 52 years, similarly found that while the learning and retrieval of the initial pairings improved into adulthood, generalization for learned pairs was present from 6 years of age and did not differ from that of adults (Braunitzer et al., 2017). Another set of studies demonstrated that 8-month-old infants can transfer a set of learned associations to a novel stimulus based on its associative similarity, providing further evidence for the early emergence of generalization, even for more complex contextually dependent associations (Werchan, Collins, Frank, & Amso, 2015; Werchan, Collins, Frank, & Amso, 2016). Concurrent near-infrared spectroscopy in these infants revealed greater recruitment of the prefrontal cortex in infants who exhibited better learning and transfer, as well as higher eye blink rate during the task, a putative measure of striatal dopamine function (Jongkees & Colzato, 2016; Karson, 1983). This evidence that generalization ability is supported by dopaminergic innervation of frontostriatal circuitry is consistent with the selective impairment in generalization ability in adult patients for whom striatal dopamine is depleted (Myers et al., 2003).

To inform goal-directed actions, learned sequential regularities or associations must be integrated with information about currently valued outcomes. One way in which this can be accomplished is by recruiting these representations to prospectively envision potential "paths" from a current state to a goal. Studies across species have identified representations of potential future trajectories encoded in hippocampal activity (Johnson & Redish, 2007), which can be recruited to simulate both experienced and novel paths toward goals (Brown et al., 2016; Gupta, van der Meer, Touretzky, & Redish, 2010; Pfeiffer & Foster, 2013). The OFC is also proposed to play a central role in representing the sequential structure or "state space" of a current task, supporting such simulations (Schuck et al., 2016). These simulations are proposed to enable online comparison of the value of alternative courses of action, facilitating goal-directed action selection (Buckner, 2010; Pezzulo, van der Meer, Lansink, & Pennartz, 2014; van der Meer,

Kurth-Nelson, & Redish, 2012). Consistent with this proposed role for prospection in goal-directed evaluation, a neuroimaging study in adults has shown that individuals whose choices reflected a more model-based evaluation process also exhibited neural signatures of mentally invoking future outcomes when making choices (Doll, Duncan, Simon, Shohamy, & Daw, 2015). No study to date, in humans or animal models, has used neural decoding approaches to characterize the capacity to prospectively simulate future states prior to adulthood. However, developmental studies using self-report methods have found that the ability to envision or simulate future events improves with age into adulthood (Atance, 2015; Coughlin, Lyons, & Ghetti, 2014, Coughlin, Robins, & Ghetti, 2017), and the ability to recruit past learning to prospectively select a goal-directed action similarly improves as children age (Redshaw & Suddendorf, 2013; Suddendorf, Nielsen, & von Gehlen, 2011). This work suggests a potential role for age-related changes in prospective simulation ability in the development of goal-directed decision-making.

Integration of learned relations to support goal pursuit can also occur in a retrospective manner. Studies in animal models suggest that receipt of reward can promote spontaneous retrieval, or "replay," of the events that preceded or are associated with an outcome (Carr, Jadhav, & Frank, 2011). This prioritized retrieval has been proposed as a mechanism for transferring value information backward across states, allowing mentally simulated experience to "retrain" learned values for remotely associated stimuli (Shohamy & Daw, 2015). Importantly, such a mechanism might enable behavioral adaptation to changes in outcome valuation or identification of a newly rewarding action (Sutton, 1991). Empirical support for this proposal comes from neuroimaging studies in adults showing that the degree to which a stimulus, previously associated with a reward predictive cue, is invoked during reward learning predicts the preference for that stimulus in a subsequent choice phase (Kurth-Nelson, Barnes, Sejdinovic, Dolan, & Dayan, 2015; Wimmer & Shohamy, 2012). As with prospective integration, to our knowledge, no studies in children or adolescents have directly examined retrospective simulation using neuroimaging approaches, and prospective and retrospective integration are often not distinguishable in many behavioral paradigms. Thus, future studies are required to better understand whether there are developmental shifts in the mechanisms by which novel value-based associations are integrated with prior learned associations.

Mounting evidence suggests that sleep plays an important role in developmental changes in memory integration and consolidation (Fischer, Wilhelm, & Born, 2007; Wilhelm, Prehn-Kristensen, & Born, 2012). Reactivation of prioritized memories, and their associated relationships, occurs not only during learning of novel reward associations but also in "offline" periods of sleep or rest following learning (Carr et al., 2011; Kudrimoti, Barnes, & McNaughton, 1999; Wilson & McNaughton, 1994). Such offline memory integration has been proposed to underpin qualitative changes in relational memory following sleep in adults (Ellenbogen, Hu, Payne, Titone, & Walker, 2007;

Stickgold & Walker, 2013). Children not only spend more time in sleeping than adults but also spend a disproportionately larger amount of time in slow-wave sleep (Ohayon, Carskadon, Guilleminault, & Vitiello, 2004), the stage during which these reactivation events occur. While no studies have directly examined sleep-dependent changes in the transfer of reward value in children, sleep-dependent facilitation of memory for emotional, relative to neutral, information is more robust in children than adults (Prehn-Kristensen et al., 2013), suggesting that sleep might have unique effects on the transformation or integration of valenced information in younger individuals. Moreover, as nearly all the studies discussed in earlier sections that have revealed developmental changes in goal-directed behavior were conducted in a single experimental session, it is unknown whether rest- or sleep-dependent memory integration and consolidation processes might facilitate subsequent goal-directed behavior.

Collectively, the findings reviewed in this section highlight developmental changes in the learning processes that play a central role in the construction of a cognitive model of a task or environment. However, while the construction of such a model is necessary for goal-directed action, it is not sufficient. *Using* this mental model further requires the ability to recognize time points at which prior knowledge could be productively leveraged to purse a current goal, as well as the selective retrieval of goal-relevant information.

Importantly, several developmental studies have observed dissociations between the ability to acquire the structural knowledge of a task necessary for goal-directed action and actual goal-directed performance on the task itself (Decker et al., 2016; Zelazo, Frye, & Rapus, 1996). Such a dissociation was present in the two-step reinforcement learning task described earlier (Decker et al., 2016). Children's habitual (model-free) choices demonstrated that they were learning stimulus—reward associations in the second stage, but they did not make model-based choices at the first stage of the task. While children did not show evidence of *using* transition structure information to make goal-directed choices, they were able to learn the task transition structure. Children, like adolescents and adults, could explicitly report the transition structure of the task (i.e., "Which planet did this spaceship usually travel to?") and were also slower to respond following rare transitions. Slower second-stage responses following rare transitions may reflect a violation of the expectation that the more frequent transition would occur and thus, reveal knowledge of the task's probabilistic transitions. In adolescents and adults, this reaction time measure of task structure knowledge correlated with the degree to which they exhibited goal-directed choices at the first-stage of each trial. However, this was not the case in children.

What cognitive factors might account for such a failure to recruit learned action—outcome knowledge to pursue a goal? The gradual development of relational integration discussed above may have hindered children's integration of learned transition structure knowledge with learned reward associations to construct a cognitive model of the task (Potter et al., 2017). However, goal-directed choice also requires the timely retrieval of this knowledge at the first stage of the task, when it can be leveraged proactively to support goal pursuit. Below, we discuss the developmental changes in the cognitive processes that support the ability to use learned knowledge about task structure to take goal-directed action.

USING COGNITIVE MODELS TO ENABLE GOAL-DIRECTED ACTION

To act to bring about a desired outcome, one must monitor the environment to determine when opportunities for goal-relevant actions arise. Such forward-looking behavior relies on cognitive control, or the ability to maintain and flexibly update a mental model of the task at hand, while preventing interference from irrelevant stimuli. Cognitive control is a key component of goal-directed behavior, as it allows for current goals to be retrieved and used to guide one's actions. A large body of work has identified substantial age-related changes in the recruitment of cognitive control (Diamond, 2006; Luna, 2009; Somerville & Casey, 2010), which likely play a critical role in the development of goal-directed decision-making.

A distinction is commonly made between two forms of cognitive control: proactive and reactive (Braver, 2012). In proactive control, a mental representation of a goal is invoked and sustained during a preparatory period prior to making a goal-directed action. Proactive control is typically engaged in anticipation of a cognitively demanding task, increasing resistance of goal representations to interference from goal-irrelevant stimuli. In contrast, reactive control occurs when a transient representation of a goal is evoked by the presence of a stimulus that either signals conflict with goal pursuit or directly activates a goal representation. For example, individuals might accomplish a goal of buying milk, eggs, and apples through two distinct cognitive control processes. Those who engage proactive control may leave their house with a mental list of groceries to buy, maintaining the goal representation during the drive to the grocery store and even when distracted at the store by goal-irrelevant stimuli such as bananas and bread. On the other hand, those recruiting reactive control may go to the grocery store and remember the items that they intended to buy only after entering the produce or dairy section of the store, which reactivates the goal representation. While both reactive and proactive control can support the pursuit of a goal, only proactive control allows for multistep planning to attain a goal. For example, in the two-step task described previously, for participants to use the information about the task structure from the previous trial to make a goal-directed choice at the first stage, proactive control is necessary to plan which spaceship should be chosen to get to the planet that has the greatest potential for reward. Reactive control does not support planning at the first stage to maximize reward but instead would be recruited at the second stage when it may be too late to select the most rewarding option.

A common task used to assess reactive versus proactive control is the AX–CPT, an adaptation of the continuous performance test (CPT). In the AX–CPT paradigm, proactive attention to contextual cues can facilitate quick correct responses when a cue is highly predictive of a target probe. On each trial, a cue ("A" or "B", which may be letters, pictures, or any other type of visual stimulus) appears, followed by a short delay period, and then a probe ("X" or "Y"). While all cue and probe combinations occur, certain cue—probe pairs ("AX") appear more frequently than others ("AY", "BX", or "BY"). A target response should be made when A is followed by X, while a nontarget response should be made for all other combinations. The disproportionate number of AX trials allows individuals to develop both an expectancy for the X probe following an A cue and a prepotent response for X probes. Critically, the pattern of performance on this paradigm can be used to distinguish which cognitive control strategy was employed. Individuals using proactive control will have worse performance on the AY trials, as the target response would have been prepared during the delay period and is more likely to be emitted than for individuals relying on reactive control. In contrast, performance in individuals using reactive control will be worse on the BX trials as the preceding B cue has not been maintained, causing them to make an erroneous target response to the X cue. In contrast, those using proactive control would have already prepared their nontarget response.

In a child-friendly adaptation of the AX-CPT task, Chatham, Frank, and Munakata (2009) found that 3-year-olds primarily relied on reactive control while 8-year-olds engaged proactive control, suggesting a developmental shift from reactive to proactive control. To investigate whether younger children are capable of invoking proactive control, a cued task-switching paradigm was adapted to create three trial types: "proactive impossible," "proactive possible," and "proactive encouraged" (Chevalier, Martis, Curran, & Munakata, 2015). For "proactive impossible" trials, only reactive control could be employed, as the cue that informed the correct action was presented simultaneously with the target, rendering proactive preparation impossible. For "proactive possible" trials, the cue preceded the probe, enabling proactive control, but remained on the screen following probe onset, facilitating reactive control. For "proactive encouraged" trials, the cue disappeared from the screen when the probe appeared, requiring cue information to be maintained in memory in order to respond correctly. Although reactive or proactive control strategies could be used for both of these trial types, reactive control was disincentivized for the "proactive encouraged" trials by making reactive control more difficult. Using this manipulation, Chevalier et al. (2015) found that whereas 10-year-old children tended to engage proactive control whenever possible, 5-year-old children engaged proactive control only when reactive control was made more challenging.

Collectively, these studies suggest that younger children tend to engage reactive control, and over the course of development, shift toward preferential engagement of proactive control (Blackwell & Munakata, 2014; Chatham et al., 2009; Chevalier et al., 2015). The age at which this developmental shift occurs depends on the complexity of the task (Chatham et al., 2009; Chevalier et al., 2015; Church, Bunge, Petersen, & Schlaggar, 2017). However, preferential engagement of reactive control early in development does not reflect an inability to engage proactive control, as younger children are able to engage proactive control if incentivized. While both cognitive control strategies can be observed early in development, the evaluation process that determines which strategy to engage may improve with age (Chevalier et al., 2015). Age-related increases in weighing the costs and benefits of each strategy according to task demands (i.e., computing the expected value of exerting each form of control; Shenhav, Botvinick, & Cohen, 2013) may support the increased reliance on proactive control (Chevalier et al., 2015). Adults may more effectively integrate value computations to inform when cognitive control should be optimally deployed (Insel, Kastman, Glenn, & Somerville, 2017).

Similar age-related trajectories are evident in working memory gating, a process that limits the information transferred into and out of working memory (Amso, Haas, McShane, & Badre, 2014; Unger, Ackerman, Chatham, Amso, & Badre, 2016). Input gating refers to the selection of task-relevant information passed into working memory, whereas output gating refers to the retrieval of information from working memory to inform action selection. A recent study suggests that children are as effective as adolescents at input gating, although they engage this strategy less often (Unger et al., 2016). Instead, children tend to rely on output gating, despite reduced efficacy in their use of this process when a subset of information must be retrieved from working memory. Selective output gating of specific items from working memory, rather than retrieving all maintained information, may impose greater cognitive demands and undergo more protracted development. Age-related improvements in output gating, along with increased reliance on input gating, may promote the deployment of working memory in a manner that better supports goal-directed behavior. Input and output gating of working memory are proposed to respectively support the engagement of proactive and reactive control. Thus, these findings corroborate the previously discussed evidence of a developmental shift toward increased reliance on proactive control.

The age-related transition toward preferential reliance on proactive control is a key component of goal-directed decision-making. In the two-step task described earlier (Decker et al., 2016), the age-related dissociation between knowledge and action may reflect developmental changes in cognitive control. Although children demonstrated verbal knowledge of the task transition structure, they were unable to translate this knowledge into goal-directed actions. This dissociation disappeared with age, potentially as individuals shifted from engaging reactive to proactive cognitive control. Development of proactive control in older participants may have supported the timely retrieval of transition structure knowledge at the first stage of the task, when it could be productively used to select the most likely path to a previously received reward. This proposal is consistent with evidence in adults that a tendency to engage proactive control is predictive of greater model-based choice in the two-step task (Otto, Skatova, Madlon-Kay, & Daw, 2015).

The development of proactive control may also improve the ability to use explicitly communicated rules to support goal-directed behavior. Rules are consistent stimulus—response mappings that apply under specific contextual conditions (Bunge, 2004; Dayan, 2007). In certain contexts, specific actions may be consistently advantageous (e.g., stopping when a traffic light is red), obviating the need for evaluation of their consequences. In such conditions, action selection can be productively informed by rules. Rules are often learned through explicit communication. For example, a child may learn to be quiet in a classroom through explicit instruction from a teacher, or a sign stating "Use your inside voice" displayed at the entrance. Use of explicitly communicated rules requires translation of a verbally encoded policy for action into a representation of both the preconditions that must be satisfied in order for the rule to apply (i.e., "What is the stimulus or context?"), as well as the action that should be performed under those conditions (i.e., "What is the response?"). Successful deployment of rules often requires verification, prior to action, that the preconditions for its application are satisfied.

Importantly, successful deployment of a rule to accomplish a goal may engage either a habitual or a goal-directed learning process, with important implications for the flexibility of the rule-guided behavior. While adults are adept at using rules to guide behavior (Cole, Laurent, & Stocco, 2013), the ability to act on the basis of explicitly communicated rules exhibits marked developmental changes. Three-year-olds are able to use explicitly instructed rules to guide behavior (e.g., "sort cards into piles based on their shape"), but they have difficulty adjusting their behavior when the task demands change and conflict with the original rules ("now sort the cards into piles based on their color"). While 3-year-olds exhibit difficulty adhering to this new rule, they show explicit knowledge of the rule, and can repeat the rule when asked, suggesting that the challenge lies in implementing this rule. By 4 years of age, children in this task are able to accomplish this switch (Zelazo et al., 1996). However, studies employing more complex tasks to assess rule-guided behavior have observed that such switch costs decrease gradually with age and persist into young adulthood (Davidson, Amso, Anderson, & Diamond, 2006). While rule use requires translation of instruction into a behavioral procedure, individuals may differ in the extent to which successful deployment of a rule harnesses a habitual learning process. To the extent that performance of a first-learned rule in younger individuals results in the formation of a habitual stimulus—response association (as opposed to an action—outcome representation), this may undermine their capacity to flexibly alter behavior in accordance with novel rules.

Proactive control and reactive control evoke dissociable patterns of neural and physiological activity. Frontal and parietal regions are broadly implicated in the selection and maintenance of goal-relevant information during the engagement of cognitive control (MacDonald, 2000; Miller & Cohen, 2001). Supporting the role of prefrontal development in proactive control, a longitudinal study in nonhuman primates found that the age-related enhancement of goal representation in prefrontal neurons, from adolescence to adulthood, was correlated with the ability to plan goal-directed actions (Zhou et al., 2016). Differences in the timing of activation in frontoparietal regions have been found to relate to the type of cognitive control engaged. Younger children exhibit greater activation in frontoparietal regions right before a goal is obtained, as is characteristic of reactive control. Greater recruitment of these regions during the preparatory period parallels age-related increases in proactive control (Andrews-Hanna et al., 2011; Church et al., 2017; Manzi, Nessler, Czernochowski, & Friedman, 2011). Similar temporal differences in pupil dilation, an index of cognitive effort, parallel developmental shifts from reactive to proactive control (Chatham et al., 2009; Chevalier et al., 2015). Younger children, who preferentially relied on reactive control, exhibited greater pupil dilation in response to the probe, whereas greater pupil dilation following the cue, in preparation of the probe, was evident in older children engaging proactive control. Developmental changes in frontoparietal circuitry and the temporal dynamics of pupil dilation reflect the shift from a tendency to rely on reactive control to greater engagement of proactive cognitive control across development (Bunge & Wright, 2007).

Collectively, this work suggests that developmental increases in proactive control, as well as effective deployment of reactive control, play an important role in the use of learned cognitive models or instructed task rules to inform action selection. The developmental shift from reactive to proactive control affords flexible, adaptive actions in which individuals are not simply reacting to goal-relevant cues but instead maintaining goal-relevant information in anticipation of the goal. Proactive control confers the obvious benefit of advanced preparation prior to pursuit of a goal (Braver, 2012). However, engaging in proactive cognitive control involves a narrowing of the focus of attention to task-relevant information, which may have the side effect of impeding the learning of information that is not currently goal-relevant (Thompson-Schill, Ramscar, & Chrysikou, 2009). Consistent with this notion, whereas adults instructed to make a suboptimal reward-driven choice learn action values that are biased by this inaccurate instruction, children's and adolescents' learning appears to more accurately reflect their experienced reward statistics (Decker, Lourenco, Doll, & Hartley, 2015). Similarly, children instructed to attend to a specific attribute of a multidimensional cue are better than adults at learning about its uninstructed attributes (Plebanek & Sloutsky, 2017). Accordingly, reliance on reactive versus proactive cognitive control may confer distinct advantages across development, with increasing engagement of proactive control promoting goal-directed behavior.

SUMMARY AND OPEN QUESTIONS

In this chapter, we examined how goal-directed decision-making changes over the course of development, adopting a conceptual framework stemming from animal learning theory, which distinguishes goal-directed from habitual instrumental action. We presented findings in human and animal models suggesting that the choices and actions of younger individuals are less sensitive to changes in outcome value or degradation of action—outcome contingency, two key assays of goal-directedness. These age-related differences in goal-directed choice were corroborated by evidence that the recruitment of model-based computations to evaluate reward-driven actions increases with age, whereas use of model-free computations appears stable from late childhood to adulthood. We discussed how developmental changes in the cognitive processes that support the construction and recruitment of a cognitive model of a task might contribute to such developmental changes in goal-directed decision-making. Our review of this literature suggests that developmental improvements in the capacity to flexibly integrate learned associations might facilitate the construction of mental models that inform goal-directed behavior. Increases in the tendency to engage cognitive control proactively, in anticipation of goal-relevant information or choices, promotes the effective use of these cognitive "maps" to guide action. We proposed that the development of structural connectivity and functional integration between the prefrontal cortex, the striatum, and the hippocampus plays a central underlying role in these cognitive and behavioral changes.

Many gaps in our understanding were highlighted in the course of this discussion that represent fruitful avenues for future research. Developmental studies directly relating assays of goal-directed behavior to neural structure and function are needed to improve our understanding of the mechanisms underlying the development of goal-directed action. Characterizing age-related changes in functional and structural connectivity between the prefrontal cortex, hippocampus, and striatum, and their functional consequences for the construction and use of cognitive models, will help to relate developmental changes in goal-directed behavior to their neural substrates. The relationship between the dopaminergic system and the control of instrumental action across development also has yet to be well characterized. An improved understanding of how memory integration and consolidation processes change over development, whether reward plays a unique role developmentally in modulating such processes, and the role of sleep in these developmental changes will help to clarify how cognitive model formation contributes to the emergence of goal-directed behavior. Investigating the neurocognitive mechanisms through which prospective representations of goals are elicited will help to elucidate how and when proactive cognitive control is recruited to support goal pursuit. Research addressing these questions would greatly augment our understanding of developmental changes in the key cognitive components of goal-directed behavior that were the focus of this chapter.

While our discussion here focused on neurocognitive developmental changes in the construction and use of cognitive models that may facilitate goal-directed action, additional processes likely influence the development of goal-directed behavior. For example, we have not addressed the question of how the arbitration between goal-directed and habitual action evaluation might shift over development. In adulthood, goal-directed and habitual evaluation processes have been proposed to operate in parallel, effectively competing for control over behavior (Balleine & O'Doherty, 2010). This competitive model is supported by studies in adult animals demonstrating that lesions to the circuitry supporting goal-directed evaluation do not eliminate instrumental behavior but instead render it habitual—insensitive to changes in outcome value or action—outcome contingency (Yin et al., 2005). Conversely, lesions to the circuitry implicated in habit learning restore the sensitivity of previously habitual actions to such changes (Yin, Knowlton, & Balleine, 2004). These findings suggest that while typically only one action evaluation system is revealed in an individual's behavior, both systems might carry out action evaluation using their respective computations and undergo some type of arbitration process to determine whether goal-directed or habitual behavior is expressed.

Alternative accounts of this arbitration process propose that the action evaluation strategy selected for behavioral expression may depend on the accuracy of its predictions (Daw et al., 2005), the costs and benefits of its computations (Kool, Gershman, & Cushman, 2017), or some combination of these factors (e.g., speed—accuracy trade-offs; Keramati, Dezfouli, & Piray, 2011). While a developmental bias toward habitual action in younger individuals could reflect difficulty in carrying out goal-directed evaluation, it might also reflect an age-related shift in such an arbitration process. For example, known developmental increases in cognitive processing speed (Kail & Salthouse, 1994; Luna, Garver, Urban, Lazar, & Sweeney, 2004) may gradually confer a competitive advantage to model-based evaluation if arbitration between these two strategies is determined via optimization of speed—accuracy trade-offs. This would be consistent with evidence that adults with higher processing speed make more model-based choices (Schad et al., 2014). The computationally intensive calculations involved in model-based evaluation (particularly for complex multistep plans) may simply take too long in children to effectively compete with the more efficient model-free computations of action value. Future studies directly probing such arbitration processes are essential for our understanding of developmental changes in the balance between these two learning systems in the control of behavior.

The distinction drawn between goal-directed and habitual action implies a dual-system perspective of behavioral control. However, there is clearly greater diversity, beyond these two "systems" in the cognitive processes that inform our motivated behavior. Through Pavlovian learning, stimuli predictive of positive or negative environmental events can acquire the capacity to elicit evolutionarily "prepared" behavioral responses. These Pavlovian reactions (e.g., rodents freezing in anticipation of a threat) can be thought of as implementing a default behavioral response in motivated contexts. Importantly, Pavlovian behaviors can facilitate or interfere with instrumental action through a process known as Pavlovian–instrumental transfer (Estes, 1943; Lovibond, 1983). For example, instrumental actions that lead to reward can be facilitated by the action invigoration typically elicited by Pavlovian reward anticipation. However, the tendency toward the inhibition of action typically elicited by Pavlovian anticipation of threat might impair performance of the same instrumental behavior. Thus, the functioning of the Pavlovian learning system is proposed to strongly modulate the performance of goal-directed instrumental action (Dayan et al., 2006). Consistent with this proposal, adults who exhibit strong Pavlovian interference with instrumental behavior also show reduced reliance on model-based evaluation (Sebold et al., 2016). Pavlovian learning emerges early in development and exhibits dynamic developmental shifts in its expression (Hartley & Lee, 2015), suggesting that age-related changes in the interaction between Pavlovian and instrumental learning may be critical for understanding the development of goal-directed decision-making.

Goal-directed and habitual instrumental learning are distinct means of action selection that both integrate over multiple past episodes to derive action values. Recent theoretical proposals and empirical data in adults suggest that value predictions derived from single episodic memories can also drive action selection in a manner that differs from both forms of instrumental evaluation (Bornstein, Khaw, Shohamy, & Daw, 2017; Gershman & Daw, 2017; Lengyel & Dayan, 2008; Ludvig, Madan, & Spetch, 2015). This mechanism for action selection has been proposed to be particularly influential in situations where individuals have little prior experience, which may be true of more choice contexts encountered at earlier developmental stages. However, developmental changes in this behavioral control process remain largely unexplored. The neurocircuitry implicated in both Pavlovian and episodic behavioral control exhibits substantive developmental changes and overlaps with the neural circuits involved in instrumental action evaluation, suggesting the potential for dynamic interactions between these evaluation systems across development. Thus, a mechanistic account of the development of goal-directed behavior must expand beyond the narrow focus presented here to examine the diverse means by which an individual can pursue a goal.

In recent years, a large body of research has focused on understanding the computational, cognitive, and neural mechanisms underlying goal-directed action. The vast majority of this work has focused on characterizing these processes in adulthood, while the changes in these processes over development have remained largely unexplored. In this chapter, we reviewed the small number of studies that have directly examined developmental changes in goal-directed behavior. Leveraging our knowledge from studies conducted in adult humans and animals, as well as our understanding of neurocognitive development more broadly, we discussed the mechanisms that might contribute to age-related increases in goal-directed action while highlighting the many gaps in our current understanding. Our hope is that in the coming years, increased efforts to characterize the development of goal-directed decision-making, across multiple levels of investigation, will begin to fill these gaps.

ACKNOWLEDGMENTS

The authors thank Kate Nussenbaum for helpful comments on the manuscript. This work was supported by the National Science Foundation (GRFP to H.A.R., CAREER Award 1654393 to C.A.H.), the National Institute on Drug Abuse (R03DA038701 to C.A.H.) and a Jacobs Foundation Early Career Fellowship to C.A.H.

REFERENCES

Amso, D., & Davidow, J. (2012). The development of implicit learning from infancy to adulthood: Item frequencies, relations, and cognitive flexibility. *Developmental Psychobiology, 54*, 664−673.

Amso, D., Haas, S., McShane, L., & Badre, D. (2014). Working memory updating and the development of rule-guided behavior. *Cognition, 133*, 201−210.

Andrews-Hanna, J. R., Mackiewicz Seghete, K. L., Claus, E. D., Burgess, G. C., Ruzic, L., & Banich, M. T. (2011). Cognitive control in adolescence: Neural underpinnings and relation to self-report behaviors. *PLoS One, 6*, e21598.

Atance, C. M. (2015). Young children's thinking about the future. *Child Development Perspectives, 9*, 178−182.

Balleine, B. W., Delgado, M. R., & Hikosaka, O. (2007). The role of the dorsal striatum in reward and decision-making. *The Journal of Neuroscience: The Official Journal of the Society for Neuroscience, 27*, 8161−8165.

Balleine, B. W., & O'Doherty, J. P. (2010). Human and rodent homologies in action control: Corticostriatal determinants of goal-directed and habitual action. *Neuropsychopharmacology: Official Publication of the American College of Neuropsychopharmacology, 35*, 48−69.

Barch, D. M., Braver, T. S., Nystrom, L. E., Forman, S. D., Noll, D. C., & Cohen, J. D. (1997). Dissociating working memory from task difficulty in human prefrontal cortex. *Neuropsychologia, 35*, 1373−1380.

Blackwell, K. A., & Munakata, Y. (2014). Costs and benefits linked to developments in cognitive control. *Developmental Science, 17*, 203−211.

Blankenship, S. L., Redcay, E., Dougherty, L. R., & Riggins, T. (2017). Development of hippocampal functional connectivity during childhood. *Human Brain Mapping, 38*, 182−201.

Bornstein, A. M., Khaw, M. W., Shohamy, D., & Daw, N. D. (2017). Reminders of past choices bias decisions for reward in humans. *Nature Communications, 8*, 15958.

Braunitzer, G., Öze, A., Eördegh, G., Pihokker, A., Rózsa, P., Kasik, L., ... Nagy, A. (2017). The development of acquired equivalence from childhood to adulthood—a cross-sectional study of 265 subjects. *PLoS One, 12*, e0179525.

Braver, T. S. (2012). The variable nature of cognitive control: A dual mechanisms framework. *Trends in Cognitive Sciences, 16*, 106−113.

Brown, T. I., Carr, V. A., LaRocque, K. F., Favila, S. E., Gordon, A. M., Bowles, B., ... Wagner, A. D. (2016). Prospective representation of navigational goals in the human hippocampus. *Science, 352*, 1323−1326.

Buckner, R. L. (2010). The role of the hippocampus in prediction and imagination. *Annual Review of Psychology, 61*, 27−48.

Bunge, S. A. (2004). How we use rules to select actions: A review of evidence from cognitive neuroscience. *Cognitive, Affective & Behavioral Neuroscience, 4*, 564−579.

Bunge, S. A., & Wright, S. B. (2007). Neurodevelopmental changes in working memory and cognitive control. *Current Opinion in Neurobiology, 17,* 243—250.

Carr, M. F., Jadhav, S. P., & Frank, L. M. (2011). Hippocampal replay in the awake state: A potential substrate for memory consolidation and retrieval. *Nature Neuroscience, 14,* 147.

Chatham, C. H., Frank, M. J., & Munakata, Y. (2009). Pupillometric and behavioral markers of a developmental shift in the temporal dynamics of cognitive control. *Proceedings of the National Academy of Sciences, 106,* 5529—5533.

Chevalier, N., Martis, S. B., Curran, T., & Munakata, Y. (2015). Metacognitive processes in executive control development: The case of reactive and proactive control. *Journal of Cognitive Neuroscience, 27,* 1125—1136.

Church, J. A., Bunge, S. A., Petersen, S. E., & Schlaggar, B. L. (2017). Preparatory engagement of cognitive control networks increases late in childhood. *Cerebral Cortex, 27,* 2139—2153.

Cole, M. W., Laurent, P., & Stocco, A. (2013). Rapid instructed task learning: A new window into the human brain's unique capacity for flexible cognitive control. *Cognitive, Affective & Behavioral Neuroscience, 13,* 1—22.

Coughlin, C., Lyons, K. E., & Ghetti, S. (2014). Remembering the past to envision the future in middle childhood: Developmental linkages between prospection and episodic memory. *Cognitive Development, 30,* 96—110.

Coughlin, C., Robins, R. W., & Ghetti, S. (2017). Development of episodic prospection: factors underlying improvements in middle and late childhood. *Child Development.* https://www.ncbi.nlm.nih.gov/pubmed/29205318.

Covington, N. V., Brown-Schmidt, S., & Duff, M. C. (2018). The necessity of the Hippocampus for statistical learning. *Journal of Cognitive Neuroscience,* 1—20.

Davachi, L., & DuBrow, S. (2015). How the hippocampus preserves order: The role of prediction and context. *Trends in Cognitive Sciences, 19,* 92—99.

Davidson, M. C., Amso, D., Anderson, L. C., & Diamond, A. (2006). Development of cognitive control and executive functions from 4 to 13 years: Evidence from manipulations of memory, inhibition, and task switching. *Neuropsychologia, 44,* 2037—2078.

Daw, N. D., Gershman, S. J., Seymour, B., Dayan, P., & Dolan, R. J. (2011). Model-based influences on humans' choices and striatal prediction errors. *Neuron, 69,* 1204—1215.

Daw, N. D., Niv, Y., & Dayan, P. (2005). Uncertainty-based competition between prefrontal and dorsolateral striatal systems for behavioral control. *Nature Neuroscience, 8,* 1704.

Dayan, P. (2007). Bilinearity, rules, and prefrontal cortex. *Frontiers in Computational Neuroscience, 1.*

Dayan, P., Niv, Y., Seymour, B., & Daw, N. D. (2006). The misbehavior of value and the discipline of the will. *Neural Networks: The Official Journal of the International Neural Network Society, 19,* 1153—1160.

Decker, J. H., Lourenco, F. S., Doll, B. B., & Hartley, C. A. (2015). Experiential reward learning outweighs instruction prior to adulthood. *Cognitive, Affective & Behavioral Neuroscience, 15,* 310—320.

Decker, J. H., Otto, A. R., Daw, N. D., & Hartley, C. A. (2016). From creatures of habit to goal-directed learners tracking the developmental emergence of model-based reinforcement learning. *Psychological Science, 27,* 848—858.

Deserno, L., Huys, Q. J. M., Boehme, R., Buchert, R., Heinze, H.-J., Grace, A. A., ... Schlagenhauf, F. (2015). Ventral striatal dopamine reflects behavioral and neural signatures of model-based control during sequential decision making. *Proceedings of the National Academy of Sciences of the United States of America, 112,* 1595—1600.

Diamond, A. (2006). The early development of executive functions. In *Lifespan cognition: Mechanisms of change* (pp. 70—95). New York: Oxford University Press.

Dickinson, A. (1985). Actions and habits: The development of behavioural autonomy. *Philosophical Transactions of the Royal Society B, 308,* 67—78.

Dolan, R. J., & Dayan, P. (2013). Goals and habits in the brain. *Neuron, 80,* 312—325.

Doll, B. B., Duncan, K. D., Simon, D. A., Shohamy, D., & Daw, N. D. (2015). Model-based choices involve prospective neural activity. *Nature Neuroscience, 18,* 767—772.

Doll, B. B., Simon, D. A., & Daw, N. D. (2012). The ubiquity of model-based reinforcement learning. *Current Opinion in Neurobiology, 22,* 1075—1081.

Ellenbogen, J. M., Hu, P. T., Payne, J. D., Titone, D., & Walker, M. P. (2007). Human relational memory requires time and sleep. *Proceedings of the National Academy of Sciences, 104,* 7723—7728.

Estes, W. K. (1943). Discriminative conditioning. I. A discriminative property of conditioned anticipation. *Journal of Experimental Psychology, 32,* 150—155.

Fischer, S., Wilhelm, I., & Born, J. (2007). Developmental differences in sleep's role for implicit off-line learning: Comparing children with adults. *Journal of Cognitive Neuroscience, 19*, 214–227.

Fiser, J., & Aslin, R. N. (2002). Statistical learning of higher-order temporal structure from visual shape sequences. *Journal of Experimental Psychology: Learning, Memory, and Cognition, 28*, 458–467.

Gabard-Durnam, L. J., Flannery, J., Goff, B., Gee, D. G., Humphreys, K. L., Telzer, E., ... Tottenham, N. (2014). The development of human amygdala functional connectivity at rest from 4 to 23 years: A cross-sectional study. *NeuroImage, 95*, 193–207.

Gershman, S., & Daw, N. (2017). Reinforcement learning and episodic memory in humans and animals: An integrative framework. *Annual Review of Psychology, 68*, 101–128.

Giedd, J. N., Blumenthal, J., Jeffries, N. O., Castellanos, F. X., Liu, H., Zijdenbos, A., ... Rapoport, J. L. (1999). Brain development during childhood and adolescence: A longitudinal MRI study. *Nature Neuroscience, 2*, 861–863.

Gläscher, J., Hampton, A. N., & O'Doherty, J. P. (2009). Determining a role for ventromedial prefrontal cortex in encoding action-based value signals during reward-related decision making. *Cerebral Cortex, 19*, 483–495.

Gogtay, N., Giedd, J. N., Lusk, L., Hayashi, K. M., Greenstein, D., Vaituzis, A. C., ... Toga, A. W. (2004). Dynamic mapping of human cortical development during childhood through early adulthood. *Proceedings of the National Academy of Sciences, 101*, 8174–8179.

Gopnik, A. (2004). A theory of causal learning in children: Causal maps and Bayes nets. *Psychological Review, 111*, 3–32.

Gottfried, J. A., O'Doherty, J., & Dolan, R. J. (2003). Encoding predictive reward value in human amygdala and orbitofrontal cortex. *Science, 301*, 1104–1107.

Grace, A. A., Floresco, S. B., Goto, Y., & Lodge, D. J. (2007). Regulation of firing of dopaminergic neurons and control of goal-directed behaviors. *Trends in Neurosciences, 30*, 220–227.

Graybiel, A. M. (2008). Habits, rituals, and the evaluative brain. *Annual Review of Neuroscience, 31*, 359–387.

Gupta, A. S., van der Meer, M. A. A., Touretzky, D. S., & Redish, A. D. (2010). Hippocampal replay is not a simple function of experience. *Neuron, 65*, 695–705.

Gu, S., Satterthwaite, T. D., Medaglia, J. D., Yang, M., Gur, R. E., Gur, R. C., & Bassett, D. S. (2015). Emergence of system roles in normative neurodevelopment. *Proceedings of the National Academy of Sciences, 112*, 13681–13686.

Halford, G. S. (1984). Can young children integrate premises in transitivity and serial order tasks? *Cognitive Psychology, 16*, 65–93.

Halford, G. S., Andrews, G., Dalton, C., Boag, C., & Zielinski, T. (2002). Young children's performance on the balance scale: The influence of relational complexity. *Journal of Experimental Child Psychology, 81*, 417–445.

Hampton, A. N., Bossaerts, P., & O'Doherty, J. P. (2006). The role of the ventromedial prefrontal cortex in abstract state-based inference during decision making in humans. *The Journal of Neuroscience: The Official Journal of the Society for Neuroscience, 26*, 8360–8367.

Hart, G., Bradfield, L. A., & Balleine, B. W. (2018). Prefrontal cortico-striatal disconnection blocks the acquisition of goal-directed action. *The Journal of Neuroscience: The Official Journal of the Society for Neuroscience*, 1311–1322.

Hartley, C. A., & Lee, F. S. (2015). Sensitive periods in affective development: Nonlinear maturation of fear learning. *Neuropsychopharmacology: Official Publication of the American College of Neuropsychopharmacology, 40*, 50–60.

Haruno, M., Kuroda, T., Doya, K., Toyama, K., Kimura, M., Samejima, K., ... Kawato, M. (2004). A neural correlate of reward-based behavioral learning in caudate nucleus: A functional magnetic resonance imaging study of a stochastic decision task. *The Journal of Neuroscience: The Official Journal of the Society for Neuroscience, 24*, 1660–1665.

Huttenlocher, P. R. (1990). Morphometric study of human cerebral cortex development. *Neuropsychologia, 28*, 517–527.

Huttenlocher, P. R., & Dabholkar, A. S. (1997). Regional differences in synaptogenesis in human cerebral cortex. *Journal of Comparative Neurology, 387*, 167–178.

Insel, C., Kastman, E. K., Glenn, C. R., & Somerville, L. H. (2017). Development of corticostriatal connectivity constrains goal-directed behavior during adolescence. *Nature Communications, 8*, 1605.

Izquierdo, A., Suda, R. K., & Murray, E. A. (2004). Bilateral orbital prefrontal cortex lesions in rhesus monkeys disrupt choices guided by both reward value and reward contingency. *The Journal of Neuroscience: The Official Journal of the Society for Neuroscience, 24*, 7540–7548.

Johnson, A., & Redish, A. D. (2007). Neural ensembles in CA3 transiently encode paths forward of the animal at a decision point. *The Journal of Neuroscience: The Official Journal of the Society for Neuroscience, 27,* 12176–12189.

Jongkees, B. J., & Colzato, L. S. (2016). Spontaneous eye blink rate as predictor of dopamine-related cognitive function—a review. *Neuroscience and Biobehavioral Reviews, 71,* 58–82.

Kail, R., & Salthouse, T. A. (1994). Processing speed as a mental capacity. *Acta Psychologica, 86,* 199–225.

Karson, C. N. (1983). Spontaneous eye-blink rates and dopaminergic systems. *Brain: A Journal of Neurology, 106,* 643–653.

Kenward, B., Folke, S., Holmberg, J., Johansson, A., & Gredebäck, G. (2009). Goal directedness and decision making in infants. *Developmental Psychology, 45,* 809–819.

Keramati, M., Dezfouli, A., & Piray, P. (2011). Speed/accuracy trade-off between the habitual and the goal-directed processes. *PLoS Computational Biology, 7.*

Klossek, U. M. H., Russell, J., & Dickinson, A. (2008). The control of instrumental action following outcome devaluation in young children aged between 1 and 4 years. *Journal of Experimental Psychology: General, 137,* 39–51.

Kool, W., Gershman, S. J., & Cushman, F. A. (2017). Cost-benefit arbitration between multiple reinforcement-learning systems. *Psychological Science, 28,* 1321–1333.

Kudrimoti, H. S., Barnes, C. A., & McNaughton, B. L. (1999). Reactivation of hippocampal cell assemblies: Effects of behavioral state, experience, and EEG dynamics. *The Journal of Neuroscience: The Official Journal of Society for Neuroscience, 19,* 4090–4101.

Kurth-Nelson, Z., Barnes, G., Sejdinovic, D., Dolan, R., & Dayan, P. (2015). Temporal structure in associative retrieval. *eLife, 4,* e04919.

Lengyel, M., & Dayan, P. (2008). Hippocampal contributions to control: The third way. In *Advances in neural information processing systems* (pp. 889–896).

Lovibond, P. F. (1983). Facilitation of instrumental behavior by a Pavlovian appetitive conditioned stimulus. *Journal of Experimental Psychology. Animal Behavior Processes, 9,* 225.

Ludvig, E. A., Madan, C. R., & Spetch, M. L. (2015). Priming memories of past wins induces risk seeking. *Journal of Experimental Psychology: General, 144,* 24–29.

Luna, B. (2009). Developmental changes in cognitive control through adolescence. In P. Bauer (Ed.), *Advances in child development and behavior* (pp. 233–278).

Luna, B., Garver, K. E., Urban, T. A., Lazar, N. A., & Sweeney, J. A. (2004). Maturation of cognitive processes from late childhood to adulthood. *Child Development, 75,* 1357–1372.

MacDonald, A. W. (2000). Dissociating the role of the dorsolateral prefrontal and anterior cingulate cortex in cognitive control. *Science, 288,* 1835–1838.

Manzi, A., Nessler, D., Czernochowski, D., & Friedman, D. (2011). The development of anticipatory cognitive control processes in task-switching: An ERP study in children, adolescents and young adults. *Psychophysiology, 48,* 1258–1275.

McClure, S. M., Berns, G. S., & Montague, P. R. (2003). Temporal prediction errors in a passive learning task activate human striatum. *Neuron, 38,* 339–346.

McDannald, M. A., Takahashi, Y. K., Lopatina, N., Pietras, B. W., Jones, J. L., & Schoenbaum, G. (2012). Model-based learning and the contribution of the orbitofrontal cortex to the model-free world. *The European Journal of Neuroscience, 35,* 991–996.

Menon, V., Boyett-Anderson, J. M., & Reiss, A. L. (2005). Maturation of medial temporal lobe response and connectivity during memory encoding. *Cognitive Brain Research, 25,* 379–385.

Miller, E. K., & Cohen, J. D. (2001). An integrative theory of prefrontal cortex function. *Annual Review of Neuroscience, 24,* 167–202.

Mills, K. L., Goddings, A.-L., Herting, M. M., Meuwese, R., Blakemore, S.-J., Crone, E. A., … Sowell, E. R. (2016). Structural brain development between childhood and adulthood: Convergence across four longitudinal samples. *NeuroImage, 141,* 273–281.

Myers, C. E., Shohamy, D., Gluck, M. A., Grossman, S., Kluger, A., Ferris, S., … Schwartz, R. (2003). Dissociating hippocampal versus basal ganglia contributions to learning and transfer. *Journal of Cognitive Neuroscience, 15,* 185–193.

Naneix, F., Marchand, A. R., Scala, G. D., Pape, J.-R., & Coutureau, E. (2012). Parallel maturation of goal-directed behavior and dopaminergic systems during adolescence. *The Journal of Neuroscience: The Official Journal of the Society for Neuroscience, 32*, 16223–16232.

Ofen, N., Chai, X. J., Schuil, K. D. I., Whitfield-Gabrieli, S., & Gabrieli, J. D. E. (2012). The development of brain systems associated with successful memory retrieval of scenes. *The Journal of Neuroscience: The Official Journal of the Society for Neuroscience, 32*, 10012–10020.

Ohayon, M. M., Carskadon, M. A., Guilleminault, C., & Vitiello, M. V. (2004). Meta-analysis of quantitative sleep parameters from childhood to old age in healthy individuals: Developing normative sleep values across the human lifespan. *Sleep, 27*, 1255–1273.

Otto, A. R., Gershman, S. J., Markman, A. B., & Daw, N. D. (2013). The curse of planning: Dissecting multiple reinforcement-learning systems by taxing the central executive. *Psychological Science, 24*, 751–761.

Otto, A. R., Raio, C. M., Chiang, A., Phelps, E. A., & Daw, N. D. (2013). Working-memory capacity protects model-based learning from stress. *Proceedings of the National Academy of Sciences, 110*, 20941–20946.

Otto, A. R., Skatova, A., Madlon-Kay, S., & Daw, N. D. (2015). Cognitive control predicts use of model-based reinforcement learning. *Journal of Cognitive Neuroscience, 27*, 319–333.

O'Doherty, J. P., Dayan, P., Friston, K., Critchley, H., & Dolan, R. J. (2003). Temporal difference models and reward-related learning in the human brain. *Neuron, 38*, 329–337.

Pennartz, C. M. A., Ito, R., Verschure, P. F. M. J., Battaglia, F. P., & Robbins, T. W. (2011). The hippocampal-striatal axis in learning, prediction and goal-directed behavior. *Trends in Neurosciences, 34*, 548–559.

Pezzulo, G., van der Meer, M. A. A., Lansink, C. S., & Pennartz, C. M. A. (2014). Internally generated sequences in learning and executing goal-directed behavior. *Trends in Cognitive Sciences, 18*, 647–657.

Pfeiffer, B. E., & Foster, D. J. (2013). Hippocampal place cell sequences depict future paths to remembered goals. *Nature, 497*, 74–79.

Plebanek, D. J., & Sloutsky, V. M. (2017). Costs of selective attention: When children notice what adults miss. *Psychological Science, 28*, 723–732.

Potter, T. C. S., Bryce, N. V., & Hartley, C. A. (2017). Cognitive components underpinning the development of model-based learning. *Developmental Cognitive Neuroscience, 25*, 272–280.

Prehn-Kristensen, A., Munz, M., Molzow, I., Wilhelm, I., Wiesner, C. D., & Baving, L. (2013). Sleep promotes consolidation of emotional memory in healthy children but not in children with attention-deficit hyperactivity disorder. *PLoS One, 8*, e65098.

Preston, A. R., Molitor, R. J., Pudhiyidath, A., & Schlichting, M. L. (2017). Schemas. In H.Eichenbaum (Series Ed.) & J. H. Byrne (Vol. Ed.) (2nd ed., *Memory systems: Vol. III. Learning and memory: A comprehensive reference* (pp. 125–132). New York: Elsevier.

Preston, A. R., Shrager, Y., Dudukovic, N. M., & Gabrieli, J. D. E. (2004). Hippocampal contribution to the novel use of relational information in declarative memory. *Hippocampus, 14*, 148–152.

Raznahan, A., Shaw, P. W., Lerch, J. P., Clasen, L. S., Greenstein, D., Berman, R., … Giedd, J. N. (2014). Longitudinal four-dimensional mapping of subcortical anatomy in human development. *Proceedings of the National Academy of Sciences, 111*, 1592–1597.

Redshaw, J., & Suddendorf, T. (2013). Foresight beyond the very next event: Four-year-olds can link past and deferred future episodes. *Frontiers in Psychology, 4*.

Roshan Cools. (2008). Role of dopamine in the motivational and cognitive control of behavior. *The Neuroscientist: A Review Journal Bringing Neurobiology, Neurology and Psychiatry, 14*, 381–395.

Saffran, J. R., Aslin, R. N., & Newport, E. L. (1996). Statistical learning by 8-month-old infants. *Science, 274*, 1926–1928.

Schad, D. J., Jünger, E., Sebold, M., Garbusow, M., Bernhardt, N., Javadi, A.-H., … Rapp, M. A. (2014). Processing speed enhances model-based over model-free reinforcement learning in the presence of high working memory functioning. *Frontiers in Psychology, 5*.

Schapiro, A. C., Gregory, E., Landau, B., McCloskey, M., & Turk-Browne, N. B. (2014). The necessity of the medial temporal lobe for statistical learning. *Journal of Cognitive Neuroscience, 26*, 1736–1747.

Schlichting, M. L., Guarino, K. F., Schapiro, A. C., Turk-Browne, N. B., & Preston, A. R. (2017). Hippocampal structure predicts statistical learning and associative inference abilities during development. *Journal of Cognitive Neuroscience, 29*, 37–51.

Schuck, N. W., Cai, M. B., Wilson, R. C., & Niv, Y. (2016). Human orbitofrontal cortex represents a cognitive map of state space. *Neuron, 91,* 1402–1412.

Schultz, W., Dayan, P., & Montague, P. R. (1997). A neural substrate of prediction and reward. *Science, 275,* 1593–1599.

Sebold, M., Schad, D. J., Nebe, S., Garbusow, M., Jünger, E., Kroemer, N. B., ... Rapp, M. A. (2016). Don't think, just feel the music: Individuals with strong Pavlovian-to-instrumental transfer effects rely less on model-based reinforcement learning. *Journal of Cognitive Neuroscience, 28,* 985–995.

Shaw, P., Kabani, N. J., Lerch, J. P., Eckstrand, K., Lenroot, R., Gogtay, N., ... Rapoport, J. L. (2008). Neurodevelopmental trajectories of the human cerebral cortex. *The Journal of Neuroscience: The Official Journal of the Society for Neuroscience, 28,* 3586–3594.

Shenhav, A., Botvinick, M. M., & Cohen, J. D. (2013). The expected value of control: An integrative theory of anterior cingulate cortex function. *Neuron, 79,* 217–240.

Shohamy, D., & Adcock, R. A. (2010). Dopamine and adaptive memory. *Trends in Cognitive Sciences, 14,* 464–472.

Shohamy, D., & Daw, N. D. (2015). Integrating memories to guide decisions. *Current Opinion in Behavioral Sciences, 5,* 85–90.

Shohamy, D., & Turk-Browne, N. B. (2013). Mechanisms for widespread hippocampal involvement in cognition. *Journal of Experimental Psychology: General, 142,* 1159–1170.

Smeets, P. M., Barnes, D., & Roche, B. (1997). Functional equivalence in children: Derived stimulus–response and stimulus–stimulus relations. *Journal of Experimental Child Psychology, 66,* 1–17.

Smittenaar, P., FitzGerald, T. H. B., Romei, V., Wright, N. D., & Dolan, R. J. (2013). Disruption of dorsolateral prefrontal cortex decreases model-based in favor of model-free control in humans. *Neuron, 80,* 914–919.

Somerville, L. H., & Casey, B. (2010). Developmental neurobiology of cognitive control and motivational systems. *Current Opinion in Neurobiology, 20,* 236–241.

Somerville, L. H., Hare, T., & Casey, B. (2011). Frontostriatal maturation predicts cognitive control failure to appetitive cues in adolescents. *Journal of Cognitive Neuroscience, 23,* 2123–2134.

Stalnaker, T. A., Cooch, N. K., & Schoenbaum, G. (2015). What the orbitofrontal cortex does not do. *Nature Neuroscience, 18,* 620–627.

Stickgold, R., & Walker, M. P. (2013). Sleep-dependent memory triage: Evolving generalization through selective processing. *Nature Neuroscience, 16,* 139.

Suddendorf, T., Nielsen, M., & von Gehlen, R. (2011). Children's capacity to remember a novel problem and to secure its future solution. *Developmental Science, 14,* 26–33.

Sutton, R. S. (1991). Dyna, an integrated architecture for learning, planning, and reacting. *ACM SIGART Bulletin, 2,* 160–163.

Tanaka, S. C., Samejima, K., Okada, G., Ueda, K., Okamoto, Y., Yamawaki, S., & Doya, K. (2006). Brain mechanism of reward prediction under predictable and unpredictable environmental dynamics. *Neural Networks: The Official Journal of the International Neural Network Society, 19,* 1233–1241.

Thompson-Schill, S. L., Ramscar, M., & Chrysikou, E. G. (2009). Cognition without control: When a little frontal lobe goes a long way. *Current Directions in Psychological Science, 18,* 259–263.

Tolman, E. C. (1948). Cognitive maps in rats and men. *Psychological Review, 55,* 189.

Tricomi, E., Balleine, B. W., & O'Doherty, J. P. (2009). A specific role for posterior dorsolateral striatum in human habit learning. *The European Journal of Neuroscience, 29,* 2225–2232.

Tricomi, E. M., Delgado, M. R., & Fiez, J. A. (2004). Modulation of caudate activity by action contingency. *Neuron, 41,* 281–292.

Unger, K., Ackerman, L., Chatham, C. H., Amso, D., & Badre, D. (2016). Working memory gating mechanisms explain developmental change in rule-guided behavior. *Cognition, 155,* 8–22.

Valentin, V. V., Dickinson, A., & O'Doherty, J. P. (2007). Determining the neural substrates of goal-directed learning in the human brain. *The Journal of Neuroscience: The Official Journal of the Society for Neuroscience, 27,* 4019–4026.

van den Bos, W., Cohen, M. X., Kahnt, T., & Crone, E. A. (2012). Striatum—medial prefrontal cortex connectivity predicts developmental changes in reinforcement learning. *Cerebral Cortex, 22,* 1247–1255.

van der Meer, M., Kurth-Nelson, Z., & Redish, A. D. (2012). Information processing in decision-making systems. *The Neuroscientist: A Review Journal Bringing Neurobiology, Neurology and Psychiatry, 18*, 342–359.

van Duijvenvoorde, A. C. K., Achterberg, M., Braams, B. R., Peters, S., & Crone, E. A. (2016). Testing a dual-systems model of adolescent brain development using resting-state connectivity analyses. *NeuroImage, 124*, 409–420.

Wahlstrom, D., White, T., & Luciana, M. (2010). Neurobehavioral evidence for changes in dopamine system activity during adolescence. *Neuroscience and Biobehavioral Reviews, 34*, 631–648.

Wendelken, C., & Bunge, S. A. (2010). Transitive inference: Distinct contributions of rostrolateral prefrontal cortex and the hippocampus. *Journal of Cognitive Neuroscience, 22*, 837–847.

Werchan, D. M., Collins, A. G. E., Frank, M. J., & Amso, D. (2015). 8-month-old infants spontaneously learn and generalize hierarchical rules. *Psychological Science, 26*, 805–815.

Werchan, D. M., Collins, A. G. E., Frank, M. J., & Amso, D. (2016). Role of prefrontal cortex in learning and generalizing hierarchical rules in 8-month-old infants. *The Journal of Neuroscience: The Official Journal of the Society for Neuroscience, 36*, 10314–10322.

Wierenga, L. M., Langen, M., Oranje, B., & Durston, S. (2014). Unique developmental trajectories of cortical thickness and surface area. *NeuroImage, 87*, 120–126.

Wilhelm, I., Prehn-Kristensen, A., & Born, J. (2012). Sleep-dependent memory consolidation — what can be learnt from children? *Neuroscience and Biobehavioral Reviews, 36*, 1718–1728.

Wilson, M. A., & McNaughton, B. L. (1994). Reactivation of hippocampal ensemble memories during sleep. *Science, 265*, 676–679.

Wilson, R. C., Takahashi, Y. K., Schoenbaum, G., & Niv, Y. (2014). Orbitofrontal cortex as a cognitive map of task space. *Neuron, 81*, 267–279.

Wimmer, G. E., & Shohamy, D. (2012). Preference by association: How memory mechanisms in the hippocampus bias decisions. *Science, 338*, 270–273.

Wunderlich, K., Smittenaar, P., & Dolan, R. J. (2012). Dopamine enhances model-based over model-free choice behavior. *Neuron, 75*, 418–424.

Yin, H. H., & Knowlton, B. J. (2006). The role of the basal ganglia in habit formation. *Nature Reviews Neuroscience, 7*, 464–476.

Yin, H. H., Knowlton, B. J., & Balleine, B. W. (2004). Lesions of dorsolateral striatum preserve outcome expectancy but disrupt habit formation in instrumental learning. *The European Journal of Neuroscience, 19*, 181–189.

Yin, H. H., Ostlund, S. B., Knowlton, B. J., & Balleine, B. W. (2005). The role of the dorsomedial striatum in instrumental conditioning. *The European Journal of Neuroscience, 22*, 513–523.

Zeithamova, D., Dominick, A. L., & Preston, A. R. (2012). Hippocampal and ventral medial prefrontal activation during retrieval-mediated learning supports novel inference. *Neuron, 75*, 168–179.

Zeithamova, D., Schlichting, M. L., & Preston, A. R. (2012). The hippocampus and inferential reasoning: Building memories to navigate future decisions. *Frontiers in Human Neuroscience, 6*.

Zelazo, P. D., Frye, D., & Rapus, T. (1996). An age-related dissociation between knowing rules and using them. *Cognitive Development, 11*, 37–63.

Zhou, X., Zhu, D., King, S. G., Lees, C. J., Bennett, A. J., Salinas, E., … Constantinidis, C. (2016). Behavioral response inhibition and maturation of goal representation in prefrontal cortex after puberty. *Proceedings of the National Academy of Sciences of the United States of America, 113*, 3353–3358.

CHAPTER 14

Social Learning: Emotions Aid in Optimizing Goal-Directed Social Behavior

Oriel FeldmanHall[1], Luke J. Chang[2]

[1]Department of Cognitive, Linguistic & Psychological Sciences, Brown University, Providence, RI, United States;
[2]Department of Psychological and Brain Sciences, Dartmouth College, Hanover, NH, United States

INTRODUCTION

Goal-directed behavior plays an enormous role in everyday human life. Goal pursuit allows humans to navigate through life in a purposeful way, facilitating the conceptualization and achievement of highly abstract and complex goals ("I want to be a doctor when I grow up"), as well as simpler, more immediate goals ("I have a hankering for a good burger, I wonder where I can find one?"). Achieving outcomes along this spectrum of goals requires a cognitive system that can translate higher level, conceptual goals into tractable, concrete actions (Gollwitzer & Moskowitz, 1996, pp. 361–399). Decades of work have been devoted to understanding the neurocognitive system that governs such goal-directed behavior. We now know much about how we represent goals and keep them in our working memory (Badre, Satpute, & Ochsner, 2012), how we update goals (Daw, Niv, & Dayan, 2005), and what happens when conflicts arise between multiple goals (Botvinick, Cohen, & Carter, 2004). However, much less is known about how goal-directed behavior unfolds in dynamic and evolving social environments. And yet, many of our most important goals—for example, being a caring parent, teacher, or citizen of the community—have enormous social and emotional qualities.

Here we propose a psychological model that captures goal-directed behavior within the social domain. Because society and social groups function better and are more effective when individuals cooperate and help others (Tyler & Blader, 2000), one long-term superordinate goal is to maintain the well-being of the group. We argue that upholding the group's welfare can be broken down into a handful of basic social needs (e.g., preventing harm to others), which serve as a core suite of goal-directed motivations. These basic goals typically act in direct opposition to the more immediate goal to self-enhance and promote one's own welfare (i.e., one's own well-being does not always bear a one-to-one correspondence with the group's well-being) (Thibaut & Kelley, 1959, 1978). The integration and subsequent resolution of these conflicting goal states is paramount to successful socialization. The affective signals (i.e., emotional prediction errors) that accompany the representations of social goals facilitate the

Goal-Directed Decision Making
ISBN 978-0-12-812098-9, https://doi.org/10.1016/B978-0-12-812098-9.00014-0

resolution of conflicts between different goals, helping to arbitrate between whether a goal is actively pursued or abandoned.

Subsequently, translating these goals into actionable outputs requires that an individual dynamically learn and update their goals through experience. Individuals constantly shift goals and strategies according to the outcomes of their previous choices and other information received from the changing environment. Since people are motivated to adapt their choices to correspond with the ever evolving social world, we posit that goal-directed behavior is flexible across contexts. In other words, social goals, and their implementation, are not stable, but rather are modulated by the context in which the goal-directed behavior arises. Finally, we conclude by proposing a social value cost function model that, depending on how each basic social need is weighted, can dictate the type of goal-directed behavior that is employed. This model facilitates a formalized testable hypothesis of the mechanisms underlying goal-directed social behavior.

Motivation and representation of social goals

Successfully navigating the social world requires that one's goals be represented in a manner consistent with promoting social well-being. Social goals are internal mental representations that relate to attaining an end state involving the welfare of others or the group (Kruglanski & Webster, 1996). Because social goals are so intimately linked with individual experiences, they can vary widely between people (Reeve, 2008). Here we argue, however, that irrespective of an individual's experience, there are a handful of goals—that manifest as needs and desires—which help to facilitate successful socialization. This set of goals are internal states that motivate behavior and goal pursuit within social environments (Ajzen & Fishbein, 1969). We suggest there are three basic goals that operate most potently within the social domain: (1) preventing harm to others, (2) social affiliation, and (3) minimizing uncertainty. This set of social goals can, at times, act in direct opposition to enhancing one's own well-being (a conflicting goal). The result is a tension between behaving in ways that facilitate successful socialization and acting in ways that augment one's own welfare.

Harm prevention

One of the most deeply held social goals is to prevent harm to others (Haidt, 2012). This desire to prevent harm has a long evolutionary lineage that can be traced back through our phylogenetic tree to ancestors common to other primate species (de Waal, 1997). Research illustrates that people are not only averse to performing harmful actions (Cushman, Gray, Gaffey, & Mendes, 2012; Greene, Sommerville, Nystrom, Darley, & Cohen, 2001; Mikhail, 2000), but will go to great lengths to avoid harming another, even in the face of a superordinate goal—such as being commanded by an army officer to kill during battle (Grossman, 1996). Indeed, observing another in pain is enough to increase one's physiological reactivity (Cushman et al., 2012), an indication of being in a highly aversive state. Individuals also report increased feelings of psychological distress in the wake of applied harm (Batson, Van Lange, Ahmad, & Lishner, 2003). This heightened aversive emotional state is generated even in the absence of actual harmful outcomes, for instance, when using a rubber knife to simulate stabbing (Cushman et al., 2012). The "avoid harm" goal manifests both as a need to refrain from impulses that may result in harm to others and as a desire to actively behave in ways that thwart harm (Bandura, 1991). Effectively, this potently experienced "avoid harm" goal helps to buoy collective social welfare and successful socialization.

Individuals who routinely exhibit behavioral patterns consistent with breaking the "avoid harm" goal (e.g., psychopaths), fail to inhibit impulses to harm, and exhibit little remorse, guilt, or empathy in the aftermath of harmful actions. Most diagnosed psychopaths do not farewell in society (Petherick, 2014), as they are three times as likely to partake in violent, criminal behavior and spend a significant portion of their adult lives behind bars (Kiehl & Hoffman, 2011). One theory posits that psychopaths do not hold the "avoid harm" goal because they exhibit generally low levels of affective physiological responsivity (Wang, Baker, Gao, Raine, & Lozano, 2012). Without sufficient levels of physiological responding, psychopaths seek out behavior that stimulates their heart rate and arousal levels, which can often take the form of violent behavior.

Affiliation

A second overarching social goal is the need to affiliate with others (Baumeister & Leary, 1995; McClelland, 1985). Examples from both inside and outside the laboratory reveal the strength by which humans feel and act on the desire to belong. Even the existence of a superficial social connection to another person (e.g., sharing a birthday) or group (e.g., finding out that you belong to an arbitrarily defined minimal group) makes the goal of needing to belong more accessible and relevant, and can ultimately bolster the motivation to affiliate (Walton, Cohen, Cwir, & Spencer, 2012).

Affiliative motivations can be so strong that an individual may conform to the behavioral patterns of others, even if the individual does not agree or approve of their own conforming behavior (Asch, 1956). In some instances, conforming to the group can be quite innocuous, such as when individuals shift their own preferences for how much they like certain musical styles (Berns & Moore, 2011; Campbell-Meiklejohn, Bach, Roepstorff, Dolan, & Frith, 2010) or how attractive they find another individual (Klucharev, Hytonen, Rijpkema, Smidts, & Fernandez, 2009; Zaki, Schirmer, & Mitchell, 2011). In other cases, conforming can have positive social consequences, such as enhancing one's own cooperative or charitable behavior after watching others behave prosocially (Fowler & Christakis, 2010; Nook, Ong, Morelli, Mitchell, & Zaki, 2016).

This canon of work illustrates that explicit social influence can readily induce compliant behavior (Asch, 1956; Cialdini & Goldstein, 2004). Emerging research reveals just how deep the need to affiliate goes: people are so sensitive to subtle social dynamics that they are even willing to alter their own behavior in the absence of overt social influence. For example, simply observing risk-taking behaviors in others increases one's own risk-taking profile (Suzuki, Jensen, Bossaerts, & O'Doherty, 2016). Within the moral domain, learning to implement the punishment preferences of a person who was the victim of a fairness violation enhances one's own desire to punish once they are affronted with a similar moral violation (FeldmanHall, Otto, & Phelps, 2018). Together, these findings indicate the strength by which individuals desire to affiliate and gain approval from others (Baumeister & Leary, 1995).

Converging evidence of affiliative behavior can also be observed in real-time laboratory interactions, such as when two or more individuals engage in economic games. In one-shot public goods games, where players can choose whether or not to put their own money into a common pool (which subsequently gets divided equally among the group members), contributions are typically high (Levitt & List, 2007). The fact that most people contribute, even when they understand that in order to maximize their own payout they should free ride, suggests that people are willing to forgo monetary rewards in order to signal a positive reputation to others. These findings are mirrored in one-shot prisoner's dilemmas, where 50%—60% of individuals choose to cooperate despite knowing they can potentially receive an even higher payoff if they were to defect (so long as their opponent chooses to cooperate, Barcelo & Capraro, 2015). Additional evidence from the priming literature illustrates that subconsciously activating the representation of a goal—in this case, priming participants with words related to "cooperation"—can also cause people to work together more in economic games (Bargh, Lee-Chai, Barndollar, Gollwitzer, & Trötschel, 2001).

The need to affiliate is also exemplified by how strongly individuals adhere to and enforce social norms, oftentimes at a cost to themselves. This type of goal setting is triggered by environmental cues that direct and motivate social goals to cooperate, socialize, trust, punish, and reciprocate (Charness & Rabin, 2002). If we take the case of punishment as an example, the social norm in Western cultures prescribes that a transgressing perpetrator should be punished, even if the individual who is enacting the punishment does not directly benefit herself. Many studies have demonstrated that people are willing to punish on behalf of others (i.e., third-party punishment), even when it is highly costly (Fehr & Gachter, 2002). Since it can be monetarily costly for an individual to punish a perpetrator on behalf of another victim, punishing can be likened to an altruistic act. Therefore, third-party punishment typically reflects the shared communal demands of the environment, revealing that behavior is oriented toward the collective goals of the social community and not simply the interests of the self. Indeed, without such situational cues dictating appropriate social behavior alongside a strong desire to affiliate, individuals would likely behave in ways that would maximize their own welfare, without regard for the greater good.

Typically, the need to belong and affiliate is strongest among one's own group members: Individuals will go to great lengths to behave in ways that accord with behavioral patterns elicited from people deemed similar (e.g., conformity with ingroup but not outgroup, Cikara & Van Bavel, 2014). This need to affiliate with one's own group can be so strong that people are even willing to endure pain (in the form of electric shocks) to prevent other group members—even previously unknown individuals—from receiving electric shocks themselves (Hein, Silani, Preuschoff, Batson, & Singer, 2010).

Minimize uncertainty

Daily social life is rife with the constant need to navigate uncertain social exchanges, including deciding who to trust and confide in, whether or not to loan money to an acquaintance, or contemplating whether to participate in a potentially dangerous activity with your friends (e.g., consume alcohol or drugs). These social choices can be risky, as it is uncertain how outcomes will unfold (e.g., Can my friend keep a secret? Will my money be returned?). Most people experience uncertainty as highly aversive (Bar-Anan, Wilson, & Gilbert, 2009; Ellsberg, 1961; Hirsh, Mar, & Peterson, 2012) and therefore have a strong desire to resolve it (Kahneman, Slovic, & Tversky, 1982). These aversive feelings that accompany the experience of uncertainty become especially acute in the social domain, which acts as a potent motivator for reducing such uncertainty.

The goal to reduce social uncertainty can be met by acquiring information about other individuals. This can happen through direct experience—repeated engagements where trust or cooperative behavior can be explicitly experienced or tested (Chang, Smith, Dufwenberg, & Sanfey, 2011; Fareri, Chang, & Delgado, 2012, 2015; King–Casas et al., 2005), or by vicariously learning about the social value of another (Olsson, Nearing, & Phelps, 2007). Either route allows an individual to gain more information and update their social value estimates of other people, which leads to a stable and restricted set of expectations about the personalities and potential emotional and cognitive states of each person. In doing so, an individual can better predict what others will do, which in turn allows them to better predict their own future states. Ultimately, an individual's social success and well-being is tied to her ability to resolve the uncertainty associated with other people and social situations, with failures to do so often manifesting in clinical mood disorders (Engelmann, Meyer, Fehr, & Ruff, 2015).

Enhance self-benefit

The basic goals described above can often act in direct opposition to the desire to enhance one's own well-being and self-benefit. The ability to enhance the self—whether through money, power, or prestige—serves as an evolutionarily primal drive that aims to optimize survival. For example, research suggests that the appeal of money can have a profoundly negative influence on people's willingness to engage in prosocial behavior. If given the opportunity to make more money through cheating and lying, individuals routinely cheat and lie (Ariely & Gino, 2011; Greene, Rand, & Nowak, 2012). Other research illustrates that when the monetary incentive is great, the number of individuals willing to break social norms, such as reciprocal trust, increases dramatically (Gneezy et al., 2011). Indeed, our own work reveals that if the monetary enhancement is sufficiently compelling (approximately $300), individuals administer electric shocks to others, willingly forgoing the "avoid harm" goal in order to enhance their own monetary self-benefit (FeldmanHall, Mobbs, Hiscox, Navrady, & Dalgleish, 2012; Feldmanhall, Dalgleish, & Mobbs, 2013).

The ease with which individuals abandon prosocial goals (e.g., affiliation, harm prevention) in the service of enhancing their own welfare, suggests that self-benefit is an equally potent and accessible goal. For example, amplifying an individual's dominance and prestige (through priming manipulations) can enhance the salience of the self-benefit goal state, which ultimately reduces the willingness to engage in prosocial acts (Guinote, 2007). Converging evidence also reveals how feeling powerful can enhance reward-seeking and disinhibited social behavior across a variety of contexts (Galinsky, Gruenfeld, & Magee, 2003; Keltner, Gruenfeld, & Anderson, 2003). In these cases, feeling powerful typically increases the rate at which an individual engages in antisocial acts, such as increased sexual aggression and harassment (Bargh, Raymond, Pryer, & Strack, 1995). In contrast, reduced power is associated with inhibited, avoidant social behavior, which traditionally aligns with group norms or the promotion of social welfare.

Translating goals to action

How do we translate social goals into actions? In order to elucidate the mechanisms governing goal-directed social behavior, we must understand the computational processes that are involved in translating superordinate goals into behavioral outputs. Quantitative disciplines such as economics, engineering, and computer science have been successful in developing normative frameworks that can describe an optimal decision policy given a set of specific goals. However, while computationally viable, these decision policies do not always capture how people *actually* behave (Camerer, 2003; Kahneman, 2003; Kahneman & Tversky, 1979). For example, standard economic theory, which assumes a rational agent is solely motivated by self-interest, is particularly poor at predicting behavior in cooperative social interactions such as bargaining (Guth, Schmittberger, & Schwarze, 1982), trust (Berg, Dickhaut, & Mccabe, 1995), and public goods games (Yamagishi, 1986). Subsequent theories that incorporate psychological motivations associated with emotions (Bell, 1982; Charness & Dufwenberg, 2006; Loewenstein, 1987, 1996; Loomes & Sugden, 1982; Mellers, Schwartz, Ho, & Ritov, 1997), concern for others' intentions (Rabin, 1993), and the collectives' outcomes

(Bolton & Ockenfels, 2000; Fehr & Gachter, 1999) have dramatically improved the ability to predict actual social behavior.

Decision policies

We review several quantitative frameworks that can be used to characterize motivated social behavior. In particular, these decision policies describe how, given an individual's social goals, selecting certain actions can maximize a stimulus' subjective value. Broadly, this process can be viewed as an optimization problem where an agent selects actions that maximize the likelihood of achieving their current social goal. We assume agents consider the expected costs and benefits associated with the outcome of a given choice and select the choice with the highest overall expected outcome for the self. There are several different types of decision policies. For example, consider a set of choice options X, where i describes a specific choice $i \in X$. Selecting the action that most aligns with the agent's goal requires applying an explicit decision policy. These policies can be deterministic, such as the greedy rule that always selects the available action a_i associated with the highest predicted expected outcome value or utility u, $a_i = argmax(u(X))$.

Alternatively, policies can select options stochastically. In these cases, choices are selected in proportion to their overall value, and the degree of stochasticity is controlled by a temperature parameter. This is traditionally modeled as a softmax function, more formally

$$a_i = \frac{e^{\frac{u_i}{\beta}}}{\sum_{j=1}^{n} e^{\frac{u_j}{\beta}}}, \tag{14.1}$$

where a_i is the probability of selecting action i, u_i is the value of action i, n is the total number of actions, and $0 < \beta < 1$ is the temperature parameter representing the stochasticity of the decision policy. Social decisions can be modeled as weighing the expected costs and benefits associated with each choice outcome and applying a decision policy to select the action with the highest overall expected utility (e.g., greedy, softmax, etc.).

Common valuation system

These various decision policies provide a principled rule for how to select an action after comparing the pros and cons of each available choice. The ability to apply a decision policy in order to carry out the desired social goal is predicated on the assumption that we can quantitatively compare the value associated with each option in the goal set. Traditionally, the pros and cons of each choice are only considered with respect to *self-interested goals* (e.g., how much money will I win or lose?). However, it remains an open question as to how we can incorporate *social goals* (e.g., harm prevention, affiliation, and social uncertainty) into a common value function, which can be compared across choices. This requires establishing a common *value metric* (Levy & Glimcher, 2012) to integrate all of the costs and benefits associated with the available options including one's social goals (Rangel, Camerer, & Montague, 2008; Ruff & Fehr, 2014).

Economics, for example, has developed a number of tools to help establish a common value metric at the behavioral level. The Weak Axiom of Revealed Preferences (WARP) establishes the existence of a convex utility function that describes a rational agent's preferences for bundles of goods by only assuming transitivity (i.e., if A > B, and B > C, then A > C, Samuelson, 1938). Importantly, this function can also be usefully applied to social contexts, including how agents value the outcomes of others (i.e., an altruistic response, Andreoni & Miller, 2002). These are commonly referred to as social or other-regarding preferences in behavioral economics.

Expected Utility Theory is another framework for describing expectations of subjective value that adds several additional axioms to WARP in addition to transitivity (i.e., completeness, independence, and continuity) (Bernoulli, 1738; von Neumann & Morgenstern, 2007). This theory provides a powerful normative framework to describe optimal decision-making strategies when making choices under uncertainty. For example, the expected utility resulting from selecting a given choice u_i can be formally described as the sum of the value of each attribute c associated with the outcome $v_{i,c}$ scaled by the expectation of the outcome being realized $p_{i,c}$,

$$u_i = \sum_{c=1}^{n} p_{i,c} \cdot v_{i,c} \qquad (14.2)$$

Though Expected Utility Theory has been very successful at providing a normative framework to understand how policies can impact economies, it has not fared as well describing how individuals make decisions. Indeed, the impressive growth and popularity of behavioral economics in the 1970s and 1980s can be attributed to an increasing realization that such normative theories of how people ought to make decisions substantially deviated from observations of how people actually behaved (Camerer, 2003; Kahneman, 2003; Kahneman & Tversky, 1979). This groundbreaking work resulted in a number of extensions to the Expected Utility Theory framework which incorporates psychological values, such as sources of value originating from social preferences (Fehr & Camerer, 2007), empathic concern for others (FeldmanHall, Dalgleish, Evans, & Mobbs, 2015), and emotional motivations (Chang & Jolly, 2017; Chang & Smith, 2015). In the following sections, we build on this framework and outline how emotional value signals arising from approaching and avoiding social goals can be integrated with self-interested value signals to impact subsequent actions.

Emotion motivates social goal-directed behavior

Emotions describe a set of phenomenological experiences that result from the intersection of our broader goals, moment-to-moment cognitive evaluations of the external world, and our internal physiological states. Similar to the somatovisceral sensations that signal internal homeostatic goal states such as hunger, thirst, and sleep (Panksepp, 1998), emotions provide motivational signals that guide us to approach resources, avoid harm (Davidson & Irwin, 1999), and navigate the complexities of the social world (Chang & Jolly, 2017; Chang & Smith, 2015; Chang et al., 2011; FeldmanHall et al., 2013, 2015). Emotions can impact the decision-making process in a variety of ways (Chang & Sanfey, 2008; Loewenstein & Lerner, 2003). At the time of the decision, immediate emotions (e.g., gut feelings associated with risk) or incidental emotions (e.g., transient moods) can bias how we interpret information and value outcomes. Emotions can also be anticipated as an affective experience resulting from the outcome of selecting an action and incorporated directly into the value function associated with the choice. In general, we tend to value things that will make us feel good and devalue things that will make us feel bad. Ultimately, these valenced motivations can serve as signals to guide us to approach or avoid outcomes depending on our broader goals (Carver & Scheier, 1990). In the following sections, we provide examples of how emotions can impact behavior at both conscious and nonconscious levels.

Conscious emotions

Incorporating emotional motivations into utility functions through counterfactual value can dramatically improve predictions of both social (Koenigs & Tranel, 2007) and nonsocial behavior (Bell, 1982; Coricelli, Dolan, & Sirigu, 2007; Lohrenz, McCabe, Camerer, & Montague, 2007; Loomes & Sugden, 1982). Imagine buying a brand-new computer only to find out that if had you waited another week it would have been discounted an additional 15%. Though we are equally satisfied with the product, we often devalue the purchase as a consequence of regretting buying the computer a week too soon. By modeling the emotion regret, researchers have been able to capture the fact that although people are motivated by maximizing their own outcomes (e.g., the goal to enhance self-benefit), they also care about minimizing making a suboptimal decision, even if such a decision is associated with an overall positive outcome (Gilovich & Medvec, 1995; Mellers & McGraw, 2001).

An additional extension to Expected Utility Theory provided by psychological game theory is the ability to incorporate belief-dependent value into utility functions (Dufwenberg & Kirchsteiger, 2004; Geanakoplos, Pearce, & Stacchetti, 1989). This framework allows utility functions to incorporate a variety of psychological motivations, such as sensitivity to fairness (e.g., reciprocating others' good or bad intentions, Dufwenberg & Kirchsteiger, 2004; Rabin, 1993) and social emotions such as guilt from disappointing a relationship partner (Battigalli & Dufwenberg, 2007; Dufwenberg & Gneezy, 2000) and anger from believing a social norm have been violated (Battigalli, Dufwenberg, & Smith, 2015; Chang & Sanfey, 2013; Chang & Smith, 2015). The marriage of these psychological motivations with formal models has provided a useful first-order approximation for how people integrate different sources of value and has been successfully leveraged to predict behavior in cooperative socially interactive contexts.

Nonconscious emotions

Goal-directed behavior is also known to be guided by nonconscious emotional mental processes (Aarts et al., 2005; Aarts, Custers, & Veltkamp, 2008; Custers & Aarts, 2010). In these cases, an individual's repertoire of readily available social goals is directly accompanied by a suite of implicit affective signals, typically conceptualized in terms of valence (e.g., positive or negative), that act as motivators or demotivators, depending on the context (Fazio, Sanbonmatsu, Powell, & Kardes, 1986).

For example, by activating the reward system, positive affect motivates the pursuit of approach-related social goals (Aarts et al., 2008; Custers & Aarts, 2010; Davidson, 1992). If a goal previously exists as a desired state, then it is already yoked to a positive affective signal that enhances how readily one pursues the goal—assuming the goal is primed. This has been shown to influence a number of social phenomena, including socializing (Aarts & Custers, 2007) and social equity concerns (Ferguson, 2007). Positive affect, however, can also be paired with a neutral goal (Aarts et al., 2005). In these situations, priming affectively valenced words outside of conscious awareness can activate evaluative conditioning processes (De Houwer, Thomas, & Baeyens, 2001), such that repeatedly pairing neutral goal concepts (e.g., drinking) with positive valenced phenomena (e.g., words such as "nice") increases the motivation to pursue the formally neutral goal of drinking.

In contrast, negative affect can act as a demotivator of social goal pursuit. For example, subliminally priming undergraduates with negatively valenced words (e.g., "war" or "trash") in conjunction with the goal of partying—a well-documented desired state (Sheeran et al., 2005)—made participants work less to attain the goal of partying (Aarts, Custers, & Holland, 2007). Dovetailing with this, pioneering work exploring the role of physiological arousal processes, indexed through galvanic skin responses, found that arousal levels bias how readily one continues to pursue rewarding or unrewarding choices (Bechara, Damasio, & Damasio, 2000; Ferguson & Bargh, 2004; Phelps, 2005; Winkielman & Berridge, 2004). Thus, even without conscious awareness, emotions can serve to shape choices to either pursue or abandon a social goal.

From emotions to social goals

These findings, which were popularized in canonical control theory accounts of motivated behavior (Carver, 1984; Carver & Scheier, 1981; Carver & Scheier, 1990), highlighted two critical aspects of the relationship between affect and goal pursuit in the nonsocial domain. First, affective signals act as a basic input in determining the motivation to pursue a goal. Second, they do so by changing the rate at which an individual pursues or avoids certain goals (Cacioppo, Gardner, & Berntson, 1999; Phelps, 2005)—a theory that has proved to be robust across time and disciplines. For example, consider how we regulate our hunger. Glucose levels in our blood are constantly assessed by the hypothalamus. When glucose starts to drop below a certain homeostatic level, we begin to feel a proportional amount of hunger, which effectively prioritizes our control system to set goals to seek out food. In the pursuit of food, we employ a decision policy to find a meal that satisfies our resource constraints (e.g., location, time, and financial budget) as well as our motivational desires (that juicy burger). Once food is found and consumed, hormones begin converting the food into energy and our hunger dissipates—which then frees up our control system to prioritize other goals to pursue.

Although much less is known about how humans flexibly adapt their behavior as they navigate through social contexts that require simultaneous pursuit of multiple goals, it stands to reason that a similar relationship between emotions and goal-directed behavior exists in the social domain as well. Indeed, the notion that the affect influences social goals has been previously proposed outside the field of psychology. In sociology, for example, it has been argued that emotions can help align both our actions and identities during social interactions (Heise, 2007; Rogers, 2015; Schroder, Hoey, & Rogers, 2016). It is likely that the computational processes supporting flexible social goal pursuit are similar to other control systems that regulate human behavior (e.g., a common valuation system), including low-level homeostatic processes (Panksepp, 2004; Robinson & Berridge, 2013) and higher-level cognitive control processes (Miller & Cohen, 2001). This analogous control system allows us to incorporate and prioritize multiple, and sometimes competing, social goals. In these cases, emotions would provide both an approach and avoid signal when monitoring progress toward one of our three fundamental social goals to affiliate, minimize harm to others, or reduce our overall uncertainty—especially when these social goals come in direct conflict with the goal to enhance one's own self-benefit.

These goal-directed behaviors are also likely to be modulated by social context. For example, if resources are scarce, one's own needs are likely to be more salient—and with it, the need to enhance self-benefit. In contrast, if resources are bountiful, one is likely able to attend to the needs of others more readily. In a similar vein, if one knows they will be repeatedly encountering a certain person, goals to affiliate and minimize harm, for example, are likely to become more salient than in situations in which one encounters a person they know they will never see again. The cues provided by social environments help establish the appropriate social goal and the attendant emotional signal. Below, we outline how affective error signals might motivate us to achieve the social goals to minimize harm to others and seek out social affiliation.

Model of goal-directed social behavior
A social-affective control model

Building off accounts that illustrate affect as a key component of many goal-directed or control theories of behavior, we propose a system for selecting actions that help pursue a social goal (Fig 14.1). Critically, we believe social goals are regulated via positive and negative emotional error signals that impact our decision policies. First, the system establishes a social goal; take for instance the desire to affiliate. Values associated with different attributes of the decision space—in this case, ensuring that affiliative or conforming actions are taken—are integrated, and a decision policy (e.g., softmax) is applied to select the next action from the set of possible choices. The system then continually monitors the environment for outcomes that result from selecting this action. By evaluating how our position changes relative to the goal of social affiliation (e.g., actions that bring one closer to successfully or unsuccessfully affiliating), progress toward achieving this goal can be monitored after every action. If the action resulted in deviations from affiliating with others, this creates an error signal in the form of an emotion (e.g., negative affect, Chang & Jolly, 2017; Montague & Lohrenz, 2007). This emotional error signal is then integrated into the value function, which ultimately biases which action is selected next. The rate of change (e.g., how quickly actions are updated in accordance with the emotional error signal) reflects how quickly we are motivated toward achieving our goal to affiliate and directly corresponds to the attendant-affective responses that monitor our ongoing progress.

This framework can be used to illustrate how behavior can be optimized to maximize multiple social goals that can dynamically change and shift as our priorities also shift across various contexts. Moreover, when goals compete with one another (e.g., enhance one's own benefit or affiliate with others), the associated affective responses that monitor actions that bring one closer to either goal will shape which goal is ultimately pursued. For example, if the negative emotions that accompany failing to pursue the goal to affiliate is stronger than the negative emotional response for failing to increase the money in one's bank account, then one should pursue the goal of affiliating.

Negative affect

One way in which we can satisfy our goal to affiliate with others is to avoid acting differently from the group norm. This can be described as minimizing the distance

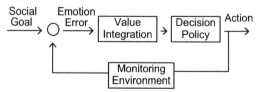

Figure 14.1 Theoretic model of how emotional error signals help us to adaptively select actions that are aligned with our social goals (e.g., preventing harm to others).

between our own behavior and shared beliefs about appropriate behavior for a specific context (i.e., social norms). Classic experiments demonstrate that we are motivated to behave in ways that consistently align with such "descriptive norms"—that is, what we believe most people would do, even when it counters our own beliefs (Asch, 1956; Cialdini & Goldstein, 2004). One instantiation of needing to affiliate with others stems from feeling intense negative affect when we are ostracized from the group or feel that we do not fit in with our peers (Leary & Richman, 2009). These negative feelings can be so powerful that the mere anticipation of being excluded, shunned, or at odds with others can generate attendant negative feelings, which in turn motivates people to act in ways that accord with the group's behavior. Within the framework of our model, the value of affiliation can be formulated as a utility function, where an agent receives utility from selecting the choice that maximizes their payoff π_i, while minimizing the deviation from the group's behavior. Here, we define a descriptive norm set by the group as the expectation of our beliefs about the likelihood of others taking certain actions i, $E[\phi_i]$ (Sanfey, Stallen, & Chang, 2014). Deviations from the group behavior generate negative affect signals.

$$u(i) = \pi_i - \alpha \cdot max(E[\phi_i] - i, 0) - \beta \cdot min(i - E[\phi_i], 0), \qquad (14.3)$$

where α and β differentially scale signed deviations from the groups' normative behavior and are constrained between [0,1].

This utility function can describe a host of affiliative behaviors, including norm adherence and enforcement in the two-person bargaining task known as the ultimatum game (UG). In the UG, Player A proposes a split of an endowment to Player B. Player B then decides whether to accept the split as proposed, or reject the offer, thereby punishing Player A—in which case both players receive nothing (Guth et al., 1982). Chang and Sanfey tested a variant of this conformity model using participants' self-reported beliefs about the type of proposals they expected to encounter in the game (Chang & Sanfey, 2013). This model provided a better account of players' decisions compared to another social preference model that proposes that players prefer equitable outcomes (Bolton & Ockenfels, 2000; Fehr & Schmidt, 1999).

In a related experiment, Xiang and colleagues provide even stronger evidence for how we track social norms (Xiang, Lohrenz, & Montague, 2013). In this study, the experimenters manipulated participants' expectations in the UG by exposing different participants to three different distributions of offers (high, medium, low). In the test phase, all groups of participants were given offers from the medium distribution. Participants decided whether to accept or reject the offers and reported their affective responses by selecting from a set of emoticons. The authors combined an ideal Bayesian observer model with a social preference utility function to track how beliefs about the social norm are updated after each offer (Chang & Sanfey, 2013; Fehr & Schmidt, 1999). This provided trial-by-trial estimates of the prediction error and variance prediction error for a given offer conditional on prior beliefs. Identical offers were rejected more frequently when participants expected offers from a high distribution compared to when participants expected offers from a low distribution, indicating that the social norm manipulation successfully changed attitudes toward fairness violations and subsequent decisions to refrain from punishing norm violators. Effectively, specific social norms can influence expectations, which in turn allow individuals to update their social goals—in this case, deciding whether to punish.

Regret

Importantly, this social-affective control system can also operate on simulated actions and does not require that the system only update after experiencing an outcome consequent to our actions (Chang & Jolly, 2017). This is a critical feature, as it allows for anticipated feelings resulting from expected outcomes to create emotional error signals (Chiu & Montague, 2008) that provide value with an associated action (Mellers & McGraw, 2001). One example of this simulated process—and one we touched on briefly before—is anticipated regret (Coricelli et al., 2007), or when an emotional error signal results from making a decision that does not align with the goal of making the best decision. If a decision turns out to have an unfavorable outcome, the deciding agent will most likely feel disappointment. However, if an agent makes a decision and learns that they could have made an even better decision regardless of outcome favorability, they will feel regret (Bell, 1982; Loomes & Sugden, 1982; Mellers et al., 1997). Thus, regret serves as an error signal that indicates deviations from the broader goal of making an optimal social choice (e.g., deciding to help someone who may subsequently return the favor, thereby eliciting better future outcomes for yourself). Importantly, regret can be anticipated at the time of decision, which can change our valuation of the choice set, and can ultimately motivate current behavior to minimize future regret (Bault, Coricelli, & Wydoodt, 2016; Coricelli et al., 2005).

Guilt

As discussed above, another important goal for agents as they navigate their social landscape is to minimize harm to others. Guilt occurs in these interpersonal contexts when one believes they have harmed or disappointed another individual (Battigalli & Dufwenberg, 2007). Guilt is considered a prosocial emotion in that agents have a tendency to take actions that repair the relationship following social transgressions (Baumeister, Stillwell, & Heatherton, 1994; Regan, Williams, & Sparling, 1972). Like regret, even the anticipation of guilt through simulating the act of committing a transgression can provide a powerful motivation for goal-directed choices to act prosocially (Massi Lindsey, 2005). This has been successfully demonstrated in the context of honoring a relationship partner's trust in the Trust Game (Berg et al., 1995). In this game, Player A invests in Player B by transferring a portion of his initial endowment to Player B. The investment amount is multiplied by a predetermined factor (typically fourfold), and Player B then decides how much, if any, of the multiplied investment to return to Player A.

If Player A invests money in Player B, Player B generally reciprocates by sending back some portion of the money, despite there being no advantage in doing so. In fact, if Player B wanted to solely maximize her monetary payout, she should keep all the money and return nothing to Player A, as there is no fear of reprisal in one-shot Trust Games. Why then is there overwhelming evidence that Player B routinely behaves in classically "irrational" ways by sending back the money to Player A? One possibility is the anticipated guilt that Player B would feel if she kept the money, thereby failing to uphold the social contract of trust. Indeed, models that consider other-regarding preferences such as warm-glow altruism (Andreoni, 1990), intention-based reciprocity (Dufwenberg & Kirchsteiger, 2004; Rabin, 1993) and inequity-aversion (Charness & Rabin, 2002; Falk, Fehr, & Fischbacher, 2008; Fehr & Schmidt, 1999) reveal that minimizing anticipated guilt provides a powerful signal to motivate honoring a partner's trust (Battigalli & Dufwenberg, 2007; Chang & Smith, 2015; Charness & Dufwenberg, 2006; Dufwenberg & Gneezy, 2000)—accounts which dramatically outperform models derived from classical economic theory (Cox, 2004). According to this model, Player B receives positive value from the money they keep and negative value from the anticipated guilt of disappointing Player A by not returning any money. Player B has a second-order belief about the amount that Player A expects them to return, and any difference between the expectations that Player A may hold, and what Player B intends to return, can create anticipatory guilt in Player B that shapes their goal to reciprocate. Formally, Player B's utility function U_B for selecting action i can be described as

$$U_B(i) = \pi_B(i) - \theta_B \cdot \max\left(E_B^2[\pi_A] - \pi_A(i), 0\right), \tag{14.4}$$

where $\pi_B(i)$ is B's payoff for action i, $\pi_A(i)$ is A's payoff for action i, $E_B^2[\pi_A]$ is B's second-order belief about what they believe A expects his payoff to be, and θ_B is a free parameter representing Player B's sensitivity to feeling guilt. In this formalization, guilt has negative value and is represented as an emotional error signal resulting from disappointing a partner's expectations.

There have been several laboratory studies providing support for guilt-aversion stimulating prosocial behavior (Charness & Dufwenberg, 2006; Nihonsugi, Ihara, & Haruno, 2015). The amount of money that Player B returns is directly proportional to their beliefs about Player A's expectations (Dufwenberg & Gneezy, 2000) such that Player B will be even more likely to reciprocate if he/she believes that his/her partner has expectations of cooperation (Chang et al., 2011; Reuben, Sapienza, & Zingales, 2009). The fact that guilt can induce strategy changes during these tasks reveals how emotional prediction errors make individuals flexible in their choice selection—able to adaptively respond during dynamic social exchanges to achieve a common social goal.

CONCLUSIONS

How humans pursue goals has been a topic of great interest for many decades. We now know much about how goals are represented and translated into action. Much less research, however, has focused on how goal-directed behavior unfolds during social interactions. And yet, many of our most important and primary goals are qualitatively social in nature. Indeed, the success of one's social experience is directly linked with how readily one achieves their social goals. Here, we argue there are three social goals—preventing harm, affiliating with others, and resolving social uncertainty—that serve as the basic building blocks of promoting social well-being. However, these social goals often come into direct conflict with an equally potent goal—the desire to enhance our own well-being.

How humans translate these oftentimes conflicting goals into concrete actions requires learning about the world and updating the relevance of each goal according to the outcomes of previously taken actions. Here, we argue that errors in pursuing these goals can manifest in the form of specific emotional feelings such as guilt, regret, or anger, and can provide a powerful motivational signal for how people update their goals and alter their actions. These emotional error signals result from monitoring how our actions help us progress toward (or away from) a goal, acting to regulate our behavior to be consistent with our social goals akin to a control theoretic system. Accordingly, our social-affective control model offers a tractable framework for understanding how these emotional prediction errors might guide individuals to undertake or circumvent certain social behaviors during social goal pursuit. Indeed, such a formal model offers both testable hypotheses about when and how these emotional prediction errors shape social goal pursuit and provides a useful roadmap for developing future experiments that can better parse the role of emotion in guiding certain social goals.

While we primarily focused on how various negative emotional phenomena act as prediction errors to guide social goal pursuit, it is likely that there are several other positive emotional experiences, such as empathy, that also provide an error signal to promote prosocial goals (FeldmanHall et al., 2015; Lockwood, Apps, Valton, Viding, & Roiser, 2016). Although there is little work that we are aware of that formally quantifies this process, we hypothesize that a prediction error likely motivates empathy in a similar way to those previously described above. Future work aimed at elucidating this process will deepen our understanding of the relationship between emotional experiences and social goal pursuit.

Finally, there is relatively little work that has explored how social contexts bias the relationship between emotions and goal pursuit. And yet, there are arguably many cases in which one's social goals to affiliate, minimize uncertainty or prevent harm are more salient than others. Take for example situations in which one is caring for a child versus entertaining friends. In these cases, the actions taken surrounding the care for a child will likely prioritize harm prevention, whereas the salient actions when engaging with friends will likely correspond with the social goal to affiliate. In other words, the desire to minimize harm is not stable across all social environments but is rather modulated by the context in which the goal-directed behavior arises. Our hope is that future work can help identify and document the various contexts that either increase or decrease the selection of certain social goals.

REFERENCES

Aarts, H., Chartrand, T. L., Custers, R., Danner, U., Dik, G., Jefferis, V., & Cheng, C. M. (2005). Social stereotypes and automatic goal pursuit. *Social Cognition, 23*, 464–489.

Aarts, H., & Custers, R. (2007). In search of the nonconscious sources of goal pursuit: Accessibility and positive affective valence of the goal state. *Journal of Experimental Social Psychology, 43*, 312–318.

Aarts, H., Custers, R., & Holland, R. W. (2007). The nonconscious cessation of goal pursuit: When goals and negative affect are coactivated. *Journal of Personality and Social Psychology, 92*, 165–178.

Aarts, H., Custers, R., & Veltkamp, M. (2008). Perception in the service of goal pursuit: Motivation to attain goals enhances the perceived size of goal-instrumental objects. *Social Cognition, 26*(6), 720–736.

Ajzen, I., & Fishbein, M. (1969). Prediction of behavioral intentions in a choice situation. *Journal of Experimental Social Psychology, 5*(4), 400–416.

Andreoni, J. (1990). Impure altruism and donations to public goods: A theory of warm-glow giving. *The Economic Journal of Nepal, 100*(401), 464–477.

Andreoni, J., & Miller, J. (2002). Giving according to GARP: An experimental test of the consistency of preferences for altruism. *Econometrica, 70*, 737–753.

Ariely, D., & Gino, F. (2011). The dark side of creativity: Original thinkers can be more dishonest. *Journal of Personality and Social Psychology, 102*, 445–459.

Asch, S. E. (1956). Studies of independence and conformity .1. A minority of one against a unanimous majority. *Psychological Monographs, 70*(9), 1–70.

Badre, D., Satpute, A. B., & Ochsner, K. N. (2012). The neuroscience of goal-directed behavior. In H. Aarts, & A. J. Elliot (Eds.), *Goal-directed behavior*. New York: Taylor & Francis Group, LLC.

Bandura, A. (1991). Social cognitive theory of moral thought and action. In *Handbook of moral behavior and development, Vol. 1: Theory*. Lawrence Erlbaum Associates, Inc.

Bar-Anan, Y., Wilson, T. D., & Gilbert, D. T. (2009). The feeling of uncertainty intensifies affective reactions. *Emotion, 9*(1), 123–127.

Barcelo, H., & Capraro, V. (2015). Group size effect on cooperation in one-shot social dilemmas. *Scientific Reports, 5*.

Bargh, J. A., Lee-Chai, A., Barndollar, K., Gollwitzer, P. M., & Trötschel, R. (2001). The automated will: Nonconscious activation and pursuit of behavioral goals. *Journal of Personality and Social Psychology, 81*(6).

Bargh, J. A., Raymond, P., Pryer, J. B., & Strack, F. (1995). Attractiveness of the underling – an automatic power → sex association and its consequences for sexual harassment and aggression. *Journal of Personality and Social Psychology, 68*(5), 768–781.

Batson, C. D., Van Lange, P. A. M., Ahmad, N., & Lishner, D. A. (2003). *Altruism and helping behavior. The sage handbook of social psychology*. Sage Publications.

Battigalli, P., & Dufwenberg, M. (2007). Guilt in games. *American Economic Review, 97*(2).

Battigalli, P., Dufwenberg, M., & Smith, A. (2015). *Frustration and anger in games*.

Bault, N., Coricelli, G., & Wydoodt, P. (2016). Different attentional patterns for regret and disappointment: An eye-tracking study. *Journal of Behavioral Decision Making*.

Baumeister, R. F., & Leary, M. R. (1995). The need to belong: Desire for interpersonal attachments as a fundamental human motivation. *Psychological Bulletin, 117*(3), 497–529.

Baumeister, R. F., Stillwell, A. M., & Heatherton, T. F. (1994). Guilt: An interpersonal approach. *Psychological Bulletin, 115*(2), 243–267.

Bechara, A., Damasio, H., & Damasio, A. R. (2000). Emotion, decision making and the orbitofrontal cortex. *Cerebral Cortex, 10*(3), 295–307.

Bell, D. E. (1982). Regret in decision-making under uncertainty. *Operations Research, 30*(5), 961–981.

Berg, J., Dickhaut, J., & Mccabe, K. (1995). Trust, reciprocity, and social-history. *Games and Economic Behavior, 10*(1), 122—142.

Bernoulli, D. (1738). *Specimen theoriae novae de mensura sortis.*

Berns, G. S., & Moore, S. E. (2011). A neural predictor of cultural popularity. *Journal of Consumer Psychology, 22.*

Bolton, G. E., & Ockenfels, A. (2000). A theory of equity, reciprocity, and competition. *American Economic Review, 90,* 166—193.

Botvinick, M. M., Cohen, J. D., & Carter, C. S. (2004). Conflict monitoring and anterior cingulate cortex: An update. *Trends in Cognitive Sciences, 8*(12), 539—546.

Cacioppo, J. T., Gardner, W. L., & Berntson, G. G. (1999). The affect system has parallel and integrative processing components: Form follows function. *Journal of Personality and Social Psychology, 76*(5), 839—855.

Camerer, C. F. (2003). *Behavioral game theory: Experiments in strategic interaction.* Princeton University Press.

Campbell-Meiklejohn, D. K., Bach, D. R., Roepstorff, A., Dolan, R. J., & Frith, C. D. (2010). How the opinion of others affects our valuation of objects. *Current Biology, 20*(13), 1165—1170.

Carver, C. S., & Scheier, M. F. (1981). A control-systems approach to behavioral self regulation. In L. Wheeler (Ed.), *Review of personality and social psychology* (Vol. 2, pp. 107—140). Beverly Hills CA: Sage.

Carver, C. S., & Scheier, M. F. (1984). A control-theory approach to behavior and some implications for social skills training. In P. Trower (Ed.), *Radical approaches to social skills training* (pp. 144—179). London/New York: Croom Helm/Methuen.

Carver, C. S., & Scheier, M. F. (1990). Origins and functions of positive and negative affect — a control-process view. *Psychological Review, 97*(1), 19—35.

Chang, L. J., & Jolly, E. (2017). Emotions as computational signals of goal error. *Nature of Emotions.*

Chang, L. J., & Sanfey, A. G. (2008). Emotion, decision-making, and the brain. *Neuroeconomics,* 31—53. D.H.K. McCabe, Elsevier.

Chang, L. J., & Sanfey, A. G. (2013). Great expectations: Neural computations underlying the use of social norms in decision-making. *Social Cognitive and Affective Neuroscience, 8*(3), 277—284.

Chang, L. J., & Smith, A. (2015). Social emotions and psychological games. *Current Opinion in Behavioral Sciences.*

Chang, L. J., Smith, A., Dufwenberg, M., & Sanfey, A. G. (2011). Triangulating the neural, psychological, and economic bases of guilt aversion. *Neuron, 70*(3), 560—572.

Charness, G., & Dufwenberg, M. (2006). Promises and partnership. *Econometrica, 74*(6), 1579—1601.

Charness, G., & Rabin, M. (2002). Understanding social preferences with simple tests. *Quarterly Journal of Economics, 117*(3), 817—869.

Chiu, P. H., Lohrenz, T. M., & Montague, P. R. (2008). Smokers' brains compute, but ignore, a fictive error signal in a sequential investment task. *Nature Neuroscience, 11*(4), 514.

Cialdini, R. B., & Goldstein, N. J. (2004). Social influence: Compliance and conformity. *Annual Review of Psychology, 55,* 591—621.

Cikara, M., & Van Bavel, J. J. (2014). The neuroscience of intergroup relations: An integrative review. *Perspectives on Psychological Science, 9*(3), 245—274.

Coricelli, G., Critchley, H. D., Joffily, M., O'Doherty, J. P., Sirigu, A., & Dolan, R. J. (2005). Regret and its avoidance: A neuroimaging study of choice behavior. *Nature Neuroscience, 8*(9), 1255.

Coricelli, G., Dolan, R. J., & Sirigu, A. (2007). Brain, emotion and decision making: The paradigmatic example of regret. *Trends in Cognitive Sciences, 11*(6), 258—265.

Cox, J. C. (2004). How to identify trust and reciprocity. *Games and Economic Behavior, 46*(2), 260—281.

Cushman, F., Gray, K., Gaffey, A., & Mendes, W. B. (2012). Simulating murder: The aversion to harmful action. *Emotion, 12*(1), 2—7.

Custers, R., & Aarts, H. (2010). The unconscious will: How the pursuit of goals operates outside of conscious awareness. *Science, 329*(5987), 47—50.

Davidson, R. J. (1992). Anterior cerebral asymmetry and the nature of emotion. *Brain and Cognition, 20,* 122—151.

Davidson, R. J., & Irwin, W. (1999). The functional neuroanatomy of emotion and affective style. *Trends in Cognitive Sciences, 3*(1), 11—21.

Daw, N. D., Niv, Y., & Dayan, P. (2005). Uncertainty-based competition between prefrontal and dorsolateral striatal systems for behavioral control. *Nature Neuroscience, 8*(12), 1704−1711.

De Houwer, J., Thomas, S., & Baeyens, F. (2001). Association learning of likes and dislikes: A review of 25 years of research on human evaluative conditioning. *Psychological Bulletin, 127*(6), 853−869.

Dufwenberg, M., & Gneezy, U. (2000). Measuring beliefs in an experimental lost wallet game. *Games and Economic Behavior, 30*(2), 163−182.

Dufwenberg, M., & Kirchsteiger, G. (2004). A theory of sequential reciprocity. *Games and Economic Behavior, 47*(2), 268−298.

Ellsberg, D. (1961). Risk, ambiguity, and the savage axioms. *Quarterly Journal of Economics, 75*(4), 643−669.

Engelmann, J. B., Meyer, F., Fehr, E., & Ruff, C. C. (2015). Anticipatory anxiety disrupts neural valuation during risky choice. *Journal of Neuroscience, 35*(7), 3085−3099.

Falk, A., Fehr, E., & Fischbacher, U. (2008). Testing theories of fairness—intentions matter. *Games and Economic Behavior, 62*(1), 287−303.

Fareri, D. S., Chang, L. J., & Delgado, M. R. (2012). Effects of direct social experience on trust decisions and neural reward circuitry. *Frontiers in Neuroscience, 6*, 148.

Fareri, D. S., Chang, L. J., & Delgado, M. R. (2015). Computational substrates of social value in interpersonal collaboration. *The Journal of Neuroscience: The Official Journal of the Society for Neuroscience, 35*(21), 8170−8180.

Fazio, R. H., Sanbonmatsu, D. M., Powell, M. C., & Kardes, F. R. (1986). On the automatic activation of attitudes. *Journal of Personality and Social Psychology, 50*(2), 229−238.

Fehr, E., & Camerer, C. F. (2007). Social neuroeconomics: The neural circuitry of social preferences. *Trends in Cognitive Sciences, 11*(10), 419−427.

Fehr, E., & Gachter, S. (1999). Collective action as a social exchange. *Journal of Economic Behavior & Organization, 39*, 341−369.

Fehr, E., & Gachter, S. (2002). Altruistic punishment in humans. *Nature, 415*, 137−140.

Fehr, E., & Schmidt, K. M. (1999). A theory of fairness, competition, and cooperation. *Quarterly Journal of Economics, 114*(3), 817−868.

FeldmanHall, O., Dalgleish, T., Evans, D., & Mobbs, D. (2015). Empathic concern drives costly altruism. *Neuroimage, 105*, 347−356.

Feldmanhall, O., Dalgleish, T., & Mobbs, D. (2013). Alexithymia decreases altruism in real social decisions. *Cortex: A Journal Devoted to the Study of the Nervous System and Behavior, 49*(3), 899−904.

FeldmanHall, O., Mobbs, D., Hiscox, L., Navrady, L., & Dalgleish, T. (2012). What we say and what we do: The relationship between real and hypothetical moral choices. *Cognition, 123*(3), 434−441.

FeldmanHall, O., Otto, A. R., Phelps, E. A. (In Press). Learning moral values: another's desire to punish enhances one's own punitive behavior. *Journal of Experimental Psychology: General.*

Ferguson, M. J., & Bargh, J. A. (2004). Liking is for doing: The effects of goal pursuit on automatic evaluation. *Journal of Personality and Social Psychology, 87*(5), 557−572.

Fowler, J. H., & Christakis, N. A. (2010). Cooperative behavior cascades in human social networks. *Proceedings of the National Academy of Sciences of the United States of America, 107*(12), 5334−5338.

Galinsky, A. D., Gruenfeld, D. H., & Magee, J. C. (2003). From power to action. *Journal of Personality and Social Psychology, 85*(3), 453−466.

Geanakoplos, J., Pearce, D., & Stacchetti, E. (1989). Psychological games and sequential rationality. *Games and Economic Behavior, 1*(1), 60−79.

Gilovich, T., & Medvec, V. H. (1995). The experience of regret − what, when, and why. *Psychological Review, 102*(2), 379−395.

Gollwitzer, P. M., & Moskowitz, G. B. (1996). *Goal effect on thought and behavior. Social psychology: Handbook of basic principles*. New York: Guilford Press.

Greene, J. D., Rand, D., & Nowak, M. A. (2012). Spontaneous giving and calculated greed. *Nature, 489*, 427−430.

Gneezy, U., Meier, S., & Rey-Biel, P. (2011). When and why incentives (don't) work to modify behavior. *Journal of Economic Perspectives, 25*, 191−210.

Greene, J. D., Sommerville, R. B., Nystrom, L. E., Darley, J. M., & Cohen, J. D. (2001). An fMRI investigation of emotional engagement in moral judgment. *Science, 293*(5537), 2105−2108.

Grossman, D. (1996). *On killing: The psychological cost of learning to kill in war and society*. Boston, Little: Brown.

Guinote, A. (2007). Power and goal pursuit. *Personality & Social Psychology Bulletin, 33*(8).

Guth, W., Schmittberger, R., & Schwarze, B. (1982). An experimental-analysis of ultimatum bargaining. *Journal of Economic Behavior & Organization, 3*(4), 367–388.

Ferguson, M. (2007). The Automatic Evaluation of Goals. *ACR North American Advances.*

Haidt, J. (2012). Chapter 7: The moral foundations of politics. In *The righteous mind: Why good people are divided by politics and religion*. Pantheon Books.

Hein, G., Silani, G., Preuschoff, K., Batson, C. D., & Singer, T. (2010). Neural responses to ingroup and outgroup members' suffering predict individual differences in costly helping. *Neuron, 68*(1), 149–160.

Heise, D. (2007). *Expressive order: Confirming sentiments in social actions.* New York: Springer.

Hirsh, J. B., Mar, R. A., & Peterson, J. B. (2012). Psychological entropy: A framework for understanding uncertainty-related anxiety. *Psychological Review, 119*(2), 304–320.

Kahneman, D. (2003). A perspective on judgment and choice: Mapping bounded rationality. *The American Psychologist, 58*(9), 697.

Kahneman, D., Slovic, P., & Tversky, A. (1982). *Judgment under uncertainty: Heuristics and biases.* Cambridge: Cambridge University Press.

Kahneman, D., & Tversky, A. (1979). Prospect theory: An analysis of decision under risk. *Econometrica, 47*, 278.

Keltner, D., Gruenfeld, D. H., & Anderson, C. (2003). Power, approach, and inhibition. *Psychological Review, 110*(2), 265–284.

Kiehl, K. A., & Hoffman, M. B. (2011). The criminal psychopath: History, neuroscience, treatment, and economics. *Jurimetrics, 51*, 355–397.

King-Casas, B., Tomlin, D., Anen, C., Camerer, C. F., Quartz, S. R., & Montague, P. R. (2005). Getting to know you: Reputation and trust in a two-person economic exchange. *Science, 308*(5718), 78–83.

Klucharev, V., Hytonen, K., Rijpkema, M., Smidts, A., & Fernandez, G. (2009). Reinforcement learning signal predicts social conformity. *Neuron, 61*(1), 140–151.

Koenigs, M., & Tranel, D. (2007). Irrational economic decision-making after ventromedial prefrontal damage: Evidence from the ultimatum game. *Journal of Neuroscience, 27*(4), 951–956.

Kruglanski, A. W., & Webster, D. M. (1996). Motivated closing of the mind: 'Seizing' and 'Freezing'. *Psychological Review, 103*(2), 263–283.

Leary, M. R., & Richman, L. S. (2009). Reactions to discrimination, stigmatization, ostracism, and other forms of interpersonal rejection. *Psychological Review, 116*(2), 365–383.

Levy, D. J., & Glimcher, P. W. (2012). The root of all value: A neural common currency for choice. *Current Opinion in Neurobiology, 22*(6), 1027–1038. https://doi.org/10.1016/j.conb.2012.06.001.

Levitt, S. D., & List, J. A. (2007). What do laboratory experiments measuring social preferences reveal about the real world. *Journal of Economic Perspectives, 21*, 153–174.

Lockwood, P., Apps, M., Valton, V., Viding, E., & Roiser, J. P. (2016). Neurocomputational mechansims of prosocial learning and links to empathy. *Proceedings of the National Academy of Sciences of the United States of America, 113*(35), 9763–9768.

Loewenstein, G. (1987). Anticipation and valuation of delayed consumption. *Economic Journal, 97*(387), 666–684.

Loewenstein, G. (1996). Out of control: Visceral influences on behavior. *Organizational Behavior and Human Decision Processes, 65*(3).

Loewenstein, G. F., & Lerner, J. S. (2003). The role of affect in decision making. In R. Davidson, H. Goldsmith, & K. Scherer (Eds.), *Handbook of affective sciences.* Oxford University Press.

Lohrenz, T., McCabe, K., Camerer, C. F., & Montague, P. R. (2007). Neural signature of fictive learning signals in a sequential investment task. *Proceedings of the National Academy of Sciences of the United States of America, 104*(22), 9493–9498.

Loomes, G., & Sugden, R. (1982). Regret theory — an alternative theory of rational choice under uncertainty. *Economic Journal, 92*(368), 805–824.

Massi Lindsey, L. L. (2005). Anticipated guilt as behavioral motivation. *Human Communication Research, 31*(4), 453–481.

McClelland, D. C. (1985). *Human motivation.* Glenview, IL: Scott, Foresman and Company.

Mellers, B. A., & McGraw, A. P. (2001). Anticipated emotions as guides to choice. *Current Directions in Psychological Science, 10*(6), 210–214.

Mellers, B. A., Schwartz, A., Ho, K., & Ritov, I. (1997). Decision affect theory: Emotional reactions to the outcomes of risky options. *Psychological Science, 8*(6), 423—429.

Mikhail, J. M. (2000). *Rawls' linguistic analogy: A study of the 'generative grammar' model of moral theory described by John Rawls in 'a theory of justice'.* Ithaca: Cornell University.

Miller, E. K., & Cohen, J. D. (2001). An integrative theory of prefrontal cortex function. *Annual Review in Neuroscience, 24,* 167—202.

Montague, P. R., & Lohrenz, T. (2007). To detect and correct: Norm violations and their enforcement. *Neuron, 56*(1).

Nihonsugi, T., Ihara, A., & Haruno, M. (2015). Selective increase of intention-based economic decisions by noninvasive brain stimulation to the dorsolateral prefrontal cortex. *Journal of Neuroscience, 35*(8).

Nook, E. C., Ong, D. C., Morelli, S. A., Mitchell, J. P., & Zaki, J. (2016). Prosocial conformity: Prosocial norms generalize across behavior and empathy. *Personality & Social Psychology Bulletin, 42*(8), 1045—1062.

Olsson, A., Nearing, K. I., & Phelps, E. A. (2007). Learning fears by observing others: The neural systems of social fear transmission. *Social Cognitive and Affective Neuroscience, 2,* 3—11.

Panksepp, J. (1998). Chapter 9: Energy is delight: The pleasures and pains of brain regulatory systems. In *Affective neuroscience: The foundations of human and animal emotions.* Oxford University Press.

Panksepp, J. (2004). *Affective neuroscience: The foundations of human and animal emotions.* Oxford University Press.

Petherick, W. (2014). *Profiling and serial crime: Theoretical and practical issues.* Waltham, MA: Elsevier Inc.

Phelps, E. A. (2005). The interaction of emotion and cognition: Insights from studies of the human amygdala. In *The new unconscious.* New York: Guilford Press.

Rabin, M. (1993). Incorporating fairness into game theory and economics. *American Economic Review, 83*(5), 1281—1302.

Rangel, A., Camerer, C., & Montague, P. R. (2008). A framework for studying the neurobiology of value-based decision making. *Nature Reviews Neuroscience, 9*(7), 545—556.

Reeve, J. (2008). *Understanding motivation and emotion.* John Wiley & Sons, Inc.

Regan, D. T., Williams, M., & Sparling, S. (1972). Voluntary expiation of guilt: A field experiment. *Journal of Personality and Social Psychology, 24*(1), 42.

Reuben, E., Sapienza, P., & Zingales, L. (2009). Is mistrust self-fulfilling. *Economics Letters, 104*(2), 89—91.

Robinson, M. J., & Berridge, K. C. (2013). Instant transformation of learned repulsion into motivational "wanting". *Current Biology, 23*(4), 282—289.

Rogers, K. B. (2015). Expectation states, social influence, and affect control: Opinion and sentiment change through social interaction. In , *Vol. 32. Advances in group processes* (pp. 65—98). Emerald Group Publishing Limited.

Ruff, C. C., & Fehr, E. (2014). The neurobiology of rewards and values in social decision making. *Nature Reviews Neuroscience, 15*(8), 549—562.

Samuelson, P. A. (1938). A note on the pure theory of consumer's behaviour. *Economica, 5*(19), 353—354.

Sanfey, A. G., Stallen, M., & Chang, L. J. (2014). Norms and expectations in social decision-making. *Trends in Cognitive Sciences, 18*(4), 172—174.

Schroder, T., Hoey, J., & Rogers, K. B. (2016). Modeling dynamic identities and uncertainty in social interactions. *American Sociological Review, 81*(4), 828—855.

Sheeran, P., Aarts, H., Custers, R., Rivis, A., Webb, T. L., & Cooke, R. (2005). The goal-dependent automaticity of drinking habits. *The British Journal of Social Psychology, 44*(Pt 1), 47—63.

Suzuki, S., Jensen, E. L., Bossaerts, P., & O'Doherty, J. P. (2016). Behavioral contagion during learning about another agent's risk-preferences acts on the neural representation of decision-risk. *Proceedings of the National Academy of Science of the United States of America, 113*(14), 3755—3760.

Thibaut, J. W., & Kelley, H. H. (1959). *The social psychology of groups.* New York: John Wiley & Sons, Inc.

Thibaut, J. W., & Kelley, H. H. (1978). *Interpersonal relations: A theory of interdependence.* London: John Wiley & Sons, Inc.

Tyler, T. R., & Blader, S. L. (2000). *Cooperation in groups: Procedural justice, social identity and behavioral engagement.* New York: Taylor & Francis Group, LLC.

von Neumann, J., & Morgenstern, O. (2007). *Theory of games and economic behavior.* Princeton University Press.

de Waal, F. (1997). *Good natured: The origins of right and wrong in humans and other animals.* Cambridge, MA: Harvard University Press.

Walton, G. M., Cohen, G. L., Cwir, D., & Spencer, S. J. (2012). Mere belonging: The power of social connections. *Journal of Personality and Social Psychology, 102*(3), 513–532.

Wang, P., Baker, L. A., Gao, Y., Raine, A., & Lozano, D. I. (2012). Psychopathic traits and physiological responses to aversive stimuli in children aged 9-11 years. *Journal of Abnormal Child Psychology, 40*(5), 759–769.

Winkielman, P., & Berridge, K. C. (2004). Unconscious emotion. *Current Directions in Psychological Science, 13,* 120–123.

Xiang, T., Lohrenz, T., & Montague, P. R. (2013). Computational substrates of norms and their violations during social exchange. *The Journal of Neuroscience: The Official Journal of the Society for Neuroscience, 33*(3).

Yamagishi, T. (1986). The provision of a sanctioning system as a public good. *Journal of Personality and Social Psychology, 51.*

Zaki, J., Schirmer, J., & Mitchell, J. P. (2011). Social influence modulates the neural computation of value. *Psychological Science, 22*(7), 894–900.

CHAPTER 15

The Balance Between Goal-Directed and Habitual Action Control in Disorders of Compulsivity

Sanne de Wit[1,2]
[1]Department of Clinical Psychology, University of Amsterdam, Amsterdam, The Netherlands; [2]Amsterdam Brain and Cognition, University of Amsterdam, Amsterdam, The Netherlands

INTRODUCTION

There is universal agreement that habits play an important role in failures to flexibly adjust behavior when goals change. In other words, habits contribute to the "intention—behavior gap" (Sheeran, 2002). In certain psychopathologies, habits may even present a serious threat to health and well-being, as well as an important target for therapeutic interventions. In line with a transdiagnostic approach to compulsivity, the present chapter focuses on an imbalance between habitual and goal-directed control as a key mechanism in disorders of compulsive behavior, focusing specifically on: substance abuse, obesity and eating disorders, and obsessive—compulsive disorder (OCD). The specificity of the goal-directed/habitual balance as a transdiagnostic factor will be discussed, as well as the relevance for clinical interventions. Recently, many articles and chapters have appeared on the role of habits in compulsivity. The specific aim of this chapter is to present the evidence in an accessible yet critical manner, with an emphasis on behavioral research with outcome devaluation paradigms, and minimal reference to the neurobiological basis, except for when this serves to link the neural dual-system architecture of goal-directed and habitual control to these different disorders. In the next section, the theoretical framework and experimental measurement of the balance between goal—directed action and habits will be briefly introduced.

The balance between goal-directed action and habit

According to a definition derived from folk psychology (which has been widely applied to animal models of instrumental action control), goal—directed action meets two criteria (Heyes & Dickinson, 1990). According to the belief criterion, goal-directed actions are mediated by knowledge of the causal action—outcome relationship. According to the desire criterion, goal-directed actions are only executed when the outcome is currently desirable, or in other words, a goal. In associative terms, performance is mediated by response—outcome (R—O) associations, which allow behavior to meet the belief criterion. This associative structure needs to be integrated with a motivational system that allows

Goal-Directed Decision Making
ISBN 978-0-12-812098-9, https://doi.org/10.1016/B978-0-12-812098-9.00015-2

current motivation to determine whether the action is performed (e.g., see associative–cybernetic model, de Wit & Dickinson, 2009). Consider the example of drinking a glass of wine at the local bar in order to relax and more freely engage in social interactions. At first, this may be a goal-directed action, but when this behavior is repeatedly performed, it can gradually turn into a habit, such that it is automatically elicited by environmental stimuli like the bar or the circle of friends with whom one usually enjoys a drink. The main conditions for habit formation are captured in Thorndike's law of effect (1911), according to which the experience of a rewarding outcome (O) of an instrumental response leads to the (positive) reinforcement of a mental association between the response (R) and the contextual stimuli (S) that were present at the time of execution of the response. As a consequence, on future occasions, the contextual stimuli will directly activate this response through the S-R association. Furthermore, (negative) reinforcement through the cancellation of a dreaded aversive event should similarly strengthen S-R links, thereby giving rise to avoidance habits. For instance, as the development of alcoholism progresses, the cancellation of withdrawal symptoms or depressive mood upon drinking may lead to a temporary feeling of relief and should thereby negatively reinforce this habit. In contrast, the experience of an aversive outcome following an instrumental response (or cancellation of an anticipated reward) should weaken the S-R association. However, because habits are mediated by direct S-R associations, they are not *directly* influenced by a change in the current desirability of the outcome. Strengthening or weakening of the S-R association can only occur gradually as a consequence of extended experience with the instrumental contingency and the outcome with its new incentive value.

To account for a gradual loss of goal-directed control as a function of behavioral repetition, dual-system theories posit that repetition shifts the balance between a goal-directed and habitual system (e.g., Balleine & O'Doherty, 2010; Dickinson, 1985). According to one such model, the associative–cybernetic model, the degree to which behavior is habitual (or goal-directed) is determined by the relative strengths of associations in an R-O system and S-R system (de Wit & Dickinson, 2009). These learning systems sometimes compete—but also cooperate—to control action (see Fig. 15.1).

Experimental assessment of the balance between goal-directed and habitual control

To investigate the balance between goal-directed and habitual control, Dickinson et al. developed the outcome devaluation paradigm. The first stage of this task consists of instrumental learning. For example, hungry rats are trained to press a lever for food pellets. Subsequently, they are removed from the learning context (i.e., the operant chamber), and an aversion is conditioned to the food pellets by pairing consumption with lithium chloride–induced nausea in half of the rats. Therefore, for these rats, the food pellets are "devalued." Subsequently, they are returned to the operant chamber where they can once again press the lever. However, during this extinction test, the food pellet

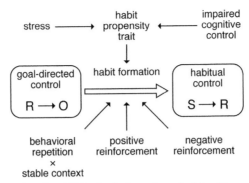

Figure 15.1 *Factors influencing the shift in the balance from habitual to goal-directed action control.*

outcome is no longer presented. This is an important aspect of the design, as presentation of the devalued outcomes contingent upon responding would be expected (according to the law of effect) to gradually weaken or inhibit S–R associations. Only if rats formed R–O associations during the initial training phase and are able to base responding on their current motivation for the outcome, should they be able to immediately and flexibly adjust performance following devaluation. To dissociate goal-directed from habitual control, it is crucial therefore that the test is conducted in extinction. Importantly, using this paradigm, Adams (1982) showed that rats that had received moderate training (100 lever presses), immediately decreased responding when the pellets were devalued. In contrast, rats that had received extensive training (500 lever presses) continued to respond for the devalued food in the extinction test. In other words, lever pressing had become a relatively inflexible habit. Importantly, reduced behavioral sensitivity to outcome devaluation in the extinction test was not due to ineffectiveness of the devaluation procedure, as the overtrained animals would quickly learn to refrain from responding for the devalued food in a reacquisition test in which the outcome was once again presented contingent on responding.

The development of insensitivity to outcome devaluation as a function of behavioral repetition has since been replicated in several animal studies (Balleine & O'Doherty, 2010) and in one human study with healthy adults (Tricomi, Balleine, & O'Doherty, 2009). In the latter study, a computer task was used to study key pressing for M&M's and Frito's by hungry participants in the presence of distinct fractal stimuli. Following instrumental training, one of the two snacks was devalued through satiation, and subsequently, participants were once again offered the opportunity to press the keys during exposure to the training stimuli. While moderately trained participants (12 trials/stimulus) were sensitive to current outcome value, the extensively trained group (72 trials/stimulus distributed over 3 days) continued to respond for the devalued outcome. However, a recent attempt to replicate those findings failed (de Wit et al., in press). Moreover, the latter study also failed to show habits as a function of behavioral repetition with two other existing outcome devaluation tasks (the slips-of-action paradigm and an avoidance task). Clearly, experimentally demonstrating overtrained habits in humans is challenging (for a more elaborate discussion, we refer the interested reader to de Wit et al., in press). However, outcome devaluation paradigms with just the moderate training condition have been used in many studies to investigate the neural basis of goal-directed and habitual control in humans. The idea behind this line of research is that habits are formed from the outset of acquisition, such that the balance between habitual and goal-directed control can already be reflected in shifting activations following moderate training and can vary between individuals and different psychopathologies.

Neural basis of goal-directed action and habits

Dual-system theories of goal-directed and habitual action control receive the strongest evidence from animal lesioning and inactivation studies, which have provided strong evidence for dissociable corticostriatal circuitries that support goal-directed action (including prelimbic cortex and dorsomedial striatum [DMS]) and habits (including infralimbic cortex and dorsolateral striatum [DLS]) (for more detailed discussion, see Balleine & O'Doherty, 2010; Lingawi, Dezfouli, & Balleine, 2016). In translational human research using the technique of functional MRI, performance on outcome devaluation paradigms has been found to be positively associated with activation of the ventromedial prefrontal cortex (vmPFC) and caudate, implicating these regions in goal-directed action (e.g., Valentin, Dickinson, & O'Doherty, 2007; de Wit, Corlett, Aitken, Dickinson, & Fletcher, 2009). On the other hand, behavioral repetition has been shown to lead to increased engagement of the posterior putamen, suggesting that this region may be involved in habit formation (Tricomi et al., 2009). Furthermore, structural MRI studies have provided evidence for neural dual-system architecture by relating estimated white matter tract strength to performance on a "slips-of-action paradigm." This is a variant of a computerized outcome devaluation paradigm, in which instrumental discrimination training with pictures serving as stimuli and as outcomes (e.g., stimulus 1: right key press - → outcome A; picture 2: right key press → outcome B) is followed by "instructed devaluation," meaning that participants are instructed that a picture that was previously worth points will suddenly lead to deduction of points (or financial credits) from a total score (e.g., O(A) but not O(B) is devalued). Subsequently, participants are shown the trigger cues (e.g., S1/S2, etc.) in rapid succession, and their task is to continue to respond for valuable outcomes (e.g., O(B)) while withholding responses for devalued outcomes (e.g., O(A)). Good performance on the slips-of-action task indicates relatively strong goal-directed control and/or weak S-R habitual control. By relating performance on this paradigm to individual differences in the estimated strength of white matter tracts, a caudate-vmPFC network has been implicated in goal-directed control, and a posterior putamen—premotor cortex network in habits (de Wit et al., 2012).

A transdiagnostic perspective: the role of goal-directed and habitual control in compulsivity

Compulsive behaviors are a central characteristic of many psychopathologies. For example, drug abuse is characterized by a strong urge to seek out and use drugs despite their harmful effects. Similarly, binge eating disorder (BED) and certain cases of obesity are associated with an urge to consume large amounts of energy-dense food, while in eating disorders, such as anorexia nervosa (AN) and bulimia nervosa, as little food as possible is consumed in order to lose weight. Finally, patients with OCD feel compelled to perform repetitive behaviors, such as checking, washing, and ordering. These outwardly very different behaviors have in common their persistence despite awareness of serious, adverse consequences regarding health, occupational, and social functioning. Several researchers have argued, therefore, that we should adopt a transdiagnostic approach to research into compulsivity across disorders like addiction, obesity and eating disorders, and OCD (e.g., Gillan, Fineberg, & Robbins, 2017; Robbins, Gillan, Smith, de Wit, & Ersche, 2012).

Transdiagnostic approaches focus on constructs of behavior (and their genetic and neural bases), that play a role across different psychopathologies, rather than on disorder-specific symptoms and diagnostic categories. A strong argument for this perspective is that there exist high degrees of comorbidity between different psychopathologies, as well as overlap in their core symptoms, as in the case of compulsivity. These are not captured by categorical distinctions in current classification systems, such as the *Diagnostic and Statistical Manual of Mental Disorders* (*DSM*) and *International Classification of Diseases*. In the context of compulsivity, several researchers have made a strong stance for investigating the balance between goal-directed and habitual control as a potentially important mechanism underlying compulsivity across different disorders (e.g., Gillan, Robbins, Sahakian, van den Heuvel, & van Wingen, 2016; Robbins et al., 2012).

The different ways in which reliance on habits could contribute to compulsive behaviors have been summarized in Fig. 15.1. Firstly, extensive behavioral repetition in stable contexts, which is usually a feature of compulsive behavior, fosters the transition from goal-directed to habitual control. Secondly, strong (positive/negative) reinforcement (for example, by the rewarding experience of drugs or the neutralization of fear in OCD patients) can further accelerate this process. Finally, a general tendency to form strong habits fast, referred to in this chapter as "habit propensity," may constitute an important transdiagnostic trait in compulsive conditions (Robbins et al., 2012). Habit propensity could be due to weak goal-directed control, strong habitual control, or a combination of the two. Unfortunately, outcome devaluation paradigms (with just the moderate training condition) do not allow one to determine the origin of impaired behavioral flexibility, although this issue can perhaps to some extent be mitigated by combining behavioral assays with neuroimaging. However, impaired goal-directed control and aberrantly strong habit formation could both render certain individuals particularly vulnerable to the development of maladaptive and even compulsive habits. Alternatively, habit propensity could be an emergent characteristic that arises, for example, as a consequence of neuroadaptations that result from substance abuse. These three different factors that determine habit strength may lead to aberrantly strong, maladaptive S-R habits in disorders of compulsivity and contribute to a sense of loss of control over behavior and to treatment resistance. Another factor that has been incorporated in Fig. 15.1 is the effect of stress on habit propensity, with acute as well as chronic stress increasing habit propensity and disrupting the balance between the underlying corticostriatal circuitries as revealed by functional and structural MRI (with the latter showing decreased caudate and increased putamen volume) (Schwabe & Wolf, 2009; Soares et al., 2012). Chronic/acute stress is highly prevalent among psychiatric disorders and is therefore an important factor to consider. Another factor concerns cognitive control capacities, including working memory and response inhibition. From a limited cognitive resources perspective, goal-directed control is more dependent on those functions being intact than habit. Impairments in these functions are ubiquitous in disorders of compulsivity (Fineberg et al., 2010; Robbins et al., 2012) and may therefore contribute to habit propensity.

DISRUPTIONS OF THE BALANCE BETWEEN GOAL-DIRECTED AND HABITUAL CONTROL IN DISORDERS OF COMPULSIVITY

Substance abuse

Many people use drugs recreationally. However, in some people, frequent goal-directed drug seeking may ultimately lead to the formation of drug habits (Tiffany, 1990), through both positive reinforcement by the powerfully rewarding properties of drugs and negative reinforcement by temporarily relief from anxiety or depression and ultimately from withdrawal symptoms (Koob, 2013). Stress may further accelerate the transition from goal-directed control to habits (Schwabe, Dickinson, & Wolf, 2011). According to the influential model of Everitt et al., habits pave the way for the development of compulsive drug seeking that persists despite severely harmful effects on one's health, problems in relationships, and impaired professional functioning (Everitt & Robbins, 2015; Everitt, Dickinson, & Robbins, 2001). By now, there is a wealth of evidence for an imbalance between goal-directed and habitual control in addiction (more than in any other clinical condition), mainly coming from animal studies that allow for experimental control over drug-seeking history.

Repeated alcohol seeking in rats has been shown to result in habit formation. In a study by Corbit, Nie, and Janak (2012), rats were trained to press a lever for unsweetened ethanol. A "minimal training group" received 14 consecutive daily training sessions, while an "extensive training group" received 56 sessions. Each rat was tested twice in extinction, once following devaluation through "satiation" (i.e., drinking a substantial amount of alcohol) and once in a nondevalued condition. The results showed that the minimal training group reduced responding in the devalued condition relative to the nondevalued, while devaluation failed to affect responding in the extensive training group. Furthermore, temporary inactivation of the DMS (through infusion of a GABA-agonist) led to habitual alcohol seeking even after minimal training, while DLS inactivation after 8 weeks of training restored goal-directed control. These results support the idea that repeated alcohol seeking gradually transitioned from DMS-driven goal-directed behavior to DLS-driven habits, in the same way that responding for natural rewards (like food) becomes habitual and DLS-dependent with overtraining (Yin, Knowlton, & Balleine, 2004).

If drugs are especially powerful reinforcers of habits, then habit formation should be accelerated with drug rewards relative to natural rewards like food. To investigate this, Dickinson, Wood, and Smith (2002) compared sensitivity to outcome devaluation in rats trained to lever press concurrently for alcohol and food pellets (e.g., right → alcohol; left → food). Subsequently, half of the rats received alcohol–LiCl pairings, and the other half received food–LiCl pairings. In support of the notion that drug habits form faster, rats pressed less for food if this had been devalued through LiCl pairings, while alcohol devaluation failed to reduce responding on the corresponding lever (but for discussion of caveats of research into drug alcohol-seeking habits, see O'Tousa & Grahame, 2014). This failure to adjust performance following devaluation was not due to a failure to devalue the alcohol, as both outcomes no longer acted as an effective reinforcer in a subsequent reacquisition test. This finding is not limited to alcohol, as a similar study provided evidence for impaired devaluation sensitivity of cocaine seeking (Miles, Everitt, & Dickinson, 2003). These studies strongly suggest that drug seeking leads to aberrantly strong habit formation.

In humans, there has been far less experimental investigation into the formation of drug-seeking habits, and studies in smokers by Hogarth and colleagues have failed to provide converging evidence. Hogarth and Chase (2011) trained smokers to press two keys, one for cigarettes and one for chocolate. On each trial, each key press had a 50% probability of being reinforced. Subsequently, the cigarette outcome was devalued by allowing participants to smoke ad libitum (i.e., through satiation). In the choice test that followed, smokers appeared goal-directed, meaning that they were able to reduce responding for cigarettes following the devaluation procedure. However, this very simple concurrent choice test that assesses choice in the absence of trigger cues does not seem optimal for the behavioral expression of habits. It would be interesting to follow this research up with a test during which external cues are presented with a strong association with the habitual response.

A related line of research has addressed whether drug abusers show a general tendency to rely on habits at the expense of flexible, goal-directed control. To investigate this possibility, Sjoerds et al. (2013) trained their participants to respond for (fruit and drink) pictures that were worth financial credits, before devaluing some of these through instruction. Alcohol-dependent patients were found to perform significantly worse than healthy matched controls in the instructed outcome devaluation test. Further support for the idea that alcohol dependents relied more on the habit system than the control group, came from the fMRI finding that they showed greater activation in the habit circuitry (posterior putamen) and weaker activation of the vmPFC during instrumental acquisition. Furthermore, vmPFC activation was inversely related to disease duration. More recently, Ersche et al. (2016) used the slips-of-action paradigm to investigate sensitivity to devaluation in cocaine abusers, with pictures of animals (that counted toward supermarket vouchers) functioning as the trigger cues and instrumental outcomes. Cocaine addicts showed more cue-triggered responding for devalued outcomes than healthy controls. These two studies provide convergent support for the idea that addiction is associated with a general tendency to rely on habits. The question remains, however, whether this habit propensity arises as a consequence of aberrantly strong habit or from impaired goal-directed control, or a combination of the two. Furthermore, cross-sectional research in humans does not inform us about causality. Habit propensity may be a consequence of chronic drug use but could alternatively constitute a predisposing trait that contributes to vulnerability to addiction.

Animal research suggests that habit propensity can i as a consequence of prolonged drug use. Corbit et al. (2012) found that rats trained to respond for sucrose still showed a devaluation effect after 8 weeks, but this was not the case in a group of rats that during the same time period was given alcohol to drink in their home cages. Similarly, Nelson and Killcross (2006) showed that sensitization to amphetamine accelerated the transition from goal-directed to habitual responding (see also Nordquist et al., 2007). As of yet, it is less clear whether habit propensity also constitutes a vulnerability factor. This question could be addressed in future research, e.g., in longitudinal investigations.

The studies reviewed so far suggest that accelerated habit formation and habit propensity play a role in drug addiction but do not directly implicate habits in *compulsive* drug use despite aversive consequences. Outcome devaluation tests are usually performed in extinction, meaning that the devalued outcome is not presented contingent on responding. As such, this paradigm can capture mindless, habitual behavior or so-called "slips of action," but it fails to provide a model of compulsive behavior. Indirect evidence for the notion of compulsive habits comes from an animal study showing that extended self-administration of cocaine leads to persistent responding despite punishment in the form of aversive, electric shocks—in a subset of ∼20% of rats (Deroche-Gamonet, 2004; see also,; Belin, Mar, Dalley, Robbins, & Everitt, 2008). In a related procedure, a seeking—taking chain schedule was adopted, in which during half of the trials the outcome of seeking responses was the opportunity to take cocaine but unpredictably on the other half of the trials, the outcome was an aversive foot shock (Pelloux, Everitt,

& Dickinson, 2007). As a result, rats seeking cocaine ran the risk of foot shock punishment. After a brief cocaine history, drug seeking was goal-directed such that rats reduced responding when the punishment contingency was introduced. In contrast, after an extended history of cocaine self-administration, a subset of rats (~20%) continued to seek cocaine despite receiving foot shocks on half of the trials (but see, Jonkman, Pelloux, & Everitt, 2012). Furthermore, in vivo optogenetic stimulation of the prelimbic cortex (the rodent functional homologue of the vmPFC) reduced drug seeking in a subset of compulsive animals, while for the remaining rats, optogenetic inhibition of this brain region led to increased cocaine seeking under the punishment schedule (Chen et al., 2013).

Fig. 15.2 summarizes the different ways in which habits could contribute to compulsive drug seeking. Of course, this figure is not exhaustive, and there are many other factors that could contribute to addiction. Importantly, next to triggering habits, drug-associated stimuli can induce drug craving or "wanting" (Berridge & Robinson, 2011). According to Berridge and Robinson, aberrantly strong "wanting" of drugs, rather than habits, is the driving force of compulsive drug seeking. They attribute a very modest role for habits, in the development of drug rituals and absent-minded drug seeking, as for example lighting a cigarette in the absence of a strong desire to smoke. One of their arguments against a central role of habits is that daily habits are not compulsive despite extensive repetition, and compulsive drug seeking is not "absent-minded" but rather is accompanied by an urge or desire. However, compulsive drug-seeking habits may differ from daily habits in these respects due to extremely strong reinforcement by drugs that hijack the dopaminergic motivational system in vulnerable individuals, thereby giving

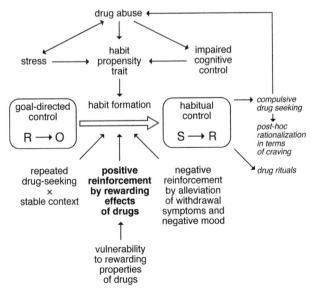

Figure 15.2 *Habits in substance abuse.* Summary of hypothesized links between different factors influencing the shift in the balance from habitual to goal-directed action control.

rise to a subjective urge to engage in drug seeking and taking. Another argument is that addicts are perfectly well capable of novel, complex behaviors if these are required to secure access to drugs. It is true that current experimental models that focus on simple motor responses fail to capture this complexity of drug-seeking behavior. It is an important outstanding question at what level of abstraction habits can be represented. For example, a concrete motor response could be turning left to get to the usual hangout of the drugs dealer, whereas a broader, abstract response representation could be to go to the drugs dealer. The latter response representation could lead to different motor responses dependent on the dealer's current whereabouts. A related question is whether the trigger cue always needs to constitute an external stimulus in the environment or whether internal cues like moods and even thoughts can also trigger habits. In the latter case, habits could readily generalize across different external contexts.

These different arguments notwithstanding, as reviewed here, there is a lot of evidence for accelerated habit formation in addiction, next to evidence for aberrantly strong craving in substance abuse. It seems plausible, therefore, that both of these processes contribute toward compulsive drug seeking, with their relative contributions changing in the course of the development of an addiction. Indeed, Everitt and Robbins (2015) proposed that craving plays a dominant role early in the development of an addiction, while habits develop later and ultimately turn into compulsive behavior. Finally, subjective reports of craving as the main driving force in addiction may *partially* arise as a posthoc rationalization of a mismatch between explicitly held goals and excessive drug-seeking habits; or in other words, a way to resolve cognitive dissonance (Everitt & Robbins, 2005).

Obesity

Obesity can have many different causes, including metabolic disturbances that are genetically rooted. However, the recent increase in the percentage of overweight (BMI ≥ 25 kg/m^2) and obese adults (BMI ≥ 30 kg/m^2) to 40% globally (Ng, Fleming, Robinson, Thomson, & Graetz, 2014) suggests that changes in the environment play a very important role. Specifically, increasing availability of palatable, high-calorie food and aggressive marketing is thought to lead to excessive consumption of energy-dense food. Furthermore, certain technological advancements may encourage a sedentary life style. The combination of high calorie intake and low calorie expenditure is thought to ultimately drive the recent rise in overweight and obesity. Obesity is associated with increased mortality, with serious health risks including diabetes, cardiovascular disease, high blood pressure, certain cancers, breathing problems, and osteoarthritis. Furthermore, it is associated with stigmatization, depression, and decreased quality of life (Jia & Lubetkin, 2010). It also is related to high medical costs and a significant loss in productivity as a result of increased sick leaves (Neovius, Neovius, Kark, & Rasmussen,

2012). Despite full awareness of these detrimental consequences, and often despite intentions to adhere to a healthier diet, many obese patients struggle to adhere to a healthier life style, which has led some researchers to reframe obesity in vulnerable individuals in terms of "food addiction" (e.g., de Jong, Vanderschuren, & Adan, 2012). A neurobiological finding that supports this idea is that obesity as well as stimulant abuse is associated with low striatal dopamine D2 receptor binding and reduced prefrontal metabolism (including orbitofrontal cortex (Volkow, Wang, Tomasi, & Baler, 2013)). In this section, we will explore whether, as in substance abuse, aberrantly strong habit formation contributes to the inability of these obese individuals to break harmful habits (see Fig. 15.3).

As discussed in the Introduction, continued food seeking leads to the formation of habits (Adams, 1982; Tricomi et al., 2009), and stress is associated with dominant habitual control over food seeking (Schwabe & Wolf, 2009). Relatedly, high stress levels are associated with obesity and can lead to "comfort eating" of palatable food (Morris, Beilharz, Maniam, Reichelt, & Westbrook, 2015). An unhealthy diet may have a negative influence on cognitive functioning, which could be speculated to foster habit propensity in obesity (Jansen, Houben, & Roefs, 2015).

So far, food-seeking habits in obesity specifically have not been investigated as a function of behavioral repetition, but there have been several human outcome devaluation studies that investigated action control after limited training. Horstmann et al. (2015) conducted the first such study to investigate habits in obesity. Obese and lean men were trained to press keys for two snack rewards (in a task akin to that used by Tricomi et al. (2009), but just with minimal training) before one of the two was devalued through

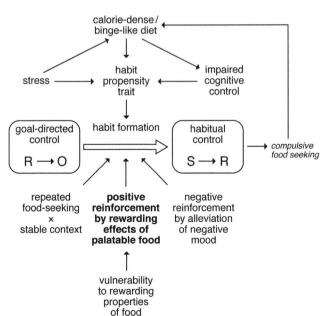

Figure 15.3 Habits in obesity. Summary of hypothesized links between different factors influencing the shift in the balance from habitual to goal-directed action control.

specific satiety. Overall, satiation successfully reduced self-reported desire to eat the snack in question. However, there was a negative correlation between BMI and behavioral sensitivity to outcome devaluation through specific satiety, supporting the notion of accelerated habit formation in overweight/obesity. In a more recent study, Janssen et al. (2017) used Hogarth's simple choice version of the outcome devaluation paradigm (Hogarth & Chase, 2011), to show that the ability to immediately reduce responding following specific satiety was impaired in (mostly female) individuals with higher obesity scores (based on BMI, waist circumference, and waist-to-hip ratio). These two studies support the hypothesis that action control is shifted toward habitual control in obesity. To rule out that differential success of outcome devaluation through satiation contributed to these correlations, future research should include a reacquisition test following the extinction test, during which the food outcomes are once again presented contingent on responding (Adams, 1982). If devaluation is equally effective in reducing food desire in participants with higher obesity scores, they should reduce responding for the devalued outcome during this reacquisition test (as opposed to the extinction test). Another related question is whether inflexible action control in obesity is restricted to responses that are reinforced with food rewards, or whether there is a general tendency to rely on habits in obesity.

To investigate general habit propensity in obesity, Dietrich, De Wit, and Horstmann (2016) used the slips-of-action paradigm with animal pictures functioning as the trigger cues and as the outcomes. They found no evidence for increased vulnerability to habitual slips of action in relation to BMI (in a group of healthy-weight, overweight, and obese [male and female] participants). A more recent study replicated this null result using the same paradigm to compare obese individuals and matched lean controls, except that food pictures instead of animal pictures were used (Watson, Wiers, Hommel, Gerdes, & de Wit, 2017). These studies suggest that habit formation is not generally accelerated in obesity. An interesting remaining possibility is that reliance on compulsive habits may be a specific issue for obese individuals with a diagnosis of BED. BED is characterized by recurrent episodes of eating large quantities of food, accompanied by a feeling of a loss of control. Some researchers have argued that the food addiction model may be most readily applicable to obese individuals with BED (de Jong et al., 2012). In support of this idea, a recent study showed that obese individuals with BED relied on inflexible, "model-free" as opposed to " model-based" decision-making (which have been proposed to capture habitual and goal-directed control, respectively; see Discussion) while those without BED performed at the same level as healthy-weight controls (Voon et al., 2015). However, an outcome devaluation study is required to investigate habitual and goal-directed control more directly in this population.

In order to gain insight into the causal link between obesity and an imbalance between habitual and goal-directed control, we again turn to animal research. Animal work shows that obesogenic diets can lead to impaired cognitive functioning, including deficits in goal-directed control over food seeking. Kendig, Boakes, Rooney, and Corbit (2013) showed that chronic restricted access to sucrose solution reduced sensitivity to outcome devaluation. Subsequently, Furlong, Jayaweera, Balleine, and Corbit (2014) investigated the effect of a 5-week high-fat/high-sugar diet of sweetened condensed milk on instrumental responding for different food rewards (food pellets and sucrose solution). They showed that intermittent, but not continuous, access to the sweetened milk diet led to reduced goal-directed control over food seeking. Furthermore, they implicated the DLS in this impairment, by providing evidence for greater activation of this region in the restricted access group and demonstrating that experimentally decreasing this activity (through infusion of an AMPA-receptor or dopamine D1-receptor antagonist into the DLS) restored goal-directed performance in the restricted group. Their results indicate, therefore, that a palatable, high-sugar/high-fat diet can have a drastic impact on the flexibility of food seeking by affecting the neural substrates of habits.

Cyclic patterns of food availability can also affect the general balance between habitual and goal-directed control over food seeking. Parkes, Furlong, Black, and Balleine (2017) investigated the effect of repeated cycles of restriction and feeding on standard lab chow on sensitivity to outcome devaluation. Firstly, all rats were trained to respond for two distinct food outcomes (grain and purified pellets). Subsequently, one group of rats received the alternating, bingelike feeding schedule during 30 days, while a control group was on restricted access during the first 20 days followed by 10 days of unrestricted access. The researchers found that rats on a bingelike feeding schedule persisted in responding for a food reward they had just been sated on (in contrast to the control rats), pointing to a general propensity to rely on habits at the expense of goal-directed control.

In summary, there are several ways in which reliance on habits may develop and play a role in compulsive food seeking in obesity (see Fig. 15.3). Food seeking has been shown to become habitual as a consequence of behavioral repetition. Furthermore, there is preliminary evidence that goal-directed control over food seeking is impaired in obesity (Horstmann et al., 2015; Janssen et al., 2017), but further investigation is warranted. Finally, animal research suggests that calorie-dense and bingelike diets can lead to habit propensity. This emergent characteristic may contribute to failures of overweight and obese individuals to change their dietary habits. Convergent (indirect) evidence comes from a large neuroimaging study (Medic et al., 2016), that provided evidence for a negative association between BMI and gray matter volume of the vmPFC, an area that has been implicated in goal-directed control. The relationship between (structural and functional) neuronal alterations in obesity and BED and action control should be investigated in future studies.

Eating disorders

AN predominantly affects adolescent girls and young women and is characterized by compulsive dietary restriction and overexercise aimed at losing weight despite serious adverse consequences, including lowered body temperature, low blood pressure, a disrupted menstrual cycle, reduced bone mineral density, a slow heart rate, and in extreme cases death. Anorexia patients report an intense fear of gaining weight, even when they are severely underweight.

Self-starvation in AN may start out as goal-directed to achieve the rewarding outcome of weight loss. To illustrate with an example, Janet Caldwell started dieting when she was 12 years old. *"At first, she was quite pleased with her weight reduction, and she was able to ignore feelings of hunger by remembering the weight loss goal she had set for herself. However, each time she lost the number of pounds she had set for her goal she decided to lose just a few more pounds… Janet felt that, in her second year of dieting, her weight loss had continued beyond her control… although there had been occasions over the past few years where she had been fairly 'down' or unhappy, she still felt driven to keep on dieting."* (Leon, 1984, pp. 179–184). As this example illustrates, extensive repetition of dieting and exercising in AN may foster the formation of compulsive behavior. At first, fixed behavioral patterns may develop regarding the purchasing, preparation, and consumption of food—with strong preferences for certain ingredients and meal consumption at fixed times and in specific orders. Similarly, rigid patterns emerge for exercising such as running a fixed route every morning or performing 100 abdominal exercises before going to bed (Treasure, Claudino, & Zucker, 2010). These habits may become aberrantly strong in AN as a consequence of the strong positive reinforcement and positive experience of "being in control" that patients experience when they achieve weight loss, which may be further boosted by heightened reward sensitivity as a consequence of starvation. There is also a negative reinforcement component in AN, as a consequence of the avoidance of consumption of energy-dense food and weight gain, which can give rise to feelings of relief. As a result, dietary and exercise habits in AN may transition into compulsive behavior. Indeed, the commonly reported experience of a gradual loss of control that occurs as rigid restrictive eating and excessive exercising patterns develop, is in line with the possibility that habit formation contributes to compulsive behavior in AN. Accordingly, several researchers have proposed that outcome-insensitive habits play an important role in AN and may contribute to treatment resistance (Godier & Park, 2014; Walsh, 2013), similarly to habits in addiction (Godier & Park, 2015) and in OCD (Steinglass & Walsh, 2006).

Chronic dietary restriction and starvation are sources of stress in AN (Park, Godier, & Cowdrey, 2014) that could further accelerate habit formation. Another effect of starvation is to impair cognitive function. However, a recent study failed to provide direct evidence for habit propensity in AN (Godier et al., 2016). Patients performed as well as controls on two outcome devaluation paradigms. Firstly, in the slips-of-action paradigm (with animal or with fruit pictures functioning as the trigger cues and outcomes), they successfully withheld responses to cues that signaled the availability of outcomes that were no longer valuable. Secondly, they performed an avoidance task in which they were trained to avoid electrical shocks, before one of the electrodes was removed, thereby effectively devaluing this aversive outcome. AN patients reduced avoidance of the removed electrode, just like healthy controls. It is noteworthy in this respect that there appears to be a discrepancy between the neural changes in AN and what would be expected for habit propensity. For example, Foerde, Steinglass, Shohamy, and Walsh (2015) showed that food choices by AN patients in a computerized task were associated with *hyper*activity in the caudate—a region previously implicated in goal–directed control (for a review of the neurobiological basis, see Steinglass et al., 2016). However, the neural basis of the balance between goal-directed and habitual control has hitherto not been directly assessed in AN.

A failure to find evidence for general habit propensity in AN does not exclude the possibility that over time their dietary and exercise behaviors become compulsive habits. Furthermore, it remains possible that AN patients are relatively prone to form habits fast in their daily lives, as a consequence of perfectionism, a character trait that people with AN tend to score high on (Egan, Wade, & Shafran, 2011). At first, the rigid rules that AN patients apply to their dieting and exercise patterns may support efficient goal achievement, but this also ensures the reinforcement of highly consistent S-R associations between stable contextual variables and behaviors, thus forging strong habits.

To conclude, few studies have investigated the role of habits in AN, but Fig. 15.4 illustrates how aberrantly strong habit formation could contribute to compulsive behavior in AN. So far, research has failed to provide evidence for general habit propensity, but it remains possible that disorder-specific weight loss behaviors become aberrantly strong habits. Lab investigations using computerized tasks may fail to capture the influence of perfectionism on consistent performance of rigid S-R rules that play an important role in the daily routines of patients, and thereby underestimate the role that habits play in

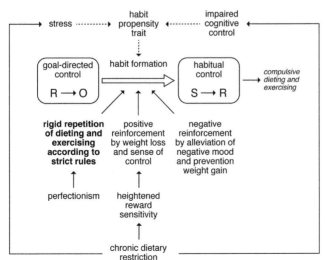

Figure 15.4 *Habits in eating disorders.* Summary of hypothesized links between different factors influencing the shift in the balance from habitual to goal-directed action control. The dotted arrows represent hypothesized links that are not supported by current empirical evidence.

AN. Another important factor that deserves further attention is the influence of starvation on the balance between goal-directed control and habitual control. Also, it would be interesting to explore the role of habits in the related eating disorder bulimia nervosa (Berner & Marsh, 2014). Finally, future neuroimaging research with outcome devaluation paradigms is required to shed light on the role of habits and corticostriatal circuitries in compulsive behavior in AN.

Obsessive–compulsive disorder

A young mother washes her hands excessively upon recurring, anxious thoughts about strangling her own daughter: "*Whenever the thought occurs, I wash my hands. It seems that I can rinse this terrible thought away causing the anxiety to decrease (…) At the moment I find washing my hands more annoying than the thought itself. It takes a long time and I become very anxious when I don't have the facilities to wash my hands.*" (Denys, 2011). The OCD patient in this example initially washes her hands to reduce distress at fear-provoking thoughts. However, after a while, she develops tolerance, leading her to wash her hands even more excessively and ultimately to washing despite little or no alleviation of stress. At this point, the behavior has turned into a compulsive act that for some patients is even more problematic than the obsessive thoughts that characterize OCD. In this regard, OCD resembles addiction (Denys, 2011).

OCD is characterized by repetitive behaviors (e.g., hand washing, ordering, checking) or mental acts (e.g., praying, counting, repeating words silently) that the individual feels driven to perform in response to an obsession or according to rules that must be applied rigidly. According to the *DSM-5*, the goal of these behaviors or mental acts is to prevent or reduce anxiety or distress, or prevent some dreaded event or situation, even when these behaviors or mental acts are not connected in a realistic way with what they are designed to neutralize or prevent, or are clearly excessive. The *DSM-5* thus offers the conventional view that compulsions are goal-directed behaviors. Cognitive therapy of OCD is based on this goal-directed account of OCD and focuses on a number of "cognitive biases" that may drive the excessive performance of compulsive acts (van Oppen & Emmelkamp, 2000). A collaborative team of researchers in the field of OCD, the Obsessive Compulsive Cognitions Working Group, identified three cognitive domains that best represent these biases: an inflated sense of responsibility and heightened threat estimation; perfectionism and intolerance of uncertainty; and importance and control of thoughts (Steketee et al., 2003).

More recently, there has been growing interest in the alternative hypothesis that compulsions result from aberrantly strong habit formation and, if anything, impaired goal-directed control (Evans, Lewis, & Iobst, 2004; Robbins et al., 2012). We will present here arguments for and against this hypothesis. First of all, the experiences of patients (and their clinicians) are often in line with the goal-directed perspective on OCD. For example, at least initially, the young mother in our example reported that she washed her hands to achieve the goal of alleviating stress. However, we should be cautious in interpreting such subjective accounts. Excessive repetition of compulsions could cause "cognitive dissonance," an unpleasant tension or conflict between one's behavior and one's explicitly held convictions. To reduce this tension, people may offer posthoc rationalizations of their compulsions. For example, "I check the door to prevent burglary" could be a goal-directed posthoc rationalization for extreme checking whether the door is locked, to the point that it becomes almost impossible to leave the house. Accordingly, Robbins et al. (2012) suggested that OCD may be better characterized as COD, with obsessions arising as a reaction to compulsions, rather than the other way around.

An argument against the goal-directed perspective is that compulsive behavior in many OCD patients is of an egodystonic nature, which means that the symptoms are inconsistent with the individual's self-perception. To illustrate, consider an OCD patient who has to check that his front door is locked three times in a row. He may be perfectly aware that this is excessive and that the behavior holds no realistic relationship with the aim of preventing burglary. Still, he finds himself unable to disengage from this checking behavior. In some cases, the absence of a realistic link is even more obvious, as in the case of someone who washes his hands to prevent that loved ones will be harmed by bad people (Rachman & De Silva, 1978). This egodystonic nature of compulsions is in line with a habitual account of OCD.

According to a habitual account of compulsive behavior in OCD, the excessive repetition of behaviors according to strict rules, and reinforced by a temporary sense of relief or stress reduction, should foster the development of habits. As a consequence, contextual stimuli will start to trigger the behavior habitually, as for example the sight of the front door evoking an urge to check the lock. While in most people habits such as these never turn into compulsions, some people may be especially vulnerable. A general propensity to rely on habits—either due to aberrantly strong habit formation or to impaired goal-directed control—could lead to these habits spinning out of control. The first direct evidence for habit propensity in OCD came from a study with the slips-of-action paradigm (with neutral fruit pictures functioning as trigger cues and outcomes). This study demonstrated that OCD patients were impaired at directing their responses toward still-valuable outcomes and away from devalued ones (Gillan et al., 2011). Furthermore, their vulnerability to slips of action was directly related to symptom severity. As compulsive behaviors in OCD are usually avoidant, Gillan et al. (2014) set out to further investigate habit propensity in an aversive context. To this end, they tested patients and matched controls on a shock avoidance paradigm in which mild shocks, administered through electrodes attached to the left and right wrist, could be avoided by pressing a foot pedal on the corresponding side. Following the learning phase, one of the electrodes was disconnected in full view of the participants, in order to "devalue" this shock. Following minimal training, both patients and controls performed less avoidance responses for the "deval-ued" shock. However, when the training duration was extended, thereby providing more opportunity for habit learning, a subset of the OCD patients showed evidence for dominant habitual control by continuing to press the pedal that avoided the "deval-ued" shock, which was associated with a self-reported urge to respond. In a subsequent fMRI study, this subjective urge was related to hyperactivity of the caudate (Gillan et al., 2015). The authors concluded that, counterintuitively, *more* activity in the goal-directed network (in the absence of changes in the habit network) relates to a goal-directed impairment in OCD, shifting the balance of control toward compulsive habits. Relat-edly, hyperactivity of components of the goal-directed network—the caudate and the vmPFC/orbitofrontal cortex—has been identified as a relatively consistent neurobiolog-ical marker of OCD (Whiteside, Port, & Abramowitz, 2004).

The relationship between habit propensity and compulsive symptomatology may not be restricted to OCD patients, but rather appears to be continuous in the general pop-ulation. In a sample of 93 healthy, young adults, scores on the Obsessive—Compulsive Inventory—Revised were related to goal-directed performance on the slips-of-action paradigm. All subscales except for Hoarding were negatively related to the ability to selectively withhold responses toward no longer-valuable outcomes (i.e., Washing, Checking, Neutralizing, Ordering, and Obsessing), although after other variables were controlled for (including stress, anxiety, and depression), only the Checking subscale remained a significant predictor of the balance between goal-directed and habitual con-trol. Interestingly, a recent metaanalysis suggested that this symptom dimension is strongly related to cognitive deficits in planning and inhibition (Leopold & Backenstrass, 2015), which could contribute to impaired goal-directed control.

Finally, a habitual account of compulsions may not necessarily be at odds with the view that cognitive biases play an important role in OCD. As mentioned before in the context of eating disorders, perfectionism in OCD (Egan et al., 2011) could foster the formation of habits by leading to strict behavioral rules and routines. Indeed, perfec-tionism has been related to symmetry and ordering behaviors that may be aimed at or reinforced by reaching the feeling of "just right" (Brakoulias et al., 2014). Memory un-certainty (and intolerance thereof) is another cognitive bias that has been hypothesized to contribute toward OCD symptoms. It is possible, however, that memory uncertainty arises as a consequence (rather than cause) of dominant S-R habitual control over avoid-ance behavior. In line with this idea, several studies have provided preliminary evidence that memory uncertainty increases with repeated checking (Boschen & Vuksanovic, 2007; Radomsky, Dugas, Alcolado, & Lavoie, 2014; Van Den Hout & Kindt, 2004). For example, participants were instructed to engage in repeated checking in an interac-tive computer animation that displayed light bulbs or gas rings. The researchers found that repeated checking reduced subjective trustworthiness, vividness, and detail of mem-ory of the checked events (while accuracy of the memory was unaffected). To the extent

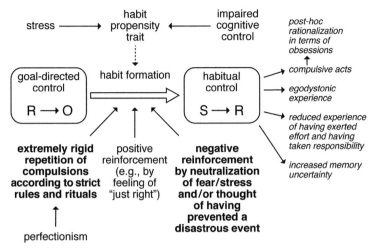

Figure 15.5 *Habits in obsessive–compulsive disorder.* Summary of hypothesized links between different factors influencing the shift in the balance from habitual to goal-directed action control.

that habits are triggered by cues and performed with less awareness, this could account for this increase in OCD-like memory uncertainty. Of course, memory uncertainty could motivate further checking, thus creating a vicious cycle. Finally, another effect of behavioral repetition is to undermine the subjective experience of having exerted effort and of having acted responsibly, which relates to another cognitive bias in OCD (Van Den Hout & Kindt, 2004).

The evidence for, and speculations regarding, a role of habits in OCD are summarized in Fig. 15.5. Firstly, as reviewed above, there is converging evidence for general habit propensity in OCD. At least, in part, this appears to be driven by impaired goal-directed control, with one study relating this to hyperactivity of the goal-directed neural circuitry. It remains an outstanding question whether aberrantly strong habitual control also contributes to the imbalance. To gain a deeper understanding of habit development and the balance with goal-directed control, this field is in urgent need of suitable animal models that allow for investigation of *causal* links between behavioral repetition and OCD symptoms, and the neural basis thereof (Camilla d'Angelo et al., 2014). Finally, the relationship between cognitive biases and avoidance habits in OCD offers an interesting avenue for future research.

Habit propensity in other disorders

In the previous section, evidence was reviewed for a role of habits in compulsive behavior in addiction, obesity, eating disorders, and OCD. However, an imbalance between goal-directed and habitual control has also been demonstrated in several other disorders, as will be reviewed below.

Gilles de la Tourette syndrome

Gilles de la Tourette syndrome (GTS) is characterized by recurrent stereotyped movements or vocalizations, which are often performed unintentionally. GTS shows overlap with OCD, with complex tics sometimes being hard to distinguish from compulsions. Like habits, these tics are exacerbated by stress. GTS patients have been shown to perform worse than healthy controls on the slips-of-action task, providing evidence for habit propensity (Delorme et al., 2016). Furthermore, the severity of their tics correlated negatively with performance, and an exploratory analysis revealed that the latter correlated with estimated white matter tract strength between the putamen and motor cortex, implicating the integrity of the habitual system in their tendency to rely on habits. Finally, tic severity in a subset of unmedicated patients correlated with tract strength between the putamen and the supplementary motor. Therefore, it appears that integrity of the habitual network is related to both habitual control and tics severity.

Autism spectrum disorder

Another clinical condition in which evidence has been found for habit propensity is autism spectrum disorder (ASD) (Alvares, Balleine, Whittle, & Guastella, 2016). Restricted and repetitive behaviors, as well as insistence on sameness, are core symptoms of ASD. To investigate whether a tendency to rely on habits underlies these symptoms, Alvares and colleagues tested adult patients on an outcome devaluation task that was based on that used by Tricomi et al. (2009; but without an extended training condition). They showed that the ability to selectively reduce responding for a prefed food reward was reduced in autism. Interestingly, however, an earlier study had failed to find evidence for habit propensity as assessed with the slips-of-action task, in children between 8 and 12 years with ASD (Geurts & de Wit, 2013). A possible explanation for this discrepancy is that the adults had a longer history of habitual behavior which may have led to stronger development of the underlying neural circuitry (Alvares et al., 2016). Future research should address this interesting hypothesis about the interaction between disease history and habit propensity in ASD (and indeed other disorders). Relatedly, healthy aging has been shown to lead to reduced goal-directed control (de Wit, van de Vijver, & Ridderinkhof, 2014). Thus, the aging brain may become increasingly vulnerable to the development of a general imbalance between goal-directed and habitual control, which may compound with disorder-related habit propensity.

Schizophrenia

Patients with schizophrenia were tested with a novel task that displayed a snack machine that could be tilted to the right or left to obtain snacks (Morris, Quail, Griffiths, Green, & Balleine, 2015). Following instrumental training, one of the snacks was devalued by showing a movie of one of the snacks infested with cockroaches. This devaluation procedure led to reduced responding in healthy controls but failed to affect responding by patients. This impairment was related to reduced caudate engagement during valued actions over devalued.

Social anxiety disorder

Social anxiety disorder (SAD) patients (Alvares, Balleine, & Guastella, 2014; Alvares et al., 2016) performed worse than matched healthy controls on an outcome devaluation (through satiation) task that was based on the design by Tricomi et al. (2009; but without an extended training condition). SAD is not typically considered to be compulsive, but still the results do not seem altogether surprising in light of the finding that SAD patients scored significantly higher on anxiety and stress than the matched controls, taken together with several demonstrations that stress impairs goal-directed control (e.g., Dias-Ferreira et al., 2009; Schwabe & Wolf, 2009). Furthermore, individual differences in general anxiety and psychological stress (as well as impulsivity) within the social anxiety group correlated negatively with test performance (Alvares et al., 2014).

Parkinson disease

Finally, Parkinson disease (PD) has also been associated with a disease severity—dependent goal-directed impairment (de Wit, Barker, Dickinson, & Cools, 2011), in line with progressive dopamine depletion of ventral corticostriatal circuitry.

To conclude, habit propensity has been demonstrated in a number of different psychopathologies. GTS is very closely related to OCD, but for the other disorders it is less obvious that compulsivity is a core characteristic. This begs the question as to how specific habit propensity is to disorders of compulsivity.

DISCUSSION

Challenges and future directions

In the previous section, evidence was presented for habit propensity not only in compulsive disorders but also in ASD, schizophrenia, SAD, and PD. Habit propensity may be fostered across these different disorders by factors that are shared by most psychiatric disorders, such as stress and cognitive control impairments. As discussed by Gillan et al. (2017), this raises the possibility that habit propensity as a mechanism for compulsivity lacks clinical specificity. Alternatively, it could be argued that categorical diagnoses obscure commonalities between these disorders, specifically relating to compulsivity. In line with the latter option, there are high rates of comorbidity between these different disorders, and arguably there is some degree of compulsivity present in all of these disorders. To address this issue, Gillan et al. (2017) propose that to further advance the field, dimensional studies should be conducted that transcend disorder-specific studies. The first such dimensional study was conducted into model-based/model-free decision-making (computational parameters that have been proposed to capture goal-directed/habitual action control) with a large sample of almost 2000 participants that completed the two-step task online and filled out several questionnaires (Gillan, Kosinski, Whelan, Phelps, & Daw, 2016). Model-based decision-making was negatively correlated with self-reported OCD symptomatology and was also related to alcohol abuse and eating disorders. Finally, a factor analysis was conducted to distill three factors from the different questionnaires that were used, one of which pertained to a loss of control over repetitive action and thought. This factor was a better predictor of model-based decision-making than any of the individual questionnaires and was dissociable from the two other factors (anxious depression and social withdrawal). This study thus provides support for the specificity of impaired model-based decision-making as a transdiagnostic mechanism for compulsivity.

A further improvement of this line of research into habits in compulsivity would be the development of tasks that tap more specifically into the mechanisms underlying the balance between goal-directed and habitual control (i.e., strong habit formation, weak goal-directed control, or both), which may reveal more subtle differences between disorders. An indirect way to achieve this is to combine the outcome devaluation task with a neurobiological measure that can assess the neural dual-system balance (e.g., Sjoerds et al., 2013). However, the most direct way to tap more directly into impairments of goal-directed control versus aberrantly strong habit formation would be to study behavioral flexibility as a function of behavioral repetition. It is a caveat of current outcome devaluation studies in patient populations (unlike several animal studies in the addiction literature) that these usually fail to manipulate the extent of training (for one notable exception, see Gillan et al., 2014). Unfortunately, at present there is no experimental paradigm available that reliably reveals habit formation as a function of behavioral repetition (de Wit et al., in press). Yet another alternative approach is offered by the aforementioned computational approaches that attempt to dissociate between two learning algorithms for goal-directed ("model-based") and habitual strategies ("model-free") in a two-step task (Daw, Gershman, Seymour, Dayan, & Dolan, 2011). Using this approach, alcohol and stimulant use have been shown to be negatively related to the model-based parameter. Furthermore, obese individuals with BED (but not without) and OCD patients also showed impaired model-based control (Voon et al., 2015; for a review, see Voon, Reiter, Sebold, & Groman, 2017). The observation that variations in the model-based parameter are related to disorders of compulsivity could suggest that habit propensity in these disorders is primarily driven by impaired goal-directed control. However, a caveat of this interpretation is that the model-based parameter appears to explain most of the variance in performance across (clinical and nonclinical) studies with this paradigm. The question arises whether the model-free parameter adequately captures habits, and indeed several researchers have recently raised this concern (Lingawi et al., 2016; Voon, Reiter, Sebold, & Groman, 2017).

An alternative behavioral paradigm that can be used to dissociate between flexible goal-directed control and stimulus-driven habits is the (outcome-specific) Pavlovian–

instrumental transfer (PIT) task. During the instrumental learning phase of the standard PIT task, subjects learn to perform instrumental responses to obtain a rewarding outcome (e.g., R1–O1; R2–O2). In the second phase, they learn that Pavlovian stimuli are predictive of these outcomes (e.g., S1–O1; S2–O2). Finally, in the critical test phase, they are given the opportunity to perform the instrumental responses while the Pavlovian stimuli are presented in the background. With this design, it has been shown that Pavlovian stimuli will bias responding toward the signaled outcome (e.g., S1 triggers R1 and S2 triggers R2, presumably through S1–O1–R1 and S2–O2–R2 associations). An elaborate discussion of this paradigm is beyond the scope of the present chapter, so the interested reader is referred to recent reviews (Cartoni, Balleine, & Baldassarre, 2016; Corbit & Balleine, 2016). Importantly, however, the PIT effect is undiminished by outcome devaluation, e.g., through satiation (Watson, Wiers, Hommel, & De Wit, 2014; but for contrasting findings, see, e.g.,; Seabrooke, Le Pelley, Hogarth, & Mitchell, 2017) and engages the habit neural circuitry (e.g., van Steenbergen, Watson, Wiers, Hommel, & de Wit, 2017). Therefore, the PIT task shows promise as a model of stimulus-driven reward seeking, in for example addiction (Corbit & Janak, 2016; Hogarth & Chase, 2011) and obesity (Watson et al., 2017). The potency of PIT effects lies in the fact that they will readily generalize to any Pavlovian stimulus that becomes associated with a certain outcome, like drugs or food. In other words, the drug-/food-seeking behavior does not need to be extensively repeated in the presence of that stimulus for it to trigger future habitual responding. PIT could therefore provide another important pathway to stimulus-driven, outcome-insensitive, and possibly even compulsive behavior.

Another important suggestion for future research concerns the consideration of individual differences in cognitive control functions. It is plausible that cognitive control functions, such as working memory and response inhibition, influence habit propensity, but this relation is still not well understood and should be investigated in future research. In light of well-documented impairments in these functions in disorders of compulsivity, future patient studies should control for this by including standardized behavioral measures of these constructs.

Finally, to link habit formation directly with the development of compulsive behavior, insensitivity to outcome devaluation in extinction in animal/human models needs to be directly related to persistence of behavior *despite the experience of negative consequences*, as has been reported in animal studies in the substance abuse literature (e.g., Pelloux, Everitt, & Dickinson, 2007).

Treatment

A transdiagnostic approach to compulsivity encourages the translation of successful treatments across diagnostic boundaries. Insight into the processes that underlie compulsivity—including potentially the balance between goal-directed and habitual

Figure 15.6 *Behavioral therapeutic interventions that may predominantly target the goal-directed versus habitual system.*

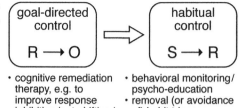

- cognitive remediation therapy, e.g. to improve response inhibition/set-shifting/working memory
- 'episodic future thinking' training

- behavioral monitoring/psycho-education
- removal (or avoidance of) habit triggers
- implementation intentions (replacing habits with more desirable ones)
- contingency management
- habit reversal therapy
- cue exposure with response prevention
- cognitive bias modification (e.g., re-training attentional/approach bias)

control—can further guide the development of effective treatments of disorders of compulsivity and formulation of treatment targets.

Impairments in goal-directed control may limit the capacity for behavioral change through standard cognitive-behavioral treatment. This possibility received support from a recent study in patients with SAD (Alvares et al., 2014), with goal-directed impairment being a negative predictor of treatment outcome. In this case, standard cognitive-behavioral treatment may be boosted by strengthening goal-directed control, by procedures that aim to break the maladaptive habit, and/or by capitalizing on the relatively strong habitual system to encourage positive behavioral change. Below, several such treatments are mentioned and summarized in the framework of dual-system theory in Fig. 15.6. By no means is this intended to be an exhaustive overview, and all of these different treatments should be regarded as promising components of a larger cognitive-behavioral therapy protocol, as opposed to being stand-alone treatments.

Strengthening goal-directed control

Clinical interventions that aim to strengthen goal-directed control (and related cognitive control functions) include cognitive remediation therapies (CRTs). Preliminary investigations in this area suggest that CRT can in some cases increase the effectiveness of standard cognitive-behavior therapy. For example, a working memory training in alcoholics appeared to be effective in decreasing alcohol intake (Houben, Wiers, & Jansen, 2011), and CRT is also thought to improve cognitive flexibility in individuals with AN (Dingemans et al., 2014; Tchanturia et al., 2007). Promising results have also been found with CRT aimed at improving response inhibition to counter overeating and obesity (for a review, see Jansen et al., 2015). Another approach is offered by episodic future thinking (EFT). EFT aims to support goal-directed control over behavior by training people to mentally simulate future events. For example, this treatment has been shown to reduce calorie intake in obese individuals (Daniel, Stanton, & Epstein, 2013).

Relatedly, efforts to reduce stress in patients could improve goal-directed behavioral control. Soares et al. (2012) showed that in college students, a period of no stress reversed the negative effects of stress (during exam period) on the balance between goal-directed and habitual control and partially reversed associated decreases in caudate and increases in putamen volume.

Identifying habit trigger cues

Habits are contextually dependent. Changing habits starts, therefore, with insight into the (external and/or internal) cues that trigger these. This can be achieved through behavioral monitoring (e.g., a food diary) and can be used for the formulation of a functional analysis that describes the three-term (stimulus: response → outcome) contingency between the triggers, maladaptive behaviors, and outcomes. Psychoeducation is an additional tool that can be used to teach patients about identifying trigger cues and creating awareness of the development of maladaptive habits.

Removal or avoidance of habit triggers

According to the "habit discontinuity hypothesis," S-R habits can be disrupted by a change in context, thereby providing a window of opportunity for adapting behavior in light of one's current goals (Verplanken, Walker, Davis, & Jurasek, 2008). Therefore, once triggers of maladaptive habits have been identified, these should—if feasible—be removed from the individual's direct environment, or otherwise avoided. Triggers can be part of a physical environment (e.g., a bar where one regularly goes to drink) or a social environment (e.g., the group of people in whose company ones uses drugs). In line with this idea, Heatherton and Nichols (1994) found that attempts to change some undesirable aspect of one's life were more successful in participants who moved to a new location around the same time. Less invasive ways of changing the context could be, for example, redecorating one's house after a return home from rehab. Finally, triggers of (maladaptive) habits can also be internal, i.e., certain moods or thoughts, in which case those need to be addressed (e.g., as part of cognitive-behavioral therapy).

Implementation intentions

Avoiding triggers of habits is often not entirely feasible. In these cases, when the stimulus—response associations that underlie a maladaptive habit have been identified, implementation intentions (IIs) can be used to replace these with a more desirable habit (which has been proposed to be more effective than simply suppressing the old habit; see, e.g., Adriaanse, van Oosten, de Ridder, de Wit, & Evers, 2011). IIs are concrete "if-then" plans that link the critical trigger cue or situation with the new behavior (Hagger et al., 2016). It has been proposed that IIs create "instant automaticity" (Gollwitzer, 1993).

Therefore, this strategy may capitalize on a relatively intact habitual system to encourage behavioral change. Implementation intentions have been shown to successfully change existing unwanted habits, such as unhealthy diets (Adriaanse, Vinkers, De Ridder, Hox, & De Wit, 2011). A recent metaanalysis suggested that this strategy is also effective in clinical samples (Toli, Webb, & Hardy, 2016), but more research is needed to assess their effectiveness in reducing compulsive behavior.

Contingency management

Contingency management (CM) means that desirable behavior is positively reinforced, which could lead to the formation of adaptive habits. For example, abstinence or refraining from an undesirable behavior is positively reinforced with vouchers, tokens, or money. CM is most regularly employed in the treatment of drug abuse (Higgins, Heil, & Sigmon, 2013).

Habit reversal therapy

Habit reversal therapy (HRT) is an intervention that is used to reduce a wide range of undesired, repetitive behaviors, including tics. It consists of several components, but an important one is that the patient is trained to link critical situations with a behavior that replaces the tic and that is incompatible with the tic at the motor level, as for example isometric tensing of the muscles opposite to the tic movement (Bate, Malouff, Thorsteinsson, & Bhullar, 2011).

Exposure response prevention

Exposure response prevention (ERP) is the behavioral treatment of choice for OCD. In ERP, OCD patients are exposed to stimuli that trigger an obsession and/or compulsion, and they practice suppressing the compulsive behavior (Abramowitz, 1996). ERP has also been used in addiction (Kaplan, Heinrichs, & Carey, 2011), and has been translated to eating behavior, with food cue ERP aiming to break the association between food-associated cues (like the sight and smell of food, or specific situations/moods or thoughts) and thereby reduce overeating and binge eating (Jansen et al., 2015). However, for a detailed discussion of the limitations of extinction treatments, the interested reader is referred to Bouton (2014).

Cognitive bias modification

Cognitive bias modification (CBM) aims to change automatic biases regarding disorder-related stimuli, as for example the sight of drugs in substance abuse. Such stimuli are thought to strongly attract the attention, thereby allowing the cue to trigger habitual responses (next to inducing craving). Attentional bias retraining aims to train attention away from the location of these stimuli (e.g., using a visual probe task) and has proven to be effective in alcohol-dependent individuals (e.g., Schoenmakers et al., 2010).

Another bias concerns the tendency to approach reward-associated cues, as for example drugs. The approach bias can be retrained with the approach-avoidance task. For example, participants are trained to use a joystick to move away from alcohol pictures (e.g., Wiers, Eberl, Rinck, Becker, & Lindenmeyer, 2011). CBM may also be effective in the context of OCD (Najmi & Amir, 2010), and food reward (Kemps, Tiggemann, Orr, & Grear, 2014), but further research is required to establish whether this is a promising avenue for treatment of these conditions.

Pharmacological and neuromodulatory treatment

The different behavioral treatments of compulsive behavior discussed here may be further enhanced by psychopharmacological and neuromodulatory interventions. Neuroscientific investigations of compulsivity across disorders and of the balance between goal-directed and habitual control are of crucial importance here to identify pharmacological targets, which requires not only human studies but, importantly, also animal research with adequate models of compulsivity.

For instance, the role of dopamine in habit formation and the dysregulation of the dopamine system across disorders of compulsivity suggest that this is a promising therapeutic target. Indeed, although OCD is typically treated with selective serotonin reuptake inhibitors (SSRIs), antipsychotic (dopamine antagonists) have also shown promise as an add-on medication (Cavedini, Bassi, Zorzi, & Bellodi, 2004). Conversely, Pelloux, Dilleen, Economidou, Theobald, and Everitt (2012) found that an SSRI reduced (over-trained) compulsive drug seeking in rats, thereby pointing to its potential to reduce compulsive cocaine seeking in patients (Ersche et al., 2011).

Deep brain stimulation (DBS) is a neuromodulatory intervention that involves implanting electrodes in the brain that send electrical impulses. The reported efficacy of DBS to the striatum (specifically, the nucleus accumbens) in OCD (Denys et al., 2010) and addiction (Muller et al., 2009) is consistent with the involvement of cortico-striatal circuits across compulsive disorders. DBS (to the nucleus accumbens) may also lead to symptom alleviation in AN patients (Wu et al., 2013). Another neuromodulatory intervention is repetitive transcranial magnetic stimulation (rTMS), a noninvasive brain stimulation technique that targets the dorsolateral PFC. There is preliminary evidence that rTMS has some efficacy across disorders, reducing cravings and consumption in substance dependence (Barr et al., 2011), decreasing compulsions and obsessions in OCD (Blom, Figee, Vulink, & Denys, 2011), and reducing anxiety and potentially the urge to exercise in AN (McClelland, Kekic, Campbell, & Schmidt, 2015). Future research should determine to what extent DBS and TMS affect the balance between goal-directed and habitual control.

Summary and conclusions

The present chapter provides a transdiagnostic perspective on compulsivity and identifies as an important underlying mechanism an imbalance between goal-directed and habitual control. Clearly, multiple processes play a role in disorders of compulsivity, such as substance abuse, obesity, eating disorders and OCD, with the main determining factors of compulsivity differing between disorders and even between individuals with the same diagnosis. However, there is evidence that in each of these disorders—to different extents and due to different reasons—an imbalance between goal-directed and habitual control plays a central role. As highlighted in this chapter, there still remain many unanswered questions in this field, exposing the need for integrated future research efforts—in animals and humans, behavioral and neurobiological—to uncover the role of habits in compulsivity.

REFERENCES

Abramowitz, J. S. (1996). Variants of exposure and response prevention in the treatment of obsessive-compulsive disorder: A meta-analysis. *Behavior Therapy, 27*(4), 583–600. http://doi.org/10.1016/S0005-7894(96)80045-1.

Adams, C. D. (1982). Variations in the sensitivity of instrumental responding to reinforcer devaluation. *The Quarterly Journal of Experimental Psychology, 34*(2), 77–98. http://doi.org/10.1080/14640748208400878.

Adriaanse, M. A., van Oosten, J. M. F., de Ridder, D. T. D., de Wit, J. B. F., & Evers, C. (2011). Planning what not to eat: Ironic effects of implementation intentions negating unhealthy habits. *Personality & Social Psychology Bulletin, 37*(1), 69–81. http://doi.org/10.1177/0146167210390523.

Adriaanse, M. A., Vinkers, C. D. W., De Ridder, D. T. D., Hox, J. J., & De Wit, J. B. F. (2011). Do implementation intentions help to eat a healthy diet? A systematic review and meta-analysis of the empirical evidence. *Appetite.* http://doi.org/10.1016/j.appet.2010.10.012.

Alvares, G. A., Balleine, B. W., & Guastella, A. J. (2014). Impairments in goal-directed actions predict treatment response to cognitive-behavioral therapy in social anxiety disorder. *PLoS One, 9*(4). http://doi.org/10.1371/journal.pone.0094778.

Alvares, G. A., Balleine, B. W., Whittle, L., & Guastella, A. J. (2016). Reduced goal-directed action control in autism spectrum disorder. *Autism Research.* http://doi.org/10.1002/aur.1613.

Balleine, B. W., & O'Doherty, J. P. (2010). Human and rodent homologies in action control: Corticostriatal determinants of goal-directed and habitual action. *Neuropsychopharmacology: Official Publication of the American College of Neuropsychopharmacology, 35*(1), 48–69. http://doi.org/10.1038/npp.2009.131.

Barr, M. S., Farzan, F., Wing, V. C., George, T. P., Fitzgerald, P. B., & Daskalakis, Z. J. (2011). Repetitive transcranial magnetic stimulation and drug addiction. *International Review of Psychiatry (Abingdon, England), 23*(5), 454–466. http://doi.org/10.3109/09540261.2011.618827.

Bate, K. S., Malouff, J. M., Thorsteinsson, E. T., & Bhullar, N. (2011). The efficacy of habit reversal therapy for tics, habit disorders, and stuttering: A meta-analytic review. *Clinical Psychology Review.* http://doi.org/10.1016/j.cpr.2011.03.013.

Belin, D., Mar, A. C., Dalley, J. W., Robbins, T. W., & Everitt, B. J. (2008). High impulsivity predicts the switch to compulsive cocaine-taking. *Science, 320*(5881), 1352–1355. http://doi.org/10.1126/science.1158136.

Berner, L. A., & Marsh, R. (2014). Frontostriatal circuits and the development of bulimia nervosa. *Frontiers in Behavioral Neuroscience, 8.* http://doi.org/10.3389/fnbeh.2014.00395.

Berridge, K. C., & Robinson, T. E. (2011). Drug addiction as incentive sensitization. *Addiction and Responsibility, 21*–54. http://doi.org/10.1080/09515089.2013.785069.

Blom, R. M., Figee, M., Vulink, N., & Denys, D. (2011). Update on repetitive transcranial magnetic stimulation in obsessive-compulsive disorder: Different targets. *Current Psychiatry Reports, 13*(4), 289–294. http://doi.org/10.1007/s11920-011-0205-3.

Brakoulias, V., Starcevic, V., Berle, D., Milicevic, D., Hannan, A., & Martin, A. (2014). *Psychiatric Quarterly, 85*(2), 133–142.

Boschen, M. J., & Vuksanovic, D. (2007). Deteriorating memory confidence, responsibility perceptions and repeated checking: Comparisons in OCD and control samples. *Behaviour Research and Therapy, 45*(9), 2098–2109. http://doi.org/10.1016/j.brat.2007.03.009.

Bouton, M. E. (2014). Why behavior change is difficult to sustain. *Preventive Medicine.* http://doi.org/10.1016/j.ypmed.2014.06.010.

Camilla d'Angelo, L.-S., Eagle, D. M., Grant, J. E., Fineberg, N. A., Robbins, T. W., & Chamberlain, S. R. (2014). Animal models of obsessive-compulsive spectrum disorders. *CNS Spectrums, 19*(1), 28–49. http://doi.org/10.1017/S1092852913000564.

Cartoni, E., Balleine, B., & Baldassarre, G. (2016). Appetitive pavlovian-instrumental transfer: A review. *Neuroscience and Biobehavioral Reviews.* http://doi.org/10.1016/j.neubiorev.2016.09.020.

Cavedini, P., Bassi, T., Zorzi, C., & Bellodi, L. (2004). The advantages of choosing antiobsessive therapy according to decision-making functioning. *Journal of Clinical Psychopharmacology, 24*(6), 628–631. http://doi.org/10.1097/01.jcp.0000144889.51072.03.

Chen, B. T., Yau, H.-J., Hatch, C., Kusumoto-Yoshida, I., Cho, S. L., Hopf, F. W., & Bonci, A. (2013). Rescuing cocaine-induced prefrontal cortex hypoactivity prevents compulsive cocaine seeking. *Nature, 496*(7445), 359–362. http://doi.org/10.1038/nature12024.

Corbit, L. H., & Balleine, B. W. (2016). Learning and motivational processes contributing to pavlovian–instrumental transfer and their neural bases: Dopamine and beyond. *Current Topics in Behavioral Neurosciences, 27*, 259–289. http://doi.org/10.1007/7854_2015_388.

Corbit, L. H., & Janak, P. H. (2016). Habitual alcohol seeking: Neural bases and possible relations to alcohol use disorders. *Alcoholism: Clinical and Experimental Research.* http://doi.org/10.1111/acer.13094.

Corbit, L. H., Nie, H., & Janak, P. H. (2012). Habitual alcohol seeking: Time course and the contribution of subregions of the dorsal striatum. *Biological Psychiatry, 72*(5), 389–395. http://doi.org/10.1016/j.biopsych.2012.02.024.

Daniel, T. O., Stanton, C. M., & Epstein, L. H. (2013). The future is now: Comparing the effect of episodic future thinking on impulsivity in lean and obese individuals. *Appetite, 71*, 120–125. http://doi.org/10.1016/j.appet.2013.07.010.

Daw, N. D., Gershman, S. J., Seymour, B., Dayan, P., & Dolan, R. J. (2011). Model-based influences on humans' choices and striatal prediction errors. *Neuron, 69*(6), 1204–1215. http://doi.org/10.1016/j.neuron.2011.02.027.

Delorme, C., Salvador, A., Valabrégue, R., Roze, E., Palminteri, S., Vidailhet, M., ... Worbe, Y. (2015). Enhanced habits formation in Gilles de la Tourette syndrome. *Brain: A Journal of Neurology.* awv307.

Delorme, C., Salvador, A., Valabreque, R., Roze, E., Palminteri, S., Vidailhet, M., ... Worbe, Y. (2016). Enhanced habit formation in Gilles de la Tourette syndrome. *Brain: A Journal of Neurology, 139*(2), 605–615. http://doi.org/10.1093/brain/awv307.

Denys, D. (2011). Obsessionality & compulsivity: A phenomenology of obsessive-compulsive disorder. *Philosophy, Ethics, and Humanities in Medicine, 6*(3). http://doi.org/10.1186/1747-5341-6-3.

Denys, D., Mantione, M., Figee, M., van den Munckhof, P., Koerselman, F., Westenberg, H., ... Schuurman, R. (2010). Deep brain stimulation of the nucleus accumbens for treatment-refractory obsessive-compulsive disorder. *Archives of General Psychiatry, 67*(10), 1061. http://doi.org/10.1001/archgenpsychiatry.2010.122.

Deroche-Gamonet, V. (2004). Evidence for addiction-like behavior in the rat. *Science, 305*(5686), 1014–1017. http://doi.org/10.1126/science.1099020.

Dias-Ferreira, E., Sousa, J. C., Melo, I., Morgado, P., Mesquita, A. R., Cerqueira, J. J., ... Sousa, N. (2009). Chronic stress causes frontostriatal reorganization and affects decision-making. *Science, 325*(5940), 621–625. http://doi.org/10.1126/science.1171203.

Dickinson, A. (1985). Actions and habits: The development of behavioural autonomy. *Philosophical Transactions of the Royal Society B: Biological Sciences, 308*(1135), 67–78. http://doi.org/10.1098/rstb.1985.0010.

Dickinson, A., Wood, N., & Smith, J. W. (2002). Alcohol seeking by rats: Action or habit? *The Quarterly Journal of Experimental Psychology, 55*(4), 331–348. http://doi.org/10.1080/0272499024400016.

Dietrich, A., De Wit, S., & Horstmann, A. (2016). General habit propensity relates to the sensation seeking subdomain of impulsivity but not obesity. *Frontiers in Behavioral Neuroscience, 10*(213), 1–14. http://doi.org/10.3389/fnbeh.2016.00213.

Dingemans, A. E., Danner, U. N., Donker, J. M., Aardoom, J. J., van Meer, F., Tobias, K., ... van Furth, E. F. (2014). The effectiveness of cognitive remediation therapy in patients with a severe or enduring eating disorder: A randomized controlled trial. *Psychotherapy and Psychosomatics, 83*(1), 29–36. http://doi.org/10.1159/000355240.

Egan, S. J., Wade, T. D., & Shafran, R. (2011). Perfectionism as a transdiagnostic process: A clinical review. *Clinical Psychology Review.* http://doi.org/10.1016/j.cpr.2010.04.009.

Ersche, K. D., Barnes, A., Simon Jones, P., Morein-Zamir, S., Robbins, T. W., & Bullmore, E. T. (2011). Abnormal structure of frontostriatal brain systems is associated with aspects of impulsivity and compulsivity in cocaine dependence. *Brain: A Journal of Neurology, 134*(7), 2013–2024. http://doi.org/10.1093/brain/awr138.

Ersche, K. D., Gillan, C. M., Jones, P. S., Williams, G. B., Ward, L. H. E., Luijten, M., ... Robbins, T. W. (2016). Carrots and sticks fail to change behavior in cocaine addiction. *Science (New York, N.Y.), 352*(6292), 1468–1471. http://doi.org/10.1126/science.aaf3700.

Evans, D. W., Lewis, M. D., & Iobst, E. (2004). The role of the orbitofrontal cortex in normally developing compulsive-like behaviors and obsessive-compulsive disorder. *Brain and Cognition.* http://doi.org/10.1016/S0278-2626(03)00274-4.

Everitt, B. J., Dickinson, A., & Robbins, T. W. (2001). The neuropsychological basis of addictive behaviour. *Brain Research Reviews, 36*(2), 129–138. pii:S0165017301000881.

Everitt, B. J., & Robbins, T. W. (2005). Neural systems of reinforcement for drug addiction: From actions to habits to compulsion. *Nature Neuroscience, 8*(11), 1481–1489. http://doi.org/10.1038/nn1579.

Everitt, B. J., & Robbins, T. W. (2015). Drug addiction: Updating actions to habits to compulsions ten years on. *Annual Review of Psychology, 67*, 23–50. http://doi.org/10.1146/annurev-psych-122414-033457.

Fineberg, N. A., Potenza, M. N., Chamberlain, S. R., Berlin, H. A., Menzies, L., Bechara, A., ... Hollander, E. (2010). Probing compulsive and impulsive behaviors, from animal models to endophenotypes: A narrative review. *Neuropsychopharmacology: Official Publication of the American College of Neuropsychopharmacology, 35*(3), 591–604. http://doi.org/10.1038/npp.2009.185.

Foerde, K., Steinglass, J. E., Shohamy, D., & Walsh, B. T. (2015). Neural mechanisms supporting maladaptive food choices in anorexia nervosa. *Nature Neuroscience, 18*(11), 1571–1573. http://doi.org/10.1038/nn.4136.

Furlong, T. M., Jayaweera, H. K., Balleine, B. W., & Corbit, L. H. (2014). Binge-like consumption of a palatable food accelerates habitual control of behavior and is dependent on activation of the dorsolateral striatum. *The Journal of Neuroscience, 34*(14), 5012–5022. http://doi.org/10.1523/JNEUROSCI.3707-13.2014.

Geurts, H. M., & de Wit, S. (2013). Goal-directed action control in children with autism spectrum disorders. *Autism: The International Journal of Research and Practice, 18*(4), 409–418. http://doi.org/10.1177/1362361313477919.

Gillan, C. M., Apergis-Schoute, A. M., Morein-Zamir, S., Urcelay, G. P., Sule, A., Fineberg, N. A., ... Robbins, T. W. (2015). Functional neuroimaging of avoidance habits in obsessive-compulsive disorder. *The American Journal of Psychiatry, 172*(3), 284–293. http://doi.org/10.1176/appi.ajp.2014.14040525.

Gillan, C. M., Fineberg, N. A., & Robbins, T. W. (2017). A trans-diagnostic perspective on obsessive-compulsive disorder. *Psychological Medicine,* 1–21. http://doi.org/10.1017/S0033291716002786.

Gillan, C. M., Kosinski, M., Whelan, R., Phelps, E. A., & Daw, N. D. (March 2016). Characterizing a psychiatric symptom dimension related to deficits in goaldirected control. *eLife, 5.* http://doi.org/10.7554/eLife.11305.

Gillan, C. M., Morein-Zamir, S., Urcelay, G. P., Sule, A., Voon, V., Apergis-Schoute, A. M., ... Robbins, T. W. (2014). Enhanced avoidance habits in obsessive-compulsive disorder. *Biological Psychiatry, 75*(8), 631–638. http://doi.org/10.1016/j.biopsych.2013.02.002.

Gillan, C. M., Papmeyer, M., Morein-Zamir, S., Sahakian, B. J., Fineberg, N. A., Robbins, T. W., & De Wit, S. (2011). Disruption in the balance between goal-directed behavior and habit learning in obsessive-compulsive disorder. *American Journal of Psychiatry, 168*(7), 718–726. http://doi.org/10.1176/appi.ajp.2011.10071062.

Gillan, C. M., Robbins, T. W., Sahakian, B. J., van den Heuvel, O., & van Wingen, G. (2016). The role of habit in compulsivity. *European Neuropsychopharmacology, 26*(5), 828−840. https://doi.org/10.1016/j.euroneuro.2015.12.033.

Godier, L. R., de Wit, S., Pinto, A., Steinglass, J. E., Greene, A. L., Scaife, J., ... Park, R. J. (2016). An investigation of habit learning in anorexia nervosa. *Psychiatry Research, 244*, 214−222. http://doi.org/10.1016/j.psychres.2016.07.051.

Godier, L. R., & Park, R. J. (2014). Compulsivity in anorexia nervosa: A transdiagnostic concept. *Frontiers in Psychology, 5*, 1−18. http://doi.org/10.3389/fpsyg.2014.00778.

Godier, L. R., & Park, R. J. (October 2015). Does compulsive behavior in anorexia nervosa resemble an addiction? A qualitative investigation. *Frontiers in Psychology, 6*. http://doi.org/10.3389/fpsyg.2015.01608.

Gollwitzer, P. M. (1993). Goal achievement: The role of intentions. *European Review of Social Psychology, 4*(1), 141−185.

Hagger, M. S., Luszczynska, A., de Wit, J., Benyamini, Y., Burkert, S., Chamberland, P.-E., ... Gollwitzer, P. M. (2016). Implementation intention and planning interventions in health psychology: Recommendations from the synergy expert group for research and practice. *Psychology & Health, 31*(7), 814−839. http://doi.org/10.1080/08870446.2016.1146719.

Heatherton, T. F., & Nichols, P. A. (1994). Personal accounts of successful versus failed attempts at life change. *Personality & Social Psychology Bulletin, 20*(6), 664−675. http://doi.org/10.1177/0146167294206005.

Heyes, C., & Dickinson, A. (1990). The intentionality of animal action. *Mind and Language, 5*(1), 87−103. http://doi.org/10.1111/j.1468-0017.1990.tb00154.x.

Higgins, S. T., Heil, S. H., & Sigmon, S. C. (2013). Voucher-based contingency management in the treatment of substance use disorders. In *APA handbook of behavior analysis, Vol. 2: Translating principles into practice* (pp. 481−500). http://doi.org/10.1037/13938-019.

Hogarth, L., & Chase, H. W. (2011). Parallel goal-directed and habitual control of human drug-seeking: Implications for dependence vulnerability. *Journal of Experimental Psychology: Animal Behavior Processes, 37*(3), 261−276. http://doi.org/10.1037/a0022913.

Horstmann, A., Dietrich, A., Mathar, D., Pössel, M., Villringer, A., & Neumann, J. (2015). Slave to habit? Obesity is associated with decreased behavioural sensitivity to reward devaluation. *Appetite, 87*, 175−183. http://doi.org/10.1016/j.appet.2014.12.212.

Houben, K., Wiers, R. W., & Jansen, A. (2011). Getting a grip on drinking behavior: Training working memory to reduce alcohol abuse. *Psychological Science, 22*(7), 968−975. http://doi.org/10.1177/0956797611412392.

Jansen, A., Houben, K., & Roefs, A. (November 2015). A cognitive profile of obesity and its translation into new interventions. *Frontiers in Psychology, 6*. http://doi.org/10.3389/fpsyg.2015.01807.

Janssen, L. K., Duif, I., van Loon, I., Wegman, J., de Vries, J. H. M., Cools, R., & Aarts, E. (2017). Loss of lateral prefrontal cortex control in food-directed attention and goal-directed food choice in obesity. *NeuroImage, 146*, 148−156. http://doi.org/10.1016/j.neuroimage.2016.11.015.

Jia, H., & Lubetkin, E. I. (2010). Trends in quality-adjusted life-years lost contributed by smoking and obesity. *American Journal of Preventive Medicine, 38*(2), 138−144. http://doi.org/10.1016/j.amepre.2009.09.043.

de Jong, J. W., Vanderschuren, L. J. M. J., & Adan, R. A. H. (2012). Towards an animal model of food addiction. *Obesity Facts, 5*(2), 180−195. http://doi.org/10.1159/000338292.

Jonkman, S., Pelloux, Y., & Everitt, B. J. (2012). Drug intake is sufficient, and conditioning is not necessary for the emergence of compulsive cocaine seeking after extended self-administration. *Neuropsychopharmacology: Official Publication of the American College of Neuropsychopharmacology, 37*(7), 1612−1619. http://doi.org/10.1038/npp.2012.6.

Kaplan, G. B., Heinrichs, S. C., & Carey, R. J. (2011). Treatment of addiction and anxiety using extinction approaches: Neural mechanisms and their treatment implications. *Pharmacology Biochemistry and Behavior.* http://doi.org/10.1016/j.pbb.2010.08.004.

Kemps, E., Tiggemann, M., Orr, J., & Grear, J. (2014). Attentional retraining can reduce chocolate consumption. *Journal of Experimental Psychology: Applied, 20*(1), 94–102. http://doi.org/10.1037/xap0000005.

Kendig, M. D., Boakes, R. A., Rooney, K. B., & Corbit, L. H. (2013). Chronic restricted access to 10% sucrose solution in adolescent and young adult rats impairs spatial memory and alters sensitivity to outcome devaluation. *Physiology & Behavior, 120*, 164–172. http://doi.org/10.1016/j.physbeh.2013.08.012.

Koob, G. F. (2013). Negative reinforcement in drug addiction: The darkness within. *Current Opinion in Neurobiology.* http://doi.org/10.1016/j.conb.2013.03.011.

Leon, G. R. (1984). *Case histories of deviant behavior.* Boston: Allyn & Bacon.

Leopold, R., & Backenstrass, M. (2015). Neuropsychological differences between obsessive-compulsive washers and checkers: A systematic review and meta-analysis. *Journal of Anxiety Disorders, 30*, 48–58. http://doi.org/10.1016/j.janxdis.2014.12.016.

Lingawi, N. W., Dezfouli, A., & Balleine, B. W. (2016). The psychological and physiological mechanisms of habit formation. In R. Murphy, & R. Honey (Eds.), *The Wiley handbook on the cognitive neuroscience of learning* (pp. 409–441). John Wiley & Sons, Ltd.

McClelland, J., Kekic, M., Campbell, I. C., & Schmidt, U. (2015). Repetitive transcranial magnetic stimulation (rTMS) treatment in enduring anorexia nervosa: A case series. *European Eating Disorders Review, 24.* http://doi.org/10.1002/erv.2414.

Medic, N., Ziauddeen, H., Ersche, K. D., Farooqi, I. S., Bullmore, E. T., Nathan, P. J., … Fletcher, P. C. (2016). Increased body mass index is associated with specific regional alterations in brain structure. *International Journal of Obesity (2005), 40*(7), 1177–1182. http://doi.org/10.1038/ijo.2016.42.

Miles, F. J., Everitt, B. J., & Dickinson, A. (2003). Oral cocaine seeking by rats: Action or habit? *Behavioral Neuroscience, 117*(5), 927–938. http://doi.org/10.1037/0735-7044.117.5.927.

Morris, M. J., Beilharz, J. E., Maniam, J., Reichelt, A. C., & Westbrook, R. F. (2015). Why is obesity such a problem in the 21st century? The intersection of palatable food, cues and reward pathways, stress, and cognition. *Neuroscience and Biobehavioral Reviews.* http://doi.org/10.1016/j.neubiorev.2014.12.002.

Morris, R. W., Quail, S., Griffiths, K. R., Green, M. J., & Balleine, B. W. (2015). Corticostriatal control of goal-directed action is impaired in schizophrenia. *Biological Psychiatry, 77*(2), 187–195. http://doi.org/10.1016/j.biopsych.2014.06.005.

Muller, U. J., Sturm, V., Voges, J., Heinze, H. J., Galazky, I., Heldmann, M., … Bogerts, B. (2009). Successful treatment of chronic resistant alcoholism by deep brain stimulation of nucleus accumbens: First experience with three cases. *Pharmacopsychiatry, 42*(6), 288–291. http://doi.org/10.1055/s-0029-1233489.

Najmi, S., & Amir, N. (2010). The effect of attention training on a behavioral test of contamination fears in individuals with subclinical obsessive-compulsive symptoms. *Journal of Abnormal Psychology, 119*(1), 136–142. http://doi.org/10.1037/a0017549.

Nelson, A., & Killcross, S. (2006). Amphetamine exposure enhances habit formation. *The Journal of Neuroscience, 26*(14), 3805–3812. http://doi.org/10.1523/JNEUROSCI.4305-05.2006.

Neovius, K., Neovius, M., Kark, M., & Rasmussen, F. (2012). Association between obesity status and sick-leave in Swedish men: Nationwide cohort study. *European Journal of Public Health, 22*(1), 112–116. http://doi.org/10.1093/eurpub/ckq183.

Ng, M., Fleming, T., Robinson, M., Thomson, B., & Graetz, N. (2014). Global, regional and national prevalence of overweight and obesity in children and adults 1980–2013: A systematic analysis. *Lancet, 384*(9945), 766–781. http://doi.org/10.1016/S0140-6736(14)60460-8.Global.

Nordquist, R. E., Voorn, P., de Mooij-van Malsen, J. G., Joosten, R. N., Pennartz, C. M., & Vanderschuren, L. J. (2007). Augmented reinforcer value and accelerated habit formation after repeated amphetamine treatment. *European Neuropsychopharmacology, 17*(8), 532–540. http://doi.org/10.1016/j.euroneuro.2006.12.005.

O'Tousa, D., & Grahame, N. (2014). Habit formation: Implications for alcoholism research. *Alcohol.* http://doi.org/10.1016/j.alcohol.2014.02.004.

van Oppen, P., & Emmelkamp, P. M. (2000). Issues in cognitive treatment of obsessive compulsive disorder. *Obsessive compulsive disorder: Contemporary issues in treatment*, 117–132.

Parkes, S. L., Furlong, T. M., Black, A. D., & Balleine, B. W. (2017). Intermittent feeding alters sensitivity to changes in reward value. *Appetite, 113*, 1–6. http://doi.org/10.1016/j.appet.2017.02.009.

Park, R. J., Godier, L. R., & Cowdrey, F. A. (2014). Hungry for reward: How can neuroscience inform the development of treatment for anorexia nervosa? *Behaviour Research and Therapy, 62*, 47–59. http://doi.org/10.1016/j.brat.2014.07.007.

Pelloux, Y., Dilleen, R., Economidou, D., Theobald, D., & Everitt, B. J. (2012). Reduced serotonin transmisison is causally involved in the developmental of compulsive cocaine-seeking in rats. *Neuropsychopharmacology: Official Publication of the American College of Neuropsychopharmacology*. https://doi.org/10.1038/npp.2012.111.

Pelloux, Y., Everitt, B. J., & Dickinson, A. (2007). Compulsive drug seeking by rats under punishment: Effects of drug taking history. *Psychopharmacology, 194*(1), 127–137. http://doi.org/10.1007/s00213-007-0805-0.

Rachman, S., & De Silva, P. (1978). Abnormal and normal obssesions. *Behaviour Research and Therapy, 16*(4), 233–248. http://doi.org/10.1016/j.brat.2006.05.005.

Radomsky, A. S., Dugas, M. J., Alcolado, G. M., & Lavoie, S. L. (2014). When more is less: Doubt, repetition, memory, metamemory, and compulsive checking in OCD. *Behaviour Research and Therapy, 59*, 30–39. http://doi.org/10.1016/j.brat.2014.05.008.

Robbins, T. W., Gillan, C. M., Smith, D. G., de Wit, S., & Ersche, K. D. (2012). Neurocognitive endophenotypes of impulsivity and compulsivity: Towards dimensional psychiatry. *Trends in Cognitive Sciences*. http://doi.org/10.1016/j.tics.2011.11.009.

Schoenmakers, T. M., de Bruin, M., Lux, I. F., Goertz, A. G., Van Kerkhof, D. H., & Wiers, R. W. (2010). Clinical effectiveness of attentional bias modification training in abstinent alcoholic patients. *Drug and Alcohol Dependence, 109*(1–3), 30–36. http://doi.org/10.1016/j.drugalcdep.2009.11.022.

Schwabe, L., Dickinson, A., & Wolf, O. T. (2011). Stress, habits, and drug addiction: A psychoneuroendocrinological perspective. *Experimental and Clinical Psychopharmacology, 19*(1), 53–63. http://doi.org/10.1037/a0022212.

Schwabe, L., & Wolf, O. T. (2009). Stress prompts habit behavior in humans. *The Journal of Neuroscience, 29*(22), 7191–7198. http://doi.org/10.1523/JNEUROSCI.0979-09.2009.

Seabrooke, T., Le Pelley, M. E., Hogarth, L., & Mitchell, C. J. (2017). Evidence of a goal-directed process in human pavlovian-instrumental transfer. *Journal of Experimental Psychology. Animal Learning and Cognition*. https://doi.org/10.1037/xan0000147.

Sheeran, P. (2002). Intention — behavior relations: A conceptual and empirical review. *European Review of Social Psychology, 12*(1), 1–36. http://doi.org/10.1080/14792772143000003.

Sjoerds, Z., de Wit, S., van den Brink, W., Robbins, T. W., Beekman, A. T., Penninx, B. W., & Veltman, D. J. (2013). Behavioral and neuroimaging evidence for overreliance on habit learning in alcohol-dependent patients. *Translational Psychiatry, 3*(12), e337. http://doi.org/10.1038/tp.2013.107.

Soares, J., Sampaio, A., Ferreira, L., Santos, N., Marques, F., Palha, J., … Sousa, N. (2012). Stress-induced changes in human decision-making are reversible. *Translational Psychiatry, 2*(131), 1–7.

van Steenbergen, H., Watson, P., Wiers, R. W., Hommel, B., & de Wit, S. (2017). Dissociable corticostriatal circuits underlie goal-directed versus cue-elicited habitual food seeking after satiation: Evidence from a multimodal MRI study. *European Journal of Neuroscience, 46*(2), 1815–1827.

Steinglass, J., & Walsh, B. T. (2006). Habit learning and anorexia nervosa: A cognitive neuroscience hypothesis. *International Journal of Eating Disorders, 39*(4), 267–275. http://doi.org/10.1002/eat.20244.

Steinglass, J. E., Walsh, B. T., Hudson, J., Hiripi, E., Pope, H., Kessler, R., … Rangel, A. (2016). Neurobiological model of the persistence of anorexia nervosa. *Journal of Eating Disorders, 4*(1), 19. http://doi.org/10.1186/s40337-016-0106-2.

Steketee, G., Frost, R., Bhar, S., Bouvard, M., Calamari, J., Carmin, C., … Yaryura-Tobias, J. (2003). Psychometric validation of the obsessive beliefs questionnaire and the interpretation of intrusions inventory: Part I. *Behaviour Research and Therapy, 41*(8), 863–878. http://doi.org/10.1016/S0005-7967(02)00099-2.

Tchanturia, K., Davies, H., Campbell, I. C., Wilson, G., Grilo, C., Vitousek, K., … Treasure, J. (2007). Cognitive remediation therapy for patients with anorexia nervosa: Preliminary findings. *Annals of General Psychiatry, 6*(1), 14. http://doi.org/10.1186/1744-859X-6-14.

Thorndike, E. L. (1911). *Animal intelligence: Experimental studies*. New York: The Macmillan company.

Tiffany, S. T. (1990). A cognitive model of drug urges and drug-use behavior: Role of automatic and nonautomatic processes. *Psychological Review, 97*(2), 147—168. Retrieved from http://www.ncbi.nlm.nih.gov/pubmed/2186423.

Toli, A., Webb, T. L., & Hardy, G. E. (2016). Does forming implementation intentions help people with mental health problems to achieve goals? A meta-analysis of experimental studies with clinical and analogue samples. *British Journal of Clinical Psychology, 55*(1), 69—90.

Treasure, J., Claudino, A. M., & Zucker, N. (2010). Eating disorders. *Lancet.* http://doi.org/10.1016/S0140-6736(09)61748-7.

Tricomi, E., Balleine, B. W., & O'Doherty, J. P. (2009). A specific role for posterior dorsolateral striatum in human habit learning. *European Journal of Neuroscience, 29*(11), 2225—2232. http://doi.org/10.1111/j.1460-9568.2009.06796.x.

Valentin, V. V., Dickinson, A., & O'Doherty, J. P. (2007). Determining the neural substrates of goal-directed learning in the human brain. *The Journal of Neuroscience, 27*(15), 4019—4026. http://doi.org/10.1523/JNEUROSCI.0564-07.2007.

Van Den Hout, M., & Kindt, M. (2004). Obsessive-compulsive disorder and the paradoxical effects of perseverative behaviour on experienced uncertainty. *Journal of Behavior Therapy and Experimental Psychiatry, 35*(2), 165—181. http://doi.org/10.1016/j.jbtep.2004.04.007.

Verplanken, B., Walker, I., Davis, A., & Jurasek, M. (2008). Context change and travel mode choice: Combining the habit discontinuity and self-activation hypotheses. *Journal of Environmental Psychology, 28*(2), 121—127. Retrieved from http://opus.bath.ac.uk/9300/.

Volkow, N. D., Wang, G.-J., Tomasi, D., & Baler, R. D. (2013). Obesity and addiction: Neurobiological overlaps. *Obesity Reviews, 14*(1), 2—18. http://doi.org/10.1111/j.1467-789X.2012.01031.x.

Voon, V., Derbyshire, K., Rück, C., Irvine, M. A., Worbe, Y., Enander, J., ... Bullmore, E. T. (2015). Disorders of compulsivity: A common bias towards learning habits. *Molecular Psychiatry, 20*(3), 345—352. http://doi.org/10.1038/mp.2014.44.

Voon, V., Reiter, A., Sebold, M., & Groman, S. (2017). Model-based control in dimensional psychiatry. *Biological Psychiatry, 82*(6), 391—400. http://doi.org/https://doi.org/10.1016/j.biopsych.2017.04.006.

Walsh, B. T. (2013). The enigmatic persistence of anorexia nervosa. *American Journal of Psychiatry, 170,* 477—484. http://doi.org/10.1176/appi.ajp.2012.12081074.

Watson, P., Wiers, R. W., Hommel, B., & De Wit, S. (2014). Working for food you don't desire. Cues interfere with goal-directed food-seeking. *Appetite, 79,* 139—148. http://doi.org/10.1016/j.appet.2014.04.005.

Watson, P., Wiers, R. J., Hommel, B., Gerdes, V. E. A., & de Wit, S. (2017). Stimulus control over action for food in obese versus healthy-weight individuals. *Frontiers in Psychology — Eating Behavior, 8.* http://doi.org/https://doi.org/10.3389/fpsyg.2017.00580.

Whiteside, S. P., Port, J. D., & Abramowitz, J. S. (2004). A meta-analysis of functional neuroimaging in obsessive-compulsive disorder. *Psychiatry Research — Neuroimaging.* http://doi.org/10.1016/j.pscychresns.2004.07.001.

Wiers, R. W., Eberl, C., Rinck, M., Becker, E. S., & Lindenmeyer, J. (2011). Retraining automatic action tendencies changes alcoholic patients' approach bias for alcohol and improves treatment outcome. *Psychological Science, 22*(4), 490—497. http://doi.org/10.1177/0956797611400615.

de Wit, S., Barker, R. A., Dickinson, A., & Cools, R. (2011). Habitual versus goal-directed action control in Parkinson disease. *Journal of Cognitive Neuroscience, 23*(5), 1218—1229.

de Wit, S., Corlett, P. R., Aitken, M. R., Dickinson, A., & Fletcher, P. C. (2009). Differential engagement of the ventromedial prefrontal cortex by goal-directed and habitual behavior toward food pictures in humans. *The Journal of Neuroscience, 29*(36), 11330—11338. http://doi.org/10.1523/JNEUROSCI.1639-09.2009.

de Wit, S., & Dickinson, A. (2009). Associative theories of goal-directed behaviour: A case for animal—human translational models. *Psychological Research, 73*(4), 463—476. http://doi.org/10.1007/s00426-009-0230-6.

de Wit, S., Kindt, M., Knot, S. L., Verhoeven, A. A. C., Robbins, T. W., Gasull-Camos, J., ... Gillan, C. M., (in press). Shifting the balance between goals and habits: Five failures in experimental habit induction. *Journal of Experimental Psychology: General.*

de Wit, S., van de Vijver, I., & Ridderinkhof, K. R. (2014). Impaired acquisition of goal-directed action in healthy aging. *Cognitive, Affective & Behavioral Neuroscience, 14*(2), 647–658. http://doi.org/10.3758/s13415-014-0288-5.

de Wit, S., Watson, P., Harsay, H. A., Cohen, M. X., van de Vijver, I., & Ridderinkhof, K. R. (2012). Corticostriatal connectivity underlies individual differences in the balance between habitual and goal-directed action control. *The Journal of Neuroscience, 32*(35), 12066–12075. http://doi.org/10.1523/JNEUROSCI.1088-12.2012.

Wu, H., Van Dyck-Lippens, P. J., Santegoeds, R., van Kuyck, K., Gabriels, L., Lin, G., ... Nuttin, B. (2013). Deep-brain stimulation for anorexia nervosa. *World Neurosurgery, 80*(3–4), S29.e1–S29.e10. http://doi.org/. https://doi.org/10.1016/j.wneu.2012.06.039.

Yin, H. H., Knowlton, B. J., & Balleine, B. W. (2004). Lesions of dorsolateral striatum preserve outcome expectancy but disrupt habit formation in instrumental learning. *European Journal of Neuroscience, 19*(1), 181–189. http://doi.org/10.1111/j.1460-9568.2004.03095.x.

CHAPTER 16

Drug Addiction: Augmented Habit Learning or Failure of Goal-Directed Control?

Teri M. Furlong[1], Laura H. Corbit[2]
[1]Neuroscience Research Australia, Randwick, NSW, Australia; [2]Department of Psychology, The University of Toronto, Toronto, ON, Canada

INTRODUCTION

A striking feature of drug addiction is the individual's continued drug use despite extreme negative consequences to long-term health and well-being. This apparent lack of flexible control over behavior has led to increased interest in the role of a habit learning process in addiction (Everitt & Robbins, 2016; Everitt et al., 2008; Ostlund & Balleine, 2008). Drugs of abuse are proposed to alter the brain to influence decision-making, reducing behavioral control of reward seeking and promoting habitual drug-seeking behaviors that continue without consideration of their consequences (Everitt & Robbins, 2013, 2016; Ostlund & Balleine, 2008). Thus, one possibility is that drug habits form like other habits as a result of repeated performance of drug-related behaviors and incremental strengthening of associations with environmental stimuli which, in time, come to trigger drug craving and seeking responses (Belin, Jonkman, Dickinson, Robbins, & Everitt, 2009). However, the habitual nature of drug abuse displayed by addicts differs from normal, adaptive habits, which can be more readily suppressed in the face of direct negative feedback. Therefore, another possibility is that drug habits form, not as a result of direct strengthening of stimulus—response associations, which are thought to underlie habit learning, but rather because flexible, goal-directed control of behavior is compromised. This could be due to direct effects of drug exposure on the brain, or, at least in humans, could relate to preexisting traits. A failure of goal-directed control would allow for premature and predominant habitual control of behavior where it would not otherwise exist, as well as a failure to suppress established habits when they no longer serve the needs or desires of the individual. The aim of this chapter is to summarize research demonstrating accelerated habitual control following exposure to drugs of abuse and to evaluate whether such habits are the result of augmented habit learning and/or failures of flexible control. By examining drug-related changes to the parallel neural circuits known to mediate goal-directed and habit learning, we can begin to understand how drug use impacts the balance between these systems. The research presented supports

Goal-Directed Decision Making
ISBN 978-0-12-812098-9, https://doi.org/10.1016/B978-0-12-812098-9.00016-4

the idea that behaviors performed following exposure to drugs of abuse are not only the result of augmented habits but also result from the failure of the goal-directed system that otherwise regulates these habits.

Experimental models of habitual versus goal-directed behavior

We perform many behaviors with a specific goal in mind. Such behaviors can be controlled flexibly; we choose to respond when the anticipated outcome of responding is desired and can also withhold responding when the outcome is unwanted. Thus, so-called goal-directed behaviors are controlled by both the contingent relationship between a response and its outcome and evaluation of that outcome (Balleine & Dickinson, 1998). This flexible, ongoing deliberation over the potential outcomes of our behaviors is cognitively demanding. In contrast, habitual actions are efficient since they are performed without consideration of their consequences (Ostlund & Balleine, 2008). Habits are instead based on reinforcement history; over time, environmental stimuli that are present when particular responses are reinforced come to elicit those same behaviors again. Such responses are adaptive; they develop, after all, out of extended experience where a particular response has been successful in producing reinforcement. Further, they allow for automated behavior and free up limited cognitive resources for more demanding tasks (Ostlund & Balleine, 2008). Associative learning accounts propose that initially actions are driven by response—outcome associations, but that with repeated performance stimulus—response associations are strengthened and, under constant conditions, eventually dominate behavioral control (Adams, 1982; Dickinson, 1985).

In the laboratory, goal-directed actions are best distinguished from habitual actions using an outcome devaluation task. In a commonly used version of this task, hungry rats are trained to press a lever for a food outcome. The value of the food is then typically reduced either by prefeeding it to induce outcome-specific satiety or by conditioning a taste aversion (Adams & Dickinson, 1981; Balleine & Dickinson, 1998). Performance of the instrumental response is then tested to assess any impact of the devaluation treatment. Rats will reduce responding if lever pressing is driven by expectation of the outcome because of an encoded response—outcome association and the current value of the outcome. However, when under habitual behavioral control, for example, following extended instrumental training, responding will continue despite the reduced desirability of the outcome (Adams, 1982). Under these conditions, behavior is argued to be stimulus driven, i.e., driven by the context and situational cues or more proximal stimuli such as the sight of the lever. Over time, these stimuli have become linked with the response resulting in a stimulus—response association that triggers responding when encountered, without consideration of the current value of the outcome. Importantly, tests of habitual versus goal-directed responding are typically conducted in extinction in order to assess which association underlies responding (response—outcome or stimulus—response) without the opportunity for new learning via feedback within the session. When feedback is provided during the session, usually by contingent delivery of the outcome, animals displaying habitual behavior will typically come to suppress responding for the devalued outcome over the course of the session (Adams, 1982; Furlong, Corbit, Brown, & Balleine, 2017; LeBlanc, Maidment, & Ostlund, 2013; Ostlund & Balleine, 2008). Thus, habits are not entirely without cognitive control as they are suppressed by negative feedback to the action, which then allows for a goal-directed strategy that is sensitive to outcome value to again regulate performance. Conceptually similar tasks have also been developed for human subjects that distinguish between goal-directed and habitual responses (as described in more detail below) (de Wit et al., 2012; Hogarth, Chase, & Baess, 2012; Tricomi, Balleine, & O'Doherty, 2009).

The impact of drugs of abuse on experimental models of habitual behavior

A vast number of studies have examined drug use in animals. However, relatively few have directly assessed whether animals flexibly control drug-seeking behavior. Several studies demonstrate that extended self-administration of cocaine, alcohol, or nicotine promotes habitual control of drug-seeking behavior (Clemens, Castino, Cornish, Goodchild, & Holmes, 2014; Corbit, Nie, & Janak, 2012; Zapata, Minney, & Shippenberg, 2010). For example, in order to demonstrate habitual cocaine seeking, animals were trained to perform a sequence of responses; a seeking response (e.g., press left lever) followed by a taking response (e.g., press right lever) that ultimately lead to response-contingent intravenous cocaine (Olmstead, Parkinson, Miles, Everitt, & Dickinson, 2000; Zapata et al., 2010). The drug-taking link of the chain was then devalued by extinction by allowing animals access to only the taking lever and withholding any cocaine. The effect of this manipulation on performance of the seeking response was then assessed after different amounts of self-administration training (Zapata et al., 2010). With limited training, extinguishing the taking lever reduced responding on the seeking lever when it was made available in a subsequent session. However, this sensitivity to devaluation was reduced by extended training, indicating that control of cocaine seeking had become habitual.

Similarly, extended but not brief, nicotine and alcohol self-administration have also been shown to produce habitual control (Clemens et al., 2014; Corbit et al., 2012). In these studies, animals were trained to self-administer intravenous nicotine (Clemens et al., 2014) or oral alcohol (Corbit et al., 2012) and were then tested for sensitivity to devaluation of the drug itself by either pairing its administration with lithium chloride injection or by allowing it to be consumed to satiety, respectively. In both cases, animals were sensitive to devaluation following limited but not extended training. These types of studies involving drug self-administration provide the most direct evidence that drug-seeking behaviors become habitual over the course of extended training. They demonstrate that drug use can start casually and be flexibly controlled, but with extended training and drug exposure, this flexibility is lost, which contributes to the transition from recreational to problem drug use.

While relatively few studies have directly examined the habitual control of drug taking itself, a larger number of studies have examined how drug exposure affects behavioral control more generally, in situations unrelated to obtaining drugs. In such studies, the experimenter administers drugs to the subjects who are then subsequently trained to perform an instrumental response for food. Outcome devaluation is then used to manipulate the value of the food reward, and sensitivity of performance is assessed. Mere exposure to a range of psychostimulants promotes habitual responding, including D-amphetamine, methamphetamine and cocaine (Corbit, Chieng, & Balleine, 2014; Furlong, Corbit, et al., 2017; LeBlanc et al., 2013; Nelson & Killcross, 2006; Nordquist et al., 2007). In each instance, relatively low doses of the drug are administered over a number of days, which is likely to better model initial drug use than chronic, heavy drug use.

Although these studies demonstrate that augmented habitual control following psychostimulant exposure resembles that resulting from extended instrumental training, such habits differ in that they are insensitive to feedback and continue in the face of negative consequences. As described in the previous section, overtraining-based habitual responses can be suppressed with response-contingent feedback, whereas it has been shown that drug-induced habitual responses are not (Furlong, Supit, Corbit, Killcross, & Balleine, 2017; Furlong, Corbit, et al., 2017; LeBlanc et al., 2013). That is, chronic pretreatment with cocaine or methamphetamine produces lever responses that are insensitive to devaluation not only when tested in extinction but also when the now unpalatable food outcome is delivered during test (Furlong, Corbit, et al., 2017; Furlong, Supit, et al., 2017; LeBlanc et al., 2013). This resistance to feedback provided by the delivery of a "punishing" food suggests that the resulting habit differs from those produced by overtraining (Furlong, Corbit, et al., 2017; Furlong, Supit, et al., 2017; LeBlanc et al., 2013; Ostlund & Balleine, 2008). In accordance, it has also been demonstrated that extended, but not limited, access to cocaine leads to drug-seeking behavior that is also insensitive to response-contingent punishment. Specifically, cocaine seeking is not suppressed by pairing cocaine delivery with a mild-footshock or with a conditioned tone stimulus that signaled footshock following extended self-administration training (Deroche-Gamonet, Belin, & Piazza, 2004; Ducret et al., 2016; Vanderschuren & Everitt, 2004). Thus, both experimenter- and self-administered psychostimulants lead to behavior that is insensitive to feedback, unlike normal adaptive habits.

Paradigms where the drugs are administered by the experimenter lack the face validity of drug self-administration paradigms. However, one advantage is that when these animals are subsequently trained to respond for food, the acute effects of the drug on locomotion, memory, appetite or other aspects of the task do not affect the interpretation of performance at test. However, because these studies do not directly assess flexible control of a drug-seeking response, the utility of these paradigms for understanding drug use itself is reduced. Nonetheless, the fact that drug history impacts learning and control of behaviors reinforced with other rewards is striking. These findings suggest that the effects of drugs on decision-making capacity are quite general and that even behaviors learned after the end of drug exposure are prone to control by the habit system, presumably because of long-lasting effects of drugs on the brain. These types of findings are likely to be equally relevant for understanding the struggles of a recovering addict.

Indeed, chronic drug users demonstrate altered ability to assess outcome value in decision-making tasks. In these tasks, symbolic rewards are often sought, such as points in the task or monetary reward upon completion, and so, as described for preclinical studies above, decision-making capacity in general rather than drug-related behaviors per se are being assessed. These studies show that chronic drug users have greater difficulty overcoming well-learned responses than healthy individuals (Ersche et al., 2016; McKim, Bauer, & Boettiger, 2016). For example, following extended training where participants earned points toward monetary gain by pressing a keypad button, cocaine addicts continued to respond to stimuli even when they were no longer rewarded (Ersche et al., 2016). In contrast, healthy controls were sensitive to this change in contingency and reduced responding to nonrewarded stimuli compared to stimuli that continued to be rewarded, thus showing sensitivity to the consequences of their responding. Similarly, individuals with a history of a substance use disorder (any drug or alcohol) demonstrate greater perseverative responding when response contingencies are changed compared to healthy controls (McKim et al., 2016). These subjects were trained to respond to particular visual stimuli by making one of four keypad choices. The response contingency was then changed so that a different keypad choice was required, and the subjects were given immediate feedback as to whether their responses were now correct or incorrect. Subjects with a history of substance abuse made more incorrect choices, which were consistent with the original contingency, than healthy controls, indicating greater perseverative responding (McKim et al., 2016). Both of these studies suggest that drug addicts are less sensitive to negative feedback to their actions, and that drug exposure alters general decision-making capacity, similar to the effects that drugs have in preclinical studies.

Neurocircuitry underlying habitual versus goal-directed behavior

Drugs of abuse have powerful effects on the central nervous system and interact with the brains' reward system. This provides the opportunity for drugs to alter learning-related plasticity, and such effects presumably underlie the ability of drugs to alter drug-seeking behaviors (Gremel & Lovinger, 2017; Wolf, 2016). There are a number of ways this might happen, and to understand how drugs might shift the balance between goal-directed and habitual control, it is important to consider the neural circuits that underlie these response systems. It is now well recognized that two distinct systems mediate goal-directed versus habitual actions. These center on the dorsomedial striatum (DMS) and dorsolateral striatum (DLS), respectively (Balleine, Liljeholm, & Ostlund, 2009; Balleine & O'Doherty, 2010; Yin & Knowlton, 2006).

The striatum can be subdivided into dorsal striatum (i.e., DMS and DLS) and ventral striatum (i.e., nucleus accumbens core and shell). The dorsal striatum is the primary input site of the basal ganglia and receives glutamatergic projections primarily from the cortex and thalamus and dopaminergic inputs from substania nigra pars compacta (Gerfen & Surmeier, 2011; Joel & Weiner, 2000; Yager, Garcia, Wunsch, & Ferguson, 2015). The dorsal striatum can be further divided across its mediolateral axis on the basis of distinct anatomical connections (Alexander, DeLong, & Strick, 1986; McGeorge & Faull, 1989). For example, cortical inputs to the dorsal striatum are topographically organized with DMS receiving inputs from association regions including prelimbic and agranular insular cortices (Reep, Cheatwood, & Corwin, 2003), and DLS receiving input from largely somatosensory and motor regions (Alloway, Lou, Nwabueze-Ogbo, & Chakrabarti, 2006; Ramanathan, Hanley, Deniau, & Bolam, 2002). Another anatomical consideration is that the dorsal striatum consists predominately of two types of inter-mingled medium spiny neurons (MSNs; approx. 90%) that differ according to their downstream projections and neurochemistry (Gerfen & Surmeier, 2011). MSNs of the so-called direct pathway project directly to the basal ganglia output nuclei (i.e., the substantia nigra pars reticulata and globus pallidus internus) and express dopamine D1 receptors and the neuropeptide dynorphin (Gerfen & Surmeier, 2011; Kreitzer & Malenka, 2008; Yager et al., 2015). On the other hand, MSNs that make up the indirect pathway project indirectly to the basal ganglia output nuclei (i.e., via the globus pallidus externus and subthalamic nucleus) and express dopamine D2 and adenosine 2A receptors and the neuropeptide enkephalin (Gerfen & Surmeier, 2011; Kreitzer & Malenka, 2008; Yager et al., 2015). A balance in activity between these pathways has been proposed to regulate a variety of striatal functions, the most well studied being motor control, including the initiation and termination of movement (Kreitzer & Malenka, 2008). However, the importance and potential dysregulation of the so-called "go" and "no-go" pathways for behavioral control in general, and in particular for drug addiction, have also been considered (Yager et al., 2015).

In addition to the anatomical differences between DMS and DLS, important distinc-tions have also been made regarding the contribution of each to learning and behavioral control. The DMS is essential for the acquisition and expression of goal-directed behaviors. When lesions or pharmacological manipulation of the DMS occur prior to instrumental training, rats can acquire an instrumental response but do not appear to encode specific response—outcome associations. For example, they demonstrate instru-mental performance that is insensitive to outcome devaluation or manipulations of the action—outcome contingency, following limited training when control rats are sensitive to such manipulations (Corbit & Janak, 2010; Yin, Ostlund, Knowlton, & Balleine, 2005). The DMS is also critical for the expression of goal-directed behavior as posttraining lesions or pharmacological inactivation impair previously established goal-directed control (Corbit & Janak, 2010; Yin et al., 2005). Furthermore, lesions of the prelimbic cortex, basolateral amygdala, or mediodorsal thalamus, which provide inputs to DMS, also reduce sensitivity to outcome value, suggesting that these regions are also part of the neurocircuitry regulating goal-directed action control (Balleine, Killcross, & Dickinson, 2003; Corbit & Balleine, 2003, 2005; Corbit, Muir, & Balleine, 2003; Killcross & Coutureau, 2003; Ostlund & Balleine, 2005). In contrast, the DLS is recognized to mediate habitual behavior as rats with lesions of the DLS fail to develop habitual responding despite extended training (Faure, Haberland, Conde, & El Massioui, 2005; Yin, Knowlton, & Balleine, 2004). Further, lesions of the infralimbic cortex or central amygdala that functionally interact with the DLS also disrupt the normal expression of habit learning and leave animals sensitive to outcome devaluation under conditions where control animals are not (Killcross & Coutureau, 2003; Lingawi & Balleine, 2012). Thus, together these findings demonstrate that there are two distinct neural circuits that mediate goal-directed and habitual behavior that center on medial and lateral divisions of the dorsal striatum.

In addition to insensitivity to devaluation under extinction conditions, lesions of the goal-directed neurocircuitry also lead to instrumental actions that are not regulated by response-contingent feedback (Ostlund & Balleine, 2008). Thus, when the DMS, basolateral amygdala or mediodorsal thalamus are lesioned, animals either remain insensitive to outcome devaluation (as they are under extinction conditions) or are slow to revert to a goal-directed strategy, when given immediate response-contingent feedback during test (Balleine et al., 2003; Corbit & Balleine, 2005; Corbit et al., 2003; Ostlund & Balleine, 2008). Thus, these findings suggest that the goal-directed neurocircuitry is required for regulating sensitivity to outcome value when feedback is provided.

Several studies also suggest that two distinct neurocircuits centered on the striatum regulate goal-directed and habitual action control in humans. When goal-directed actions are performed by healthy individuals (i.e., when there is explicit knowledge of the response—outcome contingency), imaging studies reveal activation of the anterior caudate (Tanaka, Balleine, & O'Doherty, 2008), as well as the ventromedial prefrontal cortex and medial orbital cortex (Sjoerds et al., 2013; Tanaka et al., 2008). In contrast, extensive training of the response—outcome contingencies, which leads to insensitivity to outcome devaluation, and thus a shift to habitual control, is associated with activation of the posterior putamen (Sjoerds et al., 2013; Tricomi et al., 2009). Furthermore, in healthy individuals, stronger white matter connectivity between the caudate and its cortical inputs (i.e., ventromedial prefrontal cortex) predicts goal-directed action control (i.e., responding to a still-valued outcome), whereas stronger connectivity between the posterior putamen and its cortical inputs (i.e., sensorimotor cortex) predicts habitual action control (i.e., responding for a no-longer-valuable outcome) (de Wit et al., 2012). Together these studies suggest that two distinct neurocircuits are responsible for the balance between goal-directed and habitual performance in humans, similar to those identified in the rodent. In accordance, the anterior caudate is considered to be functionally equivalent to the DMS, which mediates goal-directed action in rodents, and the posterior putamen in humans is thought to be equivalent to the habit-promoting DLS in rodents (Balleine & O'Doherty, 2010; Tanaka et al., 2008; Tricomi et al., 2009).

DRUG EFFECTS ON GOAL-DIRECTED AND HABIT SUBSTRATES

While it is well established that actions and habits rely on largely independent circuits, exactly how drugs act to promote habitual control is only beginning to be understood. For example, drugs could act to augment plasticity in the DLS and thereby directly strengthen habit learning. Alternatively, drugs could undermine normal plasticity and activity within the DMS circuitry and consequently interfere with goal-directed control. Of course, combinations of these effects are also possible. Importantly, these possibilities are difficult to disentangle behaviorally (i.e., the outcome devaluation task does not dissociate the two possibilities). However, examining the effects of drugs on the neural circuits known to control goal-directed versus habitual behaviors can give insight into how each system is affected and aid in the understanding of the nature of the observed shift in behavioral control.

Before examining the persistent effects of psychostimulants and alcohol on the dorsal striatum, it is worth considering that preexisting traits can also influence the development of habitual behavioral control. One such trait is impulsivity, which is defined as an inability to withhold responding or to act without thought (Furlong, Leavitt, Keefe, & Son, 2016; Hogarth et al., 2012; Perry & Carroll, 2008). Impulsivity is highly correlated with drug use, although for human drug abusers it is often difficult to determine which came first (Everitt et al., 2008; Perry & Carroll, 2008). However, there is evidence that impulsivity in drug users is in part genetically determined, e.g., it has been demonstrated that while impulsivity is increased in drug users compared to healthy controls, the same is true of the addicts' siblings who were nondrug users (Ersche et al., 2012; Ersche, Turton, Pradhan, Bullmore, & Robbins, 2010). Importantly, high-impulsivity traits in humans are associated with a reduction in goal-directed behavior (Hogarth, Chase, & Baess, 2012). That is, nondrug users who score highly on measures of impulsivity also show reduced sensitivity to outcome devaluation compared to individuals with low impulsivity scores (Hogarth et al., 2012). Furthermore, the strength of white matter connectivity between the putamen and its cortical inputs is associated with habitual actions in healthy individuals (as described above) (de Wit et al., 2012). These findings suggest that individual differences in trait impulsivity and brain structure promote habitual behavioral control. Therefore, a tendency toward habits is possibly not only a consequence of drug exposure but also a preexisting risk factor, as is the case for impulsivity (Everitt et al., 2008; Perry & Carroll, 2008).

Psychostimulants

Psychostimulants act acutely to increase dopaminergic signaling in dorsal striatum (Barrot et al., 1999; Bustamante et al., 2002), and repeated exposure, at doses that augment habitual behavior, sensitize the dorsal striatum so that dopamine release is subsequently potentiated, i.e., there is greater dopamine release in chronically exposed animals with subsequent stimulation compared to drug-naïve animals (Howell & Kimmel, 2008; Patrick, Thompson, Walker, & Patrick, 1991; Paulson & Robinson, 1995; Yamada et al., 1988). Thus, it has been proposed that changes in dopamine activity may promote the transition to habitual action (Nelson & Killcross, 2006; Wickens, Horvitz, Costa, & Killcross, 2007). In addition, persistent structural changes to the dorsal striatal dopamine system have also been reported following doses of psychostimulants that augment habits (i.e., a relatively low dose, repeated over a number of days, with at a short abstinence period so that findings are not confounded by the acute administration of the drug). These changes include reductions in dopamine transporter binding (DAT), dopamine D1 and D2 receptor binding and levels of dopamine's precursor enzyme, tyrosine hydroxylase (Kleven, Perry, Woolverton, & Seiden, 1990; Kousik, Carvey, & Napier, 2014; Luscher & Malenka, 2011; Maggos et al., 1998; Schwendt et al., 2009). However, findings are largely inconsistent, especially for cocaine, where no changes to DAT or dopamine receptor density have also been reported using similar dosing schedules (Boulay, Duterte-Boucher, Leroux-Nicollet, Naudon, & Costentin, 1996; Claye, Akunne, Davis, DeMattos, & Soliman, 1995; Pilotte, Sharpe, & Kuhar, 1994; Przewlocka & Lason, 1995). Thus, it is possible that in order to detect these changes to the dopamine system, longer exposure periods or moderately higher doses are required (Nader, Sinnott, Mach, & Morgan, 2002; Thanos et al., 2017), or alternatively that persistent structural changes to the dopamine system following low doses of psychostimulants are not responsible for dopamine sensitization or the development of habits.

In addition to effects on dopamine, chronic exposure to psychostimulants is well recognized to disrupt the glutamate system and produce neuroplastic changes that contribute to the development of addiction. These changes have been extensively examined and reviewed for the ventral striatum (Grueter, Rothwell, & Malenka, 2012; Kalivas, 2009; Kalivas & Volkow, 2011) and are thought to contribute to a wide-range of drug-associated behaviors implicated in the development of addiction including initiation and maintenance of drug intake, reinforcement learning, and drug relapse (Belin et al., 2009; Grueter et al., 2012; Kalivas, 2009). A much smaller literature also demonstrates persistent drug-induced changes to the glutamate system in the dorsal striatum. Although one view is that a transition in behavioral control from the ventral to the dorsal striatum may underlie the development of compulsive, habitual drug seeking (Everitt & Robbins, 2013, 2016), it is unclear whether changes to ventral striatum are a necessary first step or whether changes to ventral and dorsal striatum occur in parallel.

Doses of psychostimulants that augment habits produce persistent changes to dorsal striatum that are indicative of increased excitability. That is, repeated exposure to relatively low doses of amphetamines or cocaine are associated with persistent increases in dorsal striatal N-methyl-D-aspartate (NMDA) receptor-mediated synaptic transmission (Moriguchi, Watanabe, Kita, & Nakanishi, 2002; Nishioku, Shimazoe, Yamamoto, Nakanishi, & Watanabe, 1999) and increased density of glutamate receptors (both α-amino-3-hydroxy-5-methyl-4-isoxazolepropionic acid (AMPA) receptor subunits, as well as metabotropic glutamate receptors (Ary & Szumlinski, 2007; Furlong, Corbit, et al., 2017; Kerdsan, Thanoi, & Nudmamud-Thanoi, 2009; Loftis & Janowsky, 2000; Mao & Wang, 2001). Further, the density of vesicular glutamate transporters, which package glutamate into synaptic vesicle for release, are increased, as well as the density of glutamate transporters that are responsible for glutamate reuptake (Furlong, Corbit, et al., 2017; Nishino et al., 1996; Shirai, Shirakawa, Nishino, Saito, & Nakai, 1996). Together, these studies suggest that psychostimulants cause hyperactivity of glutamate system and increased synaptic plasticity at glutamate synapses in dorsal striatum. However, these studies largely examine the dorsal striatum as a whole and do not consider its medial and lateral subdivisions or instead focus on DLS alone while the DMS is rarely examined. Importantly, when the DMS and DLS are analyzed separately, it has been demonstrated that methamphetamine has bidirectional effects within these regions that are consistent with their differing roles in action control and decision-making (Furlong, Corbit, et al., 2017; Jedynak, Uslaner, Esteban, & Robinson, 2007).

Specifically, methamphetamine exposure has been shown to increase the density of dendritic spines on MSNs in the DLS, while reducing their density in DMS (Jedynak et al., 2007). Given that glutamatergic input to the dorsal striatum is predominately to the spines of MSNs, this finding suggests that methamphetamine increases the excitability of the habit-promoting DLS and reduces the capacity of the DMS to mediate goal-directed behavior (Jedynak et al., 2007). Further support for this idea comes from quantification of the density of glutamate receptor subunits and vesicular transport proteins located on synaptic membranes (Furlong, Corbit, et al., 2017). Rats were chronically treated with low doses of methamphetamine over a 1-month period, and then a 2-month abstinence period was imposed in order to determine the persistent effects of methamphetamine that may underlie the insensitivity to outcome value seen with this dosing schedule (Furlong, Corbit, et al., 2017). In accordance with the effects on spine density, methamphetamine exposure was found to increase the density of NMDA and AMPA receptor subunits and vesicular glutamate transporters in the DLS, while there was a reduction in the density of these proteins in the DMS. Hence, methamphetamine exposure bidirectionally impacted the glutamate system across the subregions of the dorsal striatum in a manner that suggests that methamphetamine not only facilitates the development of habitual behavior but also reduces goal-directed action control. Similar studies are therefore required to determine whether other psychostimulants, like cocaine and D-amphetamine, also differentially impact the glutamate system of the DMS and DLS, as analyses treating the entire dorsal striatum as a single structure are obviously problematic if these drugs have bidirectional effects on these regions.

Alcohol

Alcohol exposure has been shown to produce a similar hyperexcitability of the DLS to that seen following psychostimulant exposure. Extensive alcohol consumption in primates increases the density of dendritic spines in the caudoventral area of the putamen (which is homologous to the DLS) but not in the anterior caudate (which is homologous to the DMS) (Cuzon Carlson et al., 2011). The changes to spine density in the putamen were also associated with enhanced glutamate transmission. Furthermore, alcohol exposure reduced GABAergic synaptic transmission in the putamen, thus, reducing inhibitory tone on MSNs, which would allow for excitatory activity to predominate (Cuzon Carlson et al., 2011). In addition, chronic alcohol downregulates endocannabinoid-CB1 receptor signaling and CB1-dependent long-term depression (LTD) (DePoy et al., 2013) and promotes LTD at MSNs and fast-spiking interneuron inputs onto MSNs, within the DLS (Patton, Roberts, Lovinger, & Mathur, 2016). Together, these results suggest that alcohol reduces inhibition and increases the excitability of DLS, both of which would lead to increased DLS output, and in turn favor habitual behavioral control. In accordance, neuroimaging during a decision-making task where participants must make a left or right response to particular picture stimuli reveals greater activity in habit-promoting posterior putamen of alcoholics compared to healthy individuals (Sjoerds et al., 2013). Dysregulation of DMS activity has also been reported and may contribute to behavioral effects associated with alcohol exposure (Wang, Carnicella et al., 2007; Wang et al., 2015). Notably, chronic intermittent alcohol exposure in mice is associated with reduced excitatory cortical transmission to DMS, which is suggestive of hypoactive DMS functioning and reduced goal-directed behavioral control (discussed in more detail below) (Renteria, Baltz, & Gremel, 2018).

RESCUING ADAPTIVE BEHAVIOR FOLLOWING DRUG-INDUCED HABITUAL BEHAVIOR

As described in the previous sections, psychostimulants and alcohol cause persistent neuroadaptations to both the DMS and DLS, which regulate goal-directed and habitual behavior, respectively. Hence, the loss of behavioral control that results from exposure to these drugs possibly results from neuroadaptations that produce an overactive DLS and underactive or dysfunctional DMS. Consistent with this suggestion, goal-directed behavior can be restored following exposure to drugs of abuse by manipulations that either reduce activity in the DLS that alter activity in the DMS. As noted above, extended self-administration of either cocaine or alcohol produces behavior that is insensitive to devaluation. Sensitivity is restored by pharmacological inactivation of the DLS (Corbit et al., 2012; Zapata et al., 2010). Such results suggest that extended training and chronic drug exposure leads to behavioral control dependent on activity in the DLS but that when activity in this structure is suppressed, goal-directed control can be at least partially restored.

Importantly, goal-directed behavioral control can also be restored by manipulations that target the DMS rather than the DLS. Three recent studies provide compelling evidence that normalizing DMS function reinstates goal-directed control (Corbit et al., 2014; Furlong, Supit, et al., 2017; Renteria et al., 2018). One study used in vitro electrophysiology to investigate the impact of cocaine on synaptic plasticity in the dorsal striatum (Corbit et al., 2014). It was demonstrated that cocaine treatment, which accelerated the development of habitual actions in rats, was associated with increased frequency of spontaneous and miniature excitatory postsynaptic potentials (EPSCs) recorded from MSNs of the DMS. In contrast, there were no such changes in the DLS (Corbit et al., 2014). Hence, it was hypothesized that this change in activity in the DMS contributed to the loss of behavioral control by dysregulating activity in the goal-directed circuit. Therefore, the objective was to restore normal activity in the DMS and examine the subsequent impact on behavioral control. Given that increases in EPSC frequency can be accounted for by drug-induced alterations in glutamate homeostasis, N-acetylcysteine, a regulator of cystine-glutamate exchange that normalizes glutamate homeostasis, was utilized (Kalivas, 2009). It was found that the administration of N-acetylcysteine not only prevented the effects of cocaine on EPSCs in DMS but also protected goal-directed behavior in cocaine-exposed animals (Corbit et al., 2014). Therefore, it is likely that cocaine exposure also influences behavioral control through changes in DMS activity, and not just through the DLS.

In a second study, recruitment of the dorsal striatum was examined using the neuronal activity marker c-fos during the expression of habitual behavior (Furlong, Supit, et al., 2017). It was first demonstrated that testing in a methamphetamine-paired context transiently renders behavioral performance habitual. Specifically, a distinct context was paired with methamphetamine injection, and a second context that differed in its visual, tactile, and olfactory properties was paired with vehicle-control injection. Rats were then trained to press two levers, each for a different outcome, in a third context, and were tested for sensitivity to outcome devaluation in both the drug and vehicle-paired contexts. It was found that animals were goal-directed in the vehicle-paired context; however, when the same animals were tested in the drug-paired context, they demonstrated habitual behavioral performance (Furlong, Supit, et al., 2017). A similar context-mediated effect on behavioral control has also been demonstrated for alcohol-paired contexts (Ostlund, Maidment, & Balleine, 2010), thus showing the importance of drug-paired cues for mediating the expression of habitual behavior in addition to other aspects of addiction-related behavior, such as reinstatement and Pavlovian–instrumental transfer (Bossert, Marchant, Calu, & Shaham, 2013; Corbit & Janak, 2016; Crombag, Bossert, Koya, & Shaham, 2008). Additionally, when the outcome was contingently delivered in a subsequent test in the methamphetamine-paired context, animals continued to show insensitivity to outcome value and thus failed to respond appropriately to negative feedback for their actions (Furlong, Supit, et al., 2017). As described above, these kinds of "compulsive" habits, which continue in the face of negative feedback, are associated with disruption in the normal functioning of neurocircuitry that promotes goal-directed actions. In accordance, habitual actions induced by the methamphetamine context were associated with reduced c-fos in the DMS, thus indicating reduced recruitment of the DMS during habitual performance (Furlong, Supit, et al., 2017). Given that there were no such differences in c-fos in the DLS, the DMS was subsequently targeted in an attempt to attenuate habits. As the reduction in c-fos occurred selectively in direct pathway neurons (which were identified by the absence of enkephalin coexpression), activity in the indirect pathway was then reduced using an adenosine 2A receptor antagonist, to restore the balance in activity between these pathways. Adenosine 2A receptors are selectively located on indirect pathway neurons, and antagonism of these receptors inhibits the activity of these neurons (Tozzi et al., 2007), and it has previously been shown that when they are genetically knocked out of striatum in a mouse model, it prevents the development of habits (Yu, Gupta, Chen, & Yin, 2009). When the adenosine 2A antagonist was infused directly into DMS prior to test, it restored sensitivity to outcome value in the methamphetamine-paired context when feedback of action was provided (Furlong, Supit, et al., 2017). Hence, these results provide further support for the idea that drug exposure alters behavior, in part, by acting on the circuitry mediating goal-directed behavior and that behavioral control can be restored by manipulations of this circuitry.

Finally, a third study also confirmed that habitual behavior can be restored by targeting the goal-directed neurocircuitry (Renteria et al., 2018). This study first demonstrated that exposure to alcohol, which promoted habitual behavior in mice, was associated with attenuated cortical transmission to DMS. Specifically, patch-clamp recordings of direct or indirect pathway neurons of cre-transgenic mice (dopamine D1 or D2 receptors, or adenosine 2A receptors) were made 15 days after chronic intermittent inhalation of alcohol vapor or air. Compared to air vapor, exposure to alcohol vapor was associated with reduced probability of neurotransmitter release when orbital frontal cortex (OFC) neurons projecting to direct pathway neurons were optically activated. There was no such difference between treatment groups when OFC neurons projecting to indirect pathway neurons were activated. Given that these results indicate reduced cortical neurotransmission to DMS, the OFC-DMS pathway was subsequently activated using a DREADD approach during behavioral testing. It was demonstrated that increasing the activity of this projection restored goal-directed behavior in alcohol-exposed mice. Thus, this study further confirms that drug-induced alterations in the functioning of the goal-directed neurocircuitry contribute to the expression of habitual behavior.

Overall, the results of studies using inactivation or other pharmacological manipulations are consistent with the idea that the goal-directed and habit learning systems exist in parallel and that the relative dominance shifts across the course of extended training and/or as the result of adaptations following drug exposure. One of the implications of this account is that the goal-directed system is not entirely lost but that the habit system comes to dominate control if activity in the DLS is potentiated or normal activity in the DMS is reduced or dysregulated. Thus, multiple strategies that can restore balance between these parallel circuits may improve behavioral control, as a growing literature seems to suggest, yet understanding the specific effects of a particular drug will likely be important for selecting the most effective intervention.

CONCLUSIONS

In the past decade, numerous reports have shown that drugs can act to accelerate habitual control over drug-related behaviors, as well as decision-making strategies more generally. Until recently, it has been unclear whether this shift in control is due to drug effects that directly strengthen the habit system, or whether drug exposure produces deficits in goal-directed learning and the emergence of habitual control reflects compensation by, or early uncovering of, the habit system. These possibilities are difficult to disentangle behaviorally as changes in the hallmark tasks, such as outcome devaluation, do not dissociate between these accounts. However, examining the effects of drugs on the neural circuits known to control goal-directed versus habitual behaviors is beginning to reveal how each system is affected. This distinction will ultimately be important for developing the most effective treatments using either pharmacological or behavioral approaches.

REFERENCES

Adams, C. D. (1982). Variations in the sensitivity of instrumental responding to reinforcer devaluation. *The Quarterly Journal of Experimental Psychology, 34*(2), 77–98.

Adams, C. D., & Dickinson, A. (1981). Instrumental responding following reinforcer devaluation. *The Quarterly Journal of Experimental Psychology, 33*(2), 109–121.

Alexander, G. E., DeLong, M. R., & Strick, P. L. (1986). Parallel organization of functionally segregated circuits linking basal ganglia and cortex. *Annual Review of Neuroscience, 9*, 357–381. http://doi.org/10.1146/annurev.ne.09.030186.002041.

Alloway, K. D., Lou, L., Nwabueze-Ogbo, F., & Chakrabarti, S. (2006). Topography of cortical projections to the dorsolateral neostriatum in rats: Multiple overlapping sensorimotor pathways. *Journal of Comparative Neurology, 499*(1), 33–48. http://doi.org/10.1002/cne.21039.

Ary, A. W., & Szumlinski, K. K. (2007). Regional differences in the effects of withdrawal from repeated cocaine upon homer and glutamate receptor expression: A two-species comparison. *Brain Research, 1184*, 295–305. http://doi.org/10.1016/j.brainres.2007.09.035.

Balleine, B. W., & Dickinson, A. (1998). Goal-directed instrumental action: Contingency and incentive learning and their cortical substrates. *Neuropharmacology, 37*(4–5), 407–419.

Balleine, B. W., Killcross, A. S., & Dickinson, A. (2003). The effect of lesions of the basolateral amygdala on instrumental conditioning. *The Journal of Neuroscience: the Official Journal of the Society for Neuroscience, 23*(2), 666–675.

Balleine, B. W., Liljeholm, M., & Ostlund, S. B. (2009). The integrative function of the basal ganglia in instrumental conditioning. *Behavioural Brain Research, 199*(1), 43–52. http://doi.org/10.1016/j.bbr.2008.10.034.

Balleine, B. W., & O'Doherty, J. P. (2010). Human and rodent homologies in action control: Corticostriatal determinants of goal-directed and habitual action. *Neuropsychopharmacology: Official Publication of the American College of Neuropsychopharmacology, 35*(1), 48–69. http://doi.org/10.1038/npp.2009.131.

Barrot, M., Marinelli, M., Abrous, D. N., Rouge-Pont, F., Le Moal, M., & Piazza, P. V. (1999). Functional heterogeneity in dopamine release and in the expression of Fos-like proteins within the rat striatal complex. *The European Journal of Neuroscience, 11*(4), 1155–1166.

Belin, D., Jonkman, S., Dickinson, A., Robbins, T. W., & Everitt, B. J. (2009). Parallel and interactive learning processes within the basal ganglia: Relevance for the understanding of addiction. *Behavioural Brain Research, 199*(1), 89–102. http://doi.org/10.1016/j.bbr.2008.09.027.

Bossert, J. M., Marchant, N. J., Calu, D. J., & Shaham, Y. (2013). The reinstatement model of drug relapse: Recent neurobiological findings, emerging research topics, and translational research. *Psychopharmacology, 229*(3), 453–476. http://doi.org/10.1007/s00213-013-3120-y.

Boulay, D., Duterte-Boucher, D., Leroux-Nicollet, I., Naudon, L., & Costentin, J. (1996). Locomotor sensitization and decrease in [3H]mazindol binding to the dopamine transporter in the nucleus accumbens are delayed after chronic treatments by GBR12783 or cocaine. *The Journal of Pharmacology and Experimental Therapeutics, 278*(1), 330–337.

Bustamante, D., You, Z. B., Castel, M. N., Johansson, S., Goiny, M., Terenius, L., … Herrera-Marschitz, M. (2002). Effect of single and repeated methamphetamine treatment on neurotransmitter release in substantia nigra and neostriatum of the rat. *Journal of Neurochemistry, 83*(3), 645–654.

Claye, L. H., Akunne, H. C., Davis, M. D., DeMattos, S., & Soliman, K. F. (1995). Behavioral and neurochemical changes in the dopaminergic system after repeated cocaine administration. *Molecular Neurobiology, 11*(1–3), 55–66.

Clemens, K. J., Castino, M. R., Cornish, J. L., Goodchild, A. K., & Holmes, N. M. (2014). Behavioral and neural substrates of habit formation in rats intravenously self-administering nicotine. *Neuropsychopharmacology: Official Publication of the American College of Neuropsychopharmacology, 39*(11), 2584–2593. http://doi.org/10.1038/npp.2014.111.

Corbit, L. H., & Balleine, B. W. (2003). The role of prelimbic cortex in instrumental conditioning. *Behavioural Brain Research, 146*(1–2), 145–157.

Corbit, L. H., & Balleine, B. W. (2005). Double dissociation of basolateral and central amygdala lesions on the general and outcome-specific forms of pavlovian-instrumental transfer. *The Journal of Neuroscience: the Official Journal of the Society for Neuroscience, 25*(4), 962–970. http://doi.org/10.1523/JNEUROSCI.4507-04.2005.

Corbit, L. H., Chieng, B. C., & Balleine, B. W. (2014). Effects of repeated cocaine exposure on habit learning and reversal by N-acetylcysteine. *Neuropsychopharmacology: Official Publication of the American College of Neuropsychopharmacology, 39*(8), 1893–1901. http://doi.org/10.1038/npp.2014.37.

Corbit, L. H., & Janak, P. H. (2010). Posterior dorsomedial striatum is critical for both selective instrumental and Pavlovian reward learning. *The European Journal of Neuroscience, 31*(7), 1312–1321. http://doi.org/10.1111/j.1460-9568.2010.07153.x.

Corbit, L. H., & Janak, P. H. (2016). Changes in the influence of alcohol-paired stimuli on alcohol seeking across extended training. *Frontiers in Psychiatry, 7*(169). http://doi.org/10.3389/fpsyt.2016.00169.

Corbit, L. H., Muir, J. L., & Balleine, B. W. (2003). Lesions of mediodorsal thalamus and anterior thalamic nuclei produce dissociable effects on instrumental conditioning in rats. *The European Journal of Neuroscience, 18*(5), 1286−1294.

Corbit, L. H., Nie, H., & Janak, P. H. (2012). Habitual alcohol seeking: Time course and the contribution of subregions of the dorsal striatum. *Biological Psychiatry, 72*(5), 389−395. http://doi.org/10.1016/j.biopsych.2012.02.024.

Crombag, H. S., Bossert, J. M., Koya, E., & Shaham, Y. (2008). Review. Context-induced relapse to drug seeking: A review. *Philosophical Transactions of the Royal Society B: Biological Sciences, 363*(1507), 3233−3243. http://doi.org/10.1098/rstb.2008.0090.

Cuzon Carlson, V. C., Seabold, G. K., Helms, C. M., Garg, N., Odagiri, M., Rau, A. R., … Grant, K. A. (2011). Synaptic and morphological neuroadaptations in the putamen associated with long-term, relapsing alcohol drinking in primates. *Neuropsychopharmacology: Official Publication of the American College of Neuropsychopharmacology, 36*(12), 2513−2528. http://doi.org/10.1038/npp.2011.140.

de Wit, S., Watson, P., Harsay, H. A., Cohen, M. X., van de Vijver, I., & Ridderinkhof, K. R. (2012). Corticostriatal connectivity underlies individual differences in the balance between habitual and goal-directed action control. *The Journal of Neuroscience: the Official Journal of the Society for Neuroscience, 32*(35), 12066−12075. http://doi.org/10.1523/JNEUROSCI.1088-12.2012.

DePoy, L., Daut, R., Brigman, J. L., MacPherson, K., Crowley, N., Gunduz-Cinar, O., … Holmes, A. (2013). Chronic alcohol produces neuroadaptations to prime dorsal striatal learning. *Proceedings of the National Academy of Sciences of the United States of America, 110*(36), 14783−14788. http://doi.org/10.1073/pnas.1308198110.

Deroche-Gamonet, V., Belin, D., & Piazza, P. V. (2004). Evidence for addiction-like behavior in the rat. *Science, 305*(5686), 1014−1017. http://doi.org/10.1126/science.1099020.

Dickinson, A. (1985). Actions and habits: The development of behavioural autonomy. *Philosophical Transactions of the Royal Society of London, 308*, 67−78.

Ducret, E., Puaud, M., Lacoste, J., Belin-Rauscent, A., Fouyssac, M., Dugast, E., … Belin, D. (2016). N-acetylcysteine facilitates self-imposed abstinence after escalation of cocaine intake. *Biological Psychiatry, 80*(3), 226−234. http://doi.org/10.1016/j.biopsych.2015.09.019.

Ersche, K. D., Gillan, C. M., Jones, P. S., Williams, G. B., Ward, L. H., Luijten, M., … Robbins, T. W. (2016). Carrots and sticks fail to change behavior in cocaine addiction. *Science, 352*(6292), 1468−1471. http://doi.org/10.1126/science.aaf3700.

Ersche, K. D., Jones, P. S., Williams, G. B., Turton, A. J., Robbins, T. W., & Bullmore, E. T. (2012). Abnormal brain structure implicated in stimulant drug addiction. *Science, 335*(6068), 601−604. http://doi.org/10.1126/science.1214463.

Ersche, K. D., Turton, A. J., Pradhan, S., Bullmore, E. T., & Robbins, T. W. (2010). Drug addiction endophenotypes: Impulsive versus sensation-seeking personality traits. *Biological Psychiatry, 68*(8), 770−773. http://doi.org/10.1016/j.biopsych.2010.06.015.

Everitt, B. J., Belin, D., Economidou, D., Pelloux, Y., Dalley, J. W., & Robbins, T. W. (2008). Review. Neural mechanisms underlying the vulnerability to develop compulsive drug-seeking habits and addiction. *Philosophical Transactions of the Royal Society of London. Series B, Biological Sciences, 363*(1507), 3125−3135. http://doi.org/10.1098/rstb.2008.0089.

Everitt, B. J., & Robbins, T. W. (2013). From the ventral to the dorsal striatum: Devolving views of their roles in drug addiction. *Neuroscience and Biobehavioral Reviews, 37*(9 Pt A), 1946−1954. http://doi.org/10.1016/j.neubiorev.2013.02.010.

Everitt, B. J., & Robbins, T. W. (2016). Drug addiction: Updating actions to habits to compulsions ten years on. *Annual Review of Psychology, 67*, 23−50. http://doi.org/10.1146/annurev-psych-122414-033457.

Faure, A., Haberland, U., Conde, F., & El Massioui, N. (2005). Lesion to the nigrostriatal dopamine system disrupts stimulus-response habit formation. *The Journal of Neuroscience: the Official Journal of the Society for Neuroscience, 25*(11), 2771−2780. http://doi.org/10.1523/JNEUROSCI.3894-04.2005.

Furlong, T. M., Corbit, L. H., Brown, R. A., & Balleine, B. W. (2017). Methamphetamine promotes habitual action and alters the density of striatal glutamate receptor and vesicular proteins in dorsal striatum. *Addiction Biology*. http://doi.org/10.1111/adb.12534.

Furlong, T. M., Leavitt, L. S., Keefe, K. A., & Son, J. H. (2016). Methamphetamine-, d-amphetamine-, and p-chloroamphetamine-induced neurotoxicity differentially effect impulsive responding on the stop-signal task in rats. *Neurotoxicity Research, 29*(4), 569–582. http://doi.org/10.1007/s12640-016-9605-9.

Furlong, T. M., Supit, A. S., Corbit, L. H., Killcross, S., & Balleine, B. W. (2017). Pulling habits out of rats: Adenosine 2A receptor antagonism in dorsomedial striatum rescues meth-amphetamine-induced deficits in goal-directed action. *Addiction Biology, 22*(1), 172–183. http://doi.org/10.1111/adb.12316.

Gerfen, C. R., & Surmeier, D. J. (2011). Modulation of striatal projection systems by dopamine. *Annual Review of Neuroscience, 34*, 441–466. http://doi.org/10.1146/annurev-neuro-061010-113641.

Gremel, C. M., & Lovinger, D. M. (2017). Associative and sensorimotor cortico-basal ganglia circuit roles in effects of abused drugs. *Genes, Brain and Behavior, 16*(1), 71–85. http://doi.org/10.1111/gbb.12309.

Grueter, B. A., Rothwell, P. E., & Malenka, R. C. (2012). Integrating synaptic plasticity and striatal circuit function in addiction. *Current Opinion in Neurobiology, 22*(3), 545–551. http://doi.org/10.1016/j.conb.2011.09.009.

Hogarth, L., Chase, H. W., & Baess, K. (2012). Impaired goal-directed behavioural control in human impulsivity. *The Quarterly Journal of Experimental Psychology, 65*(2), 305–316. http://doi.org/10.1080/17470218.2010.518242.

Howell, L. L., & Kimmel, H. L. (2008). Monoamine transporters and psychostimulant addiction. *Biochemical Pharmacology, 75*(1), 196–217. http://doi.org/10.1016/j.bcp.2007.08.003.

Jedynak, J. P., Uslaner, J. M., Esteban, J. A., & Robinson, T. E. (2007). Methamphetamine-induced structural plasticity in the dorsal striatum. *The European Journal of Neuroscience, 25*(3), 847–853. http://doi.org/10.1111/j.1460-9568.2007.05316.x.

Joel, D., & Weiner, I. (2000). The connections of the dopaminergic system with the striatum in rats and primates: An analysis with respect to the functional and compartmental organization of the striatum. *Neuroscience, 96*(3), 451–474.

Kalivas, P. W. (2009). The glutamate homeostasis hypothesis of addiction. *Nature Reviews. Neuroscience, 10*(8), 561–572. http://doi.org/10.1038/nrn2515.

Kalivas, P. W., & Volkow, N. D. (2011). New medications for drug addiction hiding in glutamatergic neuroplasticity. *Molecular Psychiatry, 16*(10), 974–986. http://doi.org/10.1038/mp.2011.46.

Kerdsan, W., Thanoi, S., & Nudmamud-Thanoi, S. (2009). Changes in glutamate/NMDA receptor subunit 1 expression in rat brain after acute and subacute exposure to methamphetamine. *Journal of Biomedicine and Biotechnology, 2009*(329631). http://doi.org/10.1155/2009/329631.

Killcross, S., & Coutureau, E. (2003). Coordination of actions and habits in the medial prefrontal cortex of rats. *Cerebral Cortex, 13*(4), 400–408.

Kleven, M. S., Perry, B. D., Woolverton, W. L., & Seiden, L. S. (1990). Effects of repeated injections of cocaine on D1 and D2 dopamine receptors in rat brain. *Brain Research, 532*(1–2), 265–270.

Kousik, S. M., Carvey, P. M., & Napier, T. C. (2014). Methamphetamine self-administration results in persistent dopaminergic pathology: Implications for Parkinson's disease risk and reward-seeking. *The European Journal of Neuroscience, 40*(4), 2707–2714. http://doi.org/10.1111/ejn.12628.

Kreitzer, A. C., & Malenka, R. C. (2008). Striatal plasticity and basal ganglia circuit function. *Neuron, 60*(4), 543–554. http://doi.org/10.1016/j.neuron.2008.11.005.

LeBlanc, K. H., Maidment, N. T., & Ostlund, S. B. (2013). Repeated cocaine exposure facilitates the expression of incentive motivation and induces habitual control in rats. *PLoS One, 8*(4), e61355. http://doi.org/10.1371/journal.pone.0061355.

Lingawi, N. W., & Balleine, B. W. (2012). Amygdala central nucleus interacts with dorsolateral striatum to regulate the acquisition of habits. *The Journal of Neuroscience: the Official Journal of the Society for Neuroscience, 32*(3), 1073–1081. http://doi.org/10.1523/JNEUROSCI.4806-11.2012.

Loftis, J. M., & Janowsky, A. (2000). Regulation of NMDA receptor subunits and nitric oxide synthase expression during cocaine withdrawal. *Journal of Neurochemistry, 75*(5), 2040–2050.

Luscher, C., & Malenka, R. C. (2011). Drug-evoked synaptic plasticity in addiction: From molecular changes to circuit remodeling. *Neuron, 69*(4), 650–663. http://doi.org/10.1016/j.neuron.2011.01.017.

Maggos, C. E., Tsukada, H., Kakiuchi, T., Nishiyama, S., Myers, J. E., Kreuter, J., ... Kreek, M. J. (1998). Sustained withdrawal allows normalization of in vivo [11C]N-methylspiperone dopamine D2 receptor binding after chronic binge cocaine: A positron emission tomography study in rats. *Neuropsychopharmacology: Official Publication of the American College of Neuropsychopharmacology, 19*(2), 146–153. http://doi.org/10.1016/S0893-133X(98)00009-8.

Mao, L., & Wang, J. Q. (2001). Differentially altered mGluR1 and mGluR5 mRNA expression in rat caudate nucleus and nucleus accumbens in the development and expression of behavioral sensitization to repeated amphetamine administration. *Synapse, 41*(3), 230–240. http://doi.org/10.1002/syn.1080.

McGeorge, A. J., & Faull, R. L. (1989). The organization of the projection from the cerebral cortex to the striatum in the rat. *Neuroscience, 29*(3), 503–537.

McKim, T. H., Bauer, D. J., & Boettiger, C. A. (2016). Addiction history associates with the propensity to form habits. *Journal of Cognitive Neuroscience, 28*(7), 1024–1038. http://doi.org/10.1162/jocn_a_00953.

Moriguchi, S., Watanabe, S., Kita, H., & Nakanishi, H. (2002). Enhancement of N-methyl- D-aspartate receptor-mediated excitatory postsynaptic potentials in the neostriatum after methamphetamine sensitization. An in vitro slice study. *Experimental Brain Research, 144*(2), 238–246. http://doi.org/10.1007/s00221-002-1039-3.

Nader, M. A., Sinnott, R. S., Mach, R. H., & Morgan, D. (2002). Cocaine- and food-maintained responding under a multiple schedule in rhesus monkeys: Environmental context and the effects of a dopamine antagonist. *Psychopharmacology, 163*(3–4), 292–301. http://doi.org/10.1007/s00213-002-1202-3.

Nelson, A., & Killcross, S. (2006). Amphetamine exposure enhances habit formation. *The Journal of Neuroscience: the Official Journal of the Society for Neuroscience, 26*(14), 3805–3812. http://doi.org/10.1523/JNEUROSCI.4305-05.2006.

Nishino, N., Shirai, Y., Kajimoto, Y., Kitamura, N., Yamamoto, H., Yang, C. Q., & Shirakawa, O. (1996). Increased glutamate transporter (GLT-1) immunoreactivity in the rat striatum after repeated intermittent administration of methamphetamine. *Annals of the New York Academy of Sciences, 801*, 310–314.

Nishioku, T., Shimazoe, T., Yamamoto, Y., Nakanishi, H., & Watanabe, S. (1999). Expression of long-term potentiation of the striatum in methamphetamine-sensitized rats. *Neuroscience Letters, 268*(2), 81–84.

Nordquist, R. E., Voorn, P., de Mooij-van Malsen, J. G., Joosten, R. N., Pennartz, C. M., & Vanderschuren, L. J. (2007). Augmented reinforcer value and accelerated habit formation after repeated amphetamine treatment. *European Neuropsychopharmacology: the Journal of the European College of Neuropsychopharmacology, 17*(8), 532–540. http://doi.org/10.1016/j.euroneuro.2006.12.005.

Olmstead, M. C., Parkinson, J. A., Miles, F. J., Everitt, B. J., & Dickinson, A. (2000). Cocaine-seeking by rats: Regulation, reinforcement and activation. *Psychopharmacology, 152*(2), 123–131.

Ostlund, S. B., & Balleine, B. W. (2005). Lesions of medial prefrontal cortex disrupt the acquisition but not the expression of goal-directed learning. *The Journal of Neuroscience: the Official Journal of the Society for Neuroscience, 25*(34), 7763–7770. http://doi.org/10.1523/JNEUROSCI.1921-05.2005.

Ostlund, S. B., & Balleine, B. W. (2008). On habits and addiction: An associative analysis of compulsive drug seeking. *Drug Discovery Today: Disease Models, 5*(4), 235–245. http://doi.org/10.1016/j.ddmod.2009.07.004.

Ostlund, S. B., Maidment, N. T., & Balleine, B. W. (2010). Alcohol-paired contextual cues produce an immediate and selective loss of goal-directed action in rats. *Frontiers in Integrative Neuroscience, 4*, 1–8. http://doi.org/10.3389/fnint.2010.00019.

Patrick, S. L., Thompson, T. L., Walker, J. M., & Patrick, R. L. (1991). Concomitant sensitization of amphetamine-induced behavioral stimulation and in vivo dopamine release from rat caudate nucleus. *Brain Research, 538*(2), 343–346.

Patton, M. H., Roberts, B. M., Lovinger, D. M., & Mathur, B. N. (2016). Ethanol disinhibits dorsolateral striatal medium spiny neurons through activation of a presynaptic delta opioid receptor. *Neuropsychopharmacology: Official Publication of the American College of Neuropsychopharmacology, 41*(7), 1831–1840. http://doi.org/10.1038/npp.2015.353.

Paulson, P. E., & Robinson, T. E. (1995). Amphetamine-induced time-dependent sensitization of dopamine neurotransmission in the dorsal and ventral striatum: A microdialysis study in behaving rats. *Synapse, 19*(1), 56–65. http://doi.org/10.1002/syn.890190108.

Perry, J. L., & Carroll, M. E. (2008). The role of impulsive behavior in drug abuse. *Psychopharmacology,* *200*(1), 1−26. http://doi.org/10.1007/s00213-008-1173-0.

Pilotte, N. S., Sharpe, L. G., & Kuhar, M. J. (1994). Withdrawal of repeated intravenous infusions of cocaine persistently reduces binding to dopamine transporters in the nucleus accumbens of Lewis rats. *The Journal of Pharmacology and Experimental Therapeutics, 269*(3), 963−969.

Przewlocka, B., & Lason, W. (1995). Adaptive changes in the proenkephalin and D2 dopamine receptor mRNA expression after chronic cocaine in the nucleus accumbens and striatum of the rat. *European Neuropsychopharmacology: the Journal of the European College of Neuropsychopharmacology, 5*(4), 465−469.

Ramanathan, S., Hanley, J. J., Deniau, J. M., & Bolam, J. P. (2002). Synaptic convergence of motor and somatosensory cortical afferents onto GABAergic interneurons in the rat striatum. *The Journal of Neuroscience: the Official Journal of the Society for Neuroscience, 22*(18), 8158−8169.

Reep, R. L., Cheatwood, J. L., & Corwin, J. V. (2003). The associative striatum: Organization of cortical projections to the dorsocentral striatum in rats. *Journal of Comparative Neurology, 467*(3), 271−292. http://doi.org/10.1002/cne.10868.

Renteria, R., Baltz, E. T., & Gremel, C. M. (2018). Chronic alcohol exposure disrupts top-down control over basal ganglia action selection to produce habits. *Nature Communications, 9*(1), 211. http://doi. org/10.1038/s41467-017-02615-9.

Schwendt, M., Rocha, A., See, R. E., Pacchioni, A. M., McGinty, J. F., & Kalivas, P. W. (2009). Extended methamphetamine self-administration in rats results in a selective reduction of dopamine transporter levels in the prefrontal cortex and dorsal striatum not accompanied by marked monoaminergic depletion. *The Journal of Pharmacology and Experimental Therapeutics, 331*(2), 555−562. http://doi.org/ 10.1124/jpet.109.155770.

Shirai, Y., Shirakawa, O., Nishino, N., Saito, N., & Nakai, H. (1996). Increased striatal glutamate transporter by repeated intermittent administration of methamphetamine. *Psychiatry and Clinical Neurosciences, 50*(3), 161−164.

Sjoerds, Z., de Wit, S., van den Brink, W., Robbins, T. W., Beekman, A. T., Penninx, B. W., & Veltman, D. J. (2013). Behavioral and neuroimaging evidence for overreliance on habit learning in alcohol-dependent patients. *Translational Psychiatry [electronic Resource], 3*, e337. http://doi.org/10. 1038/tp.2013.107.

Tanaka, S. C., Balleine, B. W., & O'Doherty, J. P. (2008). Calculating consequences: Brain systems that encode the causal effects of actions. *The Journal of Neuroscience: the Official Journal of the Society for Neuroscience, 28*(26), 6750−6755. http://doi.org/10.1523/JNEUROSCI.1808-08.2008.

Thanos, P. K., Kim, R., Delis, F., Rocco, M. J., Cho, J., & Volkow, N. D. (2017). Effects of chronic methamphetamine on psychomotor and cognitive functions and dopamine signaling in the brain. *Behavioural Brain Research, 320*, 282−290. http://doi.org/10.1016/j.bbr.2016.12.010.

Tozzi, A., Tscherter, A., Belcastro, V., Tantucci, M., Costa, C., Picconi, B., ... Borsini, F. (2007). Interaction of A2A adenosine and D2 dopamine receptors modulates corticostriatal glutamatergic transmission. *Neuropharmacology, 53*(6), 783−789. http://doi.org/10.1016/j.neuropharm.2007.08.006.

Tricomi, E., Balleine, B. W., & O'Doherty, J. P. (2009). A specific role for posterior dorsolateral striatum in human habit learning. *The European Journal of Neuroscience, 29*(11), 2225−2232. http://doi.org/10.1111/ j.1460-9568.2009.06796.x.

Vanderschuren, L. J., & Everitt, B. J. (2004). Drug seeking becomes compulsive after prolonged cocaine self-administration. *Science, 305*(5686), 1017−1019. http://doi.org/10.1126/science.1098975.

Wang, J., Carnicella, S., Phamluong, K., Jeanblanc, J., Ronesi, J. A., Chaudhri, N., ... Ron, D. (2007). Ethanol induces long-term facilitation of NR2B-NMDA receptor activity in the dorsal striatum: Implications for alcohol drinking behavior. *The Journal of Neuroscience: the Official Journal of the Society for Neuroscience, 27*(13), 3593−3602. http://doi.org/10.1523/JNEUROSCI.4749-06.2007.

Wang, J., Cheng, Y., Wang, X., Roltsch Hellard, E., Ma, T., Gil, H., ... Ron, D. (2015). Alcohol elicits functional and structural plasticity selectively in dopamine D1 receptor-expressing neurons of the dorsomedial striatum. *The Journal of Neuroscience: the Official Journal of the Society for Neuroscience, 35*(33), 11634−11643. http://doi.org/10.1523/JNEUROSCI.0003-15.2015.

Wickens, J. R., Horvitz, J. C., Costa, R. M., & Killcross, S. (2007). Dopaminergic mechanisms in actions and habits. *The Journal of Neuroscience: the Official Journal of the Society for Neuroscience, 27*(31), 8181—8183. http://doi.org/10.1523/JNEUROSCI.1671-07.2007.

Wolf, M. E. (2016). Synaptic mechanisms underlying persistent cocaine craving. *Nature Reviews. Neuroscience, 17*(6), 351—365. http://doi.org/10.1038/nrn.2016.39.

Yager, L. M., Garcia, A. F., Wunsch, A. M., & Ferguson, S. M. (2015). The ins and outs of the striatum: Role in drug addiction. *Neuroscience, 301*, 529—541. http://doi.org/10.1016/j.neuroscience.2015.06.033.

Yamada, S., Kojima, H., Yokoo, H., Tsutsumi, T., Takamuki, K., Anraku, S., ... Inanaga, K. (1988). Enhancement of dopamine release from striatal slices of rats that were subchronically treated with methamphetamine. *Biological Psychiatry, 24*(4), 399—408.

Yin, H. H., & Knowlton, B. J. (2006). The role of the basal ganglia in habit formation. *Nature Reviews. Neuroscience, 7*(6), 464—476. http://doi.org/10.1038/nrn1919.

Yin, H. H., Knowlton, B. J., & Balleine, B. W. (2004). Lesions of dorsolateral striatum preserve outcome expectancy but disrupt habit formation in instrumental learning. *The European Journal of Neuroscience, 19*(1), 181—189.

Yin, H. H., Ostlund, S. B., Knowlton, B. J., & Balleine, B. W. (2005). The role of the dorsomedial striatum in instrumental conditioning. *The European Journal of Neuroscience, 22*(2), 513—523. http://doi.org/10.1111/j.1460-9568.2005.04218.x.

Yu, C., Gupta, J., Chen, J. F., & Yin, H. H. (2009). Genetic deletion of A2A adenosine receptors in the striatum selectively impairs habit formation. *The Journal of Neuroscience: the Official Journal of the Society for Neuroscience, 29*(48), 15100—15103. http://doi.org/10.1523/JNEUROSCI.4215-09.2009.

Zapata, A., Minney, V. L., & Shippenberg, T. S. (2010). Shift from goal-directed to habitual cocaine seeking after prolonged experience in rats. *The Journal of Neuroscience: the Official Journal of the Society for Neuroscience, 30*(46), 15457—15463. http://doi.org/10.1523/JNEUROSCI.4072-10.2010.

CHAPTER 17

Goal-Directed Deficits in Schizophrenia

Richard W. Morris[1,2,3]

[1]School of Psychology, University of New South Wales (NSW), Sydney, NSW, Australia; [2]School of Medicine, University of Sydney, Sydney, NSW, Australia; [3]Centre for Translational Data Science, University of Sydney, Sydney, NSW, Australia

INTRODUCTION

Deficits in emotion and motivation are a core feature of schizophrenia, where scientists and clinicians have observed that individuals with schizophrenia can have difficulty expressing or experiencing emotion (i.e., flat affect and *anhedonia*) while at the same time noting they are less likely to pursue courses of action and achieve goals (i.e., poor motivation or *avolition*). For instance, Bleuler coined the term schizophrenia from "splitting of the mind" after his clinical observations of a disconnect between a patient's emotional state and his/her apparent behavior. He wrote "Even in mild cases, where wishes and desires still exist, [patients] will nevertheless do nothing towards the realization of these wishes" (Bleuler, 1950). What this problem statement recognizes is the deficit is not located in the distinct hedonic functions or cognitive functions subserving motivated behavior themselves. That is, there is now evidence that the experience of pleasure upon goal attainment or reward receipt is intact in schizophrenia, as is the cognitive planning of actions. Rather, the failure is in the connection between otherwise intact hedonic concerns and the control of actions. I will argue here that it is this disconnection that produces a profound deficit in goal-directed decision-making in schizophrenia. The aim of this chapter is to illustrate how psychological theories and computational models of reinforcement learning can help us understand the form of this disconnection and its consequences.

Goal-directed decision-making requires a close integration of our hedonic concerns and our actions—a fact that has been recognized since Aristotle insisted that the distinction between "cold" cognitive reason and "hot" hedonic desire is arbitrary, and the best decisions depend upon both reason and desire. Thinking about taking a walk, for example, will not move me to actually take a walk. I must also think that taking a walk will be good or pleasant. The close cooperation between these two concerns is reflected in current accounts of a goal-directed decision as an action performed with a desire for the outcome and a belief the action will achieve this outcome (Heyes & Dickinson, 1990; de Wit & Dickinson, 2009). The importance of integrating these two distinct prerequisites for goal-directed decision-making is illustrated by considering the difficulty of maximizing the rewarding consequences of our actions. For instance, our current motivational state, such as whether one is currently hungry or thirsty, will

Goal-Directed Decision Making
ISBN 978-0-12-812098-9, https://doi.org/10.1016/B978-0-12-812098-9.00017-6

387

influence whether we will enjoy eating corn chips or drinking beer. The varying utility of our decisions is formally acknowledged in the economic principle of marginal utility: The first ice cream tastes delicious; however, by the third ice cream, we no longer enjoy that sweet-iced confection. The everyday experience that ice cream tastes less delicious after the third cone is easy to appreciate; however, planning on that contingency is considerably more difficult. Allowing our future motivational states to influence current decision-making is a challenge that is so common we often overlook it—"his eyes were bigger than his stomach" is a widely recognized problem for children. An adult example may be packing for a tropical beach holiday in the middle of a frozen winter climate; it can be difficult to resist packing heavy winter clothing even when one is familiar with the heat and humidity of the destination. This problem is more than just a failure to predict our future motivational state; instead, it represents a widely experienced difficulty that people (and especially people with schizophrenia) have connecting or *integrating* the current value of their actions with their anticipated hedonic consequences. This integration is necessary for goal-directed action selection, and it has been the focus of theories of associative learning (Chapter 1 by Dickinson), and more recently formalized in computational models of model-based reinforcement learning (MBRL) (Daw, Niv, & Dayan, 2005; Sutton & Barto, 1998).

MODEL-BASED REINFORCEMENT LEARNING: A THEORY OF GOAL-DIRECTED DECISION-MAKING

I argue there are at least three pertinent features of MBRL, which can help us understand goal-directed decision-making. In general, MBRL is concerned with the estimation of state transition probabilities in order to model the environment (Daw et al., 2005; Sutton & Barto, 1998). In the present context, *model* implies a mental model of the task or experiment, but it could be as simple as a ***map*** between particular states, the actions available in each state, and their consequent outcome states. Thus, MBRL explicitly learns the sequential contingencies of events and actions in a task, such as which outcomes (states) follow which actions (and their respective states). The resulting map is an associative network, which connects all the paths between the different states and actions in the environment. Secondly, in order to motivate action selection in each state, MBRL assumes that the final goal state will have some ***incentive value or reward*** associated with it, which the agent is trying to obtain (or *maximize*). The value of the reward may represent hedonic value (e.g., the learned incentive value of water, given a thirsty agent) or more abstract values such as information (e.g., to promote exploration of unexplored options, given an uncertain agent). Importantly, however, the agent must be able to decide between actions prior to reward, e.g., in states with no direct reward value. In order to do this, the value of the current action must be inferred from the values of its consequent states. The best action will ultimately be connected to the desired goal state via the map of transition probabilities described above. To determine the best action, MBRL calculates the value of the available actions by integrating over the value of every consequent state that lies between the current action and the ultimate state. This ***integration*** is formally described by the Bellman equations (Bellman, 1957) but can be thought of as a mental simulation of "what-ifs". An obvious disadvantage of this approach is that it quickly becomes overwhelming when simulating multiple options in large decision trees (e.g., chess playing). However, the major advantage of MBRL is that action values are dynamically updated by the addition of new information. It is the dynamic updating of the action values, which distinguishes it from model-free approaches, since model-free reinforcement learning must rely on direct experience with the reward contingency before it can update action values. For instance, in MBRL, the action values can be updated by new experiences of the consequences in the absence of the action itself, or by inference after exposure to a new reward contingency (counterfactual reasoning). All this is to say that it is the combination of *all three* of these processes: building a mental ***map***, learning where the ***reward*** is located in that map, and ***integrating*** the reward value over multiple states back to the current action, which allows MBRL to dynamically update and carry out decision-making in a goal-directed manner. The implication from this analysis with regard to understanding schizophrenia is that any observed deficit may be due to a failure in one or all of these processes.

Of course, apparent decision-making behavior, such as selecting among different value-based options or navigating a maze, may be underwritten by other forms of learning such as model-free reinforcement learning, where the learned response is a simple form of Pavlovian approach to the learned incentive value of an environmental cue. However, such behavior will not display the same dynamic features of goal-directed decision-making. For example, sequential decision tasks, such as mazes or more abstract multistep sequences, do not by themselves guarantee that a mental map (and integration of value) is involved in adaptive behavior. Even seemingly complex maze navigation may consist of a series of learned cues guiding approach behavior via Pavlovian value or conditioned reinforcement. In such cases, allowing the agent to know the reward location has changed or is empty will not produce an immediate change in behavior. Instead, the agent will need to learn by experience that taking the same actions will no longer result in reward.

WHAT IS THE HISTORICAL EVIDENCE OF A GOAL-DIRECTED DEFICIT IN SCHIZOPHRENIA?

The predominant cognitive paradigms used to investigate decision-making in schizophrenia have typically employed instructed stimulus-response tasks (e.g., Stroop task, Simon task, flanker task). Even in cases where the task involves adapting to uninstructed contingency changes (Iowa gambling task, Wisconsin Card Sorting), the response and feedback is cued and so consists of a series of stimulus-response trials. While these tasks have furthered our understanding of decision-making, they do not capture the fact that goal-directed decisions involve choosing actions on the basis of their consequences. As a consequence of focusing on stimulus-response tasks, the deficits in decision-making in schizophrenia have been explained (incompletely) in terms of a failure of response inhibition, or the salience of cues (e.g., aberrant salience), or more recently, as a failure to anticipate cued rewards (e.g., reward prediction errors). Thus, until very recently the *goal-directed nature* of decision-making has been largely ignored in patient studies.

For instance, many studies of Wisconsin Card Sorting or serial reversal tasks have observed an apparent deficit in reward learning in schizophrenia. That is, patients tend to perseverate on previously rewarded responses rather than switch to the alternate response. In the case of serial reversal tasks, the deficit has been explicitly identified with the ability to learn from omitted rewards, rather than acquiring responses to rewarded cues. Some authors have proposed that this may stem from an inability to represent absent rewards (Gold, Waltz, Prentice, Morris, & Heerey, 2008; Waltz & Gold, 2007). The representation of absent reward plays a key role in MBRL, allowing agents to infer over the model space and reason from counterfactuals (i.e., to learn from events that did not occur). Furthermore, recent reviews have proposed that tasks such as serial reversal learning can distinguish MBRL because they contain an implicit counterfactual structure based on deducing an absent reward (Doll, Simon, & Daw, 2012; Izquierdo, Brigman, Radke, Rudebeck, & Holmes, 2017). That is, a decrease in the reward contingency of the current response implies an increase in the alternate response contingency, and this can be inferred without requiring direct experience of the increased contingency. This counterfactual structure implies the formation of a mental model and the ability to infer from it when feedback is omitted. These are, of course, cardinal features of model-based RL; however, concluding that the problem lies in these processes on the basis of a performance deficit in such tasks is fraught. Apart from the fact that most versions of serial reversal tasks involve learning stimulus-response associations rather than requiring action—outcome learning as often assumed, there are other difficulties with interpreting perseverative errors as a deficit in goal-directed learning. In particular, the evidence for counterfactual reasoning is usually provided by observations that people will shift responding to the alternate response as soon as they detect the decrease in contingency on the current response. That is, they will switch responding to the alternate without needing to experience the increased contingency on that response, as if they are rigorously follow a win-stay, lose-shift strategy—and this highlights part of the problem. Are people who shift immediately at the first nonreinforced trial inferring over a mental model, or exploring uncertain response options, or following a rule to achieve the same result? Such strategies or rules can be abstracted from experience over many trials or struck upon spontaneously; however, the involvement of a mental model is uncertain in either case.

ARE NEGATIVE PREDICTION ERRORS IMPAIRED IN SCHIZOPHRENIA?

A different explanation of perseverative errors in schizophrenia may be provided by considering the specific role of negative prediction errors in learning. Classically, associative models assume negative prediction errors drive learning when an event is expected but does not occur, i.e., the unexpected omission of an event. Thus, a deficit in negative prediction errors would explain a variety of evidence from tasks showing that people with schizophrenia have difficulty learning from absent events. It would, for instance, explain why patients can acquire rewarded responses to the same level as healthy adults but tend to perseverate when the reward contingency changes (Gold et al., 2008; Murray, Cheng, et al., 2008; Murray, Corlett, et al., 2008; Weickert, et al., 2002). It would also explain why patients acquire fear conditioning as readily as healthy controls but are slower to extinguish with nonreinforcement (Holt et al., 2009). Conversely, it is also consistent with the acquisition of a conditioned response to a CS− in a conditioned discrimination task (Jenson et al., 2008), since negative prediction errors are initially necessary to discriminate between the CS+ and CS−. Similarly, we have found that people with schizophrenia cannot downregulate attention to irrelevant cues (Morris, Griffiths, Le Pelley, & Weickert, 2013), consistent with associative models, which assume negative prediction errors drive the downregulation of attention (Mackintosh, 1975). Finally, intact positive prediction errors in schizophrenia would account for patients displaying a positive bias toward rewarding stimuli (Heerey, Bell-Warren, & Gold, 2008).

Despite its parsimony, evidence for a specific deficit in negative prediction error signaling in schizophrenia is scant. Reinforcement learning tasks have been widely used in schizophrenia, and on the basis of such tasks, people have concluded that prediction errors are aberrant in schizophrenia (Gradin et al., 2011; Murray, Cheng, et al., 2008; Murray, Corlett, et al., 2008; Schlagenhauf et al., 2014; Waltz et al., 2009). However, the role of prediction errors in driving performance in such tasks has never been demonstrated in schizophrenia. To infer that performance in reinforcement learning tasks is driven by prediction errors, versus simply eliciting errors (or surprise) without any causal role in performance, requires demonstrations that learning can be blocked or enhanced by the addition of another predictive cue (Niv & Schoenbaum, 2008; Schultz & Dickinson, 2000; Tobler, O'Doherty, Dolan, & Schultz, 2006). Instead, clinical studies have relied on correlational evidence from neuroimaging data (e.g., functional Magnetic Resonance Imaging or EEG) as evidence of aberrant prediction error signaling. Yet here again, the evidence for the involvement of prediction errors in any capacity, causal or otherwise, is weak. Prediction errors represent the unexpected delivery of reward or the unexpected omission of reward, which means they co-occur with motivational events (such as rewards and punishers), as well as salient, surprising events. Nevertheless, prediction errors are distinguishable from responses to surprise and reward value (Morris, et al., 2012; Niv & Schoenbaum, 2008). To demonstrate a neural response is not due to the salience of the cue or the motivational properties of event but instead represents a bidirectional teaching signal that can adjust learning weights, it is necessary to control for the effects of reward and salience. One way to do this is to contrast unexpected rewards with expected rewards and unexpected reward omission with expected reward omission (D'Ardenne, McClure, Nystrom, & Cohen, 2008; Morris et al., 2012; O'Doherty, Dayan, Friston, Critchley, & Dolan, 2003). These contrasts hold the reward value constant in each comparison. Furthermore, by testing for opposite changes in the direction of positive and negative prediction errors, any result cannot be explained by the unidirectional effect of surprise or salience. Unfortunately, such analyses have rarely been performed in the context of schizophrenia research.

The failure to appreciate the presence of the above confounds means that early demonstrations of prediction errors in schizophrenia are confounded with differences in motivational value or salience. Many early studies compared unexpected reward deliveries and unexpected reward omission with each other, confounding differences in reward value (Koch et al., 2010; Walter, Kammerer, Frasch, Spitzer, & Abler, 2009; Waltz et al., 2009). For instance, one influential study (Waltz et al., 2009) found a group difference in BOLD (blood oxygen level dependent) responses to unexpected rewards, implying a deficit in positive prediction-error signals. But the same study also reported a group difference in BOLD response to expected rewards versus implicit baseline (they did not report the group interaction for differences in unexpected vs. expected rewards). This means that the patient deficit may be better explained as an attenuated BOLD response to rewards (both expected and unexpected). Support for this is provided by another early study, which compared reward events to neutral events, confusing the effect of reward with prediction error and so inadvertently providing consistent evidence of a deficit in reward responses rather than prediction error in schizophrenia (Murray, Cheng, et al., 2008; Murray, Corlett, et al., 2008). More recently, two other studies have separately tested for both positive and negative prediction-error signals in schizophrenia (Dowd & Barch, 2012; Morris et al., 2012). In the case of Morris et al., we were able to determine that the attenuated signals in each case were due to both hypo-activation during unexpected events and hyperactivation during expected events in the ventral striatum. On the other hand, Dowd and Barch found no evidence of a group difference in prediction error signals. This is possibly because the reward contingencies remained constant in their task, providing little opportunity to engender prediction errors. At any rate, while there is some further evidence from more recent neuroimaging studies that prediction errors are aberrant in schizophrenia (Radua et al., 2015; Schlagen-hauf et al., 2014), the evidence for a specific deficit in negative prediction errors is weak.

In order to provide evidence (of a different form) for the claim of a deficit in negative prediction errors in schizophrenia, I have reanalyzed published data in a simple computational model of prediction error learning in schizophrenia. We (Morris, et al., 2012; Morris, Purves-Tyson, et al., 2015) have previously used a reward prediction task to investigate BOLD-related ventral striatal prediction errors in schizophrenia. In this task, participants had to learn to predict the occurrence of reward given a cue (i.e., predicting whether the hand of cards they were dealt contained the "trump" card). The trump card was rewarded on 80% of trials, while nontrump cards were rewarded on 20% of trials. Furthermore, the identity of the trump card changed every five or six trials. Thus, the task was a probabilistic serial reversal task, which was designed to provide expected and unexpected rewards, as well as expected and unexpected reward omissions (for full details of the task, see Morris et al., 2012). This allowed us to ask whether predictions were more influenced by unexpected rewards or the unexpected omission of reward. The participants were 35 people with schizophrenia aged 18 to 55 on stable antipsychotic medication and living in the community, and 25 healthy adults similar in age and gender. As reported in Morris et al., the mean accuracy of the predictions of both groups were both much greater than chance and very similar ($\sim 80\%$), indicating a similar task engagement and understanding. However, a simple Q-learning model revealed the role of unexpected outcomes in updating predictions in the task was quite different. The Q-learning model fits learning rates for each individual's predictions with a separate parameter for positive changes (alpha) due to unexpected rewards and negative changes (beta) due to the unexpected omission of reward. The model also included a parameter for exploration/exploitation (tau). For each participant, maximum likelihood estimates for the learning rate parameters alpha and beta were calculated. I excluded participants whose data could not be fit at a significantly greater than chance level, i.e., relative to a random model according to a Chi-square test. After exclusions, the n were 33 and 23, respectively. The average parameter estimates were compared between groups using two-sample t-tests, an analysis strategy that effectively treats individual parameter values as a random effect (Daw, 2011). The model provided a reasonably equivalent fit to both groups as the average likelihood of the model for the healthy adults and people with schizophrenia was 0.63 and 0.64, respectively (independent-sample t-test $t < 1$). Comparison of the learning rates is shown in Fig. 17.1. The t-tests of the learning parameters revealed a significantly lower beta in schizophrenia than healthy adults, $t_{54} = 2.63$, $P = .01$, but no significant group difference in alpha values, $t_{54} = 1.40$, $P = .17$ (a 2×2 factorial ANOVA confirmed the interaction was significant, $F(1, 54) = 11.387$, $P < .001$). This implies that the amount of change in predictions associated with the unexpected omission of reward was less among people with schizophrenia than among healthy adults. This provides some initial (albeit correlational) evidence that people

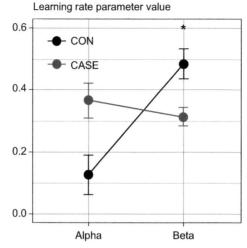

Figure 17.1 Positive (alpha) and negative (beta) learning rate parameters from a Q-learning model fitted to healthy adults (CON) and people with schizophrenia (CASE) during a probabilistic serial reversal learning task. On average, CASE had a significantly lower beta parameter, indicating that learning from the unexpected omission of reward, or negative prediction errors, was impaired in schizophrenia.

with schizophrenia have difficulty learning from the unexpected omission of reward, a deficit specifically associated with negative prediction error signaling.

More generally, most of the evidence historically used to support a deficit in goal-directed decision-making in schizophrenia has to be reassessed in light of improvements in our understanding of goal-directed learning and MBRL. Where deficits have historically been indicated in the literature, they are just as likely to be identifying problems in simple associative processes or model-free reinforcement learning, such as negative prediction error signaling.

WHAT IS THE CONTEMPORARY EVIDENCE FOR A GOAL-DIRECTED DEFICIT IN SCHIZOPHRENIA?

Establishing the involvement of a particular learning process in any decision-making task is a challenge since the likely availability of other processes to substitute for performance is difficult to eliminate. This multipath problem is explicitly recognized in dual-process theories of decision-making, which posit that habitual (or model-free) learning competes with goal-directed (or model-based) learning for control of decision-making. In other words, both processes are available and are likely to be contributing to action selection at any point in time. Accounts differ in how these processes compete (or cooperate) for control, but since both processes will contribute to performance, the emphasis shifts from isolating individual learning processes to measuring their relative contribution. Recently, one strategy has emerged, which offers to weigh the contribution of both model-based and model-free learning, in a task commonly known as the two-stage task.

The two-stage task (Daw, Gershman, Seymour, Dayan, & Dolan, 2011) is a serial reversal task over two stages, which might be best described in the form of a maze problem. In the first room, participants are given the option of entering one of two new rooms (i.e., a blue and red room), e.g., left button to get to the blue room and right button to get to the red room. Each of these colored rooms contains two doors behind which may be a reward. Participants learn the location of the reward via exploration and experience (i.e., behind the left door in the blue room). The challenge lies in the fact the location of the reward will change slowly over time. That is, it may be somewhere in the blue room for 10 trials and then in the red room for 15 trials, and so on. How participants learn about the changes in reward location also varies. That is, sometimes the location will change, and the participant will be forced to take different actions to explore the other room in order to discover where the reward is now located. When they discover the new location, their choice on the next trial should reflect that discovery. Namely, if they have just discovered the reward is now located in the blue room, they should try to return to the blue room. Under these regular conditions, they could be taking the correct action because they understand the reward is in the blue room or they could be simply repeating the most recently rewarded action sequence (with no awareness of the room or involvement of any mental map). At other times, they will be shifted against their actions to the opposite room they were intending to enter (i.e., a rare transition). Here, they may luckily discover that the reward has shifted to the new room they just entered. In this case, after the participant discovers the new location, their first choice on the next trial must be governed by their understanding that the reward is in the new room; since it was a lucky rare transition that took them to the rewarded room on the previous trial, repeating the same action will take them to the wrong room. Thus, the question of interest concerns their room choice after a lucky rare transition: Will the participant repeat the same (incorrect) choice as they did on the last trial (since this action began the sequence, which led to reward on the previous trial), or will they (correctly) choose the other action to get to the rewarded room. If the participant repeats the same action, they appear to be making their choice on the basis of the just-experienced (cached) action values, and as a result, they will end up in the incorrect room. On the other hand, if the participant correctly makes the opposite choice and selects the rewarded room, they appear to be using a mental map of the task structure along with an updated representation of the reward location in order to infer the best action. This would be an example of integrating the new state value of the rewarded room back to the current choice in order to select the best action. Thus, the two-stage task pits the cached action value against the inferred value during the first choice after a lucky rare transition. In this instance, the cached action value is acquired by a model-free process while the inferred action value is acquired by the integration of the reward value from its new location. By explicitly measuring the contribution of both to decision-making, the results offer to provide a unique insight into the resulting balance of control, which may differ between individuals as well as between clinical groups.

A variety of clinical conditions have been assessed with the two-stage task (for review, see Chapter 15 by de Wit), where it has tended to covary with transdiagnostic differences in impulsivity (Gillan et al.). Recently, Culbreth, Westbrook, Daw, Botvinich, and Barch (2016) provided the first investigation of schizophrenia with the task. They found that schizophrenia was associated with a reduced model-based learning parameter compared to healthy controls. The reduced model-based parameter means that patients were more likely to select the reinforced choice rather than the state the reward occurred in, after a lucky rare transition. In contrast, model-free estimates did not differ between groups. The equivalent model-free parameters mean that in other trials (i.e., after regular transition trials), patients were as likely to pursue a reinforced choice after it was previously rewarded as controls. There are at least three possible reasons worth discussing for why patients might display the observed deficit in goal-directed performance. The first is that patients were not motivated by the reward to the same extent as controls. The second is that patients did not form a mental map of the task during the extensive pretraining, from which they could infer the best action. The third possibility (the preferred interpretation of Culbreth et al.) is that patients formed a map but could not integrate the new reward values from the goal state back to the current choice state.

The evidence against the possibility that patients were not motivated by reward is provided by the fact that model-free learning was driven by reinforcement in patients as much as controls. This is bolstered by observations that in general, patients respond to rewards in other contexts similarly as healthy controls (Cohen & Minor, 2010; Gold et al., 2008; Kring & Moran, 2008; Strauss & Cohen, 2017; Ursu et al., 2011). Furthermore, there are precedents of equivalent performance in reinforcement learning tasks in schizophrenia and healthy controls (Deserno et al. 2017; Heerey & Gold 2007). Whether the patients learned the task structure and formed a mental map is harder to determine. While extensive instruction was provided with feedback, along with a chance to demonstrate understanding, there was no explicit assessment of whether individuals understood and remembered the transition structure for the duration of the test. The authors did find that reaction times were slower after a rare transition to an unexpected state in both groups, which they argue is evidence that patients had an expectation for the regular transition state (i.e., they learned the transition structure). However, reaction times after a rare transition to an unexpected (undesired) state may be slower because the participant no longer expected to receive any reward on that trial—the usual outcome for a rare transition trial in this case. And we have seen previously that patients can acquire reward-seeking biases, such as faster reaction times to rewarded stimuli, in the same manner as healthy people. So at present, this study by itself demonstrates a goal-directed deficit in decision-making but leaves it unclear as to the source of that deficit. The problem may be due to a difficulty with forming mental maps, or it may be due to difficulty integrating reward value over multiple states to infer action values. Future studies using this task may help clarify the source of the problem in schizophrenia, but for the moment, we must turn to other sources of evidence to answer these questions.

The other major source of evidence for goal-directed learning is provided by devaluation tests—in which adaptive action selection must utilize the updated state values after learning that the outcome is no longer valuable (Balleine & O'Doherty, 2010; Schwabe & Wolf, 2009; Valentin et al., 2007). In a typical devaluation experiment, subjects are first trained to make two different instrumental actions for distinct food rewards. After training, one of the food rewards is devalued by prefeeding subjects to satiety with that food. Prefeeding, via sensory-specific satiety, devalues the marginal utility of that reward more than any other food reward. It occurs in the absence of any access to the action, in order to avoid directly changing the action values. The subsequent choice test also occurs in the absence of any further exposure to the action—outcome contingency since that will also provide the opportunity to update the action value directly. For this reason, the choice tests are usually conducted without any feedback (i.e., extinction). If the subject adjusts their action selection to reflect the new value of the outcomes after prefeeding, this demonstrates that the action values were derived by inference over the updated value of the outcomes.

We developed a test of goal-directed decision-making for assessing people with schizophrenia. The initial participants were 18 people with schizophrenia (or schizoaffective disorder), on stable medication and living in the community, and 18 healthy adults. To begin, participants indicated their food preferences for different snack foods using a 7-point rating scale. As expected, there were no differences between our two groups in the hedonic evaluations of those snacks, consistent with other evidence of intact hedonic responses in schizophrenia. Then each participant was trained to tilt a virtual vending machine for two of those snacks (e.g., left = M&Ms, right = crackers, counterbalanced between participants). Participants correctly reported the learned action—outcome contingencies before the experiment continued. In the devaluation procedure, instead of using sensory-specific satiety (cf. Waltz et al., 2015), participants were shown a large, high-definition video of cockroaches crawling through one of the snack foods. Immediately after devaluation, participants were given the opportunity to earn snack foods by tilting the vending machine again. Importantly, no discriminative cues and no outcomes were delivered in this stage to avoid the influence of S-R learning. Participants were warned that no snack foods would appear, but they believed the snacks were still accumulating and they would have to eat them before leaving the experiment. Finally, after this choice test, we asked participants to provide hedonic ratings of the snacks as before. The results were striking—healthy adults displayed a clear choice preference for the nondevalued snack, and their hedonic ratings reflected this, indicating that they had updated their state values and successfully integrated that value back to the original action to guide selection. By contrast, the people with schizophrenia displayed no apparent preference for the nondevalued snack, yet their posttest subjective ratings clearly indicated that they now preferred that snack. Here, it appeared that state values had been successfully updated, but those values had not been integrated back to the original action to guide selection. And this cooccurred with a reduced BOLD response in the caudate nucleus of patients, consistent with reduced input from prefrontal regions thought to carry the updated state values for action selection. However, it was not that patients did not show any preferences in their action selection—their test choices were not random. The correlation between patients action selection prior to devaluation (during instrumental training) and after devaluation was very high ($r = 0.80$) (while the same correlation among the healthy adults was low $r = -0.18$, since devaluation had changed their preferences). In other words, people with schizophrenia were still choosing the same snack they preferred before devaluation, presumably on the basis of the cached action values acquired during the original instrumental conditioning. We have repeated this demonstration in a further 16 patients with the same result (Fig. 17.2. For full details, see Morris, Cyrzon, Green, Le Pelley, & Balleine, 2018). In the replication, the choice test had a nonreinforced stage as well as a reinforced stage. That is, after 30 s of the nonreinforced choices, we began displaying the snack images on-screen as they were earned. Once the onscreen feedback was provided in this reinforced test stage, people with schizophrenia began to adjust their choices to reflect their hedonic preferences in line with healthy adults.

The results of the devaluation test are in line with Culbreth's results in the two-stage decision task; however, the devaluation results provide some further precision as to the

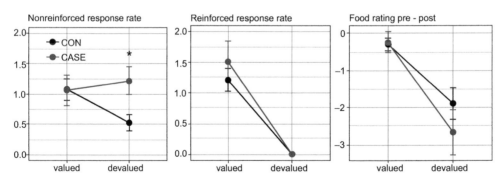

Figure 17.2 The effect of outcome devaluation on action selection and food ratings in people with schizophrenia (CASE) and healthy adults (CON). When the decisions required integration of the changed outcome values, devaluation had no impact on choices among CASE, relative to CON (left panel extinction test), *$P < 0.05$. However, when the outcomes were delivered and allowed the opportunity to update action values directly, devaluation selectively reduced actions for the devalued outcome in both groups (middle panel reinforced test). The right panel (food ratings pre—post) shows devaluation selectively reduced the value of the devalued outcome similarly in both groups. *(Adapted from Morris et al. (2018),* Translational Psychiatry.*)*

source of the deficit in schizophrenia. Both tests indicate that people with schizophrenia have a deficit in goal-directed decision-making. Plus in our hands, the fact that patients rated the devalued outcomes similarly to those of healthy adults is evidence that they successfully updated the outcome state values after devaluation. Furthermore, we explicitly determined that patients remembered the action–outcome contingencies, which implies the mental map of the task was intact at the time of test. So what remains is a potential deficit in the integration of the new outcome state values back to the current action to infer the best choice. We also associated the deficit with BOLD responses in the caudate of patients, which implicates a functional disconnection between the prefrontal cortex and the associative striatum. The neuropathology in the caudate we observed is consistent with recent PET studies, which found that schizophrenia is characterized by elevated levels of presynaptic dopamine in the associative striatum (Howes et al., 2009; Kegeles et al., 2010), rather than ventral striatum as previously thought. Others have also found reduced functional connectivity with the caudate in schizophrenia, as well as the prodromal state (Fornito et al., 2013; Quide, Morris, Shepherd, Rowland, & Green, 2013). This site receives converging inputs from the nigrostriatal path and glutamatergic inputs from the cortex, and neuroplasticity here is critical for goal-directed learning (Shiflett & Balleine, 2011). Thus, the pattern of evidence indicates the integration of state values back to actions is disrupted in schizophrenia due to neuropathology in the connections to the associative striatum.

ARE THERE OTHER SOURCES RESPONSIBLE FOR THE DEFICIT IN SCHIZOPHRENIA?

Identifying a specific deficit in schizophrenia with the integration of outcome state values implies the other two processes we have described as essential for MBRL are intact. However, this may not be the case. We also have evidence that forming the mental map between different states (stimuli, outcomes, and actions) is somewhat compromised in schizophrenia. For example, stimulus–outcome (S-O) learning can influence action selection, as demonstrated by outcome-specific Pavlovian-to-instrumental transfer (PIT). That is, after learning a predictive S-O contingency as well as a response–outcome (R-O) contingency for the same outcome, merely presenting the predictive stimulus can bias action selection for the same outcome. In other words, without learning any direct stimulus-response associations, the two associative connections (S-O, R-O) will combine to allow action selection on the basis of the common associate (i.e., O, the outcome) (Cartoni, Balleine, & Baldassarre, 2016). This would seem to require an associative map allowing inference to the common associate: a map that depends on the contingent relation among its elements (Bertran-Gonzalez, Laurent, Chieng, MacDonald, & Balline, 2013; Delamater, 1995). Furthermore, this mapping must occur without any representation of outcome value, since the ability of the stimulus to bias action selection during PIT is insensitive to outcome devaluation (Hogarth & Chase, 2012; Watson, Weirs,

Hommel, & de Wit, 2014). Thus, outcome-specific PIT appears to require the ability to infer over an associative map without imposing any integration of value, which leaves it as a unique assessment of a critical feature of MBRL.

We have assessed outcome-specific PIT in schizophrenia. In Morris, Quail, et al. (2015), we trained the same participants to expect a snack after a warning light came on the vending machine. Four different lights predicted four different outcomes: Two of the outcomes had previously been associated with the instrumental actions (e.g., tilt left → M&M, blue warning → M&M), while a third snack food outcome had never been associated with any instrumental response. That is, participants knew of no action to get this outcome and its warning light served as a general CS+ to determine the influence of food-predicting cues on total button pressing (general motivation). A fourth outcome was simply "no snack," so its warning light served as a CS−. Thus, we had two specific S-O contingencies (CS-A, CS-B) to assess the effect of the stimulus on action selection, and a CS+ and a CS− to assess stimulus effects on general motivation. When tested with the warning lights while tilting the vending machine, both healthy adults and people with schizophrenia displayed a robust influence of the specific cues (CS-A, CS-B) over action selection, i.e., outcome-specific transfer. Nevertheless, the effect was slightly attenuated among patients indicating the mapping of the S-O and R-O links via the common associate (the outcome) was somewhat compromised in schizophrenia. The effect of the CS+ and CS− on the general motivation to button press in schizophrenia was also attenuated, but press rates during the specific cues did not differ between groups. The implication of this is that the formation and use of the mental maps assumed in MBRL is compromised but achievable in schizophrenia, when it does not require any integration with value.

THE FUTURE OF MODEL-BASED REINFORCEMENT LEARNING: LEARNING CAUSAL MODELS

As discussed, MBRL is concerned with building a model of the environment, given the state caused by each action (the estimation of a transition matrix). These state transition probabilities form the mental map of the task structure. However, the transition probabilities must describe more than the probability of the outcome (O) given the action (A); i.e., $P(O|A)$. A positive contingency such as $P(O|A)$ only describes the *contiguity* between actions and outcomes; however, a negative contingency can also exist such as $P(O|\sim A)$. A negative contingency $P(O|\sim A)$ describes how the same outcome state can be reached via another path, without the action A. In order to represent the *causal relations* between states in the environment, these competing contingencies must be weighed against each other (Pearl, 2018). This is normatively described by the formalism $\Delta P = P(O|A) - P(O|\sim A)$. Consider, for instance, the case when two equally good action−outcome contingencies are provided concurrently to participants. If one of the outcomes is also provided for free (i.e., without any action), then the noncontingent outcome will selectively reduce the causal relationship of only one action and not the other. This occurs simply because the noncontingent outcome is indistinguishable from the outcomes caused by one action (i.e., both are M&Ms), but it is easily distinguishable from the outcomes caused by the other action (i.e., the other outcome is a cracker). Under these conditions, it has been amply demonstrated that both rats and people prefer to select the causal action, and in the case of people, they also judge the degraded action as less causal as well (Balleine & Dickinson, 1998; Dickinson & Mulatero, 1989; Morris, Dezfouli, Griffiths, Le Pelley, & Balleine, 2017).

Demonstrations that people and animals prefer causal actions, even when there is no difference in reward value between the actions are important because they suggest that goal-directed decision-making depends on a causal model. Consider that the value of the noncontingent outcome will diminish the reward value of both actions equally since reward can now be obtained without making either action; hence, the impact of the free reward will be equal for both actions. Also note that such demonstrations take care to equalize the rate at which free rewards and earned rewards are provided, in order to ensure there is no serendipitous differences in reward rate to explain the preference (Dickinson & Mulatero, 1989; Morris et al., 2017). Thus, any preference for the nondegraded action cannot reflect differences in reward contingencies, as is assumed in some computational models of goal-directed learning (Solway & Botvinick, 2012). Consequently, such a preference must depend upon a causal model of the task, where the transition probabilities weigh the competing contingencies between states to represent more than the simple contiguity between states.

We recently tested outcome-specific contingency degradation in schizophrenia and found evidence that learning such a causal model is severely impaired (Morris et al., 2018). Briefly, we trained participants (SZ n = 25, HA n = 25) to liberate two different snack foods from a virtual vending machine using two different action—outcome contingencies (e.g., left = M&Ms, right = crackers). In each 1-min block, participants could press freely for snacks. The positive contingency (probability of obtaining a snack) for each AO varied such that one action was reliably better than the other action and the best action changed from block to block. Since there were no negative contingencies (no free snacks delivered) in this stage, the best action in each block had the highest reward contingency and the stronger causal relationship with the outcome. Under these conditions, both groups learned the best action and rated it as more causal at similar levels. Importantly, people with schizophrenia were able to learn the best action similarly to healthy adults, even as it changed from block to block. That is, there was no evidence of reversal learning deficits or perseveration, presumably because performance was assessed over the 60-s block rather than trial to trial (Fig. 17.3, top panels. For full details, see Morris et al., 2018.). However, the next stage of testing clarified whether their performance was governed by differences in reward value or causal strength.

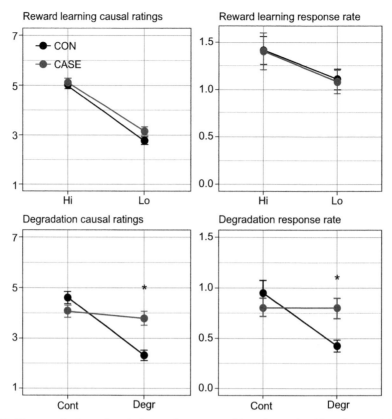

Figure 17.3 Top panels show that during an instrumental learning task with only a positive reward contingency (high vs. low), healthy adults (CON) and people with schizophrenia (CASE) distinguish the best action, on average, rating it as more causal (left) as well as preferring it (right) over the other response. Bottom panels show that people with schizophrenia do not learn to distinguish the causal action when one action–outcome contingency is degraded by noncontingent outcomes. Average causal ratings (left) and responses (right) for the degraded contingency were significantly higher among CASE than CON, *$P < 0.05$. *(Adapted from Morris et al. (2018), Translational Psychiatry.)*

In a second stage, participants learned about changes in the causal effects of their actions after one AO contingency was degraded by delivery of noncontingent outcomes. Thus, in this stage, we set the positive contingency for each AO as equal and ensured that the reward contingencies were also equal. All that differed was the negative contingency, since one action had a negative contingency (free outcomes) and the other action had no negative contingency (no free outcomes). This means that on average, the degraded action had a ΔP close to zero (equal number of free and earned outcomes for that action). Even though in the first stage, people with schizophrenia had no difficulty distinguishing the best action when it was defined by a difference in reward contingency (and it changed from block to block), this was not the case in the degradation stage. People with schizophrenia could not distinguish the best action when it was defined by a difference in causal strength in this stage, in either their actions or ratings (see Fig. 17.3, bottom panels. For full details, see Morris et al., 2018.). This implies that schizophrenia is associated with an impairment in forming a causal model of the environment. A causal model differs from models usually assumed in MBRL by virtue of the fact that it must trade-off competing contingencies to represent the causal relations between events (e.g., ΔP), rather than simply the contiguities between states. That is, a transition matrix that only represents the state-to-state contiguities is sufficient to navigate mazes and other spatial tasks for which MBRL was developed, but it is inadequate to represent the causal texture of our environment (Pearl, 2018). It is learning this form of causal model, quite different from the usual form of mental map assumed in MBRL, which seems to be impaired in schizophrenia.

CONCLUSIONS

The aim of this review is to provide a more nuanced understanding of the goal-directed deficit in schizophrenia. We have argued that MBRL as a theory of goal-directed learning has highlighted three critical features: learning a mental map of the environment, learning where reward is located within that map, and the ability to integrate the reward value of the goal state back to the current action. To date, the focus of schizophrenia research has been on determining whether an impairment exists in the integration of reward value back to the current choice. Performance deficits in the two-stage decision task and outcome devaluation tests are consistent with such an impairment; however, the role of potential impairments in the other features of MBRL may also contribute. Forming a mental map of the structure of the task is obviously a prerequisite to integrating value for decision-making. Here, the fact that people with schizophrenia can select actions according to different cues on the basis of their shared outcome implies that forming a simple map of state transitions is intact in schizophrenia. In this case, the map consists of combining R–O and S–O associations by way of the common outcome. However, forming more complex maps that require calculating competing contingencies, and which are necessary for representing causal structure, seems to be at the limit of (if not beyond) the range of function in schizophrenia. This leaves us with a slightly more nuanced view of the goal-directed decision-making in schizophrenia. For instance, on the basis of the evidence presented here, people with schizophrenia should perform sufficiently in navigation and maze tasks. Such tasks demand only a simple transition matrix of state contiguities to support performance. However, when such maze tasks depend on weighing competing routes or tasks that require competitive allocation of predictive value, such as causal learning or instrumental control, then we should start to see performance deficits. And indeed on tests involving the competitive allocation of predictive value, people with schizophrenia perform poorly (Moran, Owen, Crookes, Al-Uzri,

& Reveley, 2008; Morris et al., 2013). Looking toward the future, it seems that theoretical development of MBRL to distinguish causal models from other kinds of noncausal models will help us further characterize the deficit in schizophrenia and perhaps other disorders of goal-directed learning.

REFERENCES

Balleine, B. W., & Dickinson, A. (1998). Goal-directed instrumental action: Contingency and incentive learning and their cortical substrates. *Neuropharmacology, 37*, 48—69.

Balleine, B. W., & O'Doherty, J. P. (2010). Human and rodent homologies in action control: Corticostriatal determinents of goal-directed and habitual action. *Neuropsychopharmacology: Official Publication of the American College of Neuropsychopharmacology, 35*, 48—49.

Bellman, R. (1957). *Dynamic programming.* Princeton: Princeton University Press.

Bertran-Gonzalez, J., Laurent, V., Chieng, B. C., MacDonald, J. C., & Balline, B. W. (2013). Learning-related translocation of delta-opioid receptors in ventral striatal cholinergic interneurons mediates choice between goal-directed actions. *Journal of Neuroscience, 33*, 16060—16071.

Bleuler, E. (1950). *Dementia praecox* (Vol. 1). New York: International Universities Press.

Cartoni, E., Balleine, B. W., & Baldassarre, G. (2016). Appetitive pavlovian-instrumental tranfer: A review. *Neuroscience and Biobehavioral Reviews, 71*, 829—848.

Cohen, A. S., & Minor, K. S. (2010). Emotional experience in patients with schizophrenia revisited: Meta-analysis of laboratory studies. *Schizophrenia Bulletin, 36*, 143—150.

Culbreth, A. J., Westbrook, A., Daw, N. D., Botvinich, M., & Barch, D. M. (2016). Reduced model-based decision-making in schizophrenia. *Journal of Abnormal Psychology, 125*, 777—787.

D'Ardenne, K., McClure, S. M., Nystrom, L. E., & Cohen, J. D. (2008). BOLD responses reflecting dopaminergic signals in the human ventral tegmental area. *Science, 319*, 1264—1267.

Daw, N. D. (2011). Trial-by-trial data analysis using computational models. In *Decision making, affect, and learning: Attention and performance XXIII.* Oxford University Press.

Daw, N., Gershman, S. J., Seymour, B., Dayan, P., & Dolan, R. J. (2011). Model-based influences on humans' choices and striatal prediction-errors. *Neuron, 69*, 1204—1215.

Daw, N., Niv, Y., & Dayan, P. (2005). Uncertainty-based competition between prefrontal and dorsolateral striatal systems for behavioral control. *Nature Neuroscience, 8*, 1704—1711.

de Wit, S., & Dickinson, A. (2009). Associative theories of goal-directed behaviour: A case for animal-human translational models. *Psychological Research, 73*, 463—476.

Delamater. (1995). Outcome-selective effects of intertrial reinforcement in a Pavlovian appetitive condiitoning paradigm with rats. *Animal Learning & Behavior, 23*, 31—39.

Deserno, L., Heinz, A., & Schlagenhauf, F. (2017). Computational approaches to schizophrenia: A perspective on negative symptoms. *Schizophrenia Research, 186*, 46—54.

Dickinson, A., & Mulatero, C. W. (1989). Reinforcer specificity of the suppression of instrumental performance on a non-contingent schedule. *Behavioural Processes, 19*, 167—180.

Doll, B. B., Simon, D. A., & Daw, N. D. (2012). The ubiquity of model-based reinforcement learning. *Current Opinion in Neurobiology, 22*, 1—7.

Dowd, E. C., & Barch, D. M. (2012). Pavlovian reward prediction and receipt in schizophrenia: Relationship to anhedonia. *PLoS One, 7*, e35622.

Fornito, A., Harrison, B. J., Goodby, E., Dean, A., Ooi, C., Nathan, P. J., … Bullmore, E. T. (2013). Functional dysconnectivity of corticostriatal circuitary as a risk phenotype for psychosis. *JAMA Psychiatry, 70*, 1143—1151.

Gold, J. M., Waltz, J. A., Prentice, K. J., Morris, S. E., & Heerey, E. A. (2008). Reward processing in schizophrenia: A deficit in the representation of value. *Schizophrenia Bulletin*, 835—847.

Gradin, V. B., Kumar, P., Waiter, G., Aheard, T., Stickle, C., Milders, M., … Steele, J. D. (2011). Expected value and prediction-error abnormalities in depression and schizophrenia. *Brain: A Journal of Neurology, 134*, 1751—1764.

Heerey, E. A., & Gold, J. M. (2007). Patients with schizophrenia demonstrate dissociation between affective experience and motivated behavior. *Journal of Abnormal Psychology, 116*, 268−278.

Heerey, E. A., Bell-Warren, R. K., & Gold, J. M. (2008). Decision-making impairments in the context of intact reward sensitivity in schizophrenia. *Biological Psychiatry, 64*, 62−69.

Heyes, C., & Dickinson, A. (1990). The intentionality of animal action. *Mind and Language, 5*, 87−104.

Hogarth, L., & Chase, H. W. (2012). Evaluating psychological markers for human nicotine dependence: Tobacco choice, extinction, and pavlovian-to-instrumental transfer. *Experimental and Clinical Psychopharmacology, 20*, 213−224.

Holt, D. J., Lebron-Milad, K., Milad, M. R., Rauch, S. L., Pitman, R. K., Orr, S. P., ... Goff, D. C. (2009). Extinction memory is impaired in schizophrenia. *Biological Psychiatry, 65*, 455−463.

Howes, O. D., Montgomery, A. J., Asselin, M. C., Murray, R. M., Valli, I., Tabraham, P., ... Grasby, P. M. (2009). Elevated striatal dopamine function linked to prodromal signs of schizophrenia. *Archives of General Psychiatry, 66*, 13−20.

Izquierdo, A., Brigman, J. L., Radke, A. K., Rudebeck, P. H., & Holmes, A. (2017). The neural basis of reversal learning: An updated perspective. *Neuroscience, 345*, 12−26.

Jenson, J., Willeit, M., Zipursky, R. B., Savina, I., Smith, A. J., Menon, M., ... Kapur, S. (2008). The formation of abnormal associations in schizophrenia: Neural and behavioral evidence. *Neuropsychopharmacology: Official Publication of the American College of Neuropsychopharmacology, 33*, 473−479.

Kegeles, L. S., Abi-Dargham, A., Frankle, W. G., Gil, R., Cooper, T. B., Slifstein, M., ... Laruelle, M. (2010). Increased synaptic dopamine function in associative regions of the striatum in schizophrenia. *Archives of General Psychiatry, 67*, 231−239.

Koch, K., Schachtzabel, C., Wagner, G., Schikora, J., Schultz, C., Reichenbach, J. R., ... Schlosser, R. G. (2010). Altered activation in association with reward-related trial-and-error learning in patients with schizophrenia. *NeuroImage, 50*, 223−232.

Kring, A. M., & Moran, E. K. (2008). Emotional response deficits in schizophrenia: Insights from affective science. *Schizophrenia Bulletin, 34*, 819−834.

Mackintosh, N. J. (1975). A theory of attention: Variations in the associability of stimuli with reinforcement. *Psychological Review, 82*, 276−298.

Moran, P. M., Owen, L., Crookes, A. E., Al-Uzri, M. M., & Reveley, M. A. (2008). Abnormal prediction-error is associated with negative and depressive symptoms in schizophrenia. *Progress in Neuropsychopharmacology and Biological Psychiatry, 32*, 116−123.

Morris, R. W., Cyrzon, C., Green, M. J., Le Pelley, M. E., & Balleine, B. W. (2018). Impairments in action-outcome learning in schizophrenia. *Translational Psychiatry, 8*, 54−65.

Morris, R. W., Dezfouli, A., Griffiths, K. R., Le Pelley, M. E., & Balleine, B. W. (May 14, 2017). The algorithmic neuroanatomy of action-outcome learning. *bioRXiv*, 137851.

Morris, R. W., Griffiths, O., Le Pelley, M. E., & Weickert, T. W. (2013). Attention to irrelevant cues is related to positive symptoms in schizophrenia. *Schizophrenia Bulletin, 39*, 575−582.

Morris, R. W., Purves-Tyson, T. D., Weickert, C. S., Rothmond, D., Lenroot, R., & Weickert, T. W. (2015). Testosterone and reward prediction-errors in healthy men and men with schizophrenia. *Schizophrenia Research, 168*, 649−660.

Morris, R. W., Quail, S. Q., Griffiths, K. R., Green, M. J., & Balleine, B. W. (2015). Corticostriatal control of goal-directed action is impaired in schizophrenia. *Biological Psychiatry, 77*, 187−195.

Morris, R. W., Vercammen, A., Lenroot, R., Moore, L., Langton, J. M., Short, B., ... Weickert, T. W. (2012). Disambiguating ventral striatum fMRI-related BOLD signal during reward prediction in schizophrenia. *Molecular Psychiatry, 17*, 280−289.

Murray, G. K., Cheng, F., Clark, L., Barnett, J. H., Blackwell, A. D., Fletcher, P. C., ... Jones, P. B. (2008). Reinforcement and reversal learning in first-episode psychosis. *Schizophrenia Bulletin, 34*, 848−855.

Murray, G. K., Corlett, P. R., Clark, L., Pessiglione, M., Blackwell, A. D., Honey, G., ... Fletcher, P. C. (2008). Substantia nigra/ventral tegmental reward prediction error disruption in psychosis. *Molecular Psychiatry, 13*, 267−276.

Niv, Y., & Schoenbaum, G. (2008). Dialogues on prediction errors. *Trends in Cognitive Sciences, 12*, 265−272.

O'Doherty, J. P., Dayan, P., Friston, K., Critchley, H., & Dolan, R. J. (2003). Temporal difference models and reward-related learning in the human brain. *Neuron, 28*, 329−337.

Pearl, J. (January 2018). *Theoretical impediments to machine learning*. Tehcnical Report, R-475, submitted.

Quide, Y., Morris, R. W., Shepherd, A. M., Rowland, J. E., & Green, M. J. (2013). Task-related fronto-striatal functional connectivity during working memory performance in schizophrenia. *Schizophrenia Research, 150*, 468−475.

Radua, J., Schmidt, A., Borgwardt, S., Heinz, A., Schlagenhauf, F., McGuire, P., & Fusar-Poli, P. (2015). Ventral striatal activation during reward processing in psychosis. *JAMA Psychiatry, 72*, 1243−1251.

Schlagenhauf, F., Huys, Q. J., Deserno, L., Rapp, M. A., Beck, A., Heinze, H.-J., … Heinz, A. (2014). Striatal dysfunction during reversal learning in unmedicated schizophrenia patients. *NeuroImage, 89*, 171−180.

Schultz, W., & Dickinson, A. (2000). Neuronal coding of prediction errors. *Annual Review of Neuroscience, 23*, 473−500.

Schwabe, L., & Wolf, O. T. (2009). Stress prompts habit behavior in humans. *Journal of Neuroscience, 29*, 7191−7198.

Shiflett, M. W., & Balleine, B. W. (2011). Molecular substrates of action control in cortico-striatal circuits. *Progress in Neurobiology, 95*, 1−13.

Solway, A., & Botvinick, M. M. (2012). Goal-directed decision making as probabilistic inference: A computational framework and potential neural correlates. *Psychological Review, 119*, 120−154.

Strauss, G. P., & Cohen, A. S. (2017). A transdiagnostic review of negative symptom phenomenology and etiology. *Schizophrenia Bulletin, 43*, 712−719.

Sutton, R. S., & Barto, A. G. (1998). *Reinforcement learning: An introduction*. Cambridge, MA: MIT Press.

Tobler, P. N., O'Doherty, J. P., Dolan, R. J., & Schultz, W. (2006). Human neural learning depends on reward prediction errors in the blocking paradigm. *Jounral of Neurophysiology, 95*, 301−310.

Ursu, S., Kring, A. M., Gard, M. G., Minzenberg, M. J., Yoon, J. H., Ragland, J. D., … Carter, C. S. (2011). Prefrontal cortical deficits and impaired cognition-emotion interactions in schizophrenia. *American Journal of Psychiatry, 168*(3), 276−285.

Valentin, V., Dickinson, A., & O'Doherty, J. P. (2007). Determining the neural substrates of goal-directed learning in the human brain. *Journal of Neuroscience, 27*, 4019−4026.

Walter, H., Kammerer, H., Frasch, K., Spitzer, M., & Abler, B. (2009). Altered reward functions in patients on atypical antipsychotic medication in line with the revised dopamine hypothesis of schizophrenia. *Psychopharmacology, 206*, 121−132.

Waltz, J. A., & Gold, J. M. (2007). Probabilistic reversal learning impairments in schizophrenia: Further evidence of orbitofrontal dysfunction. *Schizophrenia Research, 93*, 296−303.

Waltz, J. A., Schweitzer, J. B., Gold, J. M., Kurup, P. K., Ross, T. J., Salmeron, B. J., … Stein, E. A. (2009). Patients with schizophrenia have a reduced neural response to both unpredictable and predictable primary reinforcers. *Neuropsychopharmacology: Official Publication of the American College of Neuropsychopharmacology, 34*, 1567−1577.

Waltz, J., Brown, J. K., Gold, J. M., Ross, T. J., Salmeron, B. J., & Stein, E. A. (2015). Probing the dynamic updating of value in schizophrenia using a sensory-specific satiety paradigm. *Schizophrenia Bulletin, 41*, 1115−1122.

Watson, P., Weirs, R. W., Hommel, B., & de Wit, S. (2014). Working for foods you don't desire. Cues interfere with goal-directed food-seeking. *Appetite, 79*, 139−148.

Weickert, T. W., Terrazas, A., Bigelow, L. B., Malley, J. D., Hyde, T., Egan, M. F., & Goldberg, T. E. (2002). Habit and skill learning in schizophrenia: Evidence of normal striatal processing with abnormal cortical input. *Learning & Memory, 9*, 430−442.

CHAPTER 18

Realigning Models of Habitual and Goal-Directed Decision-Making

Kevin J. Miller[1], Elliot A. Ludvig[2], Giovanni Pezzulo[3], Amitai Shenhav[4]

[1]Princeton Neuroscience Institute, Princeton University, Princeton, NJ, United States; [2]Department of Psychology, University of Warwick, Coventry, United Kingdom; [3]Institute of Cognitive Sciences and Technologies, National Research Council, Rome, Italy; [4]Department of Cognitive, Linguistic, and Psychological Sciences, Brown Institute for Brain Science, Brown University, Providence, RI, United States

In cognitive psychology, categories of mental behavior have often been understood through the prevailing technological and computational architectures of the day. These have spanned the distinction between types of processing (e.g., serial vs. parallel), forms of memory maintenance (e.g., short-term vs. long-term storage), and even the fundamental relationship between mind and brain (i.e., software vs. hardware). Over the past few decades, research into animal learning and behavior has similarly been informed by prominent computational architectures, especially those from the field of computational reinforcement learning (RL; themselves having drawn inspiration from research on animal behavior; e.g., Sutton & Barto, 1981, 1998). In addition to offering explicit and testable predictions for the process by which an animal learns about and chooses to act in their environment, ideas from RL have been adapted to operationalize a distinction from the animal learning literature: the distinction between habitual and goal-directed actions (Daw & O'Doherty, 2013; Daw, Niv, & Dayan, 2005; Dolan & Dayan, 2013).

THE PREVAILING TAXONOMY MAPPING ANIMAL BEHAVIOR TO REINFORCEMENT LEARNING

As described in earlier chapters, goal-directed behavior is distinguished from habits by its sensitivity to context (including motivational state), future outcomes, and the means—end relationship between the actions being pursued and the rewarding outcome expected as a result (Dickinson, 1985; Wood & Rünger, 2016). Experimentally, behavior is typically classified as *goal-directed* when an animal alters a previously rewarded action following relevant changes in the action—outcome contingencies (e.g., delivery of the outcome is no longer conditional on an action) and/or following a change in the motivational significance of the outcome expected for that action (e.g., the animal is no longer hungry; Hammond, 1980; Adams, 1982). Insensitivity to these manipulations is considered a hallmark of *habitual* behavior. These two classes of behavior are believed to be underpinned by distinct psychological processes and neural substrates—a proposal that has been borne out by evidence for dissociable patterns of neural activity (Gremel & Costa, 2013) and

Goal-Directed Decision Making
ISBN 978-0-12-812098-9, https://doi.org/10.1016/B978-0-12-812098-9.00018-8

inactivation studies showing that behavior can be made more habitual or goal-directed by selectively inactivating specific regions of the striatum and prefrontal cortex (Balleine & O'Doherty, 2010; Killcross & Coutureau, 2003; Yin & Knowlton, 2006).

Edward Tolman (1948), one of the earliest researchers into goal-directed decision-making, proposed that a distinguishing feature of goal-directed behavior was a reliance on an internal model of the environment to guide action selection, rather than action selection relying solely on the history of prior actions and associated feedback. This qualitative distinction is at the center of a parallel distinction in the RL literature between algorithms that drive an agent's action selection in a model-based or model-free manner (Kaelbling, Littman, & Moore, 1996; Littman, 2015; Sutton & Barto, 1998). Specifically, whereas *model-free* RL selects between actions based on the rewards previously experienced when performing those actions, *model-based* RL incorporates more specific information about the structure of the agent's environment and how this interacts with the agent's actions and the associated outcomes. These parallels fostered a natural alignment between the animal and machine learning literature such that goal-directed decisions were mapped onto model-based RL algorithms, and habits were mapped onto model-free algorithms (Daw et al., 2005; Dolan & Dayan, 2013). Today, these literature are so tightly interwoven that the terms model-free/habitual and model-based/goal-directed are often used interchangeably, and the linkages between them have yielded novel insights into complex decision-making phenomena, such as addiction (Lucantonio, Caprioli, & Schoenbaum, 2014; Vandaele & Janak 2017), impulsivity (Rangel, 2013), and moral judgment (Buckholtz, 2015; Crockett, 2013; Cushman, 2013). For instance, individuals with debilitating habits are thought to be driven more by model-free than model-based learning systems (Gillan, Kosinski, Whelan, Phelps, & Daw, 2016; Gillan, Otto, Phelps, & Daw, 2015), as are those who judge moral wrongdoings based on the act and its outcome rather than also taking into account other features of the situation (e.g., intention, directness of causality; Crockett, 2013; Cushman, 2013).

PROBLEMS WITH THE CURRENT TAXONOMY

Despite its popularity and intuitive foundations, key aspects of this mapping between this pair of dichotomies remain tenuous. First, the fundamental basis of RL—the idea of an agent adjusting its actions based on prior *reinforcement*—already strains against early (Hull, 1943; James, 1890; Thorndike, 1911) as well as more recent (Ouellette & Wood, 1998; Wood & Neal, 2007; Wood & Rünger, 2016) conceptualizations of habits. According to these alternate views, habits form through repetition of prior actions, *irrespective* of whether those actions were positively reinforced. In other words, the mapping between habits and model-free RL is in tension with the idea that habits may be *value-free* and therefore may not require any form of RL Miller, Shenhav, & Ludwig, (in press).

A second concern about this mapping stems from research into the neural circuitry associated with each process. In strong contrast to the relatively clean and homologous

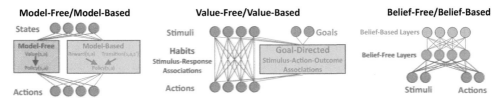

Figure 18.1 Schematic of computational architectures for habitual and goal-directed control. Left: Currently popular architecture in computational reinforcement learning (RL), in which model-free and model-based RL instantiate habitual and goal-directed control, respectively (Daw et al., 2005). Middle: Proposed architecture in which habits are implemented by direct ("value-free") connections between stimuli and the actions that are typically taken in response to those stimuli while goal-directed control is implemented by RL (Miller et al., 2016; see Fig. 2). Right: Proposed architecture in which habits are implemented by lower ("belief-free") layers in a hierarchical predictive network, while goal-directed control is implemented by higher ("belief-based") layers (Friston et al., 2016; Pezzulo et al., 2015). Left and middle panels modified from Miller et al., (2016).

neural dissociations that have been observed when distinguishing habitual and goal-directed behavior across species (Balleine & O'Doherty, 2010; Yin & Knowlton, 2006), model-free and model-based RL processes have tended to recruit largely overlapping circuits (Daw, Gershman, Seymour, Dayan, & Dolan, 2011; Doll, Simon, & Daw, 2012; Wimmer, Daw, & Shohamy, 2012; but see Lee, Shimojo, & O'Doherty, 2014; Wunderlich, Dayan, & Dolan, 2012). Moreover, the circuits implicated in both forms of RL—including regions of midbrain, ventral and dorsomedial striatum, orbital/ventromedial prefrontal cortex, and anterior cingulate cortex—overlap primarily with circuits causally implicated in goal-directed (and/or Pavlovian) behavior, rather than with the neural substrates of habitual behavior (Balleine & O'Doherty, 2010).

Together these two concerns suggest that the links between the animal and machine learning taxonomies are at the very least incomplete if not deeply misaligned. They paint a picture of (1) habits as being potentially *value-free* and therefore not mapping cleanly to either form of RL and (2) model-free and model-based RL as instead both belonging to a category of *value-based* behaviors that share mechanisms in common with goal-directed behaviors (Fig. 18.1). Habits are therefore not necessarily the product of model-free RL, and model-free RL may share more in common with model-based RL than habits do with goal-directed behavior (see also Chapter 5 by Collins).

ALTERNATIVE TAXONOMIES FOR HABITUAL AND GOAL-DIRECTED BEHAVIOR

Value-based versus value-free control

As suggested above, habitual and goal-directed behaviors may be distinguished along other dimensions and according to other schemes than those encompassed by the popular

model-free/model-based dichotomy. One alternative framework for distinguishing between habitual and goal-directed behavior focuses on the role that value does or does not
play in driving those behaviors (Miller et al., 2016). Under this proposal, goal-directed
control is understood as relying on representations of expected value: "expected discounted future reward" in the language of RL theory or "utility" in economics. Both
model-based and model-free RL agents represent expected value (i.e., both produce actions that are *value-based*) and might therefore implement different types of goal-directed
control. Habitual control, in this view, arises from a different type of agent: a *value-free*,
perseverative agent. This perseverative agent tends to repeat actions frequently taken in
the past in a particular situation, regardless of their outcomes (Fig. 18.1, middle).
Crucially, this perseverative system considers all past actions, whether they were taken
under its control or under the control of the goal-directed system. This allows behaviors
to be "passed on" from one control system to the other: If the goal-directed system tends
to frequently take the same action in a particular situation, the habitual system will learn
also to take that action (Fig. 18.2).

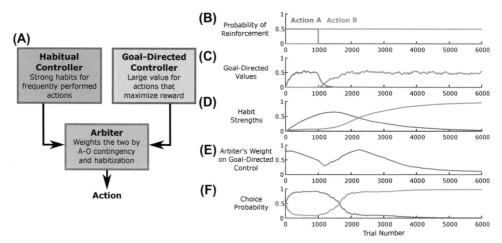

Figure 18.2 (A) Schematic of value-based/value-free architecture. Habits are implemented by a value-
free process that strengthens actions that are frequently taken in a given state, while goal-directed
control is implemented by a process that computes values based on history of reinforcement and
knowledge of the task structure. For details, see Miller et al. (2016). (B) Simulations of a reversal
learning environment: Action A is initially reinforced with higher probability (0.5) than Action B (0),
but after 1000 trials, the reinforcement probabilities reverse. (C) Soon after the reversal, the goal-
directed system learns that Action B is more valuable. (D) The habit system increasingly favors Action
A the more often it is chosen and only begins to favor Action B once that action is chosen more consistently (long after reversal). (E) The weight of the goal-directed controller gradually decreases as habits
strengthen then increases post-reversal as the global and goal-directed reinforcement rates diverge.
(F) Actions are selected on each trial by a weighted combination of the goal-directed values and the
habit strengths. Modified from Miller et al., (2016).

Such an arrangement has been shown to recapitulate classic findings in the literature on habits, including their strengthening with overtraining; their insensitivity to outcome devaluation and contingency degradation; and the widespread finding that humans and other animals perseverate on previous actions in instrumental tasks, irrespective of feedback (Miller et al., 2016). This framework also provides a natural explanation for the finding that putatively model-based and model-free representations of value and prediction error tend to colocate in brain regions associated with goal-directed control (Daw et al., 2011). More generally, this proposal is grounded in previous approaches that have incorporated Hebbian plasticity (i.e., increasing strengthening of stimulus–response associations with repetition) in other computational models of the development of automaticity (Ashby, Ennis, & Spiering, 2007; Hélie, Roeder, Vucovich, Rünger, & Ashby, 2015; Topalidou, Kase, Boraud, & Rougier, 2015).

Belief-based versus belief-free control

A second but related set of accounts distinguishes categories of behavioral control according to the role played by beliefs rather than value per se (Fig. 18.1, right; Friston et al., 2016). Under *belief-free* schemes, an agent selects actions based on stimuli or stimulus-action sets (policies). By contrast, under *belief-based* schemes, an agent maintains internal (probabilistic) estimates—or beliefs—over external states (e.g., its current or future expected locations) and uses these beliefs for action selection. Forming beliefs about the environmental state is important when the environment is partially observable (i.e., some of its parts are hidden and not directly observable, hence they need to be inferred), and the current stimulus does not unambiguously specify (for example) the agent's position or context (Box 18.1).

To better understand the difference between belief-free and belief-based schemes, consider the case of an agent in a T-maze, as depicted in Fig. 18.3. The agent can be in one of eight possible states: one of four locations (center, top-left, top-right, bottom) within one of two contexts (Fig. 18.3). In Context 1, reward is on the top-left, while in Context 2 reward is on the top-right. The agent knows its initial location (center) but does not know which context it is in. However, the agent knows that colored cues at the bottom of the maze will disambiguate the context: these cues are either blue (Context 1) or cyan (Context 2). A belief-free agent, who has no notion of state or context, would select a policy to go directly to one of the two reward sites (top-left or top-right), but will, as a result, risk missing the reward. A belief-based agent, who knows it is uncertain about the context and that the cue will reduce this uncertainty, would instead go to the cue location first (called an "epistemic action" as it aims at changing the agent's belief state and not achieving an external goal). After disambiguating its current context, the belief-based agent would go to the correct reward location with high confidence.

Box 18.1 Beliefs within a Markov Decision Process (MDP) framework

Under a belief-free scheme, an agent always directly infers its state (e.g., position in the maze) from sensory measurements (observations). This scheme is particularly attractive because an agent does not need to consider previous stimuli or actions to select an action—the current stimulus is sufficient (the so-called Markov property of MDPs). Such a scheme also assumes a one-to-one mapping between the agent's "true" state (e.g., position in the maze) and its observations. However, realistic environments complicate such inferences about one's current (or future) states in at least two ways. First, such environments are stochastic, meaning different observations can follow from the same state, such as reward being delivered probabilistically in a corner of a maze. Second, most environments involve aliasing, meaning different states generate the same observations, such as observing reward in two different corners of a maze. The former complication (stochasticity) is not a real challenge for most RL schemes, but the latter (aliasing) is more difficult to handle, as an agent cannot infer its true state from sensory measurements. The agent may thus face a credit assignment problem in assigning observations to their true causes (sometimes called hidden states). There are various ways to extend the Markov Decision Process (MDP) formulation to handle these more challenging cases. One is called belief MDP and consists in augmenting the agent's representation with a sort of memory; the hope is that even if two states are aliased and cannot be distinguished on the basis of the current observations, they can be distinguished on the basis of a trace of previous observations.

The problem described above is enriched within a framework known as a Partially Observable Markov Decision Process (POMDP; Kaelbling, Littman, & Cassandra, 1998); here, the agent is enriched with a notion of "state" that is distinct from an observation or a history of previous observations. In a probabilistic setting, an agent maintains a probability distribution or belief over its current state or even about future goal states or previous states (becoming a belief-based system). All of these beliefs are continuously updated on the basis of new observations, using mechanisms that are analogous to Kalman (or Bayesian) filters—hence, it can resolve any ambiguity about its current (or future or past) states by collecting more observations. This probabilistic approach is at the core of recent formulations of goal-directed systems that are driven by planning-by-inference (Botvinick & Toussaint, 2012), Kullback—Leibler (KL) control (van den Broek, Wiegerinck, & Kappen, 2010), and active inference (Friston, FitzGerald, Rigoli, Schwartenbeck, & Pezzulo, 2017).

In other words, a belief-free agent would reach a reward location without resolving its uncertainty first and (in this maze) fail half the time. On the contrary, a belief-based agent would perform an epistemic action first (go to the cue location) and then go to the correct reward location. This case exemplifies the similarities belief-based action selection shares with properties of goal-directed action selection, most notably, its reliance on cognitive representations (state representation and generative models), its epistemic drives (i.e., foraging for information and reducing uncertainty prior to acting), and its context sensitivity. Friston et al. (2016) proposed that a belief-free scheme is sufficient to characterize habitual systems but that a belief-based scheme is necessary to characterize goal-directed systems.

This taxonomy also illustrates that a belief-based scheme is only required when there is state uncertainty, but when uncertainty is resolved (e.g., after learning), the belief-based scheme can give way to a belief-free scheme. For instance, after engaging with the T-maze above for a sufficient number of trials to learn that the reward contingencies were stable, the agent can become sufficiently confident about them. In this case, it would not need to check context cues anymore but can go directly to the correct reward location (e.g., the top-left). In other words, with no residual uncertainty about the current context, the agent no longer needs a belief-based scheme or epistemic actions. This case is exactly when the agent can use (or learn) a classical belief-free or stimulus–response RL policy, say, to go directly to the top-left (see Friston et al., 2016 for details). Hence, belief-free action selection shares many similarities with habitual behavior—most notably, a primary dependence on stimuli to trigger actions and an inflexibility to changes in contingencies (i.e., a change in reward location). In sum, while goal-directed behavior requires a belief-based scheme, habitual behavior maps naturally to a belief-free scheme; and both can coexist within the same agent architecture. Note that while the belief-based scheme uses a notion of (expected) value, the belief-free scheme uses stimulus–response mechanisms and has no notion of value.

Other approaches

There have been several other theoretical attempts to carve the space that includes goal-directed and habitual behavior, while avoiding a strict model-based versus model-free dichotomy. One such approach leverages distinct forms of memory for past rewards: a slowly updating, long-term memory and a rapidly adjusting short-term memory (Hikosaka, Ghazizadeh, Griggs, & Amita, 2017; Silver, Sutton, & Müller, 2008). The former is thought to encode skills, which are analogous to habits in that they are automatic, precise, and inflexible, whereas the latter is responsible for flexible responding, analogous to goal-directed control. A second approach puts habits right into the goal-directed planning process, whereby planning proceeds through a typical search process, but terminates after a certain depth (Keramati, Smittenaar, Dolan, & Dayan, 2016). The value of the terminal node of the search is taken to be the habitual value and used to guide the initial choice. A third approach asserts that habits arise from "chunking" of action sequences that are initially taken under goal-directed control and are selected by the goal-directed system in situations where they are adaptive (Dezfouli & Balleine, 2013; Dezfouli, Lingawi, & Balleine, 2014). A fourth approach proposes a tripartite division between exploratory, model-based, and motor memory systems. This approach splits the model-free system into a component that generates highly variable exploratory behavior and another habitual component that sticks strictly to the learned values (Fermin et al., 2016).

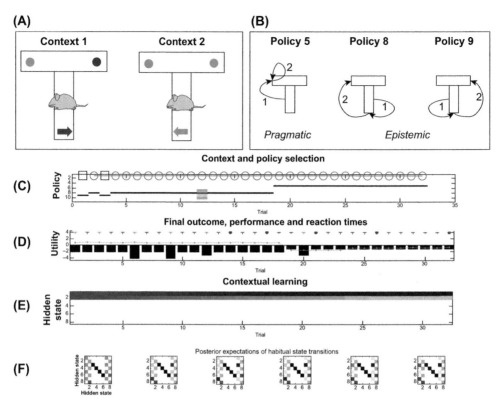

Figure 18.3 Epistemic versus pragmatic policies and belief-based versus belief-free schemes. (A) A simple choice situation that includes four spatial locations (center, top-left, top-right, bottom) and two contexts (Context 1 and Context 2), for a total of eight (four by eight) hidden states. In Context 1, the reward (*blue circle*) is located to the top-right, and a cue indicating its position (*blue arrow*) is located to the bottom. In Context 2, the reward (*cyan circle*) is located to the top-left, and a cue indicating its position (*cyan arrow*) is located to the bottom. *Red circles* are not rewarding. (B) A simulated agent starts from the center location and has to select a sequence of two actions to reach the reward site, but the agent does not know the current context. Three example policies are shown: Policy 5 (top-left, top-left) is a pragmatic policy that goes directly to the left location; Policy 8 (bottom, top-left) and Policy 9 (bottom, top-right) are epistemic policies that go to the informative (bottom) location before reaching one of the two reward locations. (C) Results of a simulation using the belief-based scheme of active inference. The panel shows the sequence of contexts (Context 1 is represented by the *green circle* and Context 2 by the *black square*) that the agent encounters for each trial and the policies it selects (*horizontal lines*). There is a clear transition between epistemic policies 8 and 9 to the pragmatic Policy 5 at trial 18, when the agent has resolved its uncertainty about its current context. (D) Outcomes (*cyan and blue dots* are rewards, *red dot* is no reward; note that outcomes are stochastic), performance (expected utility, with zero as maximum value) and reaction time for each trial. (E) Agent's belief about its initial hidden state. The values 1 and 2 indicate the center location in Context 1 and 2, respectively. Note that the agent increases its belief about being located in Context 1—which also produces the shift from an epistemic to a pragmatic policy after about 20 trials. (F) This panel shows how the belief-free component of the agent controller emerges over time. The belief-free component here corresponds to the expected transitions between the eight hidden states (columns are starting

Finally, research on category learning has described a related distinction between the competing systems: between a nonverbal, implicit system and a verbal, explicit system (Ashby & Maddox, 2011; Ashby, Alfonso-Reese, Turken, & Waldron, 1998). The former system exhibits many of the hallmarks of habitual behavior, including rapid responding and insensitivity to change, whereas the latter only emerges when there is a verbalizable rule available. Synthesizing and distinguishing these multiple taxonomies is a significant challenge for future work.

WHAT IS THE STRUCTURE OF THE GOAL-DIRECTED SYSTEM?

In the previous section, we reviewed schemes that replace model-free RL as the computational basis for habits. In this section, we consider the variety of possible schemes for the computational basis of goal-directed control, many of which merge model-based planning with model-free elements. This diversity results from the fact that optimal model-based control is impossible in realistic environments, because it would require enumerating and evaluating the full tree of possible future states, imposing computational costs that cannot be met by any physical system. Different schemes therefore represent different attempts to approximate optimal control without paying these costs (Daw & Dayan, 2014). A wide variety of such algorithms has been proposed, both within RL (Sutton & Barto, 1998) and in artificial intelligence more generally (Russell & Norvig, 2002), and the details of the algorithms used by the brain are only now beginning to be understood (Dolan & Dayan, 2013).

One strategy for reducing computational costs is to explore only parts of the search tree, whether using random rollouts (Kearns, Mansour, & Ng, 2002; Silver & Veness, 2010) or other heuristics to focus the search (Huys et al., 2015, 2012; Kocsis & Szepesvári, 2006). Evaluation of unexplored parts of the tree may be further assisted by cached values (Keramati et al., 2016). In general, these approaches entail a metacontrol problem governing how much of the tree to search (Baum, 2004; Simon, 1984). A closely related approach, termed DYNA, combines model-based and model-free RL. In this framework, a model-free value is incrementally updated through samples drawn from a world model (Gershman, Markman, & Otto, 2014; Sutton, 1991; Silver et al., 2008).

---

states; lines are end states), which are learned by "observing" one's own goal-directed behavior over time. In this particular example, an epistemic policy emerges as the agent progressively increases its expectation that it would perform a transition from state 1 (center—Context 1) to state 7 (bottom—Context 1) and from state 7 to state 3 (top-left—Context 1). When the confidence in these transitions is sufficient, the agent can shift from the belief-based scheme shown in panels (C, D) to a simpler belief-free scheme that does not require a generative model but only one of the matrices of expected transitions shown in panel (F). The latter correspond to habits and are insensitive to devaluation. Modified from Friston, et al. (2016).

Another approach to reducing the computational costs of model-based control, while remaining sensitive to at least some changes in task contingencies, is to adopt a predictive state representation, in which each state is associated with information about expected future states, but no explicit planning takes place (Dayan, 1993; Littman & Sutton, 2002). An agent using such a representation avoids many of the costs of model-based control but retains some of its flexibility, making it a plausible candidate for understanding some aspects of goal-directed behavior in humans (Momennejad et al., 2016; Russek, Momennejad, Botvinick, Gershman, & Daw, 2017).

What all these approaches have in common is that they typically do not make a sharp distinction between model-based and model-free learning. Rather, these algorithms seem to operate along a continuum, such that "model-basedness" itself forms a continuous dimension that varies in informational richness, from simple associations between states or responses and cached values to more complex associations that include information about specific outcomes and/or transition probabilities, or even conjunctive associations between states, responses, and outcomes (Alexander & Brown, 2011, 2015). This continuity between model-based and model-free algorithms stands in stark contrast to the sharp division between goal-directed and habitual mechanisms suggested by behavioral and neural data. This difference lends credence to accounts in which goal-directed control is implemented by a system with model-based and model-free aspects, whereas habitual control is implemented by separate mechanisms (Ashby et al., 2007; Dezfouli & Balleine, 2012; Friston et al., 2016; Miller et al., 2016; Pezzulo, Rigoli, & Friston, 2015; Topalidou et al., 2015).

WHAT WILL MAKE FOR A GOOD ACCOUNT OF HABITUAL AND GOAL-DIRECTED BEHAVIOR?

The previous sections have described various attempts to develop computational theories of habitual and goal-directed control, along with the challenges that each theory faces. Directly comparing the utility of these models, however, proves difficulty because of a critical limitation in this area: The different theories tend to address different aspects of the experimental literature. Indeed, it is not always clear when the various theories describe different, incompatible views of the processes by which behavior is controlled, and when they instead describe potentially compatible pieces of a larger picture which no theory yet fully encompasses. Therefore, rather than attempting such direct comparisons, in this section, we instead outline the major pieces of empirical data that future theories of habitual and goal-directed control should address.

Automaticity versus control

As discussed above, one classic criterion for distinguishing habits and goal-directed behaviors relates to the kind of action an animal takes after experiencing a degradation of contingencies or devaluation of outcomes following extended training. Habits, however, exhibit other behavioral hallmarks, including responses that are faster and more accurate (Graybiel, 2008; Wood & Rünger, 2016; e.g., Smith & Graybiel, 2013). With extended training, behavior also becomes more consistent, an observation documented most robustly in the literature on motor skill learning (Newell, 1991; Willingham, 1998; Wulf, Shea, & Lewthwaite, 2010). In other words, the inflexibility of habitual control trades off with gains in speed and consistency of action.

This set of observations collectively points to habits as being fundamentally more *automatic* than goal-directed actions. That is, relative to their goal-directed equivalents, habitual actions are selected faster and are less prone to interference from other ongoing tasks (Norman & Shallice, 1986; Posner & Snyder, 1975; Shiffrin & Schneider, 1977). This distinction was instrumental in classifying habits and goal-directed behaviors as exemplars of automatic/intuitive ("System 1") versus controlled/reflective ("System 2") processing within the dual-process literature (Wood, Labrecque, & Lin, 2014; cf. Evans, 2008). It is therefore difficult to describe a robust taxonomy of these behaviors without accounting for the relative differences in automaticity between them (Wood & Rünger, 2016) in addition to the kinds of choices an animal makes in a given setting.

Not only must a theory of these behaviors account for the automaticity of habits, it must also account for the controlled nature of goal-directed decision-making. The characteristics of decisions that involve increasing goal-directed deliberation suggest that such decisions benefit from cognitive control. For instance, increasingly goal-directed decisions are slower, more susceptible to interference from other ongoing processes, and are experienced as costly/effortful (Kool, Gershman, & Cushman, in press; Otto, Gershman, Markman & Daw, 2013; Otto, Raio, Chiang, Phelps, & Daw, 2013; Schwartz, 2004; see Chapter 7 by Kool et al. and Chapter 6 by Schmidt et al.). However, it is still unknown what type(s) of cognitive control goal-directed decisions rely on and what kinds of costs they incur. One prominent proposal suggests that goal-directed decision-making requires searching through an internal map of potential future states in order to identify the best possible future state (Kurth-Nelson, Bickel, & Redish, 2012). This search process requires selection from and maintenance of episodic and semantic memories, as well as instantiation of relevant contexts (see Chapter 6 by Schmidt et al.). The cost of goal-directed decision-making therefore may derive from the time and/or cognitive resources required for this search process (Kool et al., in press).

The development of habits

A foundational observation in the psychology of goal-directed and habitual control is that habits are slow to develop. The behavioral manifestations described above (speed, stereotypy, inflexibility, and resistance to interference) appear only after extended experience with a particular type of behavior (Adams, 1982; Dickinson, 1998; Wood & Rünger,

2016). Behavior in relatively novel environments tends to be slow, flexible, and vulnerable to distraction—the hallmarks of goal-directed control. A computational theory of goal-directed and habitual control must account for this shift, in which behavior begins under putatively goal-directed control, then over time becomes habitual. Such a theory must also account for the fact that habitization proceeds at different rates in different types of environments, for example, proceeding slowly in the case of "variable-ratio" reward schedules in which reward rate is directly proportional to the rate of performance of an action but proceeding very quickly in "variable-interval" reward schedules, which produce a reduced correlation between variability in behavior and variability in reward (Dickinson, Nicholas, & Adams, 1983; Miller et al., 2016).

Neural correlates of goals and habits

Computational theories of goal-directed and habitual control can be tested and constrained by neural data in at least two ways. First, predicted computational variables can be validated by observing corresponding neural correlates. All of the theories that we have outlined posit latent computational variables, such as the expected value associated with an action, or the associative strength between two stimuli, that are kept track of by the processes that govern behavior and that change over time. To the extent that a theory accurately describes computational mechanisms implemented by the brain, these latent variables are expected to have correlates in neural activity. Measurement of neural activity (e.g., single unit recordings, fMRI) during the performance of well-controlled tasks should reveal these neural correlates and could help adjudicate between competing models.

The second way in which neural data can inform theories of the type we consider here is by way of perturbation experiments. Specific perturbations to neural activity (e.g., lesions, pharmacology, optogenetics) have been shown to have specific effects on behavior in many tasks. A classic and robust example of this is found in the specific impairments in goal-directed or habitual control caused by lesions to specific regions of the striatum (Yin & Knowlton, 2006). A more recent example seems to demonstrate a specific role for dopamine in model-based control (see Box 18.2). A successful computational account of behavior must account for these causal mechanisms, perhaps by ascribing particular computational functions to particular structures or neurotransmitters.

Goal selection and pursuit

A rich psychological literature has characterized a set of distinct processes associated with goal-directed control, including committing to a goal (Oettingen, 2012), formulating a plan to achieve that goal (Wieber & Gollwitzer, 2017), pursuing that plan in the face of unexpected circumstances (Gollwitzer & Oettingen, 2012), and learning from one's success or failure to achieve the desired goal (Coricelli et al., 2005; Laciana & Weber, 2008).

Box 18.2 Do we need model-free control?

Many of the theories presented in the main text offer alternative computational mechanisms for habits that take the place of model-free reinforcement learning (RL). While these models allow for the possibility that model-free computations may still play a role in driving goal-directed behavior, this move nevertheless raises the question of whether stand-alone model-free algorithms are a necessary component of computational theories of decision-making.

Historically, the strongest support for model-free algorithms in the brain has come from a series of seminal studies demonstrating that firing rates of dopaminergic neurons in the midbrain show a response pattern evocative of the "prediction error" signal (Schultz, Dayan, & Montague, 1997), which plays a key computational role in model-free learning algorithms. These findings have given rise to the view that these neurons are part of a model-free control system, perhaps principally involving their strong projection to the striatum (Houk, Adams, & Barto, 1995). This picture has been complicated by recent evidence indicating that dopamine transients may be informed by model-based information (see also Chapter 11 by Sharpe and Schoenbaum). Dopamine neurons in a reversal learning task encode prediction errors that are consistent with inference (Bromberg-Martin, Matsumoto, Hong, & Hikosaka, 2010), while dopamine transients in humans encode information about both real and counterfactual rewards (Kishida et al., 2016). Perhaps most tellingly, dopamine neurons encode prediction errors indicative of model-based information (Sadacca, Jones, & Schoenbaum, 2016), and they even respond to errors in the predictions of sensory features that do not impact the value of the reward received (Takahashi et al., 2017).

These findings are congruent with a wealth of recent data from cognitive neuroscience, suggesting that dopamine plays a role in model-based rather than model-free control. Individual differences in dopamine receptor genotype correlate with the extent of model-based, but not model-free, influence on behavior (Doll et al. 2016). Model-based control is more dominant in human subjects with higher endogenous dopamine levels (Deserno et al., 2015), as well as those whose levels of dopamine have been artificially increased using drugs (Wunderlich, Smittenaar, & Dolan, 2012). Patients with Parkinson disease, in which midbrain dopaminergic neurons die in large numbers, show a smaller influence of model-based control on behavior, which is rescued if they take medication to restore systemic dopamine levels (Sharp, Foerde, Daw, & Shohamy, 2015).

This pattern of results casts doubt on the idea that dopamine neurons signal a model-free prediction error in the service of a model-free control system. The idea that such a dopaminergic model-free system underlies habitual control is further undermined by data showing that patients with Parkinson disease are relatively unimpaired at learning habits relative to goal-directed control (Hadj-Bouziane et al., 2012; de Wit, Barker, Dickinson, & Cools, 2011), and also that subjects whose levels of dopamine have been artificially depleted show fewer "slips of action," a behavioral measure of habit formation (de Wit et al., 2012). Taken together, these data suggest that dopamine may not be involved in specifically model-free computations and is unlikely to play a selective role in habitual control. More generally, these developments raise doubts about the widely accepted notion that the brain implements model-free RL algorithms, and they strongly motivate the search for alternative computational accounts of habitual behavior, such as those reviewed in this chapter.

Computational models have only begun to engage with this rich psychological phenomenology, for example, proposing that goal selection and goal pursuit map to two distinct computational processes with separate demands and neural underpinnings (O'Reilly, Hazy, Mollick, Mackie, & Herd, 2014). However, much remains to be understood concerning the psychological and phenomenological aspects of goal-directedness and the associated computational processes. These more elaborate psychological elements of goal seeking have also become an important topic for the nascent field of computational psychiatry (Montague, Dolan, Friston, & Dayan, 2012).

Goals and habits in ecological behaviors

Almost all previous theorizing about goal-directed versus habitual systems has focused on well-controlled laboratory experiments that manipulate a limited set of variables. Real-life situations, by contrast, invariably include a large state space and number of options, making some of the aforementioned strategies (e.g., exhaustive search or caching all state/action values) intractable. It has been variously proposed that dealing with real-life situations requires some form of approximation (e.g., approximate planning methods) as well as forms of abstraction, modularization, and/or hierarchization. These approximate solutions may reflect the structure inherent in the problem space (e.g., the fact that often real-life problems can be split into meaningful subproblems that can be solved one after the other, rather than solving the whole problem from start to end). Still, this is largely uncharted territory, and it is unclear whether one can find domain-general or domain-specific ways to address the full complexity of real-life situations.

The challenge of real-life complexity is particularly acute in this instance as goal-directed and habitual behavior are often distinguished by using laboratory manipulations (e.g., outcome devaluation) that emphasize aspects that are present in one condition but not the other. However, real-life situations, like shopping or planning a trip, tend to involve both habitual (e.g., stereotyped/scriptlike) components (e.g., going to the usual shop or train station) and novel challenges that have to be solved on the fly and thus need to engage a more deliberative form of reasoning (e.g., what to do if the shop is closed/the train is delayed). Some challenges posed by these tasks may be solved by reusing "cached" strategies or require some minimal form of generalization, whereas other challenges may require planning and deliberating de novo—hence, aspects of goal-directed and habitual control would plausibly need to be continuously and creatively integrated.

Recognizing that real-life behaviors are often hierarchically organized in this way, several recent proposals have adopted decision architectures that are themselves hierarchical. One such proposal comes from hierarchical RL, and proposes that behavior can be organized into abstract behavioral chunks (termed "options"), each aimed at reaching a given goal (Botvinick, 2012; Botvinick, Niv, & Barto, 2009; Sutton, Precup, & Singh, 1999). This framework provides a way to abstract away unnecessary details and plan behaviors in terms of subgoals and their associated plans/policies (e.g., go to the train station, then take the train, etc.). A related proposal suggests that the brain implements hierarchical probabilistic inference, which identifies the best ways to decompose a problem, simplifying the selection of subgoals (Balaguer, Spiers, Hassabis, & Summerfield, 2016; Donnarumma, Maisto, & Pezzulo, 2016; Maisto, Donnarumma, & Pezzulo, 2015). Other hierarchical schemes posit that simpler and more complex aspects of a plan (e.g., "go to the airport") depend on different hierarchical layers in the same network (Pezzulo et al., 2015; Pezzulo, Rigoli, & Friston, 2018). Integrating ideas like these into models of the interaction between goal-directed and habitual control offers a promising direction toward understanding the complexities of real-world behaviors.

Interactions between habitual and goal-directed control

Previous research has proposed several ways that separate habitual and goal-directed controllers could interact. One approach assumes that goal-directed and habitual mechanisms compete for control of behavior, with an arbitration mechanism that allocates control to one or another mechanism (e.g., Daw et al., 2005). The two controllers learn independently, and the arbitration mechanism can reflect the uncertainty or recent utility of each controller. As a result, behavior is alternately under control of one system or another at different times or in different contexts. This allows the agent to avoid the costs of running the goal-directed controller—whether in terms of precision (Daw et al., 2005), of time (Keramati, Dezfouli, & Piray, 2011), or of computational cost (Kool, Cushman, & Gershman, 2016)—in situations where it is not needed.

While goal-directed and habitual controllers are often modeled as learning in parallel, some proposals suggest a hierarchical organization. One family of proposals suggests that habits can be activated by goal-directed mechanisms (Aarts & Dijksterhuis, 2000), perhaps being composed of "chunked" sequences of behavior, which develop with experience (Dezfouli & Balleine, 2012; Dezfouli et al., 2014). A complementary family of proposals suggests that goals themselves may be activated by habits (Cushman & Morris, 2015; see also Chapter 7 by Kool et al.).

An alternative possibility is that both goal-directed and habitual behavior interact by forming part of a single controller, thereby cooperating to create a single integrated value. Such a cooperative architecture is incorporated into the "mixed instrumental controller" (Pezzulo, Rigoli, & Chersi, 2013). By default, this controller uses probabilistic priors on action or policy values to select action (i.e., a form of model-free control). The controller, however, also uses cost—benefit computations to decide when the prior is not sufficient. When the prior is insufficient, a second, model-based component is engaged to collect more evidence (by covertly resampling experience from the internal forward model of the task) before making a choice. This approach is closely related to the DYNA architecture (described above; Gershman et al., 2014; Sutton, 1991), in which simulated experience from a forward model is used to drive learning. Both of these schemes give rise to a continuum of choice patterns, which can stem from purely cached values or from a combination of these values and internal modeling: the more samples are drawn from the internal model, the more behavior will appear to be planned rather than model-free.

Another way to construct a continuum between habitual and goal-directed behavior is to posit that behavior at each moment results from a weighted sum of the influences of each system. One set of approaches involves an explicit arbiter that assigns weights adaptively (Lee et al., 2014; Miller et al., 2016). Another approach appeals to hierarchical predictive coding (Pezzulo et al., 2015). Here, goal-directed behavior arises when higher hierarchical layers produce long-term action predictions, and these predictions are used to "contextualize" shorter-term predictions at lower layers. Habitual behavior arises when lower layers acquire sufficient precision (a measure of inverse uncertainty in predictive coding) and become essentially insensitive to top-down messages. From this perspective, the continuum between goal-directed and habitual behavior depends on the relative strength (precision) of the top-down and bottom-up messages (predictions) in the hierarchical architecture, without an explicit arbiter.

The majority of these schemes were developed in the context of models that assert that habitual behavior relies on model-free mechanisms, and may be best suited to understanding the interactions between model-based and model-free control within the goal-directed system (see *What is the structure of goal-directed control?* section, above). Generalizing them to the case where habits are instantiated by other mechanisms remains an important direction for research.

CONCLUSIONS

A large body of evidence suggests that the brain contains separate mechanisms for goal-directed control, characterized as flexible but slow and effortful, and habitual control, characterized as inflexible but rapid and automatic (Balleine & O'Doherty, 2010; Dolan & Dayan, 2013; Yin & Knowlton, 2006). Recently, influential accounts have posited that goal-directed control is instantiated by model-based RL mechanisms, while habitual control is instantiated by model-free RL (Daw et al., 2005). These proposals have given a new theoretical foundation for investigations into the mechanisms of decision-making in general and supported new insights into many aspects of cognition, including addiction, morality, and other domains (Cushman, 2013; Lucantonio et al., 2014). At the same time, key tensions have become apparent between the model-based/model-free computational dichotomy and the theoretical and empirical literature.

Theoretical accounts of habits, both classic (Hull, 1943; James, 1890; Thorndike, 1911) and modern (Graybiel, 2008; Wood & Rünger, 2016), hold that habits are mediated by direct stimulus—response associations, which bypass any representation of expected outcome. Model-free RL, in contrast, depends critically on representations of expected value associated with each action. Empirically, the clear dissociations observed between neural structures involved in habitual and goal-directed behaviors have not been observed in tasks designed to differentiate model-based from model-free computations—instead, the regions involved have been largely overlapping (Doll et al., 2012). Together, these findings suggest that the model-based/model-free dichotomy may not map cleanly onto neural circuitry and that dominant computational models of goal-directed and habitual control may be in need of revision.

Here, we have reviewed a family of diverse alternative proposals that both arise from and engage with different portions of the literature. One element that many of these proposals share is severing the tie between habitual control and model-free RL, instead positing that habits are instantiated by an alternative computational mechanism (e.g., Dezfouli & Balleine, 2012; Friston et al., 2016; Miller et al., 2016). This realignment raises the question of whether model-free RL mechanisms in the brain are part of the goal-directed controller, or indeed whether such model-free mechanisms are at all necessary to explain human and animal behavior (see Box 18.2). More broadly, these newer proposals highlight important questions about the detailed structure of the goal-directed system and how that system resolves the inevitable trade-offs between performance and computational costs.

Finally, we have reviewed part of the broad empirical literature on goal-directed and habitual behaviors and suggested a set of phenomena that future work should seek to understand computationally. On the empirical side, this will mean developing new behavioral measures that allow for the examination of separate and interactive influences of habitual and goal-directed processes on behavior in complex environments, at different stages of habit development. On the computational side, it will be important to account not only for the observed behaviors and neural patterns within these experiments but also for the processes underlying learning and selection of habits and goal-directed behaviors, and for the real-world manifestations of these processes in both healthy and disordered populations. Such a convergence of efforts will no doubt help to adjudicate between and build on available models and work toward a full computational understanding of the neural mechanisms of goal-directed and habitual control.

REFERENCES

Aarts, H., & Dijksterhuis, A. (2000). Habits as knowledge structures: Automaticity in goal-directed behavior. *Journal of Personality and Social Psychology, 78*(1), 53—63.

Adams, C. D. (1982). Variations in the sensitivity of instrumental responding to reinforcer devaluation. *The Quarterly Journal of Experimental Psychology Section B, 34*(2), 77—98.

Alexander, W. H., & Brown, J. W. (2011). Medial prefrontal cortex as an action-outcome predictor. *Nature Neuroscience, 14*(10), 1338—1344.

Alexander, W. H., & Brown, J. W. (2015). Hierarchical error representation: A computational model of anterior cingulate and dorsolateral prefrontal cortex. *Neural Computation, 27*(11), 2354—2410.

Ashby, F. G., Alfonso-Reese, L. A., Turken, A. U., & Waldron, E. M. (1998). A neuropsychological theory of multiple systems in category learning. *Psychological Review, 105*(3), 442–481.

Ashby, F. G., Ennis, J. M., & Spiering, B. J. (2007). A neurobiological theory of automaticity in perceptual categorization. *Psychological Review, 114*(3), 632–656.

Ashby, F. G., & Maddox, W. T. (2011). Human category learning 2.0. *Annals of the New York Academy of Sciences, 1224*(April), 147–161.

Balaguer, J., Spiers, H., Hassabis, D., & Summerfield, C. (2016). Neural mechanisms of hierarchical planning in a virtual subway network. *Neuron, 90*(4), 893–903.

Balleine, B. W., & O'Doherty, J. P. (2010). Human and rodent homologies in action control: Corticostriatal determinants of goal-directed and habitual action. *Neuropsychopharmacology: Official Publication of the American College of Neuropsychopharmacology, 35*(1), 48–69.

Baum, E. B. (2004). *What is thought?* MIT Press.

Botvinick, M. M. (2012). Hierarchical reinforcement learning and decision making. *Current Opinion in Neurobiology, 22*(6), 956–962.

Botvinick, M. M., Niv, Y., & Barto, A. C. (2009). Hierarchically organized behavior and its neural foundations: A reinforcement learning perspective. *Cognition, 113*(3), 262–280.

Botvinick, M. M., & Toussaint, M. (2012). Planning as inference. *Trends in Cognitive Sciences, 16*(10), 485–488.

Bromberg-Martin, E. S., Matsumoto, M., Hong, S., & Hikosaka, O. (2010). A pallidus-habenula-dopamine pathway signals inferred stimulus values. *Journal of Neurophysiology, 104*(2), 1068–1076.

Buckholtz, J. W. (2015). Social norms, self-control, and the value of antisocial behavior. *Current Opinion in Behavioral Sciences, 3*(June), 122–129.

Coricelli, G., Critchley, H. D., Joffily, M., O'Doherty, J. P., Sirigu, A., & Dolan, R. J. (2005). Regret and its avoidance: A neuroimaging study of choice behavior. *Nature Neuroscience, 8*(9), 1255–1262.

Crockett, M. J. (2013). Models of morality. *Trends in Cognitive Sciences, 17*(8), 363–366.

Cushman, F. (2013). Action, outcome, and value: A dual-system framework for morality. *Personality and Social Psychology Review: An Official Journal of the Society for Personality and Social Psychology, Inc, 17*(3), 273–292.

Cushman, F., & Morris, A. (2015). Habitual control of goal selection in humans. *Proceedings of the National Academy of Sciences of the United States of America, 112*(45), 13817–13822.

Daw, N. D., & Dayan, P. (2014). The algorithmic anatomy of model-based evaluation. *Philosophical Transactions of the Royal Society of London. Series B, Biological Sciences, 369*(1655). https://doi.org/10.1098/rstb.2013.0478.

Daw, N. D., Gershman, S. J., Seymour, B., Dayan, P., & Dolan, R. J. (2011). Model-based influences on humans' choices and striatal prediction errors. *Neuron, 69*(6), 1204–1215.

Daw, N. D., Niv, Y., & Dayan, P. (2005). Uncertainty-based competition between prefrontal and dorsolateral striatal systems for behavioral control. *Nature Neuroscience, 8*(12), 1704–1711.

Daw, N. D., & O'Doherty, J. P. (2013). Multiple systems for value learning. In *Neuroeconomics: Decision making, and the brain*. princeton.edu. Retrieved from http://www.princeton.edu/~ndaw/do13.pdf.

Dayan, P. (1993). Improving generalization for temporal difference learning: The successor representation. *Neural Computation, 5*(4), 613–624.

de Wit, S., Barker, R. A., Dickinson, A. D., & Cools, R. (2011). Habitual versus goal-directed action control in Parkinson disease. *Journal of Cognitive Neuroscience, 23*(5), 1218–1229.

de Wit, S., Standing, H. R., Devito, E. E., Robinson, O. J., Ridderinkhof, K. R., Robbins, T. W., & Sahakian, B. J. (2012). Reliance on habits at the expense of goal-directed control following dopamine precursor depletion. *Psychopharmacology, 219*(2), 621–631.

Deserno, L., Huys, Q. J. M., Boehme, R., Buchert, R., Heinze, H.-J., Grace, A. A., ... Schlagenhauf, F. (2015). Ventral striatal dopamine reflects behavioral and neural signatures of model-based control during sequential decision making. *Proceedings of the National Academy of Sciences of the United States of America, 112*(5), 1595–1600.

Dezfouli, A., & Balleine, B. W. (2012). Habits, action sequences and reinforcement learning. *The European Journal of Neuroscience, 35*(7), 1036–1051.

Dezfouli, A., & Balleine, B. W. (2013). Actions, action sequences and habits: Evidence that goal-directed and habitual action control are hierarchically organized. *PLoS Computational Biology, 9*(12), e1003364.

Dezfouli, A., Lingawi, N. W., & Balleine, B. W. (2014). Habits as action sequences: Hierarchical action control and changes in outcome value. *Philosophical Transactions of the Royal Society of London. Series B, Biological Sciences, 369*(1655). https://doi.org/10.1098/rstb.2013.0482.

Dickinson, A. (1985). Actions and habits: The development of behavioural autonomy. *Philosophical Transactions of the Royal Society of London. Series B, Biological Sciences, 308*(1135), 67–78. The Royal Society.

Dickinson, A. (1998). Omission learning after instrumental pretraining. *The Quarterly Journal of Experimental Psychology Section B, 51*(3), 271–286. Routledge.

Dickinson, A., Nicholas, D. J., & Adams, C. D. (1983). The effect of the instrumental training contingency on susceptibility to reinforcer devaluation. *The Quarterly Journal of Experimental Psychology Section B, 35*(1), 35–51.

Dolan, R. J., & Dayan, P. (2013). Goals and habits in the brain. *Neuron, 80*(2), 312–325.

Doll, B. B., Bath, K. G., Daw, N. D., & Frank, M. J. (2016). Variability in dopamine genes dissociates model-based and model-free reinforcement learning. *The Journal of Neuroscience: The Official Journal of the Society for Neuroscience, 36*(4), 1211–1222.

Doll, B. B., Simon, D. A., & Daw, N. D. (2012). The ubiquity of model-based reinforcement learning. *Current Opinion in Neurobiology, 22*(6), 1075–1081.

Donnarumma, F., Maisto, D., & Pezzulo, G. (2016). Problem solving as probabilistic inference with subgoaling: Explaining human successes and pitfalls in the Tower of Hanoi. *PLoS Computational Biology, 12*(4), e1004864.

Evans, J. St B. T. (2008). Dual-processing accounts of reasoning, judgment, and social cognition. *Annual Review of Psychology, 59*, 255–278.

Fermin, A. S. R., Yoshida, T., Yoshimoto, J., Ito, M., Tanaka, S. C., & Doya, K. (2016). Model-based action planning involves cortico-cerebellar and basal ganglia networks. *Scientific Reports, 6*(August), 31378.

Friston, K., FitzGerald, T., Rigoli, F., Schwartenbeck, P., O'Doherty, J., & Pezzulo, G. (2016). Active inference and learning. *Neuroscience and Biobehavioral Reviews, 68*(September), 862–879.

Friston, K., FitzGerald, T., Rigoli, F., Schwartenbeck, P., & Pezzulo, G. (2017). Active inference: A process theory. *Neural Computation, 29*(1), 1–49.

Gershman, S. J., Markman, A. B., & Otto, A. R. (2014). Retrospective revaluation in sequential decision making: A tale of two systems. *Journal of Experimental Psychology: General, 143*(1), 182–194.

Gillan, C. M., Kosinski, M., Whelan, R., Phelps, E. A., & Daw, N. D. (2016). Characterizing a psychiatric symptom dimension related to deficits in goal-directed control. *eLife, 5*(March). https://doi.org/10.7554/eLife.11305.

Gillan, C. M., Otto, A. R., Phelps, E. A., & Daw, N. D. (2015). Model-based learning protects against forming habits. *Cognitive, Affective, & Behavioral Neuroscience, 15*(3), 523–536.

Gollwitzer, P. M., & Oettingen, G. (2012). Goal pursuit. In *The Oxford handbook of human motivation* (pp. 208–231).

Graybiel, A. M. (2008). Habits, rituals, and the evaluative brain. *Annual Review of Neuroscience, 31*(1), 359–387.

Gremel, C. M., & Costa, R. M. (2013). Orbitofrontal and striatal circuits dynamically encode the shift between goal-directed and habitual actions. *Nature Communications, 4*, 2264.

Hadj-Bouziane, F., Benatru, I., Brovelli, A., Klinger, H., Thobois, S., Broussolle, E., ... Meunier, M. (2012). Advanced Parkinson's disease effect on goal-directed and habitual processes involved in visuomotor associative learning. *Frontiers in Human Neuroscience, 6*, 351.

Hammond, L. J. (1980). The effect of contingency upon the appetitive conditioning of free-operant behavior. *Journal of the Experimental Analysis of Behavior, 34*(3), 297–304.

Hélie, S., Roeder, J. L., Vucovich, L., Rünger, D., & Ashby, F. G. (2015). A neurocomputational model of automatic sequence production. *Journal of Cognitive Neuroscience, 27*(7), 1412–1426.

Hikosaka, O., Ghazizadeh, A., Griggs, W., & Amita, H. (February 2017). Parallel basal ganglia circuits for decision making. *Journal of Neural Transmission*. https://doi.org/10.1007/s00702-017-1691-1.

Houk, J. C., Adams, J. L., & Barto, A. G. (1995). A model of how the basal ganglia generate and use neural signals that predict reinforcement. In J. C. Houk, J. L. Davis, & D. G. Beiser (Eds.), *Models of information processing in the basal ganglia* (pp. 249—270). MIT Press.

Hull, C. L. (1943). *Principles of behavior: An introduction to behavior theory*. Appleton-Century. http://doi.apa.org/psycinfo/1944-00022-000.

Huys, Q. J. M., Eshel, N., O'Nions, E., Sheridan, L., Dayan, P., & Roiser, J. P. (2012). Bonsai trees in your head: How the pavlovian system sculpts goal-directed choices by pruning decision trees. *PLoS Computational Biology, 8*(3), e1002410.

Huys, Q. J. M., Lally, N., Faulkner, P., Eshel, N., Seifritz, E., Gershman, S. J., ... Roiser, J. P. (2015). Interplay of approximate planning strategies. *Proceedings of the National Academy of Sciences of the United States of America, 112*(10), 3098—3103.

James, W. (1890). *The principles of psychology*. NY, US: Henry Holt and Company.

Kaelbling, L. P., Littman, M. L., & Cassandra, A. R. (1998). Planning and acting in partially observable stochastic domains. *Artificial Intelligence, 101*(1), 99—134.

Kaelbling, L. P., Littman, M. L., & Moore, A. W. (1996). Reinforcement learning: A survey. *Journal of Artificial Intelligence*. jair.org. Retrieved from http://www.jair.org/papers/paper301.html.

Kearns, M., Mansour, Y., & Ng, A. Y. (2002). A sparse sampling algorithm for near-optimal planning in large Markov decision processes. *Machine Learning, 49*(2—3), 193—208. Kluwer Academic Publishers.

Keramati, M., Dezfouli, A., & Piray, P. (2011). Speed/accuracy trade-off between the habitual and the goal-directed processes. *PLoS Computational Biology, 7*(5), e1002055.

Keramati, M., Smittenaar, P., Dolan, R. J., & Dayan, P. (October 2016). Adaptive integration of habits into depth-limited planning defines a habitual-goal-directed spectrum. *Proceedings of the National Academy of Sciences of the United States of America*. https://doi.org/10.1073/pnas.1609094113.

Killcross, S., & Coutureau, E. (2003). Coordination of actions and habits in the medial prefrontal cortex of rats. *Cerebral Cortex, 13*(4), 400—408.

Kishida, K. T., Saez, I., Lohrenz, T., Witcher, M. R., Laxton, A. W., Tatter, S. B., ... Montague, P. R. (2016). Subsecond dopamine fluctuations in human striatum encode superposed error signals about actual and counterfactual reward. *Proceedings of the National Academy of Sciences of the United States of America, 113*(1), 200—205. National Acad Sciences.

Kocsis, L., & Szepesvári, C. (2006). Bandit based Monte-Carlo planning. In , *Lecture notes in computer scienceMachine learning: ECML 2006* (pp. 282—293). Berlin, Heidelberg: Springer.

Kool, W., Cushman, F. A., & Gershman, S. J. (2016). When does model-based control pay off? *PLoS Computational Biology, 12*.

Kool, W., Gershman, S. J., & Cushman, F. A. (in press). Planning complexity registers as a cost in metacontrol. Journal of Cognitive Neuroscience.

Kurth-Nelson, Z., Bickel, W., & Redish, A. D. (2012). A theoretical account of cognitive effects in delay discounting. *The European Journal of Neuroscience, 35*(7), 1052—1064.

Laciana, C. E., & Weber, E. U. (2008). Correcting expected utility for comparisons between alternative outcomes: A unified parameterization of regret and disappointment. *Journal of Risk and Uncertainty, 36*(1), 1—17. Springer US.

Lee, S. W., Shimojo, S., & O'Doherty, J. P. (2014). Neural computations underlying arbitration between model-based and model-free learning. *Neuron, 81*(3), 687—699.

Littman, M. L. (2015). Reinforcement learning improves behaviour from evaluative feedback. *Nature, 521*(7553), 445—451.

Littman, M. L., & Sutton, R. S. (2002). Predictive representations of state. In T. G. Dietterich, S. Becker, & Z. Ghahramani (Eds.), *Advances in neural information processing systems 14* (pp. 1555—1561). MIT Press.

Lucantonio, F., Caprioli, D., & Schoenbaum, G. (2014). Transition from 'model-based' to 'model-free' behavioral control in addiction: Involvement of the orbitofrontal cortex and dorsolateral striatum. *Neuropharmacology, 76*(Pt B (January)), 407—415.

Maisto, D., Donnarumma, F., & Pezzulo, G. (2015). Divide et impera: Subgoaling reduces the complexity of probabilistic inference and problem solving. *Journal of the Royal Society Interface, 12*(104), 20141335. The Royal Society.

Miller, K., Shenhav, A., & Ludvig, E. (in press). Habits without values. *Psychological Review.*

Momennejad, I., Russek, E. M., Cheong, J. H., Botvinick, M. M., Daw, N., & Gershman, S. J. (2016). The successor representation in human reinforcement learning. *bioRxiv.* https://doi.org/10.1101/083824.

Montague, P. R., Dolan, R. J., Friston, K. J., & Dayan, P. (2012). Computational psychiatry. *Trends in Cognitive Sciences, 16*(1), 72–80.

Newell, K. M. (1991). Motor skill acquisition. *Annual Review of Psychology, 42,* 213–237.

Norman, D. A., & Shallice, T. (1986). Attention to action. In *Consciousness and self-regulation* (pp. 1–18). Boston, MA: Springer.

Oettingen, G. (2012). Future thought and behaviour change. *European Review of Social Psychology, 23*(1), 1–63. Routledge.

Otto, A. R., Gershman, S. J., Markman, A. B., & Daw, N. D. (2013). The curse of planning: Dissecting multiple reinforcement-learning systems by taxing the central executive. *Psychological Science, 24*(5), 751–761.

Otto, A. R., Raio, C. M., Chiang, A., Phelps, E. A., & Daw, N. D. (2013). Working-memory capacity protects model-based learning from stress. *Proceedings of the National Academy of Sciences of the United States of America, 110*(52), 20941–20946.

Ouellette, J. A., & Wood, W. (1998). Habit and intention in everyday life: The multiple processes by which past behavior predicts future behavior. *Psychological Bulletin, 124*(1), 54. American Psychological Association.

O'Reilly, R. C., Hazy, T. E., Mollick, J., Mackie, P., & Herd, S. (2014). Goal-driven cognition in the brain: A computational framework. *arXiv [q-bio.NC].* arXiv. Retrieved from http://arxiv.org/abs/1404.7591.

Pezzulo, G., Rigoli, F., & Chersi, F. (2013). The mixed instrumental controller: Using value of information to combine habitual choice and mental simulation. *Frontiers in Psychology, 4*(March), 92.

Pezzulo, G., Rigoli, F., & Friston, K. (2015). Active inference, homeostatic regulation and adaptive behavioural control. *Progress in Neurobiology, 134*(November), 17–35.

Pezzulo, G., Rigoli, F., & Friston, K. (2018). Hierarchical active inference: A theory of motivated control. *Trends in Cognitive Sciences, 22.*

Posner, M. I., & Snyder, C. R. R. (1975). Attention and cognitive control. In R. L. Solso (Ed.), *Information processing and cognition: The Loyola symposium.*

Rangel, A. (2013). Regulation of dietary choice by the decision-making circuitry. *Nature Neuroscience, 16*(12), 1717–1724.

Russek, E. M., Momennejad, I., Botvinick, M. M., Gershman, S. J., & Daw, N. D. (2017). Predictive representations can link model-based reinforcement learning to model-free mechanisms. *PLoS Computational Biology, 13*(9), e1005768.

Russell, S., & Norvig, P. (2002). *Artificial intelligence: A modern approach (International edition).* Pearson US Imports & PHIPEs.

Sadacca, B. F., Jones, J. L., & Schoenbaum, G. (2016). Midbrain dopamine neurons compute inferred and cached value prediction errors in a common framework. *eLife, 5.* https://doi.org/10.7554/eLife.13665.

Schultz, W., Dayan, P., & Montague, P. R. (1997). A neural substrate of prediction and reward. *Science, 275*(5306), 1593–1599.

Schwartz, B. (2004). *The paradox of choice: Why less is more.* New York: Ecco.

Sharp, M. E., Foerde, K., Daw, N. D., & Shohamy, D. (2015). Dopamine selectively remediates 'model-based' reward learning: A computational approach. *Brain: A Journal of Neurology,* awv347. Oxford Univ Press.

Shenhav, A., Musslick, S., Lieder, F., Kool, W., Griffiths, T. L., Cohen, J. D., & Botvinick, M. M. (2017). Toward a rational and mechanistic account of mental effort. *Annual Review of Neuroscience, 40*(July), 99–124.

Shiffrin, R. M., & Schneider, W. (1977). Controlled and automatic human information processing: II. Perceptual learning, automatic attending and a general theory. *Psychological review, 84*(2), 127.

Silver, D., Sutton, R. S., & Müller, M. (2008). Sample-based learning and search with permanent and transient memories. In *Proceedings of the 25th international conference on machine learning, ICML '08* (pp. 968–975). New York, NY, USA: ACM.

Silver, D., & Veness, J. (2010). Monte-carlo planning in large POMDPs. In J. D. Lafferty, C. K. I. Williams, J. Shawe-Taylor, R. S. Zemel, & A. Culotta (Eds.), *Advances in neural information processing systems 23* (pp. 2164–2172). Curran Associates, Inc.

Simon, H. A. (1984). *Models of bounded rationality: Economic analysis and public policy*. MIT Press.

Smith, K. S., & Graybiel, A. M. (2013). A dual operator view of habitual behavior reflecting cortical and striatal dynamics. *Neuron, 79*(2), 361–374.

Sutton, R. S. (1991). Dyna, an integrated architecture for learning, planning, and reacting. *SIGART Bulletin, 2*(4), 160–163. New York, NY, USA: ACM.

Sutton, R. S., & Barto, A. G. (1981). Toward a modern theory of adaptive networks: Expectation and prediction. *Psychological Review, 88*(2), 135–170.

Sutton, R. S., & Barto, A. G. (1998). *Reinforcement learning: An introduction* (Vol. 1). Cambridge: MIT Press.

Sutton, R. S., Precup, D., & Singh, S. (1999). Between MDPs and semi-MDPs: A framework for temporal abstraction in reinforcement learning. *Artificial Intelligence, 112*(1), 181–211.

Takahashi, Y. K., Batchelor, H. M., Liu, B., Khanna, A., Morales, M., & Schoenbaum, G. (2017). Dopamine neurons respond to errors in the prediction of sensory features of expected rewards. *Neuron, 95*(6), 1395–1405.e3.

Thorndike, E. L. (1911). *Animal intelligence: Experimental studies*. Macmillan.

Tolman, E. C. (1948). Cognitive maps in rats and men. *Psychological Review, 55*(4), 189–208.

Topalidou, M., Kase, D., Boraud, T., & Rougier, N. P. (2015). The formation of habits in the neocortex under the implicit supervision of the basal ganglia. *BMC Neuroscience, 16*(1), P212.

van den Broek, B., Wiegerinck, W., & Kappen, B. (2010). Risk sensitive path integral control. In P. Grünwald, & P. Spirtes (Eds.), *Proceedings of the 26th conference on uncertainty in artificial intelligence*. AUAI Press.

Vandaele, Y., & Janak, P. H. (June 2017). Defining the place of habit in substance use disorders. *Progress in Neuro-Psychopharmacology & Biological Psychiatry*. https://doi.org/10.1016/j.pnpbp.2017.06.029.

Wieber, F., & Gollwitzer, P. M. (2017). Planning and the control of action. In , *Knowledge and spaceKnowledge and action* (pp. 169–183). Cham: Springer.

Willingham, D. B. (1998). A neuropsychological theory of motor skill learning. *Psychological Review, 105*(3), 558–584.

Wimmer, G. E., Daw, N. D., & Shohamy, D. (2012). Generalization of value in reinforcement learning by humans. *The European Journal of Neuroscience, 35*(7), 1092–1104.

Wood, W, Labrecque, J. S., Lin, P. Y., & Runger, D. (2014). Habits in dual process models. In JW Sherman, B Gawronski, & Y Trope (Eds.), *Dual Process Theories of the Social Mind* (pp. 371–385). New York: Guilford.

Wood, W., & Neal, D. T. (2007). A new look at habits and the habit-goal interface. *Psychological Review, 114*(4), 843–863.

Wood, W., & Rünger, D. (2016). Psychology of habit. *Annual Review of Psychology, 67*, 289–314.

Wulf, G., Shea, C., & Lewthwaite, R. (2010). Motor skill learning and performance: A review of influential factors. *Medical Education, 44*(1), 75–84.

Wunderlich, K., Dayan, P., & Dolan, R. J. (2012). Mapping value based planning and extensively trained choice in the human brain. *Nature Neuroscience, 15*(5), 786–791.

Wunderlich, K., Smittenaar, P., & Dolan, R. J. (2012). Dopamine enhances model-based over model-free choice behavior. *Neuron, 75*(3), 418–424.

Yin, H. H., & Knowlton, B. J. (2006). The role of the basal ganglia in habit formation. *Nature Reviews Neuroscience, 7*(6), 464–476.

CHAPTER 19

The Motivation of Action and the Origins of Reward

Bernard W. Balleine
Decision Neuroscience Lab, School of Psychology, UNSW Sydney, Kensington, NSW, Australia

INTRODUCTION

This chapter is divided into two parts: The first surveys the currently accepted view of goal-directed actions, their neural bases, and their boundary conditions in non-goal-directed actions or habits particularly with regard to the reward- and reinforcement-related feedback processes that guide their acquisition. The second part focuses more specifically on the structure of these feedback processes, as it emerges from consideration of general theories of motivation.

From a neural perspective, although there is a long and extensive literature linking executive functions, such as goal-directed action, to the prefrontal cortex (Fuster, 2000; Goldman-Rakic, 1995), more recent studies suggest that these functions depend on reward-related circuitry linking the prefrontal, premotor, and sensorimotor cortices with the striatum (Dolan & Dayan, 2013; Hikosaka, Kim, Yasuda, & Yamamoto, 2014). Evidence from a range of species suggests that this corticostriatal network controls functionally heterogeneous processes involving the following: (1) actions that are flexible or goal-directed, sensitive to rewarding feedback and mediated by discrete regions of association cortices particularly medial, orbitomedial, premotor, and anterior cingulate cortices together with their targets in caudate/dorsomedial striatum (DMS) and (2) actions that are stimulus bound, relatively automatic or habitual and mediated by sensorimotor cortices and dorsolateral striatum (DLS)/putamen (cf. Balleine & O'Doherty for review). Indeed, changes in basal ganglia function produced by neurodegenerative disorders, stroke, injury, or disease can produce pathological changes in action control (Lee, 2013), and the nature of these changes depends on the locus of the damage. Damage to the associative, DMS (or the caudate nucleus) produces deficits in volitional or goal-directed action and can produce intrusive, involuntary, or compulsive actions (Chambers, Garavan, & Bellgrove, 2009; Levy & Dubois, 2006), whereas damage to the motor, DLS (or putamen) produces a loss of skilled or habitual motor movement (Pramstaller & Marsden, 1996) and can produce intrusive thoughts and cognitive demands, deficits in attention, and concentrated effort (Ellison-Wright, Ellison-Wright, & Bullmore, 2008).

Goal-Directed Decision Making
ISBN 978-0-12-812098-9, https://doi.org/10.1016/B978-0-12-812098-9.00019-X

Importantly, these two forms of action control have been argued to depend on distinct learning rules controlled by different forms of feedback; i.e., "reward" in the case of goal-directed action and "reinforcement" in the case of habits (Balleine, Liljeholm, & Ostlund, 2009). Although, historically, terms like "reward" and "reinforcement" were thought to refer to synonymous processes, recent research has found evidence of significant differences in these forms of feedback at both a behavioral and a neural level. This research is described in the first part of this chapter, whereas the second part will consider the motivational bases of these feedback processes more generally.

ACTIONS AND HABITS

It is now widely accepted that choice between different actions, e.g., pressing a lever or pulling a chain when these actions earn different food rewards, is determined by an animal encoding the association between its actions and their specific consequences and the relative value of those consequences; thus, choice is sensitive both to degradation of the action—outcome contingency and to outcome revaluation treatments (Balleine & Dickinson, 1998a). In contrast, when actions are *overtrained*, decision processes become more rigid or habitual and performance is no longer sensitive to degradation and devaluation treatments and is controlled by a process of sensorimotor association. Indeed, whereas action—outcome encoding appears to be mediated by a form of error correction learning, the development of habits is determined by event contiguity rather than contingency (Dickinson, 1994).

Assessments of the detailed circuitry associated with these forms of decision process in rodents have found that cell body lesions of the prefrontal cortex, particularly the prelimbic (Balleine & Dickinson, 1998a; Corbit & Balleine, 2003), but not the infralimbic (Killcross & Coutureau, 2003), region abolish the acquisition of goal-directed actions resulting in actions being controlled by a process of sensorimotor association alone. Prelimbic involvement in goal-directed learning is, however, limited to a period early in acquisition (Ostlund & Balleine, 2005; Tran-Tu-Yen, Marchand, Pape, Di Scala, & Coutureau, 2009) suggesting that the learning-related plasticity mediating goal-directed action is localized to an efferent structure. The prelimbic cortex projects densely to both DMS and accumbens core (Gabbott, Warner, Jays, Salway, & Busby, 2005) and, although the latter region appears not to be involved (Corbit, Muir, & Balleine, 2001), evidence has consistently implicated the DMS in goal-directed learning. Thus, within the posterior part of dorsomedial striatum (pDMS) both pre- and posttraining lesions, muscimol-induced inactivation, the infusion of the NMDA-antagonist AP-5, or drugs that block the phosphorylation of kinases associated with plasticity, such as mitogen-activated protein kinase (MAPkinase), all abolish goal-directed learning and render choice performance insensitive to both contingency degradation and outcome devaluation treatments and so leave choice habitual (Shiflett & Balleine, 2011; Yin, Knowlton, & Balleine, 2005; Yin, Ostlund, Knowlton, & Balleine, 2005).

Importantly, evidence suggests that a parallel corticostriatal circuit involving DLS together with the sensorimotor and infralimbic cortices (ILs) mediates habitual actions (Barnes, Kubota, Hu, Jin, & Graybiel, 2005; Yin, Knowlton, & Balleine, 2004). However, whereas the IL has been argued to mediate aspects of the reinforcement signal controlling sensorimotor association (Balleine & Killcross, 2006), changes in DLS appear to be training related (Costa, Cohen, & Nicolelis, 2004; Tang, Pawlak, Prokopenko, & West, 2007) and to be coupled to changes in plasticity, as behavioral processes become less flexible (Shan, Christie, & Balleine, 2015). Correspondingly, whereas overtraining causes performance to become insensitive to outcome devaluation, lesions of DLS reverse this effect rendering performance sensitive to devaluation treatments (Yin et al., 2004). Likewise, muscimol-induced inactivation of DLS renders otherwise habitual actions goal-directed (Yin, Knowlton, & Balleine, 2006) consistent with the claim that distinct corticostriatal networks control different forms of action control (Balleine & O'Doherty, 2010). This circuitry is illustrated in Fig. 19.1A.

Reward and reinforcement

An important way in which these control processes differ lies in the feedback that they use to update the specific learning rules controlling goal-directed and habitual actions. Whereas goal-directed actions depend on feedback associated with an outcome-mediated **reward** signal derived from the emotional significance of the goal, habitual actions depend on an affective **reinforcement** signal via which a specific stimulus—response (S-R) association is strengthened (Dickinson, 1994).

The essential psychological distinction between reward and reinforcement lies in their consequences: In common usage, rewards are things that are given in recognition of effort or achievement. Rewards are, therefore, given for actions over which some control can be exerted: that can be increased or decreased, withheld, released, or modified in order to achieve *a specified change in the environment that satisfies some need or desire*. In contrast, reinforcers do not encourage actions that alter the environment in specified ways but rather they work to *reinforce or stamp in a particular motor movement as a response to a specific situation or stimulus*. These elicited responses are not discretionary but reflexive, and rewarding them would be like rewarding a brick for standing still or a leaf for falling from a tree.

Movements tend to be repeated if they generate a signal capable of strengthening their association with antecedent stimuli. It has been interesting to consider the nature of this signal; whether it reflects a change in affective tone, such as a contiguous increase in positive affect (or satisfaction as envisaged within Thorndike's "law of effect" (Thorndike, 1898)), or a precise phasic change related to the predictive status of the

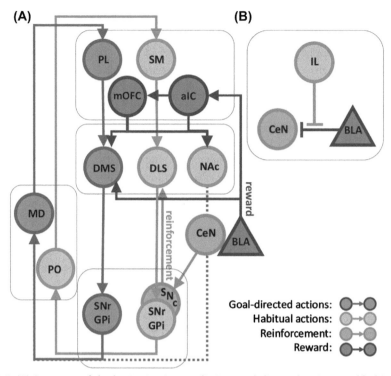

Figure 19.1 (A) Summary of the basic circuitry mediating goal-directed actions and habits and their feedback processes associated with reward and reinforcement, respectively. (B) Summary of the role of IL in the disinhibition of the BLA inhibition of CeN the substrate mediating the interaction of reward and reinforcement processes. IL manipulations should, therefore, affect habit learning and performance to the extent that these depend on the interaction between the BLA and CeN. See text for details. *alC,* anterior insular cortex; *BLA,* basolateral amygdala; *CeN,* amygdala central nucleus; *DLS,* dorsolateral striatum; *DMS,* dorsomedial striatum; *GPi,* globus pallidus interna; *IL,* infralimbic cortices; *MD,* mediodorsal thalamus; *mOFC,* medial orbitofrontal cortex; *NAc,* nucleus accumbens core; *PL,* prelimbic cortex; *PO,* posterior thalamus; *SM,* sensorimotor cortex; *SNc,* substantia nigra pars compacta; *SNr,* substantia nigra pars reticulata.

stimulus. In the latter case, which is commonly supposed by reinforcement learning accounts of habits (Daw, Niv, & Dayan, 2005), unpredicted events productive of a positive affective response, such as food for a food–deprived rat, generate a strong phasic response, whereas well-predicted food generates a weaker or zero phasic response (Schultz & Dickinson, 2000). This latter view has the virtue of predicting that the associative strength of the S-R association will asymptote; as the environment comes to predict food, the error signal will decline, meaning the S-R association will reach a point after which it will no longer be strengthened. However, experiments that have altered the predictive status of these stimulus conditions, for example, by only delivering food when a response is withheld on an omissions schedule (Davis & Bitterman, 1971), have found that such manipulations do not alter the performance of habits (Dickinson, 1998). Although reversing the contingency should provoke a strong prediction error, habits do not adjust to this variation in contingency, although such variations strongly reduce the performance of goal-directed actions. It seems unlikely, therefore, that the reinforcement signal can be reduced to an error signal (Dezfouli & Balleine, 2012).

In contrast, research into the nature of reward has revealed one of the most striking properties of goal-directed actions: that they are controlled by the *experienced value* of the outcome generated by some specified action or other. Unlike habits or other reflexes, therefore, goal-directed actions are not directly affected by the immediate energetic and affective consequences of shifts in primary motivation and are only affected when the influence of those changes on outcome value is directly experienced (Dickinson, Balleine, Watt, Gonzalez, & Boakes, 1995—See Dickinson, Chapter 1, for a more complete description of the motivational control of habits). For example, hungry rats trained to perform two actions, one delivering a solid food, the other a liquid food, will not alter their choice between actions when made thirsty until they have first consumed the liquid food in the thirsty motivational state and discovered, based on their experienced emotional responses, that it is valuable in that state (Dickinson & Dawson, 1988). This is also true of shifts between different levels of deprivation; rodents do not immediately choose an action associated with a more (or less) calorific outcome when deprivation is increased (or decreased) (Balleine, 1992; Dickinson & Balleine, 1994). Rather they have first to learn about the new reward value of the outcome before the shift in motivation will influence performance, an effect we have called *incentive learning.*

Incentive learning is a ubiquitous feature of changes in reward value and has been found to influence changes in choice after (1) taste aversion-induced outcome devaluation (Balleine & Dickinson, 1991); (2) specific satiety-induced outcome devaluation (Balleine & Dickinson, 1998b); (3) shifts from water deprivation to satiety (Lopez, Balleine, & Dickinson, 1992); (4) changes in outcome value mediated by drug states (Balleine, Ball, & Dickinson, 1994; Furlong, Supit, Corbit, Killcross, & Balleine, 2017; Ostlund, Maidment, & Balleine, 2010); (5) changes in the value of thermoregulatory rewards (Hendersen & Graham, 1979); (6) sexual rewards (Everitt, 1990; Woodson & Balleine, 2002); and (7) changes in the value of heroin in a state of withdrawal (Hutcheson, Everitt, Robbins, & Dickinson, 2001) (see Balleine, 2005 for review).

Neural feedback processes
Reward
Incentive learning is required for changes in reward value because the proximal cause of such changes is the integration of the sensory properties of the rewarding event with a change in emotional feedback, i.e., the emotional response contiguous with detection of the rewarding event (Balleine, 2001, 2005). Considerable evidence suggests that the basolateral amygdala (BLA) plays a critical role in this integrative process: Thus, lesions

(Balleine, Killcross, & Dickinson, 2003), inactivation (Ostlund & Balleine, 2008), μ-opiate receptor blockade (Wassum, Ostlund, Maidment, & Balleine, 2009), or protein synthesis inhibition (Wang, Ostlund, Nader, & Balleine, 2005) in the BLA all render rats insensitive to changes in the reward value of an outcome. However, it has become clear that, although the BLA plays a key role in encoding changes in reward value, it does not itself store those changes which, current evidence suggests, involves the anterior insular cortex, a structure strongly connected with the BLA (Parkes & Balleine, 2013; Wassum et al., 2009)—see Fig. 19.1A.

For example, we, Parkes and Balleine (2013) found that, in hungry rats trained to press two levers for distinct outcomes, whereas inactivation of the BLA prior to specific satiety-induced outcome devaluation abolished the effect of the change in outcome value on choice in an extinction test, inactivation after the outcome devaluation treatment did not. In contrast, inactivation of the insular cortex, whether conducted prior to or after the devaluation manipulation, abolished the effect of outcome devaluation on choice. We hypothesized that these effects of insular inactivation suggest that it is involved in retrieving *changes* in outcome value to guide choice and, to test this, we used an asymmetrical, temporal inactivation procedure. Hungry rats were again trained to press two levers for distinct outcomes prior to specific satiety-induced outcome devaluation and a choice extinction test. However, prior to the devaluation, the BLA in one hemisphere was inactivated leaving the contralateral BLA free to become involved in encoding the change in value. Then, after the devaluation treatment, we inactivated the insular cortex contralateral to the BLA inactivation. If the change in value involves BLA inputs to the insular cortex then this order of the inactivation treatments should abolish sensitivity to outcome devaluation on test. A control group received the reverse treatment, i.e., inactivation of the insular cortex in one hemisphere prior to devaluation and of the contralateral BLA prior to test; a treatment that should allow the insular cortex in one hemisphere to encode the change in value and so alter performance on test. This is precisely what we found; the group that had the BLA inactivation prior to the insular inactivation showed no devaluation on test, whereas the control group showed a significant devaluation effect in their choice performance (Parkes & Balleine, 2013). This result provides strong evidence for the argument that the BLA is involved in changes in reward value because of its influence on encoding those changes in the anterior insular.

We have also found that the BLA—insular cortex encoding of the change in value interacts with regions of the striatum to mediate changes in performance after outcome devaluation. So, for example, disconnection of the insular cortex from the nucleus accumbens core (NAco) after outcome devaluation abolishes changes in choice performance on test (Parkes, Bradfield, & Balleine, 2015). Likewise, lesion-induced disconnection of the BLA from the pDMS has been found to abolish sensitivity to outcome devaluation, whether conducted prior to or after training (and so prior to devaluation in both cases) (Corbit, Leung, & Balleine, 2013). Thus, as described above, prior research has established that bilateral inactivation of either the pDMS or the BLA, using the GABA-A agonist muscimol, reduced the sensitivity of actions to outcome devaluation (Ostlund & Balleine, 2008; Yin, Ostlund, et al., 2005). We have found the same is true after a unilateral lesion of the BLA and inactivation of either the contralateral BLA or the contralateral pDMS. Rats trained to press two levers for different outcomes then had one of the two outcomes devalued by specific satiety before a choice extinction test conducted on the levers. Control groups given a unilateral BLA lesion plus infusion of saline vehicle into either the BLA or pDMS prior to devaluation showed normal outcome devaluation: The rats reduced their choice of the action that in training delivered the now-devalued outcome compared to the other action. In contrast, devaluation was abolished by inactivation of contralateral BLA or, more importantly, by inactivation of contralateral pDMS prior to devaluation since the latter treatment functionally disconnects the BLA from the pDMS (Corbit et al., 2013).

These experiments demonstrate that a circuit involving the BLA, the insular cortex, and the NAco interacts with the pDMS to control the way changes in outcome value affect goal-directed action. These nodes form an integrated circuit that links the dorsal stream encoding the learning processes through which goal-directed actions are encoded, with a ventral stream encoding the reward values that control the performance of such actions (see Hart, Leung, & Balleine, 2014 for a review of this thesis).

There is one final node in this network that deserves mention and that is the medial orbitofrontal cortex (mOFC). Although the immediately adjacent cortical regions—the prelimbic cortex, the IL, and the ventral and lateral orbital cortices—play no role in the way *changes* in outcome value affect goal-directed action, when animals have to retrieve changes in value from memory then lesions or inactivation specifically of the mOFC blocks that capacity (Bradfield, Dezfouli, van Holstein, Chieng, & Balleine, 2015). What is the role of the medial orbital cortex relative, say, to the anterior insular region? We have previously found evidence that after the reward value of an outcome has been encoded, applying treatments that would otherwise affect that value were the outcome contacted have no impact on the way the value is retrieved on test. For example, cholecystokinin can serve as a satiety factor and animals given cholecystokinin prior to reexposure to food when they are hungry fail to revalue that food, as revealed in a subsequent choice test. However, conducting the test under cholecystokinin does not affect the value that is retrieved on test; i.e., the effect of cholecystokinin on the increase in value of a specific food during reexposure when hungry does not extend to the retrieval of other foods on test, *suggesting that the mechanisms involved in encoding experienced values and retrieving those values differ* (Balleine & Dickinson, 1994; Balleine, Davies, & Dickinson, 1995). One possibility is that the concrete changes in value induced by direct experience and involving the anterior insular cortex engender a further abstract encoding of those changes in value in the mOFC via the cortico-cortical connections it maintains with that structure. This suggests that the way in which the BLA-insular circuit gains access to the striatum could be not only via connections with the accumbens core but also via connections with the mOFC conveying both somatic and more abstract values, respectively (Balleine, 2005). However, to explain the effects of mOFC manipulations, the representation of abstract values must succeed and also largely replace the somatic values.

Reinforcement

In contrast to goal-directed actions, the neural basis of the reinforcement signal that controls habits appears to be relatively straightforward, being thought to reduce to dopamine release in the DLS (White, 1989). The DLS receives direct and converging afferents from sensorimotor and motor cortices as well as a dense projection from the substantia nigra dopamine neurons that a number of studies suggest is critical for plasticity in this region and mediates the reinforcement of habits (Reynolds, Hyland, & Wickens, 2001). Of particular importance in this process is the release of dopamine in the striatum produced by sustained activation of these midbrain neurons (Da Cunha, Gomez-A, & Blaha, 2012). Inputs to the DLS arise in the substantia nigra pars compacta (SNc), and one input to the SNc that has been of interest in this regard is that arising in the central nucleus of the amygdala (CeN).

It has been known for some time that the acquisition of simple sensory motor—orienting responses is mediated by the CeN—SNc connection (Holland, Han, & Gallagher, 2000; Lee et al., 2005) and, although lesions of the CeN have no effect on the acquisition of goal-directed actions, they do attenuate the general motivational effects of stimuli associated with reward (Corbit & Balleine, 2005; Hall, Parkinson, Connor, Dickinson, & Everitt, 2001). In fact, recent evidence suggests that the acquisition of habitual actions depends on a circuit involving the DLS and the CeN (Dezfouli, Lingawi, & Balleine, 2014; Lingawi & Balleine, 2012). As noted above, to demonstrate habits, rats are usually overtrained. Early in training actions are sensitive to devaluation and hence are goal directed. After extended training, however, their control shifts from being goal-directed to being habitual, i.e., their performance becomes insensitive to the effects of outcome devaluation (Dickinson, 1994). Bilateral lesions of the DLS reduce habit learning and render overtrained actions goal-directed (Yin et al., 2004). The same is true of bilateral lesions of the anterior (but not of the posterior) CeN (Lingawi & Balleine, 2012). Furthermore, whereas a unilateral lesion of the CeN and ipsilateral DLS does not affect the rats' insensitivity to outcome devaluation, we found that a unilateral lesion of the CeN plus contralateral DLS rendered performance goal-directed and generated a reliable outcome devaluation effect. This latter finding has important implications for our understanding of the circuitry mediating habits. Anatomical tracing studies have found evidence that the anterior CeN projects to the SN (Gonzales & Chesselet, 1990). The SN has two divisions: the reticulata (SNr) and the SNc. The SNr is largely composed of GABAergic projection neurons to the thalamus, which maintain collateral projections to neurons in the SNc (Groenewegen, 2003; Joel & Weiner, 2000). In contrast, the SNc is composed primarily of a bank of dopaminergic neurons that are the source of dopamine to the dorsal striatum, and it is this dopaminergic projection to the DLS that has been implicated in the reinforcement signal mediating learning in the DLS (Reynolds et al., 2001).

Interestingly, the processes that mediate goal-directed and habitual actions appear to interact; albeit likely in different ways, at different times. During the *performance* of instrumental actions, for example, the speed of action selection, which is enabled by the stimulus control of habits, allows habits to outcompete goal-directed control. Nevertheless, habits can be inhibited, and it is likely circuitry involved in goal-directed control that accomplishes that process. Recent evidence suggests that feedback connections within the basal ganglia, most notably those within the indirect pathway associated with connections from the subthalamic nucleus to the globus pallidus and thence to the dorsal striatum via the arkypallidal neuronal projection, accomplish this process; being inhibitory, these projections could quite readily alter the balance between dorsomedial and dorsolateral areas in a manner reflecting such an interaction between the performance of goal-directed and habitual actions (Bogacz, Martin Moraud, Abdi, Magill, & Baufreton, 2016; Mallet et al., 2016).

In contrast, both processes appear, generally, to be engaged in parallel during acquisition; damage to or inactivation of the pDMS renders what would otherwise be goal-directed actions habitual (Yin, Ostlund, et al., 2005), whereas inactivation of the DLS renders what would otherwise be habitual actions goal-directed (Yin et al., 2004). Nevertheless, it is interesting to consider the possibility that the processes mediating goal-directed and habitual actions interact very early during acquisition at which point, thinking adaptively, it would seem prudent to establish whether a newly acquired action were going to be of some enduring value before starting the process of stamping it in as a long-term habit. Although it is not known whether, say, the goal-directed system inhibits habit learning during initial acquisition, some evidence suggests that it does. For example, pretraining excitotoxic lesions of the BLA appear to accelerate habit learning; rather than requiring extensive training to induce habits, they emerge rapidly after only a few sessions and, by and large, with the rate of performance largely unaffected (Balleine et al., 2003). Indeed, based on recent studies, it is possible that this interaction between goal-directed and habitual control is due to some form of interaction involving amygdala subnuclei, between which the feed-forward excitatory and inhibitory connections involving the BLA and CeN have been described (Busti et al., 2011; Herry et al., 2010). However, if habit learning is delayed by BLA-mediated inhibition of the CeN then this cannot persist forever; otherwise habits would never emerge at all. Obviously, any such inhibition must itself be modulated by some third structure, but, given this scenario, which structure could that be?

Evidence that the IL is involved in habit learning has important implications in this regard (Coutureau & Killcross, 2003; Killcross & Coutureau, 2003); as with manipulations of the DLS, both excitotoxic lesions and pharmacological inactivation of the IL have been reported to impair habit learning. Although this may be taken to imply that habits depend on a circuit involving the DLS and the IL, anatomically, they are not connected with one another; however, the IL does maintain strong connections with the CeN, particularly with its lateral capsular area (McDonald, Shammah-Lagnado, Shi, & Davis, 1999; Pinard, Mascagni, & McDonald, 2012) suggesting that the IL might interact with the CeN to regulate the acquisition of habits. In fact, as illustrated in Fig. 19.1B, one intriguing hypothesis is that, rather than modulating BLA or CeN directly, the IL may modulate the inhibitory interaction between the BLA and CeN. Interaction between the IL and amygdala has been of interest recently in the context of the neural bases of the extinction of Pavlovian fear conditioning, where BLA inputs to central nucleus are thought to be regulated by IL inputs to the region, particularly to the intercalated cells (ITCs) that stand between BLA and CeN (Paré, Quirk, & Ledoux, 2004). Although the role of ITCs has recently been questioned (Pinard et al., 2012), the evidence for regulation of the BLA output by the IL is quite strong and suggests, in the current context, that, during the course of instrumental training, the IL functions in the habit circuit to inhibit the BLA inhibition of CeN to allow the CeN reinforcement signal to initiate the acquisition of habit learning. Likewise, posttraining inactivation of the IL should be predicted to result in a return of BLA inhibition of the CeN and, with it, a loss habitual performance.

Some implications

Taken together these data on the behavioral and neural bases of goal-directed and habitual action confirms that they constitute distinct control processes with very different goals: a change in the external environment in the case of goal-directed actions and in a specified motor movement in the case of habits. They have different associative structures, learning rules, and feedback processes that regulate that learning. Importantly, a single event, e.g., a morsel of food when hungry or a drop of fluid when thirsty, can serve both to reward a goal-directed action and to reinforce a habit. This parsing of a single event into multiple feedback signals to control diverse learning processes appears to be one of the critical functions of the amygdala, at least insofar as appetitive learning is concerned (although the same may be true for aversive learning, it is beyond the scope of this chapter; see Balleine & Killcross, 2006 for a review of these issues). I have considered the hypothesis that some interaction occurs between these feedback processes; that at some point they are competitive, with the reward system inhibiting the reinforcement system under the modulatory control of the IL. Generally, although other forms of interaction can occur during *performance*, e.g., the goal-directed system appears able rapidly to curtail the *performance* of habits (unless the latter can outpace that inhibitory process), during *learning* these systems appear to function relatively autonomously to establish the forms of action control with which they are associated. Although it is clear, therefore, that distinct forms of feedback exist, their origin within the general motivational processes of the animal has not often been considered, and I turn to this issue in part 2.

FROM MOTIVATION TO ACTION

Although reward and reinforcement have distinct functions in goal-directed and habitual action, how they achieve their independent influence is still an open question. From a motivational perspective, it seems unlikely that these processes have independent origins; both have clear links with regulatory processes through which events that relieve, for example, hunger, thirst, heat, cold, pain, fear, or illness may gain value. And, just as such events can function to strengthen habits, they can also provoke the acquisition of new actions. How does the contribution of motivational processes to these functions differ?

Affective change and reinforcement

Historically, motivational analyses were closely aligned to energetic conceptions such as homeostasis or drive (Cannon, 1932; Richter, 1927). There were several major proponents of drive theory and the theories took several forms but perhaps the most ambitious, certainly the most thoroughly articulated, was Clark Hull's theory of general drive (Hull, 1943). According to this view, behavior is driven by S-R associations stamped in by a process of reinforcement, the latter produced by drive reduction. Accordingly, an increase in drive could provoke, initially, an increase in general activity that would persist until drive reduction occurred, resulting in a strengthening of the relationship between the situation and whatever response preceded the drive reduction. As a consequence, when that S arose again, its associated R would be more likely to occur. To ensure the R resulted in something reasonably biologically useful, both the S and the drive reduction were linked to the animal's internal conditions; so drive stimuli could arise that were reasonably specific, say a state of hunger, to which some R could be linked that would result in drive reduction, presumably due to contact with food, along with a concomitant reduction in the stimulus/drive conditions. However, for Hull, although there could be different drive stimuli, there were no specific drives, just different sources of drive and, therefore, drive could be reduced by a reduction in any of those sources.

This view implies that different sources of drive can both **summate with** and **substitute for** other sources of drives, a position which had one strength but many failings. The failings were empirical; many studies tested the ability of drives to summate and to substitute one for another and, although summation and substitution were sometimes observed, these predictions were not generally supported (Bolles, 1967). However, the one strength of the general drive perspective should not be underemphasized: It ensured that the concept of reinforcement had an objective, observable content that did not depend on unobserved (from Hull's perspective, unobservable) connections within the animal and, therefore, was not circular. Subsequent positions that aimed to fix the problems with the general drive concept by divorcing reinforcement from drive reduction were forced to define it in relational terms; which ultimately reduced the definition of reinforcement to "the presentation of a reinforcer" and of a reinforcer to "a thing that reinforces" or, as if more words could help, to "a stimulus change that increases the strength of the preceding response" (Meehle, 1950; Paniagua, 1985)!

These statements held out against claims that reinforcement was determined by connections inside the animal, by some form of central motivational state, but, as these attempts make clear, that was hardly an advantage. Nor was it likely to persist in practice; in fact, some of the most interesting and persistent ideas from that period come from attempts to instantiate general drive theory in neural terms; for example, in terms of the effects generated by the ascending reticular activating system (Lindsley, 1957). So much became clear with the advent of incentive theories which, for both empirical and theoretical reasons, had significant advantages over drive theory (Bindra, 1959). Incentives are essentially reinforcers: Money, for example, is often held up as the primary example of a positive incentive. However, unlike reinforcers, incentives do not necessarily reinforce; the notion that one could establish the degree to which they do, when they do and when they don't, meant that the relationship between incentive and reinforcement was at least an empirical one. Furthermore, considering the properties of the reinforcing event and its effects on the animal generated important insights: Sheffield's sex and saccharine studies, in which, after running in a runway, male rats were allowed access to events designed to induce rather than reduce drive, e.g., to intromit but not ejaculate or, when hungry, to drink nonnutritive saccharine, appeared nevertheless to acquire runway performance suggesting these events served to reinforce that performance (Sheffield & Roby, 1950; Sheffield, Wulff, & Backer, 1951). These and other data motivated alternative theories of reinforcement according to which consummatory or hedonic responses were the necessary component, consistent with incentive views (Sheffield, Roby, & Campbell, 1954; Young, 1949). Nevertheless, some data naggingly persisted in demonstrating a role for drive. For example, Miller and Kessen (1952) were able to reinforce responses in a T-maze by injecting milk directly into the stomach of hungry rats through a fistula. Even though rats given oral milk learned much faster (even with a long delay), the impressive effects of direct stomach injection suggest a form of drive reduction was in play. Furthermore, Webb, Bolles, and others found that up to a point, responses established with food deprivation could be maintained with water deprivation and even with mild footshock (Bolles, 1961; Webb, 1949). Beyond that point, and with water deprivation, it was a limit of about 7 h deprivation, the substitution collapsed; indeed, thirsty rats given water responded as if they were hungry suggesting that at deprivation levels above 7 h thirst was actively inhibiting not substituting for hunger (Grice & Davis, 1957).

Nevertheless, drive theory gave way to incentive theories and so to the view that behavior reflects the anticipation of events that generate affective responses and that it is the prediction or delivery of, say, a food reward, rather than a reduction in hunger, that generates the appetitive activity necessary to strengthen the S-R association (Dickinson & Balleine, 2002). Incentive theories have been successful; they can explain the rapid changes in behavior associated with changes in the quality or quantity of the incentive used to reinforce an action and so can give a more reasoned description of the influence of reinforcers as the targets of performance and not just as they influence learning. However, because they were developed in opposition to drive theories, they typically do not, and perhaps cannot, fully stipulate the sources of reinforcement. And yet, quite clearly, specific drives modulate incentive-related activity; the level of hunger determines activity (whether one means behavioral activity or activity in a specific central motivational state—such as a nutrient system) associated with food and food anticipation; thirst modulates fluid-related activity, and so on. With regard to aversive events, similar relationships can be drawn—e.g., a drive related to pain may be thought to modulate the influence of fear- (or threat) inducing events, malaise to modulate disgust-inducing events. The activity of these incentive processes would, one imagines, increase aversive affect and engage something like the negative law of effect to diminish rather than strengthen S-R associations, whereas a reduction in this activity would be appetitive, resulting in negative reinforcement and so increase S-R association (LeDoux & Daw, 2018). It would appear necessary, therefore, as was argued by Spence (Spence, 1956), to retain both drive and incentive constructs, allowing the former to define the sources of the specific needs that are the target of the latter. Indeed, Bindra, perhaps the quintessential incentive theorist, developed a position that had at its core exactly such integration (Bindra, 1974). And such a position, in which drive and incentive are integrated, helps to overcome difficulties or confusions at the heart of motivational accounts, e.g., it is sometimes argued that hunger is an aversive state that nutrients reduce, leading to confusion as to whether nutrient commodities are positive or negative reinforcers (Betley et al., 2015). Nutrients do, of course, reduce hunger and so inhibit that source of drive but, from the perspective of incentive theory, nutrients and the stimuli that predict them activate a nutrient incentive system and so generate positive affect, i.e., they are positive reinforcers.

Collectively, therefore, this combined drive-incentive perspective supposes multiple incentive processes with attendant modulatory "drives," what might be seen as peripheral changes in state, that alter the likelihood that one of a number of central incentive systems are activated—see Fig. 19.2. Feedback from that activity diminishes the drive, whereas activity of the incentive system changes the internal state, generates state- and sensory-specific motivational associations and causes an affective output to the appetitive or aversive affective system, depending on the incentive system. At a simple neural level, if the appetitive activity reflected in midbrain dopamine neurons and their modulation by CeN provides the proximal signal through which situation—response associations

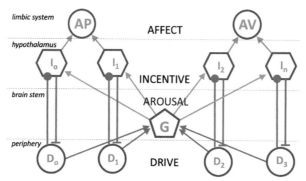

Figure 19.2 A cartoon describing the hypothetical peripheral—brainstem—hypothalamic interactions mediating the drive-incentive system, including its fixed connections with the affective processes of the limbic system. This is developed for multiple drives, **D**, incentive systems, **I**, processing appetitive (e.g., nutritive, fluidic, etc.) and aversive (e.g., threatening, disgusting, etc.) incentives, and appetitive (AP) and aversive (AV) affective systems. On this view, and to account for evidence of partial summation and substitution between sources of drive, a source of general arousal, "G," is hypothesized that distributes the activational effects of drive based on their specific modulatory impact (based, say, on a competitive threshold) on their respective incentive systems. Shown is feedback inhibition by each specific incentive system on their respective sources of drive; not shown are potential interactions between drives or incentive systems (e.g., as postulated by Konorski, 1967).

are reinforced—as described above and as illustrated in Fig. 19.3, and if, as proposed above, such affective activity is mediated by activity in the individual incentive systems, then one should anticipate that interconnections between the CeA and brain stem and hypothalamic regions associated with these systems to play a role in this reinforcement

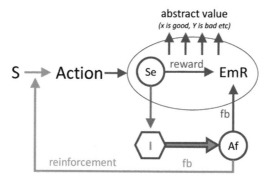

Figure 19.3 Sources of affective feedback and their influence on instrumental conditioning. In the presence of specific situational stimuli (S), actions are performed resulting in events that have two kinds of effect: One effect is driven by direct feedback via the incentive and affective systems to reinforce the stimulus—action (S-R) association. The second effect is indirect; the action resulting in a specific stimulus event produces, via the incentive and affective systems, emotional feedback, in the form of an emotional response, allowing an experienced value (and, incidentally, an abstract value) to be assigned to those consequences and an action value to be calculated based on the action—outcome contingency.

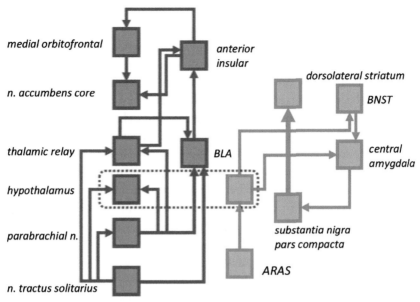

Figure 19.4 Putative neural bases of the reward and reinforcement signals supporting instrumental conditioning. The left panel shows various brain stem inputs to the hypothalamus and thalamic sensory relay targeting the basolateral amygdala through which the reward value of the instrumental outcome is encoded in the insular and medial orbital cortices. It is retrieved into the nucleus accumbens to guide the performance of goal-directed actions. In the right panel, a broad arousal signal to the hypothalamus targets the central and extended amygdala that controls dopaminergic inputs to the dorsolateral striatum to influence the acquisition of habits. *BLA*, basolateral amygdala; ARAS, ascending reticular activating system; BNST, bed nucleus of the stria terminalis.

process. This connectivity has been well documented in previous (Paredes, Winters, Schneiderman, & McCabe, 2000) and more recent mapping studies (Douglass et al., 2017), with strong inputs from the brain stem visceromotor nuclei as well as from a broad range of hypothalamic inputs both directly to the CeA and relayed via the bed nucleus of the stria terminalis. Such signals would appear consistent with a summed activation of inputs to these structures, with little if any sensory specificity to these various inputs consistent with the nonspecific activation of dopamine neurons and with a resulting general affective signal appropriate for the reinforcement of habitual actions—see Fig. 19.4.

Emotional feedback and reward

Incentive theories have, naturally enough, been concerned as much with reward as with reinforcement—although, historically, these processes were not usually differentiated. My concern in this section is with how such theories account for reward and how rewards are instantiated at a neural level. Although a general affective signal is sufficient for reinforcement, this will not provide the specificity necessary to account for the functions of reward; as described above, it is clear that, in goal-directed action, animals acquire a highly specific representation of the consequences of their actions both to inform their encoding of action—outcome contingencies and to constrain the assignment and modulation of reward values. As a consequence, one of the primary distinctions between rewards and reinforcers is the necessary involvement of specific sensory information in the representation of rewarding events. Furthermore, studies assessing instrumental incentive learning suggest that the basis for such learning is not simply affective activity but an emotional response; that a simple affective signal is not sufficient to produce a change in reward value and that such changes depend on *direct consummatory experience*. Subsequent analyses have focused on this latter requirement, proposing that consummatory experience is required because it allows the animal to experience the sensory properties of a reward in conjunction with the emotional response such contact evokes and that it is this conjunction that provides the necessary conditions for the assignment of any reward value (Balleine, 2005).

If this experience is necessary to alter the value of the goals of goal-directed actions, and so for animals to modify the performance of such actions after shifts in primary motivation, then the effect of such shifts on the performance of other responses (i.e., those not goal-directed) should not depend on this process. And by and large, that is what experiments have found. Thus, habitual instrumental actions and Pavlovian-conditioned reflexes, such as magazine approach, are directly influenced by shifts in primary motivation, and for slightly different reasons (Balleine, 1992; Dickinson et al., 1995). Whereas motivational shifts, such as a reduction in food deprivation appear to influence the performance of habits by altering the animals' level of general arousal, changes in the Pavlovian conditioned reflex (CR) appear to be due to alterations in the ability of conditioned stimuli to activate the representation of the unconditioned stimulus (Dickinson & Balleine, 2002—see also Implications for theories of conditioning section).

Careful consideration of these kinds of effect suggests that the influence of specific incentive processes on affective centers provokes a number of distinct responses: such as approach and withdrawal, changes in arousal or general activity but also likely distinct forms of emotional response. Behaviorally, considerable research suggests that fixed action patterns, unconditioned and conditioned reflexes, are produced by tastes or other stimuli associated with palatable or unpalatable outcomes (Berridge, 2000), and, along with William James, it is tempting to regard these responses as the emotional response itself. However, evidence from studies assessing the neural bases of both incentive learning and changes in taste reactivity suggests that they can be doubly dissociated. In one study, infusing the μ-opioid receptor antagonist naloxone into either the nucleus accumbens shell or the ventral pallidum blocked changes in taste reactivity produced by a sucrose solution following an increase in food deprivation (Wassum et al., 2009). Nevertheless, these naloxone infusions failed to block an increase in the incentive value of the sucrose; reexposure to the sucrose after the upshift in food deprivation increased subsequent goal-directed action even though no change in taste reactivity was observed during that exposure. Hence, it is clear that motivational shifts can have immediate effects on many classes of reflexive response, including unconditioned reflexes, conditioned reflexes, and habitual instrumental actions, and can do so by causing changes in the ability of primary incentive processes to engage the affective system. And, in one way or another, it is clear that such changes are distinct from the emotional feedback that these incentive processes can also produce. Indeed, as described in the first section, emotional feedback clearly engages the BLA and incentive learning appears critically to involve the sensory and visceral/emotion-related inputs to that structure to encode those changes in value in the insular cortex (Livneh et al., 2017).

How this is achieved at an emotional level is not yet fully understood; however, as implied by this analysis, considerable recent research has confirmed that affective processes do have two quite distinct neural effects, one associated with the discrimination of different forms of affective or emotional state, particularly, in humans, those expressed by other people and involving the well-described somatotopic representation in the somatosensory cortices, whereas the other appears to be associated with experienced emotional responses per se and involves the amygdala as well as the insular and orbito-frontal cortices (McGlone, Wessberg, & Olausson, 2014). It is these latter that appear to be critical for incentive learning and, indeed, that appear to be essential for any initial changes in value to be induced. With regard to changes in emotional response associated with changes in primary motivation, the ascending fiber pathways from the periphery to the solitary tract, particularly inputs from the gut and mouth via the vagal, fascial, and glossopharyngeal nerves, and thence to brain stem structures particularly the parabrachial complex, which provides the most caudal topographic map of the body, allowing the assembly of what has been argued to be a fundamental interoceptive map (Critchley & Harrison, 2013; Damasio & Carvalho, 2013). Subsequently, projections from this brain stem affective representation extend to thalamic and hypothalamic structures and densely to the amygdala as well as to the insular cortex both directly and via the thalamic relay (Venkatraman, Edlow, & Immordino-Yang, 2017). It has been hypothesized that the emotional responses that form the basis of incentive learning are generated within this circuitry via coordinated inputs to the BLA of visceral and sensory afferents to encode the specific sensory features of events contiguous with those responses. It is worth noting that insular cortex involvement in encoding the product of incentive learning, i.e., specific incentive values, could be taken to suggest that it plays a role in emotional experience. However, on the current view, the anterior insular cortex serves to encode incentive values and so is sensitive to the effect of emotional processing rather than itself providing the source of that processing, consistent with other recent views (e.g., Damasio & Carvalho, 2013).

It is also worth noting that touch-related sensation and discrimination has previously appeared to provide an exception to this model. Such inputs depend on spinal inputs from C and C tactile (CT) fibers, which send rapid signals to somatosensory cortices and play a key role in sensorimotor functions. However, recent research describing slow, touch-sensitive CT fibers that are specifically involved in pleasant touch suggests that touch-related emotional responses may fall into much the same category as other emotional responses and, importantly, unlike the rapid inputs to somatosensory cortices, these slow CT fibers appear to project to much the same forebrain areas as the ascending visceral sensory projections described above, terminating in insular and orbitofrontal cortices (McGlone et al., 2014).

Generally, therefore, a simple model driven by a certain amount of redundancy and following prior accounts could be generated that proposes that sensory-incentive associations underlie the ability of specific events to drive affective activity associated with the following: (1) the reinforcement signal used in the acquisition of habits; (2) the affective responses associated with Pavlovian preparatory conditioned reflexes; and (3) the emotional feedback necessary for incentive learning. What differentiates these processes on this account is not their source, which is in all cases determined by the drive-incentive system, but the specific output of the affective system, i.e., on affective feedback, arousal and emotional responding, such that each response can be measured and their effects within the neural systems associated with unconditioned stimulus (US) processing, reinforcement, and reward, detected by specific behavioral tests that discriminate between these processes—cf. Fig. 19.3.

Perhaps the most interesting, if not the most surprising, implication of this analysis is that, like reinforcement, reward is really a quite abstract signal. However, rather than being limited to internal activity in the ascending SNc dopamine projection, it is determined by an emotional response, reflecting the bodily reaction to the motivational/incentive properties of some specific event with which it is subsequently associated. However, note that, on this account, assigned reward value to an event means to associate the sensory properties of that event with an emotional response, whether those sensory events are in any way connected with the generation of the emotional response or not. It appears, therefore, that, if simply pairing a stimulus with an emotional response is all that is required, then the reward value of any event can readily be modified, even of artificial compounds, such as drugs of various kinds, that our nervous system did not evolve to regulate. Likewise, one can see a ready route for the development of specific psychopathology.

IMPLICATIONS FOR THEORIES OF CONDITIONING

Having commented on the origins of reward and reinforcement in the functions of the affective system, it is important to recognize that specifying the source of activity in this system as arising in the various incentive systems has other implications, most of which remain to be tested. Perhaps the most important implication is for "US processing" theories of Pavlovian conditioning, particularly the most influential such theory advanced by Rescorla and Wagner (Rescorla & Wagner, 1972).

That early account assumed within the model that the "value" of the representation of the US (λ) is fixed, as is its salience (β). However, on the current account, the US representation is acquired during the formation of sensory connections with the incentive systems, and so the value cannot be fixed at the start of conditioning. Likewise, the salience, what might more accurately be called the associability, of the US may also be assumed to vary much as conditioned stimulus (CS) processing theories contend that the associability of the CS does. Although this quite radically changes the assumptions of the Rescorla—Wagner model, some phenomena are difficult to explain in any other way. For example, experiments on Pavlovian conditioning have found considerable evidence for nonassociative learning phenomena, for example, exposure to a US (or other salient event) can result in an increase in unconditioned responding to that event, i.e., in *sensitization*, and sometimes to other events, i.e., *cross-sensitization* or *pseudoconditioning* (Overmier, 2002). Such phenomena accord well with the idea that USs are better processed or are more accurately represented over the course of exposure. Furthermore, such effects appear to depend on a background level of motivational arousal; Davis found that sensitization of startle to a loud noise emerged when background white noise was greater that 60 db whereas against a background of low motivational arousal such exposure resulted in habituation (Davis, 1974). These data are consistent with the notion that activity in the incentive system modulates US processing (see also Swithers, 1996).

Direct evidence for this claim in the appetitive domain comes from studies assessing the effect of a newly established motivational state on responding to the US. In "conditioned reflexes" (Pavlov, 1927, p. 23), Pavlov casually describes an experiment in which young dogs weaned from their mothers' milk are made hungry for the first time and shown meat. Despite their hunger, however, they did not salivate at the sight of the meat. Nevertheless, having eaten the meat, they did salivate when shown it for a second time. A more recent replication of this study in rats found a comparable effect, now on approach to food or water when rat pups were first made hungry or thirsty (Changizi, McGehee, & Hall, 2002; Myers & Hall, 2001). In both cases, the rat pups failed to show any more approach responses than sated animals when first made hungry or thirsty and exposed to food or water. Having experienced the food or water in these states, however, they immediately approached the site of these events when subsequently made hungry or thirsty. These experiments confirm the observations reported by Pavlov and, together with the effects of aversive arousal on startle sensitization, support the general claim that changes in responding induced by repeated exposure to a US produces a change in US processing.

Careful consideration of these effects suggests that phenomena often thought to provide support for CS processing accounts of Pavlovian conditioning could instead reflect fluctuations in US processing. For example, sometime ago, Simon Killcross and I found evidence to suggest that latent inhibition, the phenomenon that preexposure to a CS slows the subsequent conditioning of that CS, is specific to the motivational state in which the CS is presented (Killcross & Balleine, 1996). Thus, exposure to one CS, S1, when hungry and a different CS, S2, when thirsty, slowed conditioning more to S1 than S2 when the rats were made hungry and thirsty and both S1 and S2 were paired with food; whereas it slowed conditioning to S2 more than S1 when both were paired with saline. Although we interpreted these results at the time as reflecting a form of learned irrelevance, an alternative US processing account of these results can be developed according to which, by virtue of exposure to S1 when hungry (or S2 when thirsty), S1 (or S2) becomes associated with the nutritive (or fluidic) incentive system and so interferes with (perhaps overshadows or blocks) subsequent processing of the US. As a consequence, normal acquisition of the CS—US association should be delayed until the details of the US predicted by the CS are sorted out.

Some aspects of Pavlovian conditioning accord with this account; for example, sensory preconditioning—the phenomenon that a first stimulus, S1, trained to predict a second stimulus, S2, shows evidence of conditioning after S2 has been paired with a US—is also sensitive to motivational state. Although a learned irrelevance account of latent inhibition would appear to predict that exposure to S1 → S2 when the rat is hungry should reduce the relevance of both S1, S2 and their association for food, in fact the opposite has been reported: Adamec and Melzack (1970) exposed cats to pairings of a clicker stimulus and a flash stimulus when the cats were food and water deprived or when they were undeprived. In a second phase, the cats remained or were made food and water deprived, and the flash was conditioned by pairing it with milk delivery. Finally, in a test, the cats were presented with the clicker and their willingness to lick at the spout from which milk was delivered was recorded. If deprivation increases the irrelevance of the exposed stimuli to the target motivational event, in this case a nutritive fluid such as milk, then the clicker and the flash should both be less likely to generate CRs in the group exposed to these events when food and water deprived than when undeprived. In fact, the opposite results were found. In line with the view that exposure to the stimuli in the deprived state caused them to become associated with the relevant incentive systems, the cats responded more to the clicker when it was associated with the flash in the deprived state than in the undeprived state; and indeed, responded more to the clicker than did a deprived control group exposed to the click and the flash unpaired.

If this account is correct, then this kind of US processing needs to be added to formal theories of Pavlovian conditioning to fully specify what the animal learns in even quite basic situations where a simple stimulus predicts a US. One way of achieving this, summarized in Fig. 19.5, is to regard Rescorla and Wagner's lambda (λ) as essentially a

(A)   **(B)**

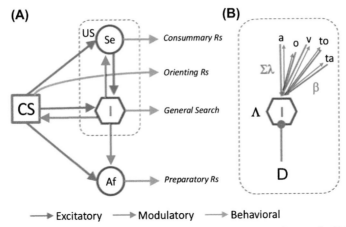

→ Excitatory   → Modulatory   → Behavioral

Figure 19.5 The hypothesis considered here regarding the motivational control of US processing in Pavlovian conditioning. (A) The conditioned stimulus can form associations with the stimulus and the motivational/incentive components of the US and with the affective system that it activates. Each of these associations produces distinct conditioned reflexes facilitated by the effects of incentive activity. (B) US processing involves the formation of associations between the specific sensory components of the US and the incentive system, and responses to those sensory events are facilitated by feedback from the incentive system and will generate sensitization.

prediction of the true lambda (Λ) that changes such that $\delta\lambda = \beta(\Lambda - \Sigma\lambda)$, with conditioning reflecting this changing value of the US; viz: $\Delta V = \alpha\beta((\Sigma\lambda) - \Sigma V)$. Alternatively, or perhaps additionally, it is possible to consider the merits of using US exposure to change the associability of the US, (β), much as has been claimed for the CS by CS processing theorists (Mackintosh, 1975; Pearce & Hall, 1980). Thus, for example, orienting to specific perceptual features of the environment appears to be strongly modulated by the animals' prevailing motivational state (Aarts, Dijksterhuis, & De Vries, 2001; Changizi & Hall, 2001) and, as such, it appears reasonable to suppose that the likelihood of a particular US being processed (β) could change as a result of experience, such the value of β on trial n is $\beta^n = |\Lambda^{n-1} - \lambda^{n-1}|$, where $\Delta\lambda^n = i\beta\Lambda^n = i \mid \Lambda^{n-1} - \lambda_0^{n-1} \mid \Lambda^n$, where i is the intensity of sensory representation of the US. Or perhaps it could be assumed $\Delta\beta_0$ is positive if $|\Lambda - \lambda_0| < |\Lambda - \lambda_X|$, where $\lambda_0 = \beta_0(\Lambda - \lambda_0)$; refer Fig. 19.5.

There are clearly many alternatives available at present, and many experiments are called for to clarify these ideas; however, the point of this analysis is not to provide the final word but merely to consider the implications of providing a fully elaborated motivational account of Pavlovian and instrumental conditioning. Although, as summarized here, we have, in the past, provided many of the details for the latter; the former has been neglected and is, one feels, a major gap in current theorizing.

ACKNOWLEDGMENTS

The preparation of this chapter was supported by a Senior Principal Research Fellowship from the NHMRC Australia, GNT#1079561.

REFERENCES

Aarts, H., Dijksterhuis, A., & De Vries, P. (2001). On the psychology of drinking: Being thirsty and perceptually ready. *British Journal of Psychology (London, England: 1953), 92*(Pt 4), 631–642.

Adamec, R., & Melzack, R. (1970). The role of motivation and orientation in sensory preconditioning. *Canadian Journal of Psychology, 24*(4), 230–239.

Balleine, B. (1992). Instrumental performance following a shift in primary motivation depends on incentive learning. *Journal of Experimental Psychology. Animal Behavior Processes, 18*(3), 236–250.

Balleine, B. W. (2001). Incentive processes in instrumental conditioning. In R. Klein, & S. Mowrer (Eds.), *Handbook of contemporary learning theories* (pp. 307–366). Hillsdale, NJ: LEA.

Balleine, B. W. (2005). Neural bases of food-seeking: Affect, arousal and reward in corticostriatolimbic circuits. *Physiology & Behavior, 86*(5), 717–730. https://doi.org/10.1016/j.physbeh.2005.08.061.

Balleine, B., Ball, J., & Dickinson, A. (1994). Benzodiazepine-induced outcome revaluation and the motivational control of instrumental action in rats. *Behavioural Neuroscience, 108*(3), 573–589.

Balleine, B., Davies, A., & Dickinson, A. (1995). Cholecystokinin attenuates incentive learning in rats. *Behavioural Neuroscience, 109*(2), 312–319.

Balleine, B., & Dickinson, A. (1991). Instrumental performance following reinforcer devaluation depends upon incentive learning. *The Quarterly Journal of Experimental Psychology Section B, 43*(3b), 279–296. https://doi.org/10.1080/14640749108401271.

Balleine, B., & Dickinson, A. (1994). Role of cholecystokinin in the motivational control of instrumental action in rats. *Behavioural Neuroscience, 108*(3), 590–605.

Balleine, B. W., & Dickinson, A. (1998a). Goal-directed instrumental action: Contingency and incentive learning and their cortical substrates. *Neuropharmacology, 37*(4–5), 407–419.

Balleine, B. W., & Dickinson, A. (1998b). The role of incentive learning in instrumental outcome revaluation by sensory-specific satiety. *Animal Learning & Behavior, 26*(1), 46–59. https://doi.org/10.3758/BF03199161.

Balleine, B. W., & Killcross, S. (2006). Parallel incentive processing: An integrated view of amygdala function. *Trends in Neurosciences, 29*(5), 272–279. https://doi.org/10.1016/j.tins.2006.03.002.

Balleine, B. W., Killcross, A. S., & Dickinson, A. (2003). The effect of lesions of the basolateral amygdala on instrumental conditioning. *The Journal of Neuroscience, 23*(2), 666–675.

Balleine, B. W., Liljeholm, M., & Ostlund, S. B. (2009). The integrative function of the basal ganglia in instrumental conditioning. *Behavioural Brain Research, 199*(1), 43–52. https://doi.org/10.1016/j.bbr.2008.10.034.

Balleine, B. W., & O'Doherty, J. P. (2010). Human and rodent homologies in action control: Corticostriatal determinants of goal-directed and habitual action. *Neuropsychopharmacology, 35*(1), 48–69. https://doi.org/10.1038/npp.2009.131.

Barnes, T. D., Kubota, Y., Hu, D., Jin, D. Z., & Graybiel, A. M. (2005). Activity of striatal neurons reflects dynamic encoding and recoding of procedural memories. *Nature, 437*(7062), 1158–1161. https://doi.org/10.1038/nature04053.

Berridge, K. C. (2000). Measuring hedonic impact in animals and infants: Microstructure of affective taste reactivity patterns. *Neuroscience and Biobehavioral Reviews, 24*(2), 173–198.

Betley, J. N., Xu, S., Cao, Z. F. H., Gong, R., Magnus, C. J., Yu, Y., & Sternson, S. M. (2015). Neurons for hunger and thirst transmit a negative-valence teaching signal. *Nature, 521*(7551), 180–185. https://doi.org/10.1038/nature14416.

Bindra, D. (1959). *Motivation: A systematic reinterpretation*. New York: Ronald Press.

Bindra, D. (1974). A motivational view of learning, performance, and behavior modification. *Psychological Review, 81*(3), 199–213.

Bogacz, R., Martin Moraud, E., Abdi, A., Magill, P. J., & Baufreton, J. (2016). Properties of neurons in external globus pallidus can support optimal action selection. *PLoS Computational Biology, 12*(7), e1005004. https://doi.org/10.1371/journal.pcbi.1005004.

Bolles, R. C. (1961). The interaction of hunger and thirst in the rat. *Journal of Comparative and Physiological Psychology, 54*, 580−584.

Bolles, R. C. (1967). *Theory of motivation.* New York: Harper & Row.

Bradfield, L. A., Dezfouli, A., van Holstein, M., Chieng, B., & Balleine, B. W. (2015). Medial orbitofrontal cortex mediates outcome retrieval in partially observable task situations. *Neuron, 88*(6), 1268−1280. https://doi.org/10.1016/j.neuron.2015.10.044.

Busti, D., Geracitano, R., Whittle, N., Dalezios, Y., Mańko, M., Kaufmann, W., ... Ferraguti, F. (2011). Different fear states engage distinct networks within the intercalated cell clusters of the amygdala. *The Journal of Neuroscience, 31*(13), 5131−5144. https://doi.org/10.1523/JNEUROSCI.6100-10.2011.

Cannon, W. B. (1932). *The wisdom of the body.* New York: Norton.

Chambers, C. D., Garavan, H., & Bellgrove, M. A. (2009). Insights into the neural basis of response inhibition from cognitive and clinical neuroscience. *Neuroscience and Biobehavioral Reviews, 33*(5), 631−646. https://doi.org/10.1016/j.neubiorev.2008.08.016.

Changizi, M. A., & Hall, W. G. (2001). Thirst modulates a perception. *Perception, 30*(12), 1489−1497. https://doi.org/10.1068/p3266.

Changizi, M. A., McGehee, R. M. F., & Hall, W. G. (2002). Evidence that appetitive responses for dehydration and food-deprivation are learned. *Physiology & Behavior, 75*(3), 295−304.

Corbit, L. H., & Balleine, B. W. (2003). The role of prelimbic cortex in instrumental conditioning. *Behavioural Brain Research, 146*(1−2), 145−157.

Corbit, L. H., & Balleine, B. W. (2005). Double dissociation of basolateral and central amygdala lesions on the general and outcome-specific forms of pavlovian-instrumental transfer. *The Journal of Neuroscience, 25*(4), 962−970. https://doi.org/10.1523/JNEUROSCI.4507-04.2005.

Corbit, L. H., Leung, B. K., & Balleine, B. W. (2013). The role of the amygdala-striatal pathway in the acquisition and performance of goal-directed instrumental actions. *The Journal of Neuroscience, 33*(45), 17682−17690. https://doi.org/10.1523/JNEUROSCI.3271-13.2013.

Corbit, L. H., Muir, J. L., & Balleine, B. W. (2001). The role of the nucleus accumbens in instrumental conditioning: Evidence of a functional dissociation between accumbens core and shell. *The Journal of Neuroscience, 21*(9), 3251−3260.

Costa, R. M., Cohen, D., & Nicolelis, M. A. L. (2004). Differential corticostriatal plasticity during fast and slow motor skill learning in mice. *Current Biology, 14*(13), 1124−1134. https://doi.org/10.1016/j.cub.2004.06.053.

Coutureau, E., & Killcross, S. (2003). Inactivation of the infralimbic prefrontal cortex reinstates goal-directed responding in overtrained rats. *Behavioural Brain Research, 146*(1−2), 167−174.

Critchley, H. D., & Harrison, N. A. (2013). Visceral influences on brain and behavior. *Neuron, 77*(4), 624−638. https://doi.org/10.1016/j.neuron.2013.02.008.

Da Cunha, C., Gomez-A, A., & Blaha, C. D. (2012). The role of the basal ganglia in motivated behavior. *Reviews in the Neurosciences, 23*(5−6), 747−767. https://doi.org/10.1515/revneuro-2012-0063.

Damasio, A., & Carvalho, G. B. (2013). The nature of feelings: Evolutionary and neurobiological origins. *Nature Reviews. Neuroscience, 14*(2), 143−152. https://doi.org/10.1038/nrn3403.

Davis, M. (1974). Sensitization of the rat startle response by noise. *Journal of Comparative and Physiological Psychology, 87*(3), 571−581.

Davis, J., & Bitterman, M. E. (1971). Differential reinforcement of other behavior (DRO): A yoked-control comparison. *Journal of the Experimental Analysis of Behavior, 15*, 237−241.

Daw, N. D., Niv, Y., & Dayan, P. (2005). Uncertainty-based competition between prefrontal and dorsolateral striatal systems for behavioral control. *Nature Neuroscience, 8*(12), 1704−1711. https://doi.org/10.1038/nn1560.

Dezfouli, A., & Balleine, B. W. (2012). Habits, action sequences and reinforcement learning. *The European Journal of Neuroscience, 35*(7), 1036−1051. https://doi.org/10.1111/j.1460-9568.2012.08050.x.

Dezfouli, A., Lingawi, N. W., & Balleine, B. W. (2014). Habits as action sequences: Hierarchical action control and changes in outcome value. *Philosophical Transactions of the Royal Society of London. Series B, Biological Sciences, 369*(1655). https://doi.org/10.1098/rstb.2013.0482.

Dickinson, A. (1994). Instrumental conditioning. In N. J. Mackintosh (Ed.), *Animal cognition and learning* (pp. 4–79). London: Academic Press.

Dickinson, A. (1998). Omission learning after instrumental pretraining. *The Quarterly Journal of Experimental Psychology Section B, 51*(3), 271–286. https://doi.org/10.1080/713932679.

Dickinson, A., & Dawson, G. R. (1988). Motivational control of instrumental performance: The role of prior experience of the reinforcer. *The Quarterly Journal of Experimental Psychology Section B, 40*(2), 113–134. https://doi.org/10.1080/14640748808402313.

Dickinson, A., & Balleine, B. (1994). Motivational control of goal-directed action. *Animal Learning & Behavior, 22*(1), 1–18. https://doi.org/10.3758/BF03199951.

Dickinson, A., & Balleine, B. W. (2002). The role of learning in motivation. In C. R. Gallistel (Ed.) (3rd ed., *Steven's handbook of experimental psychology: Vol. 3. Learning, motivation & emotion* (pp. 497–533). New York: John Wiley & Sons.

Dickinson, A., Balleine, B., Watt, A., Gonzalez, F., & Boakes, R. A. (1995). Motivational control after extended instrumental training. *Animal Learning & Behavior, 23*(2), 197–206. https://doi.org/10.3758/BF03199935.

Dolan, R. J., & Dayan, P. (2013). Goals and habits in the brain. *Neuron, 80*(2), 312–325. https://doi.org/10.1016/j.neuron.2013.09.007.

Douglass, A. M., Kucukdereli, H., Ponserre, M., Markovic, M., Gründemann, J., Strobel, C., … Klein, R. (2017). Central amygdala circuits modulate food consumption through a positive-valence mechanism. *Nature Neuroscience, 20*(10), 1384–1394. https://doi.org/10.1038/nn.4623.

Ellison-Wright, I., Ellison-Wright, Z., & Bullmore, E. (2008). Structural brain change in attention deficit hyperactivity disorder identified by meta-analysis. *BMC Psychiatry, 8*, 51. https://doi.org/10.1186/1471-244X-8-51.

Everitt, B. J. (1990). Sexual motivation: A neural and behavioural analysis of the mechanisms underlying appetitive and copulatory responses of male rats. *Neuroscience and Biobehavioral Reviews, 14*(2), 217–232.

Furlong, T. M., Supit, A. S. A., Corbit, L. H., Killcross, S., & Balleine, B. W. (2017). Pulling habits out of rats: Adenosine 2A receptor antagonism in dorsomedial striatum rescues meth-amphetamine-induced deficits in goal-directed action. *Addiction Biology, 22*(1), 172–183. https://doi.org/10.1111/adb.12316.

Fuster, J. M. (2000). Executive frontal functions. *Experimental Brain Research, 133*, 66–70.

Gabbott, P. L. A., Warner, T. A., Jays, P. R. L., Salway, P., & Busby, S. J. (2005). Prefrontal cortex in the rat: Projections to subcortical autonomic, motor, and limbic centers. *The Journal of Comparative Neurology, 492*(2), 145–177. https://doi.org/10.1002/cne.20738.

Goldman-Rakic, P. S. (1995). Architecture of the prefrontal cortex and the central executive. *Annals of the New York Academy of Sciences, 769*, 71–83.

Gonzales, C., & Chesselet, M. F. (1990). Amygdalonigral pathway: An anterograde study in the rat with Phaseolus vulgaris leucoagglutinin (PHA-L). *The Journal of Comparative Neurology, 297*(2), 182–200. https://doi.org/10.1002/cne.902970203.

Grice, G. R., & Davis, J. D. (1957). Effect of irrelevant thirst motivation on a response learned with food reward. *Journal of Experimental Psychology, 53*, 347–352.

Groenewegen, H. J. (2003). The basal ganglia and motor control. *Neural Plasticity, 10*(1–2), 107–120. https://doi.org/10.1155/NP.2003.107.

Hall, J., Parkinson, J. A., Connor, T. M., Dickinson, A., & Everitt, B. J. (2001). Involvement of the central nucleus of the amygdala and nucleus accumbens core in mediating Pavlovian influences on instrumental behaviour. *The European Journal of Neuroscience, 13*(10), 1984–1992.

Hart, G., Leung, B. K., & Balleine, B. W. (2014). Dorsal and ventral streams: The distinct role of striatal subregions in the acquisition and performance of goal-directed actions. *Neurobiology of Learning and Memory, 108*, 104–118. https://doi.org/10.1016/j.nlm.2013.11.003.

Hendersen, R. J., & Graham, J. (1979). Avoidance of heat by rats: Effects of thermal context on rapidity of extinction. *Learning and Motivation, 10*(3), 351–363.

Herry, C., Ferraguti, F., Singewald, N., Letzkus, J. J., Ehrlich, I., & Lüthi, A. (2010). Neuronal circuits of fear extinction. *The European Journal of Neuroscience, 31*(4), 599–612. https://doi.org/10.1111/j.1460-9568.2010.07101.x.

Hikosaka, O., Kim, H. F., Yasuda, M., & Yamamoto, S. (2014). Basal ganglia circuits for reward value-guided behavior. *Annual Review of Neuroscience, 37*, 289–306. https://doi.org/10.1146/annurev-neuro-071013-013924.

Holland, P. C., Han, J. S., & Gallagher, M. (2000). Lesions of the amygdala central nucleus alter performance on a selective attention task. *The Journal of Neuroscience, 20*(17), 6701–6706.

Hull, C. L. (1943). *Principles of behavior.* New York: Appleton-Century-Crofts.

Hutcheson, D. M., Everitt, B. J., Robbins, T. W., & Dickinson, A. (2001). The role of withdrawal in heroin addiction: Enhances reward or promotes avoidance? *Nature Neuroscience, 4*(9), 943–947. https://doi.org/10.1038/nn0901-943.

Joel, D., & Weiner, I. (2000). The connections of the dopaminergic system with the striatum in rats and primates: An analysis with respect to the functional and compartmental organization of the striatum. *Neuroscience, 96*(3), 451–474.

Konorski, J. (1967). *Integrative Activity of the Brain. An Interdisciplinary Approach.* Chicago: University of Chicago Press.

Killcross, S., & Balleine, B. (1996). Role of primary motivation in stimulus preexposure effects. *Journal of Experimental Psychology. Animal Behavior Processes, 22*(1), 32–42.

Killcross, S., & Coutureau, E. (2003). Coordination of actions and habits in the medial prefrontal cortex of rats. *Cerebral Cortex (New York, N.Y.: 1991), 13*(4), 400–408.

LeDoux, J., & Daw, N. D. (2018). Surviving threats: Neural circuit and computational implications of a new taxonomy of defensive behaviour. *Nature Reviews Neuroscience.* https://doi.org/10.1038/nrn.2018.22.

Lee, D. (2013). Decision making: From neuroscience to psychiatry. *Neuron, 78*(2), 233–248. https://doi.org/10.1016/j.neuron.2013.04.008.

Lee, H. J., Groshek, F., Petrovich, G. D., Cantalini, J. P., Gallagher, M., & Holland, P. C. (2005). Role of amygdalo-nigral circuitry in conditioning of a visual stimulus paired with food. *The Journal of Neuroscience, 25*(15), 3881–3888. https://doi.org/10.1523/JNEUROSCI.0416-05.2005.

Levy, R., & Dubois, B. (2006). Apathy and the functional anatomy of the prefrontal cortex-basal ganglia circuits. *Cerebral Cortex (New York, N.Y.: 1991), 16*(7), 916–928. https://doi.org/10.1093/cercor/bhj043.

Lindsley, D. B. (1957). Psychophysiology and motivation. *Nebraska Symposium on Motivation, V*, 44–105.

Lingawi, N. W., & Balleine, B. W. (2012). Amygdala central nucleus interacts with dorsolateral striatum to regulate the acquisition of habits. *The Journal of Neuroscience, 32*(3), 1073–1081. https://doi.org/10.1523/JNEUROSCI.4806-11.2012.

Livneh, Y., Ramesh, R. N., Burgess, C. R., Levandowski, K. M., Madara, J. C., Fenselau, H., ... Andermann, M. L. (2017). Homeostatic circuits selectively gate food cue responses in insular cortex. *Nature, 546*(7660), 611–616. https://doi.org/10.1038/nature22375.

Lopez, M., Balleine, B., & Dickinson, A. (1992). Incentive learning and the motivational control of instrumental performance by thirst. *Animal Learning & Behavior, 20*(4), 322–328. https://doi.org/10.3758/BF03197955.

Mackintosh, N. J. (1975). A theory of attention: Variations in the associability of stimuli with reinforcement. *Psychological Review, 82*(4), 276–298.

Mallet, N., Schmidt, R., Leventhal, D., Chen, F., Amer, N., Boraud, T., & Berke, J. D. (2016). Arkypallidal cells send a stop signal to striatum. *Neuron, 89*(2), 308–316. https://doi.org/10.1016/j.neuron.2015.12.017.

McDonald, A. J., Shammah-Lagnado, S. J., Shi, C., & Davis, M. (1999). Cortical afferents to the extended amygdala. *Annals of the New York Academy of Sciences, 877*, 309–338.

McGlone, F., Wessberg, J., & Olausson, H. (2014). Discriminative and affective touch: Sensing and feeling. *Neuron, 82*(4), 737–755. https://doi.org/10.1016/j.neuron.2014.05.001.

Meehle, P. E. (1950). On the circularity of the law of effect. *Psychological Bulletin, 47*, 52–75.

Miller, N. E., & Kessen, M. L. (1952). Reward effects of food via stomach fistula compared with those of food via mouth. *Journal of Comparative and Physiological Psychology, 45*, 555–564.

Myers, K. P., & Hall, W. G. (2001). Effects of prior experience with dehydration and water on the time course of dehydration-induced drinking in weanling rats. *Developmental Psychobiology, 38*(3), 145–153.

Ostlund, S. B., & Balleine, B. W. (2005). Lesions of medial prefrontal cortex disrupt the acquisition but not the expression of goal-directed learning. *The Journal of Neuroscience, 25*(34), 7763–7770. https://doi.org/10.1523/JNEUROSCI.1921-05.2005.

Ostlund, S. B., & Balleine, B. W. (2008). Differential involvement of the basolateral amygdala and mediodorsal thalamus in instrumental action selection. *The Journal of Neuroscience, 28*(17), 4398–4405. https://doi.org/10.1523/JNEUROSCI.5472-07.2008.

Ostlund, S. B., Maidment, N. T., & Balleine, B. W. (2010). Alcohol-paired contextual cues produce an immediate and selective loss of goal-directed action in rats. *Frontiers in Integrative Neuroscience, 4.* https://doi.org/10.3389/fnint.2010.00019.

Overmier, J. B. (2002). Sensitization, conditioning, and learning: Can they help us understand somatization and disability? *Scandinavian Journal of Psychology, 43*, 105–112.

Paniagua, F. A. (1985). The relational definition of reinforcement: Comments on circularity. *The Psychological Record, 35*(2), 193–202. https://doi.org/10.1007/BF03394925.

Paredes, J., Winters, R. W., Schneiderman, N., & McCabe, P. M. (2000). Afferents to the central nucleus of the amygdala and functional subdivisions of the periaqueductal gray: Neuroanatomical substrates for affective behavior. *Brain Research, 887*(1), 157–173.

Paré, D., Quirk, G. J., & Ledoux, J. E. (2004). New vistas on amygdala networks in conditioned fear. *Journal of Neurophysiology, 92*(1), 1–9. https://doi.org/10.1152/jn.00153.2004.

Parkes, S. L., & Balleine, B. W. (2013). Incentive memory: Evidence the basolateral amygdala encodes and the insular cortex retrieves outcome values to guide choice between goal-directed actions. *The Journal of Neuroscience, 33*(20), 8753–8763. https://doi.org/10.1523/JNEUROSCI.5071-12.2013.

Parkes, S. L., Bradfield, L. A., & Balleine, B. W. (2015). Interaction of insular cortex and ventral striatum mediates the effect of incentive memory on choice between goal-directed actions. *The Journal of Neuroscience, 35*(16), 6464–6471. https://doi.org/10.1523/JNEUROSCI.4153-14.2015.

Pavlov, I. P. (1927). *Conditioned reflexes: An investigation of the physiological activity of the cerebral cortex.* Oxford, England, UK: Oxford University Press.

Pearce, J. M., & Hall, G. (1980). A model for pavlovian learning: Variations in the effectiveness of conditioned but not of unconditioned stimuli. *Psychological Review, 87*(6), 532–552.

Pinard, C. R., Mascagni, F., & McDonald, A. J. (2012). Medial prefrontal cortical innervation of the intercalated nuclear region of the amygdala. *Neuroscience, 205*, 112–124. https://doi.org/10.1016/j.neuroscience.2011.12.036.

Pramstaller, P. P., & Marsden, C. D. (1996). The basal ganglia and apraxia. *Brain, 119*(Pt 1), 319–340.

Rescorla, R. A., & Wagner, A. R. (1972). A theory of pavlovian conditioning: Variations in the effectiveness of reinforcement and non reinforcement. In A. H. Black, & W. F. Prokasy (Eds.), *Classical conditioning II: Current research and theory* (pp. 64–99). New York: Appleton-Century-Crofts.

Reynolds, J. N., Hyland, B. I., & Wickens, J. R. (2001). A cellular mechanism of reward-related learning. *Nature, 413*(6851), 67–70. https://doi.org/10.1038/35092560.

Richter, C. P. (1927). Animal behavior and internal drives. *Quarterly Review of Biology, 2*, 307–343.

Schultz, W., & Dickinson, A. (2000). Neuronal coding of prediction errors. *Annual Review of Neuroscience, 23*, 473–500. https://doi.org/10.1146/annurev.neuro.23.1.473.

Shan, Q., Christie, M. J., & Balleine, B. W. (2015). Plasticity in striatopallidal projection neurons mediates the acquisition of habitual actions. *The European Journal of Neuroscience, 42*(4), 2097–2104. https://doi.org/10.1111/ejn.12971.

Sheffield, F. D., & Roby, T. B. (1950). Reward value of a non-nutritive sweet taste. *Journal of Comparative and Physiological Psychology, 43*, 471–481.

Sheffield, F. D., Roby, T. B., & Campbell, B. A. (1954). Drive reduction versus consummatory behavior as determinants of reinforcement. *Journal of Comparative and Physiological Psychology, 47*, 349–354.

Sheffield, F. D., Wulff, J. J., & Backer, R. (1951). Reward value of copulation without sex drive reduction. *Journal of Comparative and Physiological Psychology, 44*, 3–8.

Shiflett, M. W., & Balleine, B. W. (2011). Contributions of ERK signaling in the striatum to instrumental learning and performance. *Behavioural Brain Research, 218*(1), 240–247. https://doi.org/10.1016/j.bbr.2010.12.010.

Spence, K. W. (1956). *Behavior theory and conditioning*. New Haven: Yale University Press.

Swithers, S. E. (1996). Effects of oral experience on rewarding properties of oral stimulation. *Neuroscience and Biobehavioral Reviews, 20*(1), 27–32.

Tang, C., Pawlak, A. P., Prokopenko, V., & West, M. O. (2007). Changes in activity of the striatum during formation of a motor habit: Changes in striatal activity during habit learning. *European Journal of Neuroscience, 25*(4), 1212–1227. https://doi.org/10.1111/j.1460-9568.2007.05353.x.

Thorndike, E. L. (1898). Some experiments on animal intelligence. *Science, VII*.

Tran-Tu-Yen, D. A. S., Marchand, A. R., Pape, J.-R., Di Scala, G., & Coutureau, E. (2009). Transient role of the rat prelimbic cortex in goal-directed behaviour. *The European Journal of Neuroscience, 30*(3), 464–471. https://doi.org/10.1111/j.1460-9568.2009.06834.x.

Venkatraman, A., Edlow, B. L., & Immordino-Yang, M. H. (2017). The brainstem in emotion: A review. *Frontiers in Neuroanatomy, 11*(15). https://doi.org/10.3389/fnana.2017.00015.

Wang, S.-H., Ostlund, S. B., Nader, K., & Balleine, B. W. (2005). Consolidation and reconsolidation of incentive learning in the amygdala. *The Journal of Neuroscience, 25*(4), 830–835. https://doi.org/10.1523/JNEUROSCI.4716-04.2005.

Wassum, K. M., Ostlund, S. B., Maidment, N. T., & Balleine, B. W. (2009). Distinct opioid circuits determine the palatability and the desirability of rewarding events. *Proceedings of the National Academy of Sciences of the United States of America, 106*(30), 12512–12517. https://doi.org/10.1073/pnas.0905874106.

Webb, W. B. (1949). The motivational aspect of an irrelevant drive in the behavior of the white rat. *Journal of Experimental Psychology, 39*, 1–13.

White, N. M. (1989). Reward or reinforcement: what's the difference? *Neuroscience and Biobehavioral Reviews, 13*(2–3), 181–186.

Woodson, J. C., & Balleine, B. W. (2002). An assessment of factors contributing to instrumental performance for sexual reward in the rat. *The Quarterly Journal of Experimental Psychology Section B, Comparative and Physiological Psychology, 55*(1), 75–88. https://doi.org/10.1080/02724990143000199.

Yin, H. H., Knowlton, B. J., & Balleine, B. W. (2004). Lesions of dorsolateral striatum preserve outcome expectancy but disrupt habit formation in instrumental learning. *The European Journal of Neuroscience, 19*(1), 181–189.

Yin, H. H., Knowlton, B. J., & Balleine, B. W. (2005). Blockade of NMDA receptors in the dorsomedial striatum prevents action-outcome learning in instrumental conditioning. *The European Journal of Neuroscience, 22*(2), 505–512. https://doi.org/10.1111/j.1460-9568.2005.04219.x.

Yin, H. H., Ostlund, S. B., Knowlton, B. J., & Balleine, B. W. (2005). The role of the dorsomedial striatum in instrumental conditioning. *The European Journal of Neuroscience, 22*(2), 513–523. https://doi.org/10.1111/j.1460-9568.2005.04218.x.

Yin, H. H., Knowlton, B. J., & Balleine, B. W. (2006). Inactivation of dorsolateral striatum enhances sensitivity to changes in the action-outcome contingency in instrumental conditioning. *Behavioural Brain Research, 166*(2), 189–196. https://doi.org/10.1016/j.bbr.2005.07.012.

Young, P. T. (1949). Food-seeking drive, affective process and learning. *Psychological Review, 56*, 98–121.

INDEX

'Note: Page numbers followed by "f" indicate figures, "t" indicate tables, and "b" indicate boxes.'